EXPLOSION HAZARDS IN THE PROCESS INDUSTRIES

EXPLOSION HAZARDS IN THE PROCESS INDUSTRIES

Second Edition

ROLF K. ECKHOFF
University of Bergen, Norway

ELSEVIER

Amsterdam • Boston • Heidelberg • London
New York • Oxford • Paris • San Diego
San Francisco • Singapore • Sydney • Tokyo
Gulf Professional Publishing is an imprint of Elsevier

G P P

Gulf Professional Publishing is an imprint of Elsevier
50 Hampshire Street, 5th Floor, Cambridge, MA 02139, USA
The Boulevard, Langford Lane, Kidlington, Oxford, OX5 1GB, UK

First Edition 2005
Second Edition 2016

British Library Cataloguing-in-Publication Data
A catalogue record for this book is available from the British Library

Library of Congress Cataloging-in-Publication Data
A catalog record for this book is available from the Library of Congress

ISBN: 978-0-12-803273-2

For information on all Gulf Professional Publishing
visit our website at https://www.elsevier.com/

Publisher: Joe Hayton
Senior Acquisition Editor: Katie Hammon
Senior Editorial Project Manager: Kattie Washington
Production Project Manager: Sruthi Satheesh
Designer: Maria Ines Cruz

Typeset by TNQ Books and Journals

DEDICATION

I dedicate this second edition of this book to my dear wife, Astrid, for her everlasting love and support, to our eight grandchildren, Gunhild, Sigurd, Johanne, Elias, Maren, Simen, Martin, and Filip, and to the precious memory of the two, Åshild and Nora, whom we so sadly lost. The words in Isaiah 42:16 have continued to give me hope and courage.

Rolf K. Eckhoff

CONTENTS

ABOUT THE AUTHOR

Rolf K. Eckhoff was born in Stavanger, Norway, in 1937. He obtained his MSc in chemical engineering at the Norwegian University of Science and Technology in Trondheim in 1961. He obtained the degree dr. techn. from the same university in 1976 and the degree DSc (Eng) (higher doctorate) from the University of London, UK, in 1992. He is currently a professor emeritus of Process Safety Technology at the University of Bergen, Norway, and a concurrent professor at the Northeastern University, Shenyang, PR China. Prior to joining the University of Bergen in 1996, Eckhoff worked for 33 years as a researcher, senior researcher, and chief scientist at Chr. Michelsen Institute (presently, GexCon AS) in Bergen. His main areas of research have included powder technology, dust explosions, and gas explosions. He is the author of the textbook, *Dust Explosions in the Process Industries*, first published 1991, second edition in 1997, and third edition in 2003. The fourth edition, to appear in 2016, is co-authored by T. Skjold. Eckhoff is also the author/co-author of about 150 technical/scientific publications/papers and more than 130 research reports. He has advised industry on numerous occasions, and lectured on courses, conferences, and seminars in a number of countries in Europe, and in China, Bahrain, Israel, India, Malaysia, Thailand, Singapore, Australia, Canada, United States, and Brazil.

PREFACE TO FIRST EDITION

My very first contact with the area of industrial explosion hazards was during my postgraduate studies at King's College, University of London, 1966—68. In the early 1970s, the Norwegian process industries experienced a series of serious dust explosions, which confronted me with the challenge to establish dust-explosion prevention and mitigation as a new field of research and consultancy in Norway, based at the Chr. Michelsen Institute (CMI) in Bergen. A few years later, when the oil and gas industry on the Norwegian continental shelf evolved at great pace, I was faced with the further challenge to start up CMI's comprehensive research program on prevention and mitigation of accidental gas explosions.

In 1991, I approached the University of Bergen (UoB) to discuss the possibility of establishing process safety technology as a new field of teaching and research there. I was appointed part-time professor in 1992 and full-time professor in 1996. During the building up of my own activity at UoB, I was also taking a central role in establishing a common new framework program of process technology at large within our faculty of mathematics and natural sciences. We were convinced that our candidates should build their process safety studies on a foundation of general process technology (chemical engineering). The result of a joint effort by many people is a complete 5-year study program in process technology, comprising a 3-year bachelor program providing the general basis, followed by a 2-year master program providing the specialities. Process safety technology is one of several options for the master studies.

During my time as part-time professor, I developed a short course on "electrical apparatuses for hazardous areas" published as a 160-page basic textbook in Norwegian (Tapir, Trondheim, 1996). After joining UoB full time, I used this book as a central element in a wider basic bachelor-level course on explosion hazards in the process industries, but I then had to hand out supplementary material on case histories and prevention/mitigation. I therefore decided to write a completely new textbook, comprising all the material needed for this course. Because the book might also useful outside Norway, I chose to write it in English.

The book covers accidental gas/vapor, spray/mist, dust, and explosive/pyrotechnic/propellant explosions, whereas various types of accidental physical explosions, boiling liquid expanding vapor explosions (BLEVEs), and explosions due to chemical runaway reactions are not included.

It is my intention to produce a Powerpoint presentation covering the entire book, with some of the photographic material in color, and with additional illustrative material. This will primarily be for my own use, but such a presentation may also be helpful for other teachers using my book in their classes. We are also building up a set of about 10 laboratory-scale experiments on initiation and propagation of gas and dust explosions, which the students shall have to carry out and report on in writing as part of the course.

I am deeply grateful to the companies, research institutions, standardization organizations, publishers, etc., which granted permissions to use some of their valuable material for the book, and to the many skilled persons, including some of my students, for providing valuable information and advice.

Rolf K. Eckhoff
Bergen, June 2005

PREFACE TO SECOND EDITION

The first edition of this book was published in 2005 by Gulf Publishing Company, Houston, Texas, United States. A corrected reprint appeared in 2008. In 2013, the copyright was transferred to Elsevier Science. They decided to produce a second edition of the book to appear in 2015—16, ie, just about one decade after the appearance of the first edition.

I have, of course, reviewed all the original chapters of the book and added, modified, and deleted material when this was considered appropriate.

However, the most significant change from the first to the second edition is the addition of three new chapters. The first is Chapter 3 on "Boiling Liquid Expanding Vapor Explosions (BLEVEs)," a topic that I deliberately left out when writing the first edition, but which I later realized would add to the usefulness of the book. The same applies to the new chapter on "Water-Vapor Explosions" (Chapter 4).

The third, entirely new Chapter 9 on "Destructive and Harmful Effects by Pressure, Solid Fragments, and Heat From Accidental Explosions". I have for a long time realized that such a chapter had to be included in the second edition of this book. Finally, I have totally rewritten the previous very brief and incomplete Chapter 8 on hazards and risk analysis. This chapter now appears as a considerably expanded and updated Chapter 11 on "Review of Some Methods of Hazard and Risk Analysis."

I wish to express my sincere indebtedness to Professor Isabelle Sochet, École Nationale Supérieure d'Ingénieurs de Bourges, France, for her kind help and advice in preparing the new Chapter 9. I am also greatly indebted to five anonymous reviewers of my proposal for the book revision. All five offered many valuable suggestions, which I have largely followed, and which have no doubt improved the quality of the book. Finally I wish to thank all my students for valuable feedback on various aspects of the book, and for their enthusiasm and interest in the subject.

Rolf K. Eckhoff
Bergen, March 2016

CHAPTER ONE

Introduction

1.1 PROCESS SAFETY—A PERSISTENT CHALLENGE

Right from the start of the development of the oil and natural gas industry on the Norwegian continental shelf very high safety standards were established. This was a matter of both attitude in industry and official national policy. Until the Piper Alpha oil production platform explosion catastrophe in the North Sea in 1988, it was felt by some people that Norway was overdoing safety issues in its offshore industry. However, after this accident, the imposition of strict safety requirements in offshore oil and gas industries gained wide international acceptance. The more recent blowout, explosion, and fire catastrophe on the oil platform Deepwater Horizon in the Gulf of Mexico in 2010 has undoubtedly amplified this trend. It has also been pointed out that substantial benefits would result if the high standards of safety in the offshore industries could be adapted to industry onshore and to society at large.

However, high safety levels cannot be established once and for all by a single all-out effort. Deterioration results if the high level once attained is not actively secured by continuous maintenance and renewal. This applies to both safety technology and human factors.

Education has a key role in the continuous maintenance and renewal process. This ranges from short practical training courses to in-depth long-term education. Universities and colleges have responded to the challenge by establishing courses of study on a wide range of safety aspects. In the case of process safety, relevant topics include reliability and risk analysis; the physics, chemistry, and technology of processes and hazards; and means of accident prevention and mitigation. Much emphasis has been put on methods of reliability and risk analysis, which are indeed very important. However, it is sometimes felt by the process industry itself that education in the "hard" aspects, ie, the physics, chemistry, and technology of processes and process hazards, has been somewhat left behind. This situation presents a special challenge to universities and colleges. In 1996 the University of Bergen established

Explosion Hazards in the Process Industries
ISBN: 9780128032732
http://dx.doi.org/10.1016/B978-0-12-803273-2.00001-3

a course of study in process safety technology, with particular emphasis on the scientific and technological aspects.

In process safety, the prevention of fires and explosions, and the mitigation of their effects, is a central concern. In the case of combustible gases/vapors, loss of confinement of the combustible substance is the first step in the accidental chain of events. The next steps are generation and ignition of the flammable cloud, resulting in explosion and/or fire, which may in turn cause loss of life and limb, and damaged process plant, adjacent process areas, and even more remote building structures. As pointed out in Section 1.6 below, understanding the processes of accident escalation is an important aspect of process safety technology.

Quantitative risk analysis plays an increasing role in the effort to improve offshore process safety. In the offshore oil and gas industries highly packed, congested process plants, with compact living quarters as close neighbors, ask for very systematic and thorough analysis of all possible risk factors, including the human elements. A concise and constructive official authority policy constitutes an important basis for ensuring the necessary high level of safety.

This book has been written with intent to be a useful source of basic information on the origin, course, prevention, and mitigation of accidental explosions in the process industries. Potential readers/users of the book are expected to include people both from a wide range of process industries, from official authorities, from engineering companies, and not least the students in technical colleges and universities.

1.2 WHAT IS AN EXPLOSION?

The concept of *explosion* is not unambiguous. Various encyclopedias give varying definitions that mainly fall into two categories. The first focuses on the noise or "bang" due to the sudden release of a strong pressure wave, or blast, into the ambient atmosphere. The origin of this pressure wave, whether a chemical or mechanical energy release, is then of secondary concern. This definition of an explosion is in accordance with the basic meaning of the word ("sudden outburst").

The second main category of explosion definitions is confined a *sudden release of chemical energy as heat.* This includes explosions of gases, dust clouds, and solid explosives. The emphasis is then put on the chemical energy release itself, and an explosion is defined accordingly. One possible definition could then be "An explosion is an exothermal chemical process that, when occurring at constant volume, gives rise to a sudden and significant pressure rise."

In the present book, the definition of an explosion will shift pragmatically between the two alternatives, by focusing at either cause or effect, depending on the context. It should be noted that genuine, physical water vapor explosions (see Chapter 4) require a separate definition, because in these explosions the driving force is physical heat transfer, not exothermal chemical reaction.

1.3 GAS/VAPOR AND DUST EXPLOSIONS—REAL HAZARDS IN THE PROCESS INDUSTRIES

In the present book, the two largest chapters are devoted to gas/vapor explosions and dust explosions. This is justified by the fact that these two accidental explosion categories seem to represent the majority of accidental explosions in the process industries. The particular industries/facilities in which gas/vapor or dust explosion hazards may exist include:

Oil and natural gas industries/activities

- Oil and natural gas production installations on- and offshore
- Oil and gas refineries
- Systems for transportation of oil and gas (pipelines, ships, trains, cars, etc.)

Petrochemical, chemical, and metallurgical process industries

- Petrochemical industries producing chemicals and polymers
- Plants producing pharmaceuticals, pesticides, organic pigments, etc.
- Paint-production plants
- Pulverized-metal production (aluminum, magnesium, silicon, silicon alloys, etc.)
- Chemical food and feed production
- Production of cellulose, paper, etc., from wood

Mechanical processing

- Grain and feed storage
- Flour mills
- Sugar refineries
- Mechanical wood refining (hardboard, etc.)

Special processes

- Production, storage, and handling of explosives, pyrotechnics, and propellants

1.4 HOW AND WHERE ACCIDENTAL EXPLOSIVE GAS/VAPOR AND DUST CLOUDS ARE GENERATED IN THE PROCESS INDUSTRIES: BASIC DIFFERENCES

1.4.1 Similar Ignition and Combustion Properties of Clouds Generated

Before discussing the obvious basic differences in the modes of generation of explosive gas/vapor clouds and dust clouds, a possible reason for the similarity should be mentioned. As pointed out by Eckhoff (2006, 2009), explosive gas mixtures and explosive clouds of mists/sprays and dusts, once existing, exhibit very similar ignition and combustion properties, such as:

- flammability/explosibility limits
- laminar burning velocities and quenching distances
- the response of the burning velocity to cloud turbulence
- detonation phenomena
- adiabatic constant-volume explosion pressures of similar magnitudes
- well-defined minimum ignition energies, and
- minimum ignition temperatures for given experimental conditions

Recognition of these similarities may have contributed to the idea that the hazards of accidental gas/vapor and dust explosions could be regarded essentially identical. However, as discussed in Section 1.4.3 (in the following), this idea is only partly correct. One basic difference between gases/vapors and dusts is in the ranges of hazardous fuel concentrations. For combustible gas/vapor clouds, flame propagation is only possible when the fuel/air mixing ratios lie between the lower and the upper flammability limits. However, in the case of dusts, flame propagation is not limited only to the flammable dust concentration range of clouds. Settled dust as layers/deposits presents an additional singular regime of flame propagation. This is because, contrary to combustible gases/vapors, settled dusts are porous and will trap air in the voids between the particles. This makes it possible for sustained, although often very slow, combustion to propagate throughout such deposits.

1.4.2 Influence of Inertial Forces on the Movement of Dust Particles in a Dust Cloud

Once a combustible gas has been homogeneously mixed with air, the mixture will tend to stay homogeneous due to random molecular motion.

In clouds of dust particles, however, the fuel particles are generally so much larger than the molecules of the air (often in the range 1–100 μm) that their movement within the air is controlled by inertial forces, including gravity, rather than by random molecular motion. The role of inertial forces increases systematically with increasing particle size and increasing density of the particle material. Turbulence and other convective movement of the air can prolong the time over which the particles will stay in suspension.

1.4.3 Fundamental Differences Between the Ways Explosive Clouds Are Generated

The fundamental differences between explosive clouds of gases/vapors and dust clouds are in the ways and circumstances in which clouds of the two fuel categories are generated and sustained. The basic physics of generation and sustainment of airborne clouds of dusts and of premixed clouds in air of gas/vapor are substantially different. These differences have a major impact on the choice of means by which accidental explosions are prevented and mitigated. The paramount question is whether an explosive cloud occurs in the first place.

In the case of *gases/vapors*, the fuel mixes with the air at the molecular level. Explosive clouds are generated readily when combustible gases and vapors are accidentally released into the atmosphere. Gas/vapor cloud explosions can also be initiated inside process equipment, but most often the initiation is in accidental clouds generated outside process equipment, following loss of confinement due to leaks or equipment failure.

In the case of *dusts*, however, primary explosive clouds are practically exclusively found inside process equipment, such as mills, mixers/blenders, bucket elevators, pneumatic conveying systems, silos/hoppers, cyclones, and filters. Inside these process units the dust/powder can be kept in suspension more or less continually by rotation of the whole unit, movement of inserts, or an air flow. Therefore, explosive dust clouds may exist more or less continually in normal operation due to the basic nature of the operation.

Dust clouds generated outside process equipment by leaks from that equipment are rare and do not play any significant role as far as primary dust explosions are concerned. The duration of the processes of cloud generation is normally very short (eg, pouring/discharging operations, accidental bursting of sacks and bags). The dust particles will start to settle out of suspension as soon as the cloud generation process terminates, and typical total lifetimes of explosive clouds outside process equipment will be on the order of seconds.

One exception would be long-duration dust leaks from, eg, flanges in pressurized pneumatic transport lines, but the dust cloud volumes produced in such cases would normally be quite small and present a nuisance problem rather than a primary explosion hazard. However, dust layers from such leaks having accumulated outside process equipment present a hazard of secondary dust explosions when the layers are thrown into suspension by blasts from a primary explosion initiated inside process equipment.

1.4.4 Migration of Dust Particles Through Narrow Holes and Gaps in Enclosure Walls

Because dust particles are so much bigger than gas molecules, they will not travel through narrow holes and slots on the order of 1 mm diameter and smaller in the same way as gas molecules will do. In principle, dust particles and liquid droplets may be carried through narrow passages by the air flow generated by a moderate pressure difference across the passage. However, both dust particles and liquid droplets will easily adhere to the area around the passage entrance and to the passage walls and eventually block the passage. In the case of liquid droplets they will most often coalesce and form a liquid film.

Dust particles or liquid droplets that have been able to pass through narrow holes/gaps in this way will not remain suspended in the air inside the enclosure nor eventually form an explosive cloud. Instead, they will settle out of suspension and form a dust layer or a liquid film. In the case of dusts it is also difficult to envisage any mechanical process inside typical electrical apparatus enclosures, which could possibly redisperse such dust layers into explosive dust clouds within the enclosure. In the case of liquid droplets, liquid films will be formed.

1.5 EUROPEAN DEFINITION OF "EXPLOSIVE ATMOSPHERES"

The two European Union directives traditionally named "ATEX 118a" (1999) and "ATEX 100a" (2014), define explosive atmospheres as follows: "Mixtures with air, under atmospheric conditions, of flammable substances in the form of gases, vapours, mists or dusts in which, after ignition has occurred, combustion spreads to the entire unburnt mixture."

Unfortunately, the two directives are unclear regarding the basic differences in the ways in which explosive clouds of gases/vapors and dusts are generated in industrial practice. Discussions of this deficiency were given by Eckhoff (2006, 2009). Bringing these differences to light in European standardization is essential for providing technically and scientifically based differentiated guidance on selection of suitable means for preventing and mitigating accidental gas/vapor explosions on the one hand and accidental dust explosions on the other in the process industries. This definitely applies to the selection of appropriate electrical apparatuses (see Chapter 10).

1.6 DOMINO/ESCALATION EFFECTS FROM ACCIDENTAL EXPLOSIONS

As already mentioned in Section 1.1, an accidental explosion in the process industry may be just the initiating event that triggers one or more successive even more destructive events. In the case of gas/vapor explosions (see Chapter 2), an initial explosion event may tear apart piping and process equipment carrying combustible liquids and/or gases and give rise to major fires. The fires may, in turn, cause weakening and collapse of major building structures.

In the case of dust explosions (see Chapter 7), the primary event is most often a dust explosion in a process unit like a mill, a mixer, a cyclone, a dust filter, or a silo. If the process unit bursts, the blast wave emitted into the surroundings may whirl up dust layers that had accumulated on floors, process equipment, beams, etc., into the explosive dust cloud. If these clouds are then ignited by the flame ejected from the initial process unit, major catastrophes may result. Further escalation may result if the secondary explosions give rise to major fires.

In the comprehensive paper by Cozzani et al. (2006), improved criteria were developed for reliable prediction of potential domino/escalation effects following a primary accidental event. They based their work on recent advances in modeling of explosion and fire damage to process equipment due to heat radiation, overpressure, and missile/solid fragment ejection. They proposed a set of revised threshold criteria for triggering of the next step in the domino/escalation chain.

Cozzani et al. (2009) proposed an inherent safety–based approach for preventing domino/escalation accidents. By defining a set of indexes, they were able to calculate process and layout hazards related to these events. The indexes were based on assessment of inherent-safety distances being calculated by using specific threshold values of various parameters for escalation effects to take place.

Development of computational tools for predicting and controlling domino/escalation effects in relation to accidental explosions in the process industry remains a demanding task.

1.7 THE "HUMAN FACTOR" IN PROCESS SAFETY

The buildup and maintenance of an integrated system for preventing and mitigating explosions in the process industries has to contain a strong element addressing human relations and human behavior. Experience has shown over and over again that the root causes of accidental explosions in these industries very often are various kinds of human failure. It is crucial, therefore, that the human factor be regarded just as important and significant as the technical factors in all the good work done for reducing the risks of accidental explosions.

A systematic approach to human error analysis (HEA) is outlined in Section 11.13 in Chapter 11.

CHAPTER TWO

Gas and Vapor Cloud Explosions

2.1 COMBUSTION OF GASES AND VAPORS

2.1.1 Diffusion Combustion and "Premixed" Combustion

Burning of combustible gases is well known from daily life. Propane (C_3H_8), for example, is often used for lighting, heating, and cooking, if electricity is not easily available. When discussing the burning of combustible gases, it is important to distinguish between diffusion-controlled burning and burning of premixed gas/air. Fig. 2.1 illustrates the flame of a common gas lighter, with butane (C_4H_{10}) as the fuel. The burning is limited to a narrow zone just outside the butane outlet, where the gas makes contact and mixes with the ambient air, and hence with oxygen. The burning occurs comparatively slowly, being controlled by the rate of diffusion of fuel and oxygen molecules into the reaction zone. The same type of burning takes place in a candle flame and in the flame of an oil lamp.

In gas explosions, the situation is different. In this case, fuel gas and air have been mixed in combustible proportions in advance, before ignition occurs. Therefore, combustion, once initiated, will propagate readily

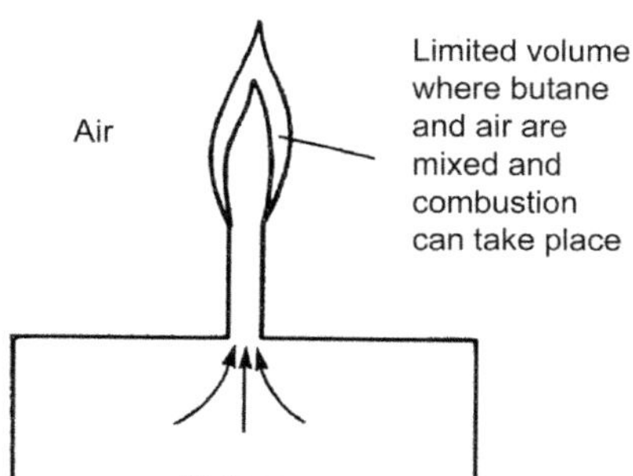

Figure 2.1 Butane lighter flame. Example of diffusion controlled burning of a combustible gas in air.

Explosion Hazards in the Process Industries
ISBN: 9780128032732
http://dx.doi.org/10.1016/B978-0-12-803273-2.00002-5

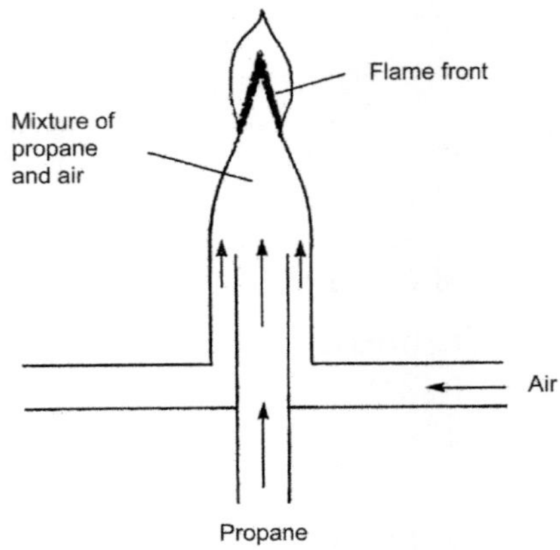

Figure 2.2 Burning of premixed propane/air in a Bunsen burner.

throughout the mixture. This kind of combustion is named "*pre-mixed*" and can be studied eg, in Bunsen burners as illustrated in Fig. 2.2. It is important then that the mass flows of propane and air are adjusted to yield a propane concentration within the combustible range (see Section 2.3 following). Furthermore, the overall mass flow of the mixture must be within certain limits. If the flow is too high, the flame will be blown away upward and extinguish; if it is too low, the flame will propagate downward into the interior of the burner.

2.1.2 Laminar Burning of Premixed Gas/Vapor and Air

Fig. 2.3 gives experimentally determined relationships between the so-called laminar burning velocity and the concentration of combustible gas in the mixture with air, for four common combustible gases.

The ideal laminar burning velocity is the lowest velocity at which a flame front can propagate through a given gas mixture (given ratio of gas/air, given pressure and temperature). The flame zone is then a smooth, planar sheet of thickness on the order of 1 mm, traveling at constant velocity through the quiescent homogeneous gas/air mixture. The heat-transfer mechanism driving the flame propagation is mainly molecular diffusion.

As Fig. 2.3 shows, the differences between S_u for the various alkanes (saturated linear hydrocarbons) are relatively small, all four maximum values falling in the range 0.42–0.47 m/s. However, as illustrated by Fig. 2.4, unsaturated hydrocarbons can have considerably higher maximum values.

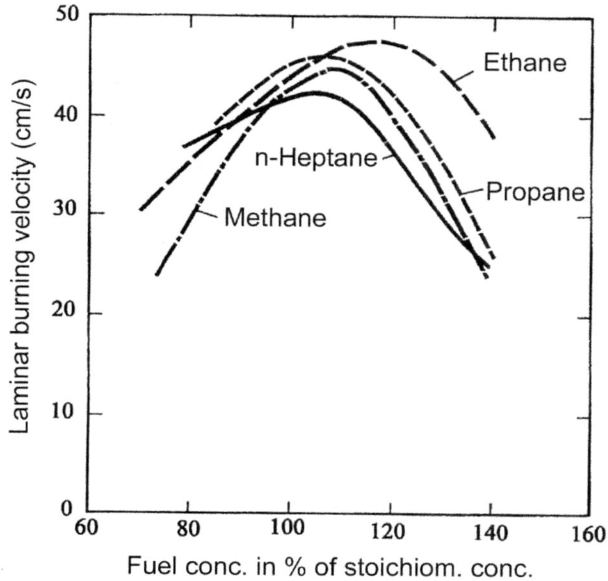

Figure 2.3 Laminar burning velocities S_u at atmospheric pressure and normal temperature for mixtures of various hydrocarbon gases and air.
From Zabetakis, M.G., 1965. Flammability characteristics of combustible gases and vapors. US Bureau of Mines, Bulletin 627, 45.

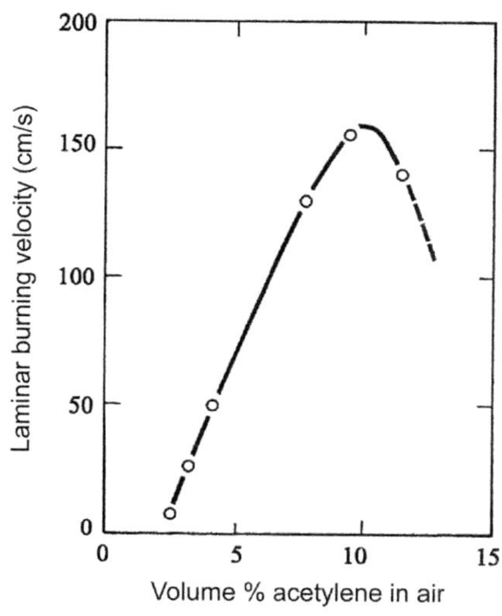

Figure 2.4 Laminar burning velocities S_u for different mixing ratios of acetylene and air, at atmospheric pressure and normal temperature.
From Zabetakis, M.G., 1965. Flammability characteristics of combustible gases and vapors. US Bureau of Mines, Bulletin 627, 59.

Table 2.1 Maximum laminar burning velocities S_u for premixed fuel/air at atmospheric pressure and normal temperature for some fuels

Fuel	Maximum S_u (cm/s)
Alkanes	40–50
Natural gas (unspecified)	40
Acetone	43
Butylene	43
Cyclohexane	44
Benzene	48
Propylene	51
Carbon monoxide	52
Methanol	57
Propylene oxide	67
Ethylene	75
Ethylene oxide	90
Acetylene	155
Hydrogen	325

From Kuchta, J.M., 1985. Investigation of fire and explosion accidents in the chemical, mining and fuel related industries. A Manual. US Bureau of Mines, Bulletin 680, 38.

Both Figs. 2.3 and 2.4 illustrate that for any combustible gas, S_u has a maximum value when the mixing ratio fuel/air is close to the stoichiometric ratio. At the stoichiometric fuel concentration, the ratio of fuel/oxygen is exactly what is required for all the oxygen to transform all the fuel to CO_2 and H_2O. When the fuel/air ratio moves away from the stoichiometric value, whether in the direction of leaner or richer mixtures, the value of S_u becomes systematically smaller.

Table 2.1 provides maximum experimental S_u values for some combustible gases in air.

2.1.3 Flammable Concentration Ranges for Premixed Gas/Vapor and Air

2.1.3.1 General

As the fuel/air ratio is increased or decreased sufficiently, departing from the stoichiometric value, two finite limits are eventually reached beyond which the mixture is no longer able to propagate a flame. These limits bracket the flammable (or explosive) range of a given combustible gas or vapor in air. Fig. 2.5 illustrates the flammability ranges in air for some gases. Because acetylene, being a very reactive compound, is able to decompose exothermally in the absence of oxygen, its upper explosion limit is in fact 100%. The first

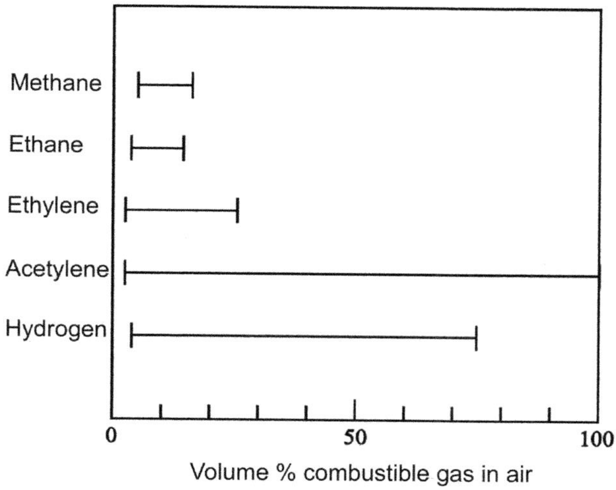

Figure 2.5 Flammable ranges for mixtures of some combustible gases and air at atmospheric pressure and normal temperature (see also Table 2.2).

two columns of Table 2.2 give numerical values for the lower and upper flammability limits of a number of combustible gases and vapors in air, including those in Fig. 2.5.

It should be pointed out that the experimental determination of flammability limits is not straightforward, and results obtained using different methods may scatter. The method used in the United States is described in American Society for Testing and Materials (ASTM) (2015a). The fuel/air mixture to be tested is placed in a 5 L spherical glass flask equipped with an internal spark gap near the flask center for electric spark ignition of the mixture. The flask is filled with a uniform mixture of the desired fuel/air ratio, followed by activation of the ignition source, and visual observation of the extent of flame propagation. Flame propagation both upward and sideways is deemed sufficient for the mixture to be regarded as flammable, whereas flame propagation upward only, is not deemed sufficient. Downward flame propagation is not regarded a necessary requirement for the mixture to be classified as flammable. In a comment, Ural (2004) pointed out that the lower and upper flammability limits (LFL, UFL, respectively) observed in the ASTM test are not necessarily identical with lower and upper explosibility limits (LEL, UEL, respectively) determined by other methods. It is clear, therefore, that experimental determination of flammability limits of mixtures of combustible gases/vapors and air is by no means straightforward.

Table 2.2 Combustibility and ignitability parameters of some combustible gases and vapors in air at atmospheric pressure and normal temperature

Fuel	Flammable limits [vol. % in air]		Flash point (°C)	Min. ign. temp. (°C)	MESG (mm)	Temp. class	Gas group code
	Lower	Upper					
Acetone	2.5	13.0	<−20	539	1.01	T1	IIA
Acetylene	2.3	100.0	?	305	0.37	T2	IIC
Ammonia	15.0	33.6	?	630	3.18	T1	IIA
Benzene	1.2	8.6	−11	498	0.99	T1	IIA
Butadiene (1,3-)	1.4	16.3	?	420	0.79	T2	IIB
Butane (n-)	1.4	9.3	?	372	0.98	T2	IIA
Butyl alcohol(1-)	1.4	12.0	35	343	0.91	T2	IIA
Cyclohexane	1.0	8.0	−17	244	0.94	T3	IIA
Butene(1-)	1.6	10.0	?	345	0.94	T2	IIA
Decane	0.7	5.6	46	235	1.05	T3	IIA
Di-ethyl ether	1.7	36.0	−45	160	0.87	T4	IIB
Di-methyl ether	2.7	32.0	?	240	0.84	T3	IIB
Ethane	2.4	15.5	−29	515	0.91	T1	IIA
Ethanol	3.1	19.0	12	400	0.89	T2	IIB
Ethene (ethylene)	2.3	36.0	?	440	0.65	T2	IIB
Heptane	0.85	6.7	−7	204	0.91	T3	IIA
Hexane	1.0	8.9	−22	225	0.93	T3	IIA
Hydrogen	4.0	77.0	?	560	0.29	T1	IIC
Hydrogen sulphide	4.0	45.5	?	260	0.83	T3	IIB
Carbon disulphide	0.6	60.0	−30	90	0.34	T6	IIC
Carbon monoxide	10.9	74.0	?	607	0.84	T1	IIB
Methane	4.4	17.0	−223	600	1.12	T1	I/IIA[a]
Naphtha	0.9	6.0	<−18	289	?	T3	IIA
Propane	1.7	10.9	−104	450	0.92	T2	IIA
Styrene	1.0	8.0	30	490	?	T1	IIA
Toluene	1.0	7.8	4	530	1.06	T1	IIA

[a]For methane Code I applies to coal mines only, whereas Code IIA applies to all other situations.

The other data in Table 2.2 will be considered when the respective parameters are discussed later in this chapter.

2.1.3.2 Flash Point of a Combustible Liquid

The flash point of a combustible liquid, as defined by The Institute of Petroleum (2002), is the lowest temperature at which, in contact with air at atmospheric pressure, a flammable liquid will give off a sufficiently rich vapor for ignition to occur in the presence of an ignition source. On the condition that the ignition source is sufficiently effective, the flash point is the lowest temperature the liquid can have for the vapor/air mixture just above the liquid surface to be able to propagate a flame. There is then a direct coupling between the lower flammability limit of vapors and the flash point of their liquid phases. This is because the flash point is the temperature at which the vapor pressure of the liquid equals the partial pressure of the actual vapor at its lower flammability limit.

Flammable liquids are grouped in hazard classes on the basis of their flash points. The Institute of Petroleum (IP) classification system uses a subdivision of the main Classes II and III according to whether the condition of handling is at a temperature below or above the flash point. This is in agreement with the conception that an explosive gas (vapor) atmosphere generally cannot exist if the flash point is significantly above the maximum temperature of that fluid.

However, as will be discussed in Chapter 5, if a pressurized combustible liquid at a temperature far below its flash point is released to the atmosphere as a spray of small droplets in air, the spray can become ignited and propagate a flame just like a premixed homogeneous explosive gas cloud.

The third column of Table 2.2 gives flash points for some combustible liquids in air. The other four parameters included in Table 2.2 will be discussed later in this chapter in their relevant contexts.

2.1.3.3 Classification of Flammable Fluids According to Their Flash Points

The classification system for flammable fluids recommended by The Institute of Petroleum (2002) is given in Table 2.3. The main classification conforms to the flash-point classification systems of the European Directives. The subdivision of Classes II and III has been standardized by The Institute of Petroleum (2002) to aid the handling of petroleum materials throughout bulk storage, loading, conveyance, discharge, and distribution. It should, however, be recognized that there is an upper temperature limit, though

Table 2.3 Classification of flammable fluids according to their flash points

Fluid class	Description
Class 0	Liquefied petroleum gases (LPG)
Class I	Liquids that have flash points below 21°C
Class II(1)	Liquids that have flash points from 21°C up to and including 55°C, handled below flash point
Class II(2)	Liquids that have flash points from 21°C up to and including 55°C, handled at or above flash point
Class III(1)	Liquids that have flash points above 55°C up to and including 100°C, handled below flash point
Class III(2)	Liquids that have flash points above 55°C up to and including 100°C, handled at or above flash point
Unclassified	Liquids that have flash points above 100°C

From The Institute of Petroleum, August 2002. Area classification code for installations handling flammable fluids, second ed. The Institute of Petroleum, London.

unlikely to be encountered, in Classes I, II(2), and III(2) classification applications above which they should not be extended; this upper limit is the temperature at which the liquid would boil.

Typical examples of petroleum fluids are shown in Table 2.4.

When the flammable fluid is a liquid, its volatility in relation to the conditions of temperature and pressure at which a potential release might take place is an essential factor, because it will determine the extent of rapid vapor formation from that release. In many of the commonly encountered types of process plant, the flammable fluid can be classified satisfactorily by its flash point, eg, in accordance with the procedure of The Institute of Petroleum (2002). Further subdivision of the main classes (II and III) into subclasses (1) and (2) is done according to whether the liquids are handled at temperatures above or below the flash point.

2.1.4 Maximum Pressures Generated from Constant-Volume Adiabatic Combustion of Premixed Gas/Vapor and Air

If the combustion occurs at constant volume, a pressure rise will result. The adiabatic temperature rise at constant volume can be expressed as

$$T_2 - T_1 = \Delta E / C_v \qquad [2.1]$$

in which ΔE is the liberated combustion heat per mole at constant volume, C_v the mean specific heat of the gaseous system at constant volume, T_1 the

Table 2.4 Classification of commonly encountered combustible petroleum materials

Class of petroleum based on flash point	Examples of petroleum materials	Flash point (°C)	Boiling range (°C)
Class 0	LPGs, ethylene, propylene	Not applicable	Propane −42 Butane −1 Isobutane −12
Class I			
<21°C	Gasoline (petrol)	−45	ca. 20−205
	Stabilized crude oil	<21	ca. −1 to 380+
	Avtage wide-cut jet fuel (JP4; Jet B)	−25	ca. 0−220
	Benzene	−11	80
	Toluene	4	110
	Naphtha	−2 to 10	30−177
	Methanol	11	65
Class II			
21−55°C	Avtur/Jet A Turbofuel	38 (min.)	150−240
	Kerosene		
	(a) Premium grade	43 (min.)	160−280
	(b) Regular	38 (min.)	150−280
Class III			
55−100°C	Gas oil/distillate		
	Heating oil	55+	250−360
	Automotive diesel fuel	55+	180−360
Unclassified	Atmosphere residues		
>100°C	Heavy fuel oils	>100	>350

From The Institute of Petroleum, August 2002. Area classification code for installations handling flammable fluids, second ed. The Institute of Petroleum, London.

initial gas mixture temperature (prior to combustion), T_2 the final temperature of the combustion products and the nitrogen (from the air), after complete combustion. Ideal T_2 values computed using Eq. [2.1] are most often somewhat higher than experimental values, due to inevitable heat losses and nonideal chemical conversion in real experiments. However, assuming that an adequate T_2-value can be obtained (see Table 2.6), the corresponding constant-volume adiabatic pressure rise can be estimated, using the equation of state (Eq. [2.2]), assuming that the gases/vapors are ideal.

$$P_2/P_1 = n_2T_2/n_1T_1 \quad [2.2]$$

Here, P is pressure (absolute), T temperatures (K), and n total number of moles of gas in the constant volume considered. A more detailed analysis is given by Kuchta (1985). Tables 2.5 and 2.6 give some experimental P_2-values. The data in the two tables originate from different sources and deviate somewhat. Different fuel/air ratios in the experiments may be one

Table 2.5 Highest maximum explosion pressure (gauge) for combustion of premixed fuel gas/air in a closed 5-L test vessel. The fuel gas/air ratio was varied until the highest explosion pressure had been found for each fuel. Normal atmospheric initial conditions

Fuel	Bar (g)	Fuel	bar (g)
Acetylene	10.3	i-Butyl alcohol	7.6
Ethylene oxide	9.9	Ethyl alcohol	7.5
Benzene	9.0	n-Butyl alcohol	7.5
Acetone	8.9	n-Decane	7.5
Ethylene	8.9	Hydrogen	7.4
n-Hexane	8.7	Carbon monoxide	7.3
n-Pentane	8.7	Methane	7.2
n-Butane	8.6	Toluene	6.8
Cyclohexane	8.6	Vinyl chloride	6.8
n-Heptane	8.6	Methyl butadiene	6.6
Propane	8.6	Styrene	6.6
Propylene	8.6	Ammonia	6.0
i-Octane	8.1	Acetic acid	5.4
Carbon disulfide	7.8	Hydrogen sulfide	5.4

From Freytag, H.H., 1965. Handbuch der Raumexplosionen. Verlag Chemie, Weinheim, Germany, p. 152.

Table 2.6 Combustion properties of some fuel gas/air mixtures. Normal atmospheric initial conditions

		Constant volume	Constant pressure	Adiabatic flame temperature	
Fuel	**Vol.% fuel in air**	**Max. pressure (bar (g))**	**Expansion ratio E (−)**	**Constant volume (K)**	**Constant pressure (K)**
Acetylene	7.75	8.80	8.38	2925	2542
Hydrogen	29.60	7.02	6.88	2755	2384
Ethylene	6.53	8.32	7.48	2740	2372
Butane	3.13	8.43	8.06	2639	2272
Propane	4.0	8.34	7.98	2633	2266
Ethane	5.67	8.21	7.87	2627	2261
Methane	9.51	7.60	7.25	2591	2227

From Lee, J.H.S., 1984. Physics of Explosions. Lecture Notes. McGill University, Montreal, p. 249.

reason for the discrepancies. Note: Initial temperatures T_1 and pressures P_1, deviating from normal atmospheric conditions, give correspondingly different P_2 values.

As a rule, the maximum adiabatic explosion pressure (abs.) produced by a given fuel/air mixture at constant volume, is proportional to the initial pressure (abs.) prior to combustion. Hence, a gas mixture giving 8 bar (abs.) at atmospheric initial pressure, will yield 16 bar (abs.) at 2 bar (abs.) initial pressure (1 bar overpressure).

In the case of premixed gas explosions in a system of closed volumes in series, coupled via comparatively narrow ducts or pipes, an explosion in one part of the system can give rise to considerable pressure rise in the still unburnt gas mixture in other parts ("pressure piling"). When the flame front reaches and ignites such precompressed gas volumes, very high-pressure transients can result, despite the fact that the entire system was at atmospheric pressure prior to the first ignition (see Section 2.4.5.2.1 in the following).

2.1.5 The "Expansion Ratio" in Combustion of Premixed Gas/Vapor and Air

Fig. 2.6 illustrates idealized adiabatic (no heat loss to the tube, no buoyancy, no interference of the wall with the gas flow) planar, laminar combustion of premixed quiescent combustible gas/air in a one-end-open straight tube/duct, at constant pressure.

If the gas mixture is ignited in a plane across the open end of the tube (Fig. 2.6a), the combustion products will expand freely into the ambient atmosphere, whereas the still unburnt gas further into the tube will remain

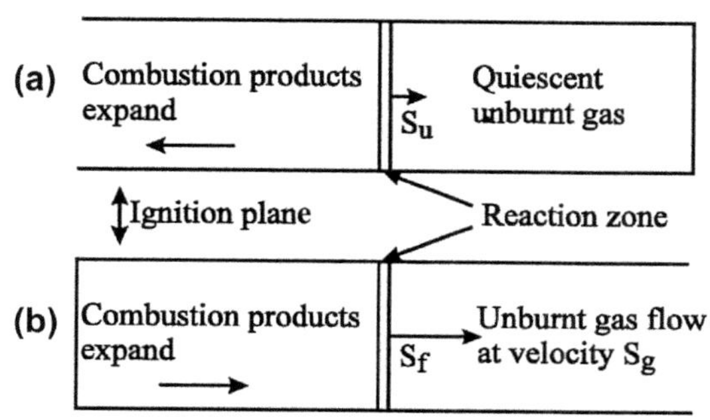

Figure 2.6 Premixed fuel gas/air burning at constant pressure with a plane, laminar flame in a one-end-open tube. (a) Ignition at open tube end. (b) Ignition at closed tube end.

quiescent. The observed flame speed in relation to the tube wall will then be identical with the laminar burning velocity S_u of the actual gas/air mixture. If ignition occurs at the closed tube end (Fig. 2.6b), however, the expansion of the combustion products is forced to occur in the same direction as that of flame propagation. Therefore, the unburnt gas mixture ahead of the flame will be pushed toward the open tube end. In this case, the observed flame speed S_f in relation to the tube wall is the sum of S_u and the flow velocity S_g of the unburnt gas. Hence, Eqs. [2.3] and [2.4] describe how the liberated combustion heat, in the case of constant pressure, gives rise to expansion.

$$S_f = S_u + S_g \qquad [2.3]$$

and, therefore,

$$S_f = S_u \cdot E \qquad [2.4]$$

Here, E is the "expansion ratio," defined by equation Eq. [2.3] as $(1 + S_g/S_u)$ or by Eq. [2.4] as S_f/S_u.

In other words, E is the ratio of the volume that a given quantity of premixed gas has attained after complete, ideal adiabatic combustion at constant pressure and its volume before combustion. For a given fuel, E varies with the fuel/air ratio and has its maximum around the stoichiometric ratio or somewhat higher. Maximum E-values for most hydrocarbons in air are in the range 7.5–8.0. Some values are given in Table 2.6. The reason for this substantial expansion is the large amount of heat liberated when the fuel and oxygen combine and form combustion products (CO_2 and H_2O for complete combustion of hydrocarbons). Note that *burning velocity* is by definition the relative linear velocity by which the combustion reaction is "eating" itself into the unburnt gas/air mixture, whereas *flame speed* denotes the linear velocity of the flame front in relation to a stagnant observer or structure/apparatus. If the combustion occurs under conditions in which expansion cannot occur freely, the liberated heat may give rise to a substantial pressure rise, which, in turn, may cause substantial damage to industrial equipment, plants, buildings, etc. (see Tables 2.5 and 2.6).

Fig. 2.7 illustrates ideal laminar *spherical* combustion of a premixed gas/air cloud, following ignition at a point in the cloud. In this case, assuming a very thin flame and negligible buoyancy, the unburnt gas/air will always be pushed in the direction of flame propagation and Eqs. [2.3] and [2.4], with the same numerical values as for the linear case in Fig. 2.6b, apply even in this case.

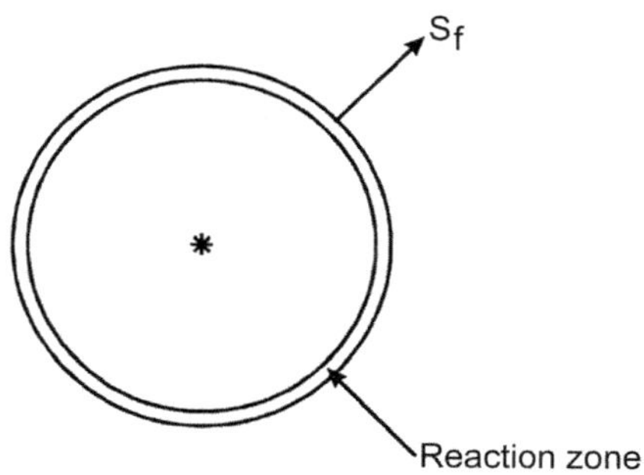

Figure 2.7 Ideal laminar spherical burning of quiescent premixed fuel gas/air, following ignition at a point.

2.1.6 Turbulent Combustion of Premixed Gas/Vapor and Air

In real premixed accidental combustion of gas clouds in an industrial plant, the cloud will not be quiescent, but turbulent. Flame propagation in turbulent gas is illustrated in Fig. 2.8.

Skjold (2003) conducted an extensive study of the influence of turbulence on the burning properties of premixed propane/air in 20 L closed vessels. In turbulent clouds the flame front is no longer a continuous, planar sheet, but folded and torn, with an effective surface area that can be substantially larger than the nominal projected area. This in turn gives rise to nominal turbulent burning velocities that can be considerably higher than the laminar value for the gas mixture in question. Fig. 2.9 gives some results from measurements in a burner similar to the one illustrated in Fig. 2.2.

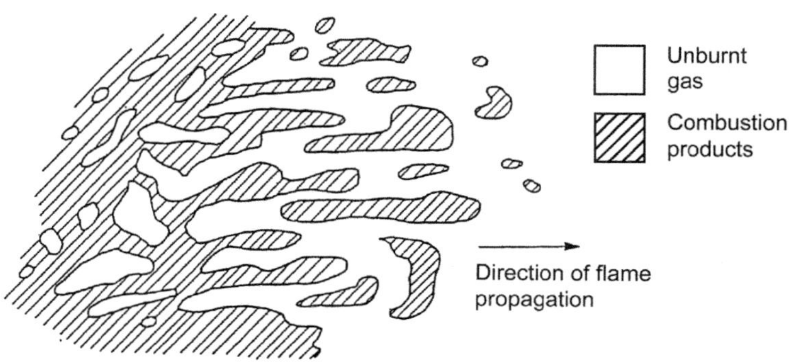

Figure 2.8 Illustration of flame-front structure in turbulent premixed gas.

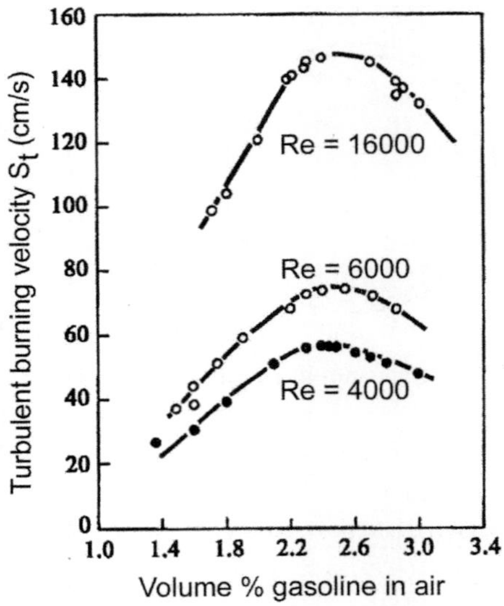

Figure 2.9 Turbulent burning velocity as a function of Reynolds number Re. Premixed gasoline/air in steady-state burner of diameter 16 mm.
From Freytag, H.H., 1965. Handbuch der Raumexplosionen. Verlag Chemie, Weinheim, Germany, p. 70.

When discussing accidental gas explosions, one distinguishes between two kinds of turbulence, viz. "initial" or "preexplosion" turbulence, and turbulence generated by the explosion itself. The latter is caused by the transient flow of the unburnt gas ahead of the flame, generated by the expansion of the combustion products behind the flame. The production of this kind of turbulence is particularly exaggerated in areas packed with process equipment, pipes, and other kinds of obstructions to the flow, as illustrated in Fig. 2.76.

Fig. 2.10 illustrates the dramatic effects that can result from gas explosions in obstructed geometries.

These experiments were performed in a one-end-open tube of diameter 2.5 m, length 10 m, and volume 50 m^3. Ring-shaped turbulence-generating steel baffles could be mounted at various positions in the tube. Because the tube was fully open at one end, the rate of combustion had to be very high for high explosion overpressures to be generated inside the tube. As the table in Fig. 2.10 shows, the influence of introducing the flow obstructing rings on the maximum explosion overpressure in the tube was dramatic, no rings giving 0.15 bar (g), 1 ring 0.6 bar (g), 3 rings 3.5 bar (g), and 6 rings 8.0 bar (g). In other words, the variations in

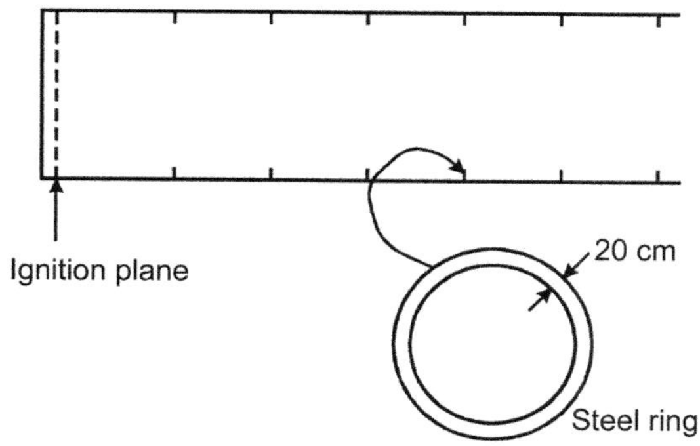

Max explosion pressures in tube for 9.5 vol. % methane in air	
Ring configuration	[bar(g)]
No ring in tube	0.15
1 ring 2 m from ign.plane	0.60
3 rings	3.50
6 rings (as shown in fig.)	8.00

Figure 2.10 Experiment performed in Norway to determine the influence of explosion-generated turbulence on the violence of gas explosions, using turbulence-generating rings (baffles).
From Moen, I.O, Lee, J.H.S., Hjertager, B.H., Fuhre, K., Eckhoff, R.K., 1982. Pressure development due to turbulent flame propagation in large-scale methanol air explosions. Combustion and Flame 47, 31–52.

overpressure were by more than a factor of 50! These experiments urged the development of a computer code (FLame ACceleration Simulator (FLACS), developed by GexCon AS in Bergen, Norway) that can predict the explosion overpressures to be expected from a given fuel gas/air system, as a function of geometry of the explosion environment, and the location of the ignition point (see Sections 2.3 and 2.4).

Section 7.2.5.3 in Chapter 7 discusses the acceleration of dust explosions in elongated systems, right up to detonation (see Section 2.1.7). This discussion also applies to explosions in premixed gases.

2.1.7 Detonation of Premixed Gas/Vapor and Air

Detonation is an extreme singular mode of flame propagation in premixed gas. In this case, the combustion zone follows immediately behind, and drives a leading shock front moving into the quiescent, unburnt gas mixture at supersonic speed. The mechanism of flame propagation is then no longer thermal diffusion, but extremely rapid compression, whereby the temperature in the compression zone is raised to such a high level that ignition occurs almost instantaneously. The necessary condition for self-sustained detonation propagation is that the shock wave is sufficiently strong for the volume inside it to become ignited and react chemically before the shock wave has traveled a significant distance away. In this way, the shock wave and the chemical reaction zone remain closely coupled, and the shock wave speed and strength are maintained. As shown in Table 2.7, typical detonation velocities in premixed hydrocarbon gas/air, at normal initial pressure, temperature, and optimum fuel concentrations, are 1800–1900 m/s, ie, on the order of five times the velocity of sound in the unburnt, uncompressed premixed gas/air. For this reason, the unburnt mixture ahead of the detonation front cannot receive any gas dynamic signal from the approaching detonation front until being caught by the front itself. Therefore, reducing the maximum explosion pressure of a detonation by venting is impossible.

Detonations can be initiated directly by means of a sufficiently strong explosive charge or electric spark. In accidental explosions in industry, deflagration-to-detonation (DDT) is the common initiation mechanism. This can, eg, occur in long ducts, when turbulence and burning rates increase to such high levels that the compressed unburnt gas, trapped between the flame front and the leading shock, ignites spontaneously. Extremely high transient-pressure peaks can then be generated before stable detonation conditions are attained (See also Section 7.2.5.3 in Chapter 7.).

In addition to typical detonation front velocities, Table 2.7 also gives corresponding detonation peak pressures. Hydrogen deviates from the hydrocarbons because the density of 30% hydrogen in air is only 18/29 of

Table 2.7 Detonation properties of some fuel gases premixed with air at atmospheric pressure and temperature. Stoichiometric mixtures

Fuel	Detonation velocity (m/s)	Overpressure in detonation front (bar (g))
Methane	1800	16.1
Ethane	1800	17.0
Propane	1800	17.3
Butane	1800	17.4
Ethylene	1860	17.7
Acetylene	1870	18.0
Hydrogen	1960	14.6

From Lee, J.H.S., 1984. Physics of Explosions. Lecture Notes. MeGill University, Montreal, p. 249

the density of air, and the velocity of sound is therefore 418 m/s as opposed to 330 m/s (see Eq. [2.5]).

Fig. 2.11 shows the result of an experiment in which stoichiometric premixed acetylene/air detonated in a 46-m long tube of diameter 0.6 m.

Some simple relations are useful. The first is

$$P_{CJ}/P_1 = \left[1 + \gamma_1 (V^*/C_1)^2\right] \Big/ [1 + \gamma_2] \qquad [2.5]$$

in which P_{CJ} is the Chapman–Jouguet detonation front pressure; P_1 the initial preignition pressure; γ_1 and γ_2 the C_p/C_v ratio (specific heats at constant pressure and volume) for the preignition gas mixture and the combustion products, respectively; V^* the detonation front velocity in the unburnt mixture;, and C_l the velocity of sound in the preignition mixture. V^*/C_l is the Mach number, which is >>1 for detonation fronts in hydrocarbon/air mixtures.

The following "rule of thumb" gives useful guidance:

$$P_{CJ} \approx 2P_2 \qquad [2.6]$$

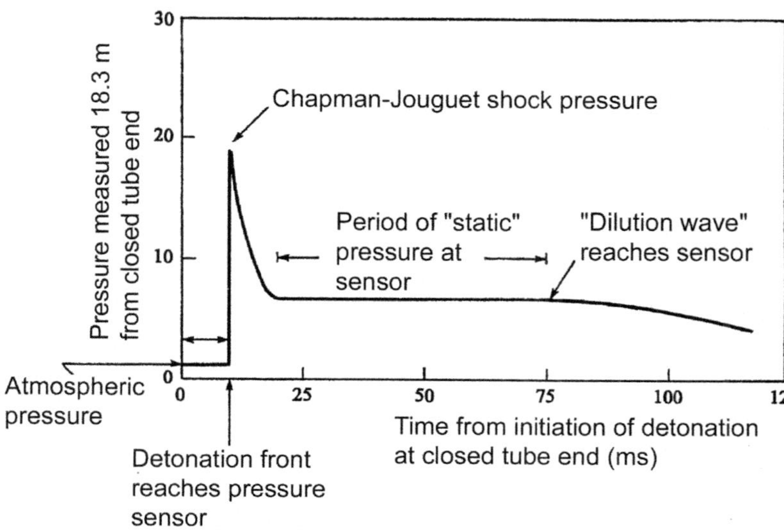

Figure 2.11 Detonation of a stoichiometric mixture of acetylene and air in a one-end-open tube of diameter 0.6 m. The detonation was initiated at the closed tube end by means of an explosive charge. The pressure sensor was located 18.3 m from the closed tube end.

From Kuchta, J.M., 1985. Investigation of fire and explosion accidents in the chemical, mining and fuel related industries. A Manual. US Bureau of Mines, Bulletin 680.

Here, P_2 is the maximum adiabatic explosion pressure at constant volume (Tables 2.5 and 2.6).

For any given ideal gas the velocity of sound C is independent of pressure, but dependent on temperature. The following relation applies:

$$C = [(\gamma P_o T)/(273\rho_o)]^{0.5} \qquad [2.7]$$

in which γ is C_p/C_v (≈ 1.4 for two-atomic gases), P_o and ρ_o the gas pressure and density at the same reference temperature, and T the absolute temperature of the gas. In other words, at any given T, the velocity of sound C is inversely proportional to $M^{0.5}$, in which M is the mole weight of the gas. This is why C, at any given temperature, is considerably higher in hydrogen than in air.

Fig. 2.12 illustrates the dramatic span (logarithmic scale!) of possible burning velocities of fuel/air mixtures, using propane/air as an example. The concentration ranges for detonation are somewhat narrower than those for turbulent combustion, which are in turn somewhat narrower than those for laminar burning. Table 2.8 gives some figures illustrating this.

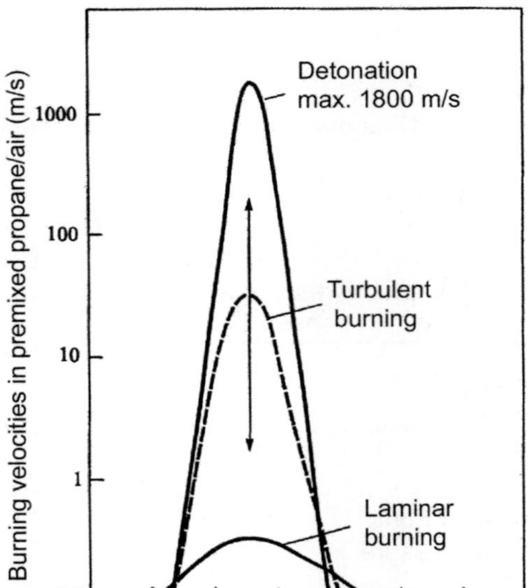

Figure 2.12 Burning velocities in premixed propane/air at atmospheric pressure and normal temperature, for different combustion modes (laminar, turbulent, or detonation).

Table 2.8 Comparison of composition limits for laminar burning and detonation for two combustible gases mixed with air

	Limits for laminar flames (vol.% in air)		Detonability limits (vol.% in air)	
Fuel	**Lower**	**Upper**	**Lower**	**Upper**
Acetylene	2.5	100	4.2	50
Hydrogen	4.0	75	18.3	59

Data from Table 2.2 and Freytag, H.H., 1965. Handbuch der Raumexplosionen. Verlag Chemie, Weinheim, Germany, p. 74.

2.2 IGNITION OF PREMIXED GAS/VAPOR AND AIR

2.2.1 Introduction

This section concentrates on initial cloud conditions of normal atmospheric pressure and temperature, and on ignition of premixed hydrocarbon gases and air. Experimental evidence elucidating the basic features of various ignition processes and practical guidelines for prevention of ignition in industrial process plants are considered.

The term ignition source is used as a general generic term embracing all categories of heat sources that may in principle give rise to ignition. The term effective ignition source indicates that a particular source will actually cause ignition if brought into contact with the specific explosive atmosphere of concern.

Numerous standards and guidelines, national as well as international, are currently in use for preventing accidental ignition of explosive atmospheres. In Europe, a comprehensive standard was drafted by CEN (1997). This document also addresses ignition prevention, and in this context refers to and builds on a series of European and fully international standards on various specific aspects of the ignition prevention problem. The guidelines for industrial practice given in the various following sections are mainly based on this document. In addition to Eckhoff (1996), the paper by Eckhoff and Thomassen (1994) is a main information source for this section.

2.2.2 What Is Ignition? Basic Theory of "Thermal Runaway"

Explosive gas mixtures can be ignited by a variety of ignition sources including:

- Open flames (matches, welding, and soldering flames, etc.)
- Glowing/smoldering materials
- Hot solid surfaces

- Burning metal particles and "thermite" flashes from impacts, grinding, etc.
- Electrical and electrostatic sparks, arcs, and other discharge forms
- Jets of hot combustion gases
- Adiabatic compression
- Light radiation, eg, light conveyed through optical fibers/cables

The question then arises whether ignition by all these quite different sources can be described qualitatively by one common global conceptual model. The answer is yes, and the model is the classical thermal explosion theory formulated by Frank-Kamenetskii (1955). In the following the essence of this theory will be outlined qualitatively with reference to Fig. 2.13.

In principle, the theory applies to any combustible system exposed to a potential ignition source or process. In the context of the present book it applies with equal validity to explosive clouds of combustible gases, liquid sprays/mists, or dusts, and to combustible dust layers/deposits.

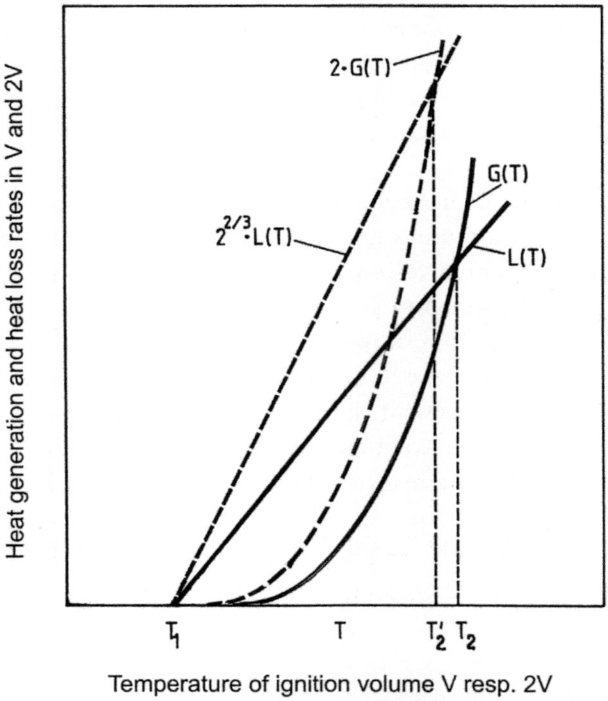

Figure 2.13 Simplified schematic illustration of the basic idea of the "thermal explosion theory" by Frank-Kamenetskii (1955).

Consider an explosive gas mixture, eg, 4 vol.% propane in air, initially at room temperature. Then consider a small "ignition volume" inside the bulk of this mixture. Assume that the "ignition volume" can be heated to any desired temperature by a small heating source, eg, an electrically heated resistance wire that has been placed inside it. Then make the simplifying assumptions that the temperature throughout the "ignition volume" is uniform at any time during the heating-up process, with a very sharp temperature drop to ambient temperature at the boundary, and that the "ignition volume" does not expand during heating. Fig. 2.13 then illustrates the increase of the heat generation rate inside the "ignition volume" due to the chemical reaction of the gas mixture there, as well as the increase of the conductive heat loss rate from this volume to the ambient atmosphere, with increasing temperature in the "ignition volume."

First, consider the solid lines representing an "ignition volume" V. The curved line then represents the rate of heat generation G(T) within the ignition volume due to the exothermal reaction between the fuel and the air, as a function of the temperature in this volume. According to the classical Arrhenius theory, this relationship is exponential for a zero-order chemical reaction. The straight solid line represents the rate of heat loss L(T), which increases linearly with the temperature drop between the "ignition volume" and the ambient gas. Fig. 2.13 illustrates that at modest temperatures T the rates of heat generation by combustion are normally substantially lower than the heat loss rates, and it is thus impossible for the temperature in the "ignition volume" to rise above ambient by the slow heat-producing chemical reaction only.

However, if T is raised by activating the "external" heating source, G(T) will start to rise exponentially. If the "external" heating is brought to an end while the temperature of the ignition volume is still below the critical temperature for ignition T_2, the rate of heat loss L(T) will still exceed the heat generation rate by chemical reaction G(T), and the temperature in the "ignition volume" will drop back to ambient temperature again. If, however, T is raised further to a point at which G(T) exceeds L(T), a positive "feed-back loop" is established by which further temperature rise is accomplished by the combustion reaction itself. This process will eventually lead to ignition. The critical temperature T_2 for ignition is the one at which:

$$G(T) = L(T) \qquad [2.8]$$

and

$$dG(T)/dT > dL(T)/dT \qquad [2.9]$$

Now consider the dotted lines in Fig. 2.13, representing an "ignition volume" of 2V. In this case the value of G(T) will be twice that for the "ignition volume" V for any value of T, simply because the amount of reactive mixture has been doubled. However, because the heat loss rate from the "ignition volume" increases proportionally with the surface area of this volume, and not with the volume itself, L(T) at any T has only increased by a factor of $2^{2/3}$ compared with that for the "ignition volume" V. Because of this, the critical temperature $T_{2'}$ at the crossing point between G(T) and L(T) will be lower in the case of 2V than in the case of V.

As a first approximation, this means that the minimum temperature that an ignition source must have to cause ignition will decrease with increasing size of the "ignition volume" to be associated with that source. In reality, the shape of the "ignition volume" will also play an important role.

2.2.3 Ignition by Open Flames and Hot Gases

Open flames are gaseous combustion reactions at temperatures of at least 1000°C. Flames and their hot gaseous reaction products, even at very small volumes, are among the most effective sources of ignition for explosive gas clouds.

2.2.4 Ignition by Hot Surfaces

2.2.4.1 Overview

The minimum hot-surface temperature for igniting a given mixture of combustible gas in air has sometimes been regarded as a fundamental constant for that mixture. However, this is a false perception. In general terms, ignition is a dynamic process in which chemical heat generation and physical heat loss compete in a complex manner in the potential ignition region, and in which the former eventually overtakes the latter. This also applies to hot-surface ignition.

The size of the hot surface and the relative movement between the combustible gas mixture and the hot surface are two key parameters controlling the minimum ignition temperature, T_{min} of a given gas mixture. Classical investigations of these effects are illustrated in Figs 2.14 and 2.15. Both these classical experiments should be compulsory background in any study of hot-surface ignition processes. Some results by Silver (1937) and Paterson (1939, 1940) are shown in Fig. 2.14.

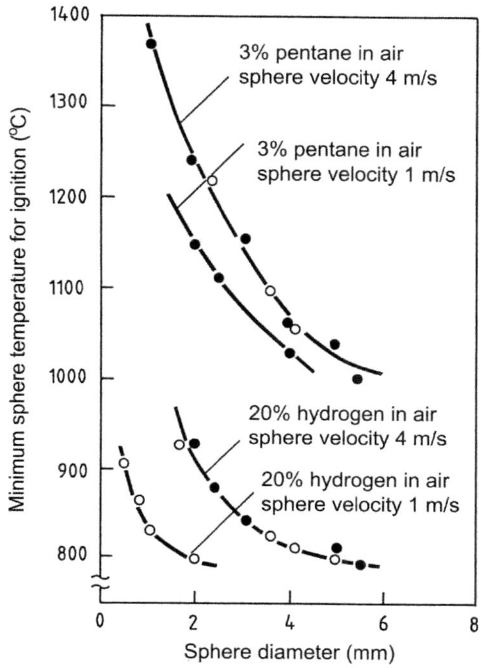

Figure 2.14 Ignition of explosive gas mixtures by hot solid spheres injected into the gas, showing the influence of sphere diameter and velocity on minimum sphere temperature for ignition: (●) quartz spheres; (o) spheres of aged platinum.
From Silver (1937), Paterson (1939) and Paterson (1940).

Two systematic trends can be extracted from Fig. 2.14:

- T_{min} decreases with increasing sphere diameter
- T_{min} decreases with decreasing sphere velocity.

There was no significant difference between spheres of quartz and aged platinum.

A similar set of classical results obtained by Mullen et al. (1949) is shown in Fig. 2.15.

In this case, the explosive pentane/air mixture was flowing past a stationary hot-metal rod at comparatively high velocities. The minimum rod temperature for ignition was recorded as a function of rod diameter and gas velocity. Fig. 2.15 reveals the same main trends as Fig. 2.14. T_{min} decreases systematically with increasing size of the hot surface and with decreasing relative velocity between the hot surface and the gas. It should be noted that the range of relative velocities is considerably higher in Fig. 2.15 than in Fig. 2.14, by a factor of about 10.

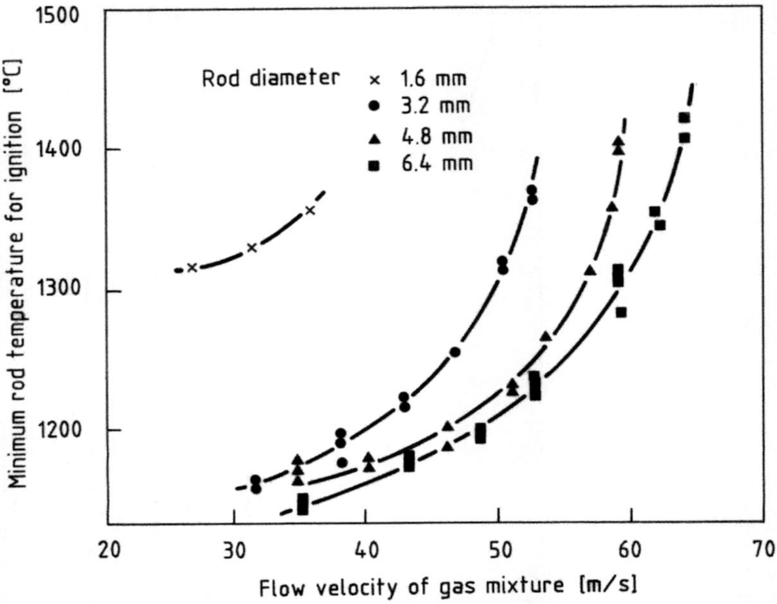

Figure 2.15 Ignition of flowing stoichiometric premixed pentane/air by hot stationary metal rods inserted into a gas flow, showing the influence of the rod diameter, and the gas velocity in relation to the rod, on the minimum rod temperature for ignition. *From Mullen, J.W., Fenn, J.B., Irby, M.R., 1949. The ignition of high velocity streams of combustible gases by heated cylindrical rods. In: Proceedings of The Combustion Institute, 3rd International Symposium on Combustion and Flame and Explosion Phenomena, pp. 317–329, Williams and Walkins, Baltimore, USA.*

This strong dependence of T_{min} on actual experimental circumstances is exposed further when comparing values for identical gas mixtures determined in different laboratory test apparatus. Müffling (1946) performed such a comparison and grouped the apparatus in eight categories as follows:

a. The explosive gas mixture is passed through a tube of known internal wall temperature

b. The mixture is admitted to a vessel of known internal wall temperature

c. The mixture is compressed adiabatically, and T_{min} is calculated from the lowest compression ratio that gives ignition

d. The combustible gas and the air are preheated separately to the desired test temperature and subsequently mixed

e. A jet of the combustible gas is injected into a vessel containing air preheated to the desired test temperature
f. Cool explosive mixture is admitted to a soap bubble surrounding a hot platinum wire of known surface temperature
g. A hot solid body of known surface temperature is dropped or ejected into cool explosive mixture
h. A hot solid rod of known surface temperature is inserted into cool mixture.

Müffling (1946) reported some results for 7 vol.% hexane in air, showing that method (c) gave a T_{min} of only 300°C, whereas method (d) gave a value of 630°C. For 4–5 vol.% heptane in air the corresponding values were 280°C and 580°C, respectively. The difference of about 300°C in both cases is substantial.

Laurendeau (1982) summarized some more recent published experimental and theoretical work on hot–surface ignition of various fuel/air mixtures, with the main emphasis on methane/air. The general validity of the overall trends exhibited by Figs 2.14 and 2.15 was confirmed.

In a subsequent investigation, Alfert and Fuhre (1988) ignited propane/air mixtures in the apparatus shown in Fig. 2.16.

After covering the top of the box by a sheet of aluminum foil, making provision for two gas outlet tubes, the box was flushed gently with a propane/air mixture of the desired composition, entering the box through the inlet in the bottom of the box. The rise of fuel concentration in the box was monitored continuously. The hot surface was generated when the overall fuel concentration approached the inlet concentration.

When the hot-surface temperature reached 600°C, the heating rate was reduced to generate a surface temperature increase of a few degrees per minute. Corresponding values of the hot-surface temperature and the propane concentration in the box were recorded at the moment of ignition, the latter being in the range of 4.5–5.0 vol.% in most of the experiments.

The results are summarized in Fig. 2.17 and compared with data obtained using the two standard test methods ASTM (2015b) and IEC (1975a), and a closed bomb method described by Kong et al. (1995).

The results from the experiments in the apparatus shown in Fig. 2.17 were from 300°C to 500°C higher than those obtained in the three conservative flask/bomb tests, including the current standard ASTM (2015b) and IEC (1975a) tests for minimum ignition temperature. This discrepancy calls for a reconsideration of the current methods of standardized testing for T_{min}.

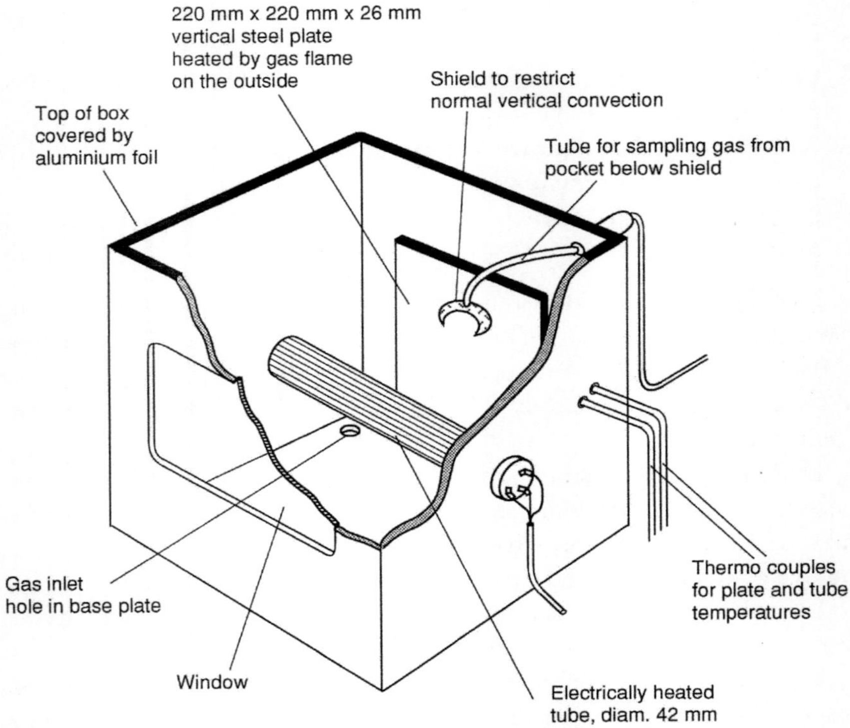

Figure 2.16 Sketch of 50 L explosion box used for determining minimum ignition temperatures for propane/air under various conditions. In addition to the gas inlet, holes for gas concentration, and gas temperature measurement probes were also provided in the base plate.
From Alfert, F., Fuhre, K., 1988. Ignition of gas clouds by hot surfaces. Which maximum temperature is relevant and acceptable for equipment in classified areas? CM1 Report No. 873301-1, Christian Michelsen Research, (from 2001: GexCon AS), Bergen, Norway.

2.2.4.2 Minimum Ignition Temperatures of Multicomponent Fuels in Air

In offshore oil and gas production, minimum ignition temperatures of mixtures of multicomponent fuels and air are often of interest. For example, natural gas is a multicomponent fuel, containing higher hydrocarbons in addition to the methane. For mixtures of methane, propane, and air, Kong et al. (1995) investigated the dependence of the ratio of propane/methane on T_{min} of the mixtures. Fig. 2.18 shows a significant nonlinear decrease of T_{min} with increasing propane/methane ratio. In the hot-bomb method used, a very high equivalence ratio Φ of 2.3 gave a significantly lower T_{min} than for $\Phi = 1$, ie, for stoichiometric mixtures.

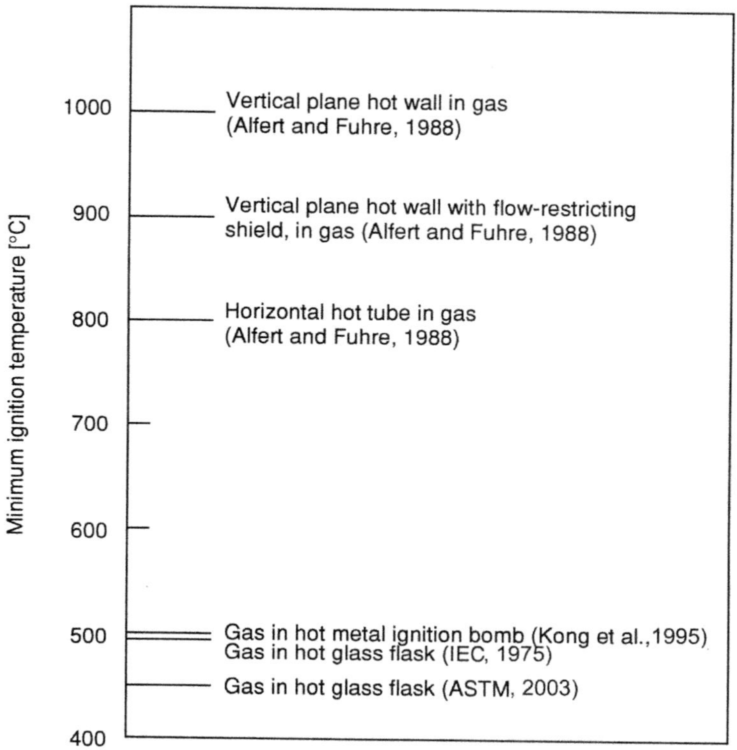

Figure 2.17 Minimum hot surface ignition temperatures for a propane/air mixture at atmospheric pressure, determined by different methods.

2.2.4.3 Solution for the Future: Dynamic Computer Simulation Models of Hot Surface Ignition

There is a need to reconsider current practices for assessing T_{min} of explosive gaseous fuel/air mixtures for the purpose of process design. A differentiated approach should be developed, which allows critical ignition conditions to be predicted for both mono- and multicomponent fuels, for different geometric hot-surface configurations, and for the conditions of heat and gas flow that actually occur in the industrial situation of concern.

The likely approach for the future will be development of comprehensive numerical models containing submodels of both chemical kinetics and transport processes. Because of the rapid development of fast computers and advanced methods of measurements, reaction kinetics modeling is progressing at great pace. Simulation of fuel oxidation by considering elementary reaction steps has shown good agreement with experimental results.

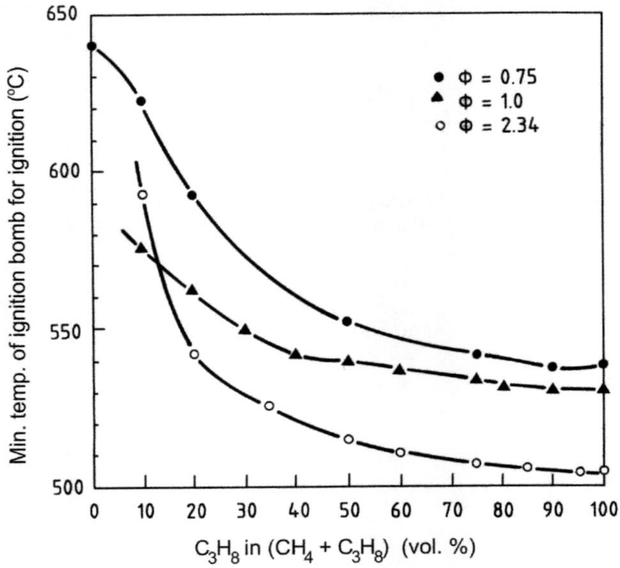

Figure 2.18 Influence of fuel composition on the minimum ignition temperature of CH_4/C_3H_8/air mixtures, for three different overall equivalence ratios Φ, in which $\Phi = 1$ for a stoichiometric fuel/air mixture, and $\Phi < 1$ and $\Phi > 1$ represent lean and rich mixtures, respectively.
From Kong, D., Eckhoff, R.K., Alfert, F., 1995. Auto-ignition of CH_4/air, C_3H_8/air, CH_4/C_3H_8/air and CH_4/CO_2/air using a 1 litre ignition bomb. Journal of Hazardous Materials 40, 69–84.

Integration of a full chemical kinetics package into a mathematical model for predicting T_{min} in practical process situations may be too ambitious at the outset. Using semiempirical approximations of the influence of chemistry could be a more realistic point of departure. From this perspective, the work of Oberhagemann (1989) is interesting. He developed a semiempirical model by which T_{min} for single organic components in air can be computed on the basis of the molecular structure of the fuel. Using this model he computed T_{min} for 380 different organic fuels and correlated with experimental values obtained by the standard IEC (1975a) method. The coefficient of correlation was 0.97, corresponding to a mean deviation between experiment and theory of about 10°C. Oberhagemann extended his model to mixtures of two, three, and four different fuels and air, and obtained good correlation with experimental values even in these cases.

2.2.4.4 Standard Test Methods for T_{min}

The current international standard test apparatus used for determining the T_{min} value of explosive gas mixtures for practical explosion prevention is shown in Fig. 2.19.

As illustrated in Fig. 2.17 the two standard methods IEC (1975a) and ASTM (2015b) are quite conservative. However, one can envisage practical accidental situations in which ignition conditions may be as conservative as in the tests.

Despite IEC (1975a) and ASTM (2015b) being quite conservative, they may not produce the absolute minimum values that can be found for any

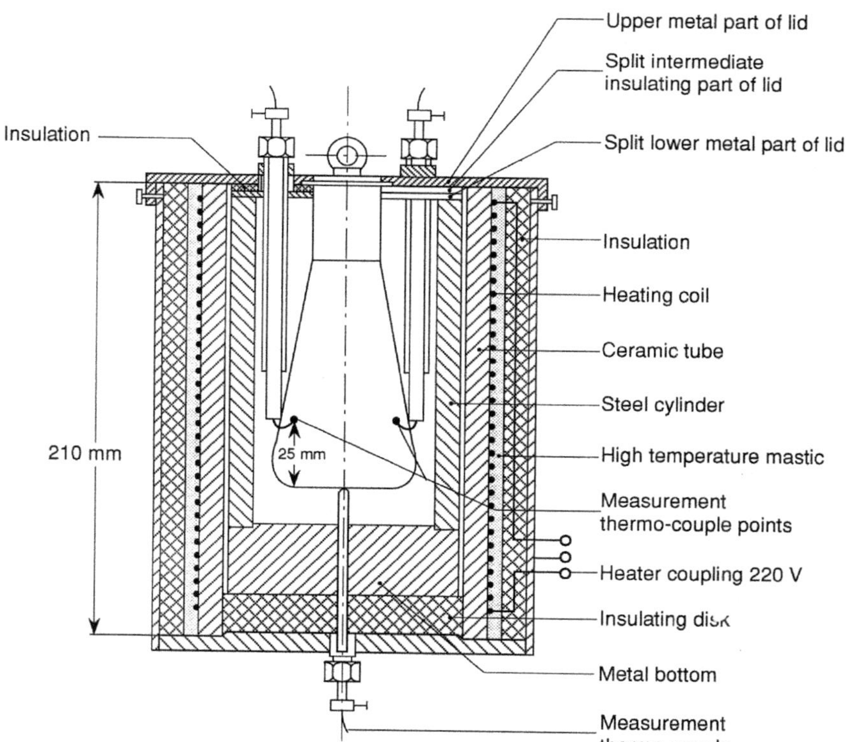

Figure 2.19 The IEC standard 200 cm^3 glass flask apparatus for determining the T_{min} values of explosive gas mixtures and vapors.
Based on IEC, 1975a. Electrical Apparatus for Explosive Gas Atmospheres. Part 4: Method of Test for Ignition Temperature, second ed. IEC Publication 79-4, Central Office of International Electrotechnical Commission, Geneva, Switzerland.

given fuel/air mixture. This is because T_{min} determined by closed/semi-closed vessel methods of this type decreases somewhat with increasing vessel size. Hence, tests in significantly larger vessels than those used in the two standard methods may yield somewhat lower T_{min} than the values obtained with the two standard methods.

2.2.5 Ignition by Burning Metal Particles, "Thermite" Reactions, and Transient "Hot-Spots"

2.2.5.1 Introductory Overview

In the past, the initiation of accidental gas explosions was sometimes attributed to "friction sparks" without any further explanation. However, this imprecise term covers a multifarious category of potential ignition sources, comprising mechanical hot-work operations such as grinding and cutting, repeated impacts on one spot, and single impacts. However, in all these cases the ignition source is generated by transformation of mechanical energy (impact and friction) into heat.

Existing knowledge suggests that the probability of ignition of a combustible gas-in-air mixture by a mechanical impact is determined by six basic parameters characterizing the impact and five parameters characterizing the gas mixture. Key parameters of the impact include:

- Chemical composition of colliding bodies
- Surface topography of colliding bodies at areas of contact
- Contact area
- Contact pressure
- Sliding velocity parallel to the contact surface
- Sliding distance (or contact time)

Key parameters of the gas include:

- Chemical composition of the fuel gas
- Concentrations of fuel, oxygen, and inert gas
- Dynamic state of the gas (turbulence and systematic velocity components, which are also influenced by the impact process itself.)
- Temperature
- Pressure

Depending on the circumstances, the actual ignition source produced by the impact/friction process is generally either small hot particles released during the process or hot spots generated on one or both of the colliding bodies during the impact. In special situations in which metals such as aluminum or titanium are involved in the impact together with rust, the impact can generate very incendiary "thermite" flashes due to exothermic transfer of oxygen from the rust to the aluminum or titanium.

2.2.5.2 Ignition by Small Burning Metal Particles from Single Impacts

Small flying hot particles from single impacts between chemically inert materials (rock and other inert minerals) cannot ignite explosive gas mixtures. This is because the mechanically generated temperature rise will not reach the high levels required for such small particles to become an ignition source. However, if metals are involved in the impact, small metal particles may be torn or cut away from the main bulk and heated mechanically to such an extent that they start to burn spontaneously while flying through the air. Then the particle temperature will increase substantially due to the release of chemical energy. Such small burning metal particles are the genuine "friction sparks."

Much work has been carried out to assess the possibility of igniting methane/air ("firedamp") in coal mines by coal-picking equipment impacting on rock. This is essentially a single-impact process. Research over many years has revealed that, if ignition occurs, the source is not the burning metal sparks (steel/hard metal), but the transient hot spot generated by the pick on the rock. In his extensive review paper, Powell (1984) concluded that small burning metal particles from mechanical impacts are not capable of igniting methane/air, and possibly not even higher alkanes/air (group IIA), unless the particle temperature exceeds 2000°C. This means that steel sparks from single impacts are unlikely to ignite natural gas/air. However, burning particles of titanium, zirconium, magnesium, and aluminum can ignite such gases. According to Powell (1984), the low probability of igniting methane/air and group IIA gases/air with steel sparks from single impacts is in accordance with American and British recommendations saying that "nonsparking" tools are superfluous in areas where such gases may be present.

Pedersen and Eckhoff (1986) studied the ignition of propane/air and acetylene/air by heat generated in tangential impacts between tips of different steel qualities or of titanium and a rusty or sandblasted steel plate. Tangential impacts against the steel plate were generated by a rigid spring-loaded arm carrying the test tip. The apparatus is illustrated in Fig. 2.20.

The strength of the impact ("net impact energy") was expressed in terms of loss of kinetic energy of the impacting arm during the impact. Some results are shown in Table 2.9.

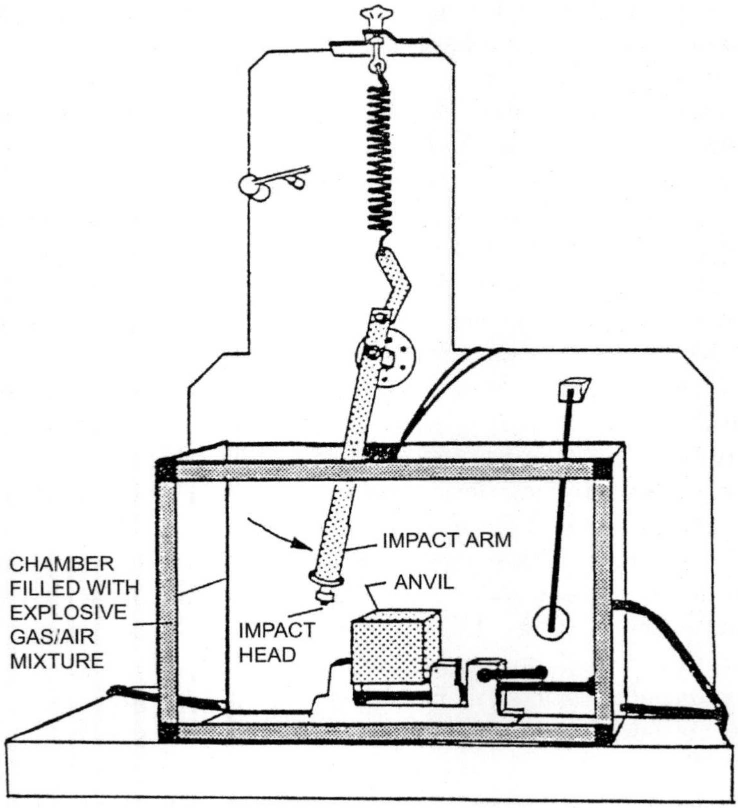

Figure 2.20 Impact-spark ignition apparatus for studying the ability of burning metal particles and "thermite flashes" from single impacts to ignite explosive gas/air mixtures. *From Pedersen, G.H., Eckhoff, R.K., December1986. Initiation of Gas Explosions by Heat Generated During Single Impact Between Solid Bodies, CMI Report No. 863302-1. Chr. Michelsen Institute (now GexCon AS), Bergen, Norway.*

Table 2.9 Results from impact ignition experiments using different tip materials impacting on a target of naturally rusted steel in an explosive mixture of 4.6 vol.% propane in air

Tip material	Number of visible sparks	Net impact energy (J)	Ignition
St 37 steel	20–50	8–10	No
Chrome–vanadium steel	10–30	6	No
Unbraco screw	~20	8	No
Acid-resistant screw	~5	8	No
Nonsparking tool	0	13–14	No
Titanium	10–1000	~10–15	Yes

From Pedersen, G.H., Eckhoff, R.K., December1986. Initiation of Gas Explosions by Heat Generated During Single Impact Between Solid Bodies, CMI Report No. 863302-1. Chr. Michelsen Institute (now GexCon AS), Bergen, Norway.

Within the experimental range of net impact energies up to 20 J, it was not possible to ignite 4.6 vol.% propane/air with sparks or hot spots from single impacts between different steel qualities and rusty or sandblasted steel. This agrees with the general conclusion of Powell (1984) on ignition of groups I and IIA gases, mentioned previously.

However, as Table 2.9 shows, titanium sparks were able to ignite propane/air. In about half of the experiments with titanium that gave ignition, one or several specific flying sparks could be identified by high-speed video as the ignition source(s). Such ignition mostly took place 50–90 ms after the impact. By this time, the velocity of the sparks had dropped to 2–5 m/s, which presumably allowed a sufficient residence time in a given gas volume for ignition to occur. In some cases, single-spark ignition took place after the spark had collided with the wall of the explosion chamber and lost most of its velocity. Similar observations were made by Ritter (1984).

Fig. 2.21 shows some more detailed results from impacts with titanium against rust in 4.6 vol.% propane in air.

It is interesting to observe that there seems to be a worst-case range of tangential impact velocities that favors ignition. At lower velocities fewer

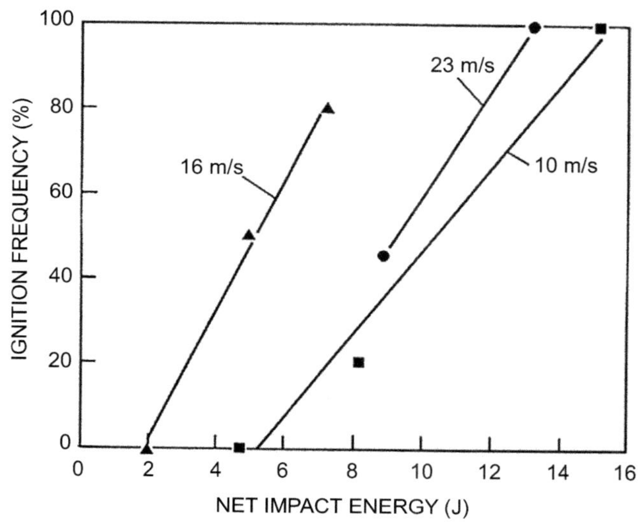

Figure 2.21 Results from tangential impact experiments with titanium against rust in a mixture of 4.6 vol.% propane in air, using the apparatus illustrated in Fig. 2.20.
From Pedersen, G.H., Eckhoff, R.K., December1986. Initiation of Gas Explosions by Heat Generated During Single Impact Between Solid Bodies, CMI Report No. 863302-1. Chr. Michelsen Institute (now GexCon AS), Bergen, Norway.

burning metal particles are produced, and ignition is less probable. At higher velocities more burning metal particles may be produced, but this is more than compensated for by the violent disturbance of the local gas cloud by the movement of the impact arm, which makes ignition less probable. Table 2.10 gives some results from ignition of acetylene/air by burning metal particles.

As can be seen, the acetylene/air mixture was readily ignited by sparks from steel qualities that are known to be even less able to release incendiary sparks than standard construction steels. High-speed video recordings revealed that ignition was caused by single flying steel sparks in all the cases covered by Table 2.10. Ignition occurred after the sparks had traveled sufficiently far for the spark velocity to become <5 m/s. Ignition of acetylene/air by the hot spot generated at the point of impact on the anvil plate was not observed in these experiments.

When using a "nonsparking" material as the test object, only one single visible spark was observed altogether. This spark became visible 45 ms after the impact, and remained visible for 90 ms. During this period, the spark velocity decreased from about 1 m/s to 0.4 m/s. It is not clear whether this weak spark originated from the "nonsparking" material or from the anvil material, which was rusty steel. Although the spark observed was very weak, its occurrence indicates that impacts involving so-called "nonsparking" materials as one of the colliding partners may in fact generate visible sparks, but it seems highly unlikely that such sparks can ignite hydrocarbon/air mixtures. It is not clear, however, whether impacts between a "nonsparking" material and light metals, eg, titanium, may produce incendiary light-metal sparks.

Table 2.10 Results from impact ignition experiments using different tip materials impacting on a target of naturally rusted steel in an explosive mixture of 7.7 vol.% acetylene in air

Tip material	Number of visible sparks	Net impact energy (J)	Ignition
Chrome-vanadium steel	~10	~6	Yes
Unbraco screw (stainless)	~20	~8	Yes
Acid-resistant steel	3–5	~8	No
Nonsparking tool alloy	0–1	8–10	No

From Pedersen, G.H., Eckhoff, R.K., December1986. Initiation of Gas Explosions by Heat Generated During Single Impact Between Solid Bodies, CMI Report No. 863302-1. Chr. Michelsen Institute (now GexCon AS), Bergen, Norway.

2.2.5.3 Ignition by Thermite Flashes

The experiments in propane/air with impacts of titanium on rusty St37 steel (see Table 2.9) revealed two different modes of ignition. In about half of the tests in which ignition occurred, it was observed to take place at or very close to the point of impact on the anvil plate, immediately after or even during impact. In such cases, an extremely luminous hemispherical volume was observed in the region of impact. The onset of flame propagation could be referred specifically to neither one single metal spark nor the hot spots generated at the points of impact. It is not fully clear what this luminous hemisphere consisted of, but the very high numbers of "sparks" (on the order of 1000) that were observed in these cases were located within this volume. The hemisphere was only observed when the peripheral velocity of the tip holder exceeded about 15 m/s. It is believed that the luminous volume was a "thermite flash" corresponding to the overall formal exothermic reaction:

$$3\ \mathrm{Ti} + 2\ \mathrm{Fe_2O_3} \rightarrow 3\ \mathrm{TiO_2} + 4\ \mathrm{Fe} + \mathrm{heat} \qquad [2.10]$$

Similar reactions would be expected between rust and other metals having greater affinities for oxygen than iron. In the case of aluminum, the softness of the metal may prevent sufficient heat generation during impact to produce a thermite flash. However, according to Kornai et al. (1994), impacts involving harder aluminum alloys may generate thermite flashes, as well as incendiary metal particles. Also, impacts on rusty surfaces covered by aluminum paint of high pigment content have generated thermite flashes. Gibson et al. (1968) demonstrated that smears of aluminum on rusty steel produced thermite flashes capable of igniting methane/air when struck by strikers made from almost any metal, including steel, brass, bronze, and even copper—beryllium.

2.2.5.4 Ignition by Transient Hot-Spots

As already pointed out, impacts not only create burning metal particles or thermite flashes, but also create transient hot surfaces ("hot spots") on the two colliding bodies. Powell and Quince (1972) applied the classical theory of frictional impacts to calculate the maximum hot-spot temperatures generated in such impacts. Eckhoff and Pedersen (1988) discussed this theory in relation to impact ignition hazards on offshore oil and gas installations. Cutler (1974, 1978) conducted experiments in which methane/air and propane/air were ignited by artificial transient hot spots generated electrically on tungsten strips. Fig. 2.22 shows how the values of T_{min} were influenced by the strip width.

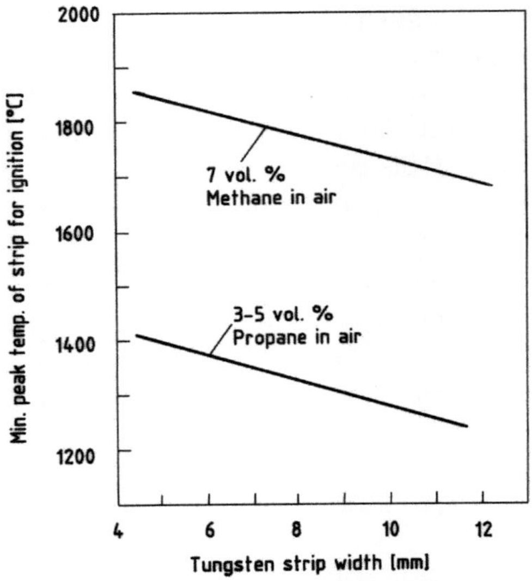

Figure 2.22 Influence of the width of a transiently heated tungsten strip on the minimum peak temperature of the strip for igniting methane/air and propane/air. *From Cutler, D.P., 1974. The ignition of gases by rapidly heated surfaces. Combustion and Flame 22, 105–109; Cutler, D.P., 1978. Further studies of the ignition of gases by transiently heated surfaces. Combusion and Flame 33, 85–91.*

As pointed out earlier, experiments with net impact energies of up to 20 J in acetylene/air revealed that ignition never occurred at the hot spot on the anvil plate or on the impacting metal tip. Therefore, net impact energies which are substantially higher than 20 J would be required for impact generation of hot spots capable of igniting natural gas/air.

2.2.6 Ignition by Electric Sparks, Arcs, and Electrostatic Discharges

2.2.6.1 Electric Sparks Between Two Conducting Electrodes

Electric sparks are produced when the strength of the electric field in the gap between two conducting electrodes exceeds what the dielectric medium in the gap can resist. For a given dielectric medium, eg, air at atmospheric conditions, the relationship between the gap distance, and the critical gap voltage for gap breakdown depends on the electrode shape and the electrode material.

Fig. 2.23 gives some results obtained in two different laboratories from measurement of electrical breakdown voltages in air at atmospheric conditions using various electrode configurations.

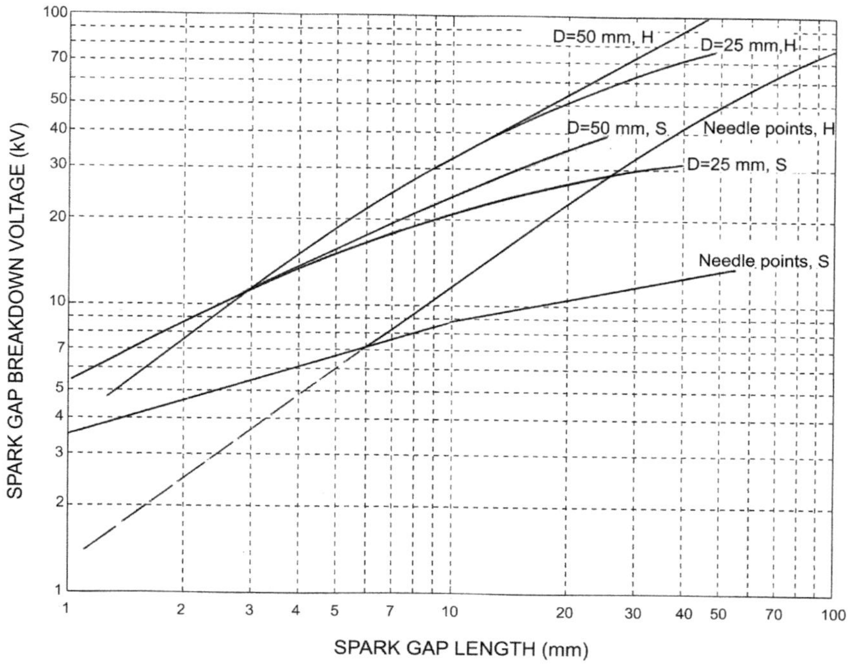

Figure 2.23 Spark-gap breakdown voltages in air at normal atmospheric conditions as functions of spark-gap length and electrode shape. "H" values are data from Handbook of Chemistry and Physics (1959/60) and "S" values are data from Smithsonian Physical Tables (1959). Details of electrode materials and electrode surface structures are not known.

All the data confirm that, for one specific electrode system, the electrical breakdown voltage increases systematically with the length of the gap between the electrodes. In addition, the data confirm that for a given gap length the electrical breakdown voltage is considerably lower with needle-point electrodes than with rounded electrodes having radii of curvature of at least a few mm. The discrepancies between the absolute breakdown voltages obtained in the two laboratories for apparently similar conditions illustrate that the actual breakdown voltage also depends on other experimental conditions than just global electrode shape and gap distance.

Extrapolation of the various curves in Fig. 2.23 into the region of voltages on the order of 100 V and less indicate that spark-over will then be possible only at very small gap distances <<1 mm. This in turn means that any ignition of an explosive gas mixture under such circumstances occurs under partly quenched conditions. This means that the required spark energy for ignition can be considerably higher than the minimum ignition energies obtained under unquenched conditions (see discussion of Fig. 2.26 following).

Electric sparks can be capacitive or inductive. Capacitive sparks, associated with the discharge of a capacitor across a given spark gap, can occur in electric circuits, or they can be caused by triboelectrically generated electrostatic charges. The theoretical spark energy, neglecting losses, is $\frac{1}{2}CU^2$, in which C is the capacitance, and U is the voltage across the spark gap just prior to gap breakdown. Inductive sparks are associated with discharge of inductive energy across gaps that are formed when live circuits are broken. In this case, the theoretical spark energy is $\frac{1}{2}Li^2$ in which L is the inductance, and i is the current just before formation of the break. Any given explosive mixture of a combustible gas and air is associated with a given minimum value of electric spark energies E_{min} that can ignite the mixture. Fig. 2.24 illustrates this for the alkanes, which constitute the main components in natural gas, eg, from the North Sea.

It is seen that the values of the equivalence ratio Φ at which the minimum ignition energy for each gas occurs increase systematically with increasing molecular weight of the fuel gas, from 0.85 for methane (MW = 16) to 1.80 for heptane (MW = 100). It is also seen that the E_{min} value of methane (0.28 mJ) is slightly higher than those of the higher alkanes (0.23–0.25 mJ).

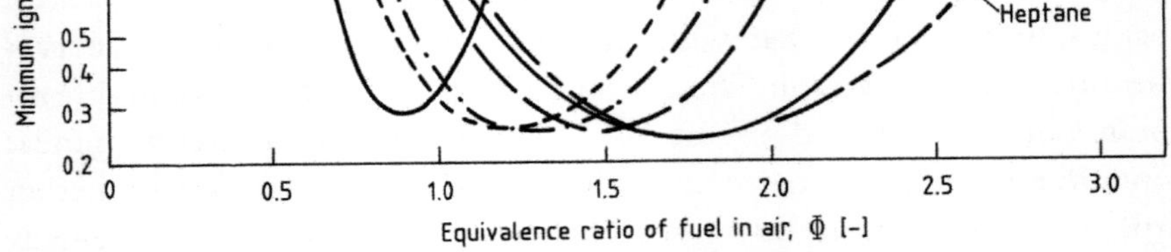

Figure 2.24 Minimum capacitive electrical spark ignition energies of mixtures of various gaseous alkanes and air as a function of the volumetric ratio of fuel and air. For $\Phi = 1$ the ratio of fuel to oxygen is stoichiometric. For $\Phi < 1$ the mixture is lean, and for $\Phi > 1$ it is rich.

From Kuchta, J.M., 1985. Investigation of fire and explosion accidents in the chemical, mining and fuel related industries. A Manual. US Bureau of Mines, Bulletin 680.

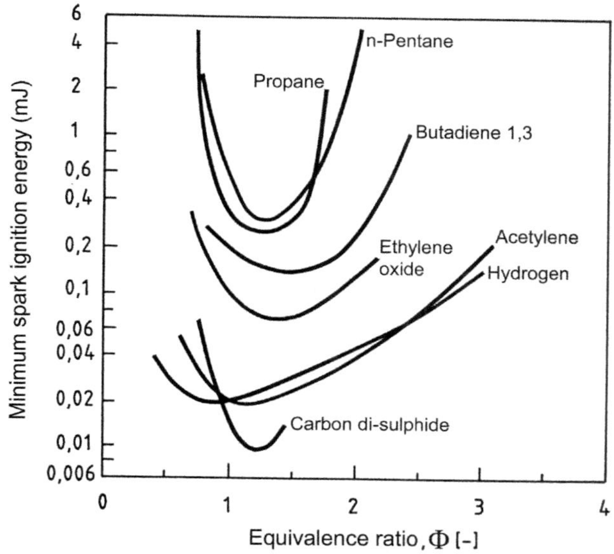

Figure 2.25 Minimum capacitive electrical spark ignition energies of mixtures of various gaseous fuels and air as a function of the volumetric ratio of fuel and air. For $\Phi = 1$ the ratio of fuel to oxygen is stoichiometric. For $\Phi < 1$ the mixture is lean, and for $\Phi > 1$ it is rich.
From Freytag, H.H., 1965. Handbuch der Raumexplosionen. Verlag Chemie, Weinheim, Germany.

Fig. 2.25 shows curves corresponding to those in Fig. 2.24, for a range of different explosive gas/air mixtures. As can be seen, E_{min} for the most ignition-sensitive gases hydrogen, acetylene, and carbon di-sulfide, are about two orders of magnitude lower than the values for the alkanes in Fig. 2.24.

All the E_{min} values in Figs 2.24 and 2.25 were determined by capacitive sparks under unquenched conditions. This implies that the fixed spark gaps used in the test were slightly larger than the quenching distances (QD) for the gas mixtures tested. This important concept is illustrated in Fig. 2.26. True E_{min} values are obtained only if the length of the electrode gap exceeds QD. As soon as the gap length becomes <QD, the minimum spark energy for ignition can increase quite significantly. The increase will be more or less abrupt, depending on the physical dimensions of the electrodes, as implied in Fig. 2.26.

Fig. 2.27 illustrates the close correlation between minimum electric spark ignition energies (E_{min}) and quenching distance (QD).

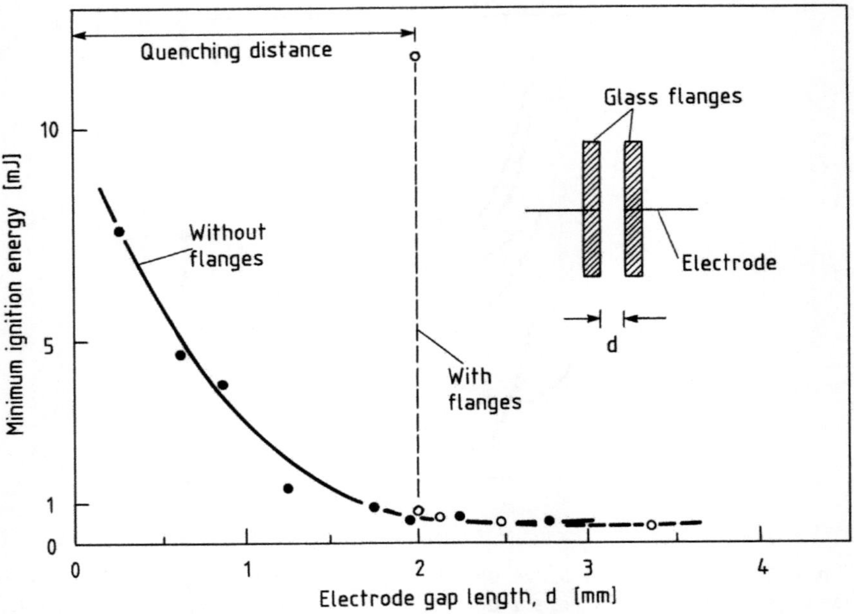

Figure 2.26 Minimum capacitive electric spark energies for ignition of a stoichiometric mixture of natural gas and air, as a function of length of gap between electrodes, with and without glass flanges around the electrode tips.
From Freytag, H.H., 1965. Handbuch der Raumexplosionen. Verlag Chemie, Weinheim, Germany.

It is generally assumed that the minimum ignition energies obtained for capacitive sparks by optimizing spark-gap length and fuel/air ratio are the true minima. However, this may not be entirely true. Fig. 2.28 gives some results from a study by Parker (1985), which illustrates that minimum ignition energy may also depend on the basic features of the spark discharge. Using constant-power sparks of mutually independent energy and duration, Parker found that the minimum ignition energy of a lean propane/air mixture increased from about 0.2 mJ for 0.1 μs discharge duration to about 2 mJ for 100 μs duration. The reason for this quite pronounced effect may be found in the fundamental process of ignition, in which the production of chemically active radicals in the spark, which is enhanced by the high spark temperatures generated at short discharge times, plays a central role. In a standard determination of E_{min} for propane/air, using a direct capacitive spark discharge, the discharge time will most probably not exceed 0.1 μs.

Various electrical apparatus can give rise to electric sparks and arcs that can cause ignition of explosive gas mixtures. Electric sparks can be generated when electric circuits are opened and closed, or by stray currents. The use

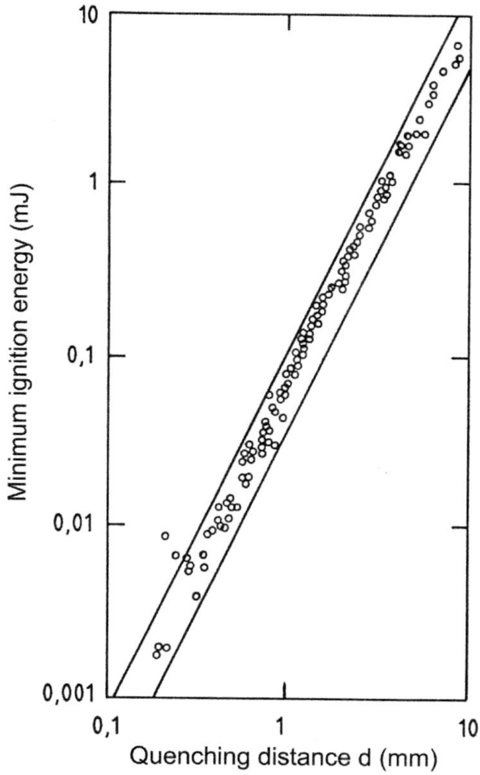

Figure 2.27 Correlation of minimum electric spark ignition energies for unquenched ignitions, and corresponding quenching distances.
From Kuchta, J.M., 1985. Investigation of fire and explosion accidents in the chemical, mining and fuel related industries. A Manual. US Bureau of Mines, Bulletin 680.

of low-voltage equipment (maximum = 50 V) for protecting personnel against electric shock does not eliminate the explosion hazard. Even lower voltages than this can produce energies which can ignite explosive gases. It has been found useful, therefore, to introduce an international standard method for experimental determination of the combinations of electrical circuit parameters that are critical for producing incendiary sparks/arcs from electrical circuitry. The IEC standard apparatus is illustrated in Fig. 2.29.

Fig. 2.30 shows experimental "ignition curves" (solid lines) for propane/air (gas group IIA), ethylene/air (gas group IIB), and hydrogen/air (gas group IIC) determined in the apparatus illustrated in Fig. 2.29. Ignition was possible at capacitor voltages even below 10 V, but for spark-over to occur the spark gap has then to be <<1 mm. This in turn means that the

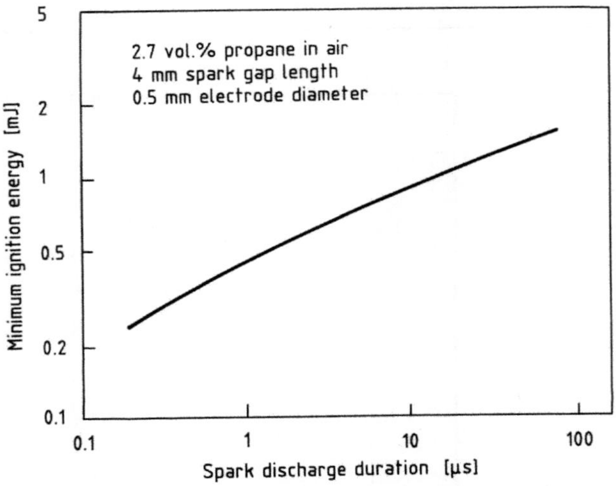

Figure 2.28 Influence of spark discharge duration on minimum ignition energy of a lean (2.7 vol.%) mixture of propane/air at normal temperature and pressure.
From Parker, S.J., 1985. Electric Spark Ignition of Gases and Dusts (Ph.D. thesis) City.

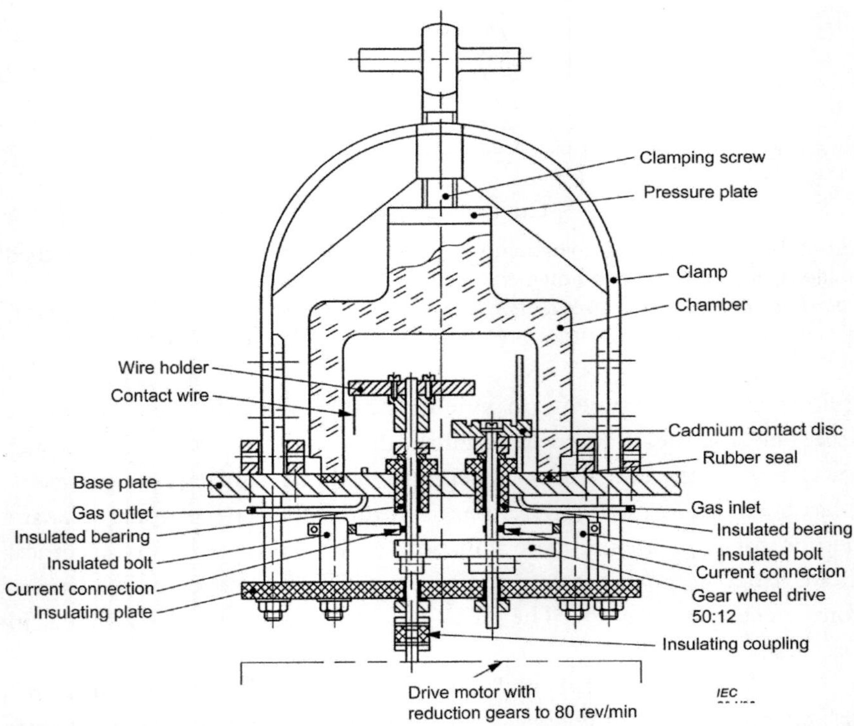

Figure 2.29 Standard IEC apparatus for experimental determination of ignition curves, and for testing of the ability of electrical circuits to produce sparks that can ignite gas mixtures.
From IEC, February 1999. Electrical Apparatus for Explosive Gas Atmospheres. Part 11: Intrinsic Safety "i", 4th ed. IEC International Standard 60079-11. Central Office of International Electrotechnical Commission, Geneva, Switzerland.

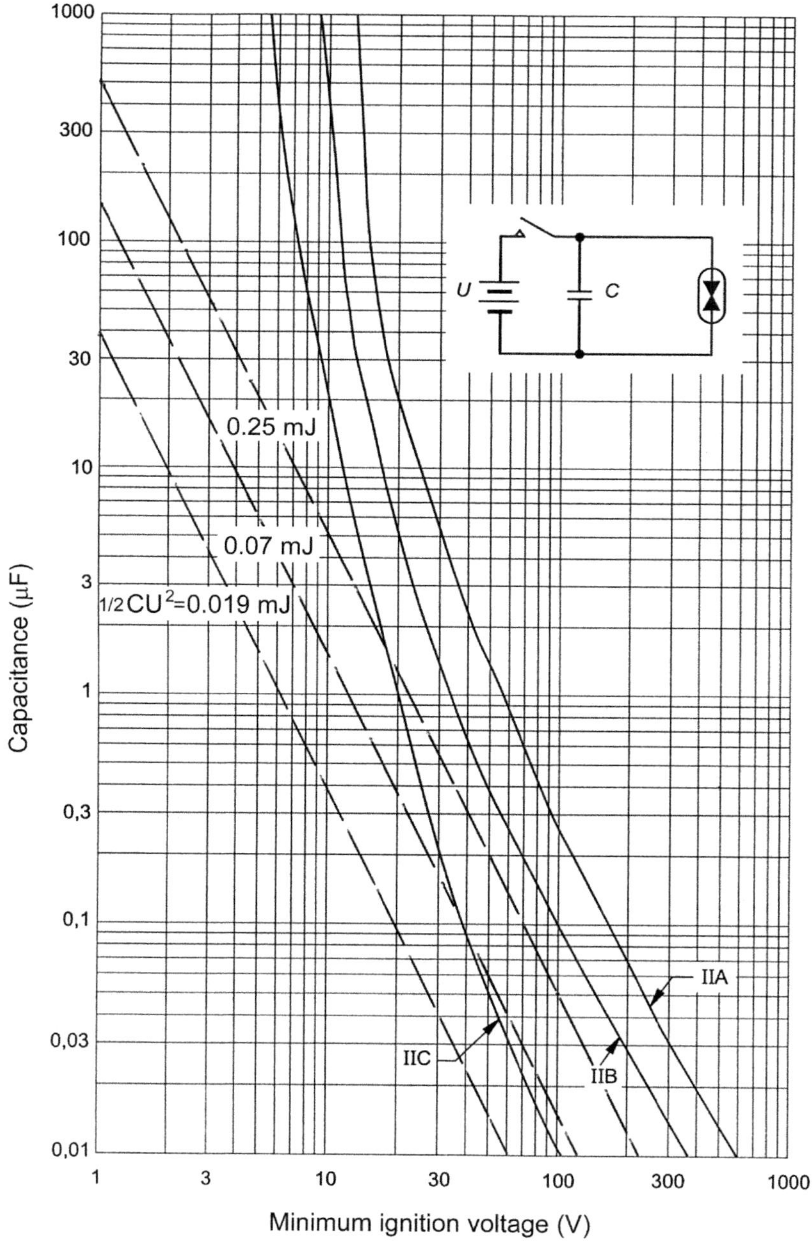

Figure 2.30 Experimental standard "ignition curves" for propane/air (IIA), ethylene/air (IIB), and hydrogen/air (IIC) for capacitive spark discharges, determined by the standard IEC break flash apparatus shown in Fig. 2.29. The dotted straight lines represent all combinations of voltage and capacitance that yield $\frac{1}{2}CU^2$ values equal to the experimental minimum ignition energies for the three respective gas mixtures. Standard ignition curves from IEC (1999).

sparks were delivered to the explosive mixture under quenched conditions (see Fig. 2.26). It is not surprising, therefore, that the dotted lines, representing the minimum $\frac{1}{2}CU^2$ for igniting the three different gas mixtures with spark gaps larger than the quenching distances, fall significantly below the experimental solid lines. It must be pointed out, however, that these dotted lines are idealized hypothetical worst cases, which cannot be obtained in reality for the voltage range on the horizontal axis in Fig. 2.30. This is because quenching is unavoidable at the short spark gaps required to obtain spark-over at these voltages.

As shown in Fig. 2.31, a corresponding effect is found for inductive circuits, in which the lines for $\frac{1}{2}Li^2 = E_{min}$ are also significantly below the corresponding minimum break flash ignition curves (minimum ignition current as a function of circuit inductance).

The apparatus illustrated in Fig. 2.29 is also used for assigning the appropriate gas group to any given combustible gas/vapor by employing the concept of "minimum ignition current" (MIC). Then an inductive electrical circuit containing a 95 mH inductance and operating at 24 V d.c. is used. The "MIC ratio" of the gas of concern, is defined as the actual MIC of that gas, when premixed with air in the most easily ignited proportions, and the corresponding MIC of the standard gas, which is a laboratory-quality methane. It is customary to assign combustible gases/vapors having MIC ratios >0.8 to the least easily ignitable gas group IIA. The intermediate group IIB comprises gases/vapors having $0.45 \leq \text{MIC} \leq 0.8$, whereas gases/vapors of MIC ratios <0.45 belong to the most easily ignitable group IIC.

Besides in electrical apparatuses, spark discharges between two conducting electrodes can also arise from triboelectric charging of nonearthed electrically conducting items. Table 2.11 indicates the levels of capacitances C and voltages U that may typically be associated with electrostatic charging of nonearthed electrical components in industry. The resulting stored energies $\frac{1}{2}CU^2$ represent the maximum spark energies that can be generated from discharge of the various capacitive components to earth.

Table 2.11 also includes the charging of a human being. Charging can, eg, occur whenever a person wearing electrically insulating shoes is walking across a floor. Charge transfer occurs every time the shoes are lifted/separated from the floor. Fig. 2.32 illustrates the discharge of an electrostatically charged person to earth. In electrical terms the human body can be regarded as a capacitor on the order of 100–300 pF, with a given internal ohmic resistance. It is commonly assumed that during discharge about half the energy $\frac{1}{2}CU^2$ stored in the capacitor is dissipated in the spark and the other half in

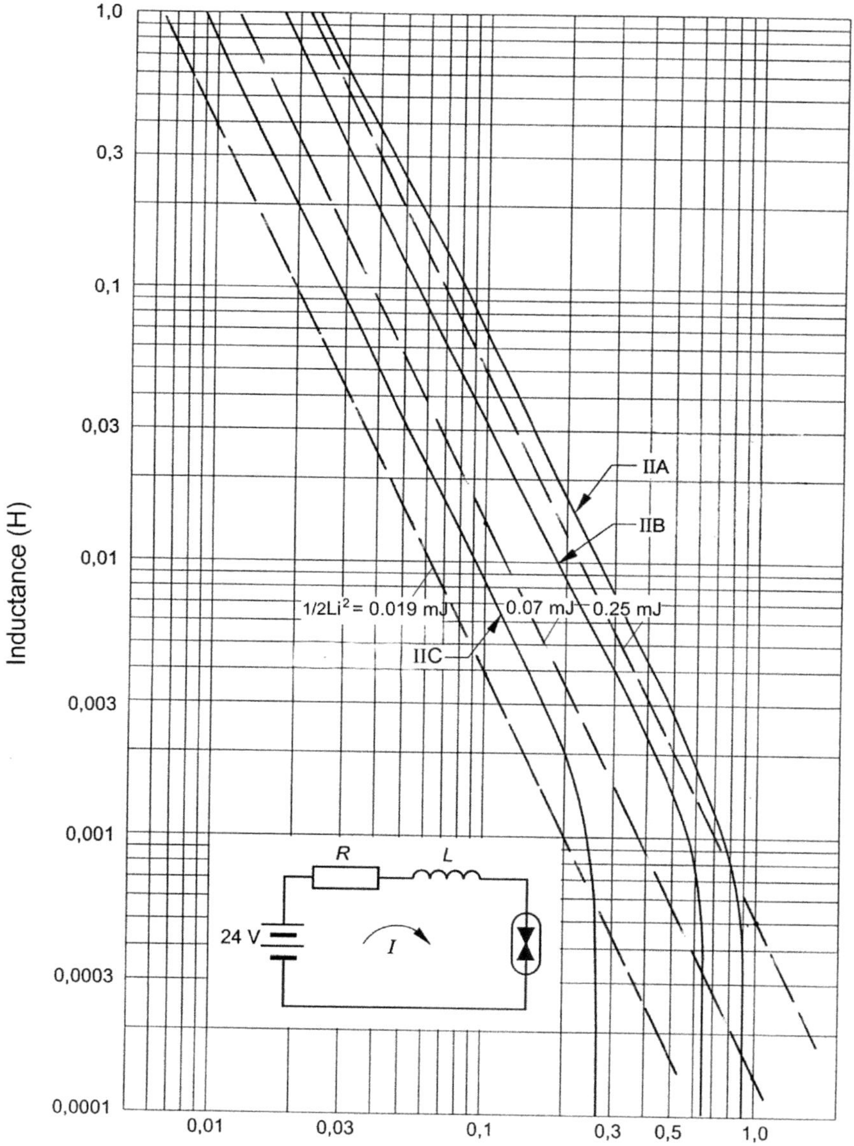

Figure 2.31 Experimental standard "ignition curves" for propane/air (IIA), ethylene/air (IIB), and hydrogen/air (IIC) for inductive spark discharges, determined by the standard IEC break flash apparatus shown in Fig. 2.29. The dotted straight lines represent all combinations of current and inductance that yield ½Li² values equal to the experimental minimum ignition energies for the three respective gas mixture standard ignition curves from IEC (1999).

Table 2.11 Examples of combinations of capacitances, voltages, and resulting spark energies, in industrial practice

Charged object	Capacitance (pF)	Potential (kV)	Energy (mJ)[a]
Single screw	1	5	0.01
Flange, nominal width = 100 mm	10	10	0.5
Shovel	20	15	2
Small container (~50 L)	50	8	2
Funnel	50	15	6
Person	300	10	15
Drum (200 L)	200	20	40
Road tanker	1000	15	100

[a]Approximate values.
From Lüttgens, G., Glor, M., 1989. Understanding and Controlling Static Electricity, Expert Verlag, Ehningen bei Böblingen, Germany.

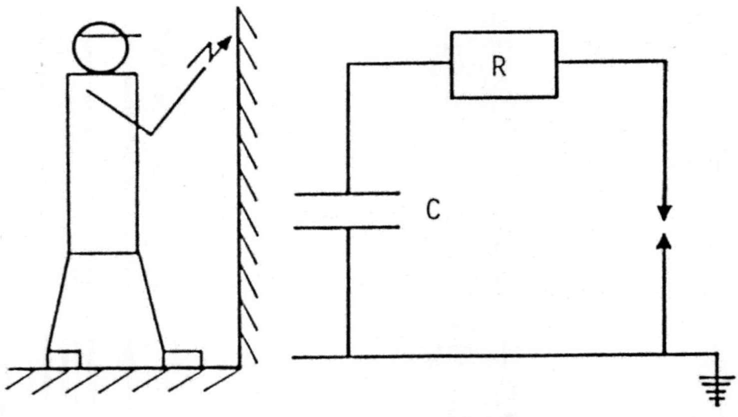

Figure 2.32 Discharge of an electrostatically charged person to earth. Approximate equivalent electrical circuit of a human being, with internal resistance and capacitance in series.
From Glor, M., 1988. Electrostatic Hazards in Powder Handling. Research Studies Press Ltd., John Wiley & Sons Inc.

the internal body resistance. As Table 2.11 shows, spark energies well above E_{min} for most gases and vapors can be generated when a charged person is discharged to earth. In many process operations it is, therefore, necessary to take active measures to make sure that electrostatic charging of persons cannot occur, or to ensure that any charged person is reliably discharged before entering an area where explosive gas atmospheres may occur.

The most important measure preventing electrostatic spark discharges is earthing of all conductive parts that could become dangerously charged (see Table 2.11). However, this protective measure is insufficient if nonconductive materials are present and become electrostatically charged to hazardous levels. In this case electrostatic one-electrode discharges can occur which may also ignite explosive gas/air mixtures. Then nonconductive materials, including nonconductive liquids, must be avoided. This may imply that only earthed metals and earthed antistatic nonmetals are permissible in, eg, process equipment.

2.2.6.2 Various Forms of One-Electrode Electrostatic Discharges from Charged Nonconductors: Concept of "Equivalent Energy"

A one-electrode discharge occurs when electrostatic charge accumulated on the surface of a nonconductor by a charge separation process is drained to earth via a conducting electrode approaching the charged nonconductor. It is common to distinguish between three types of one-electrode discharges:

- Corona discharge
- Brush discharge
- Propagating brush discharge.

Lüttgens and Glor (1989) have discussed the nature of the three types of discharges in detail and have illustrated their possible appearance in practice by examples.

Corona discharges, illustrated in Fig. 2.33, occur when the tip of a pointed conducting electrode (radius of curvature <1 mm) approaches a charged

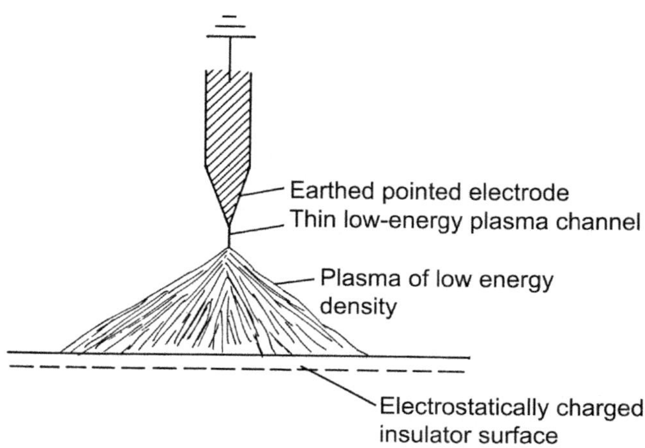

Figure 2.33 Illustration of a corona discharge from a charged insulator surface to earth.

nonconductor. This type of discharge does not occur abruptly within a short time interval as a spark, but rather as a continuous charge leakage, and only gas mixtures of very low minimum ignition energies can be ignited by corona discharges.

However, when the radius of curvature of the conducting electrode tip exceeds a few mm, the discharge will occur more abruptly, and a stem of high energy density will appear close to the electrode. Because of this appearance, this type of discharge illustrated in Fig. 2.34 is called a brush discharge.

Substantially stronger and more incendiary discharges than ordinary brush discharges can be obtained if a charged double layer of opposite polarities is generated rather than just a single charge layer. Figs 2.35 and 2.36 illustrate two different situations in which propagating brush discharges can occur. In Fig. 2.36 the strong electric field across the insulating film breaks the film material down whereby the charge is drained to earth through one or more breakdown channels through the film. Alternatively, the discharge can occur across the gap between the internal charged surface of the insulating film and earth at the pipe exit.

For basic reasons, the energy of the incendiary part of a corona, brush, or propagating brush discharge is not easy to measure directly. The concept of "equivalent energy" is useful in partly overcoming this problem. This concept, which was first introduced by Gibson and Lloyd (1965), is defined as follows: An electric discharge has the equivalent energy W if it is just capable of igniting an explosive mixture of $E_{min} = W$. The equivalent energy of any reproducible but not quantifiable electric discharge is

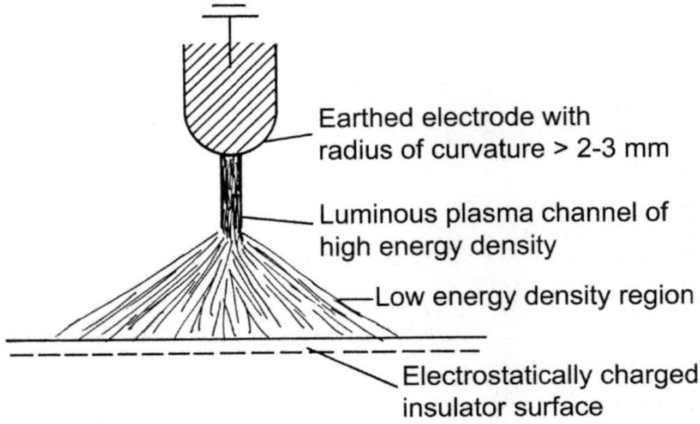

Figure 2.34 Illustration of a brush discharge from a charged insulator surface to earth.

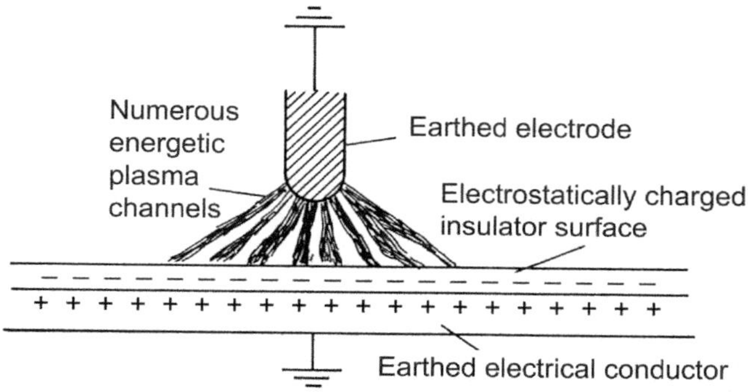

Figure 2.35 Illustration of a propagating brush discharge when an earthed electrode is brought close to a charged insulator surface in close contact with an earthed conducting surface.

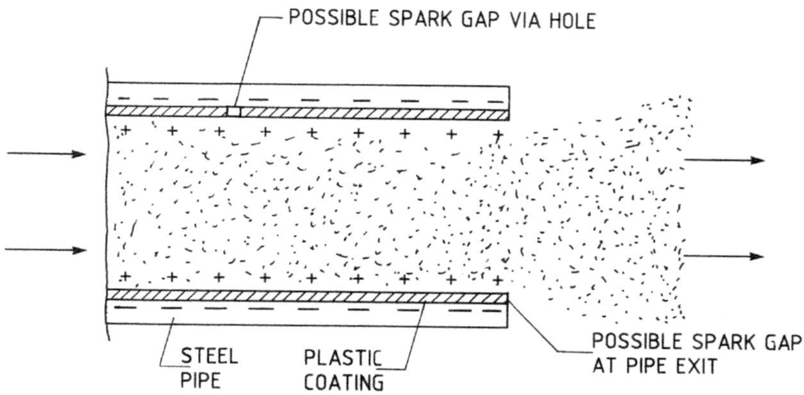

Figure 2.36 Illustration of a propagating brush discharge occurring in a steel pipe lined with an insulating film.

determined in two steps. The first is to determine the composition of a premixed fuel/air/nitrogen mixture, which can just be ignited by the unknown discharge. The second step is to E_{min} of that particular mixture, which will per definition be equal to W. The equivalent energies for the three types of one-electrode discharges previously discussed are shown in Table 2.12.

The minimum ignition energy for alkanes in air is on the order of 0.25 mJ. Therefore, it is unlikely that such mixtures, including natural gas in air, can be ignited by corona discharges. However, brush discharges may cause ignition, and propagating brush discharges can generate energies that are substantially higher by orders of magnitude than those required for ignition.

Table 2.12 Equivalent energies of three types of electrostatic one-electrode discharges

Type of discharge	Equivalent energy (mJ)
Corona	$\leq 2 \times 10^{-2}$
Brush	≤ 3
Propagating brush	$\leq 10^5$

Data from Lüttgens, G., Glor, M., 1989. Understanding and Controlling Static Electricity, Expert Verlag, Ehningen bei Böblingen, Germany.

2.2.7 Ignition by a Jet of Hot Combustion Products

2.2.7.1 The Basic Process

The situation is illustrated in Fig. 2.37.

The central question is whether the hot jet will ignite an explosive gas cloud outside the enclosure, having the same composition as the primary cloud inside. Systematic research has been carried out to investigate the influence on the probability of reignition/explosion transmission of hole diameter/slot width, length of hole/slot, volume of primary enclosure, volume of secondary enclosure, location of ignition point inside the primary enclosure, and composition of the gas mixture.

2.2.7.2 Grouping of Ignition Sensitivity of Premixed Gas/Air According to MESG

Various test apparatuses have been used for determining the maximum experimental safe gap (MESG) for a given explosive gas mixture (see Fig. 2.44). Experiments have revealed a correlation between MESG values from such tests and QD (see Fig. 2.26). A rough rule of thumb is that

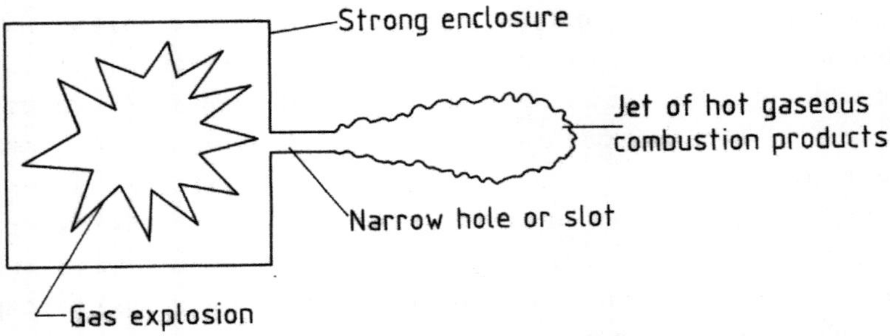

Figure 2.37 Illustration of the ejection of a jet of hot gaseous combustion products through a narrow hole or slit in the wall of a strong enclosure in which a gas explosion takes place.

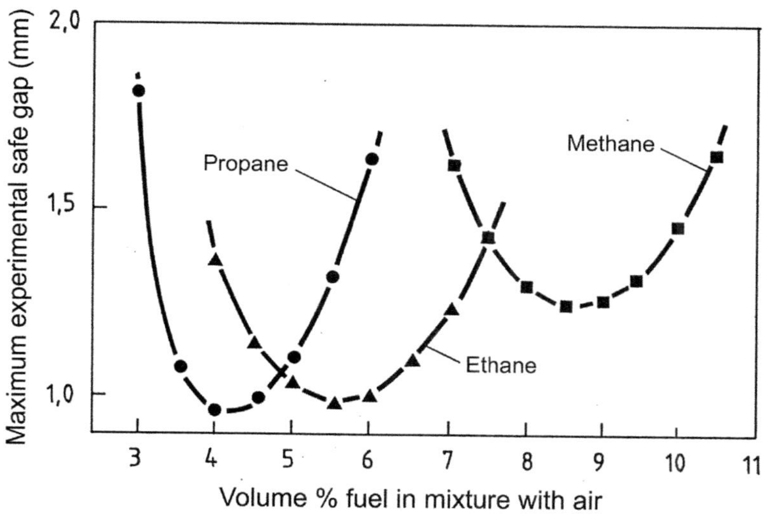

Figure 2.38 Influence of fuel concentration in mixtures with air on MESG for the three lower alkanes.
Data from Alfert, F., 1985. Sicherheitstechnische Kenngrössen von polynären Brennstoff/Luft-Gemischen und selbstzerfallenden Stoffen (Dr.-Ing. thesis). University of Dortmund, Germany.

QD $\approx$ 2·MESG. Keeping in mind that there is a close positive correlation between the quenching distance of a gas mixture and its minimum ignition energy, one would expect MESG to vary with gas type and fuel/air ratio in a manner similar to that in Fig. 2.24 for the minimum ignition energy. As shown in Fig. 2.38, this is indeed the case.

Fig. 2.39 shows the apparatus used in some gas explosion transmission experiments conducted by Larsen and Eckhoff (2000). The vertical cylindrical system consisted of two concentric chambers, one primary and one secondary, separated by a disc with the cylindrical transmission hole at its center. Before an experiment both chambers were flushed with premixed propane/air of the desired composition, until this concentration was obtained throughout the system. The electric spark ignition source in the primary chamber was positioned at the chamber axis, and the distance X_i from the ignition point to the entrance of the transmission hole was varied.

Fig. 2.40 gives a set of results showing the critical distance X_i for explosion transmission as a function of the hole diameter D.

As can be seen, the critical tube diameter for flame transmission with $X_i = 0$ was about 3 mm. In this case the conditions were probably close

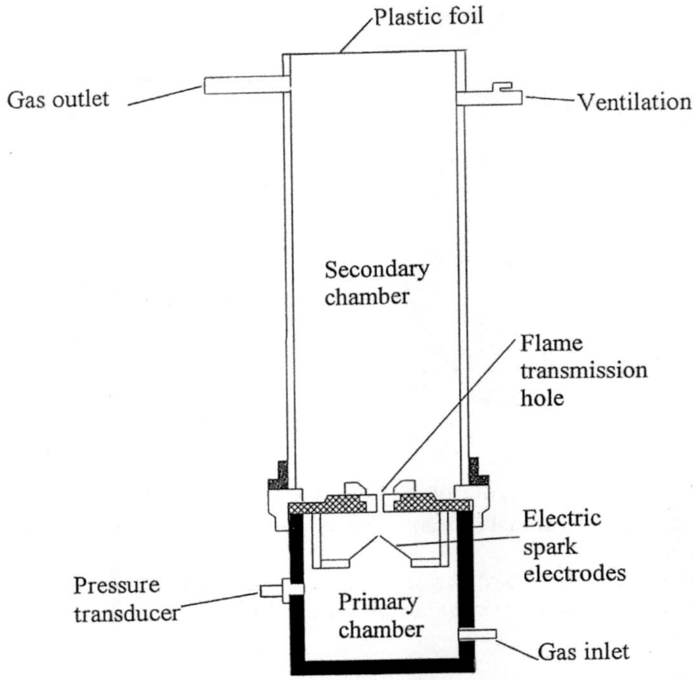

Figure 2.39 Cross section of apparatus used by Larsen and Eckhoff (2000) for experimental determination of critical hole diameters for transmission of gas explosions. Volume of primary chamber = 1 L

to laminar, and what was measured was the critical laminar quenching tube diameter. However, as X_i was gradually increased, a significant amount of gas had to burn before the flame front reached the entrance of the transmission hole. This means that at the time of flame-front arrival at the hole entrance the pressure in the primary chamber was significantly higher than the ambient pressure downstream of the hole. Hence, when the flame front arrived at the hole, hot combustion gases were pushed through the hole, producing a jet into the unburnt explosive gas in the secondary chamber. The question was then no longer whether a laminar flame front would be able to travel through the transmission hole, but rather whether reignition would occur downstream of the hole, where the hot combustion gases forced through the hole mixed with the unburnt mixture. As Fig. 2.40 shows, it appeared that the minimum critical hole diameter for such reignition was only about half the critical diameter from laminar flame transmission. However, with a further increase of X_i the critical hole

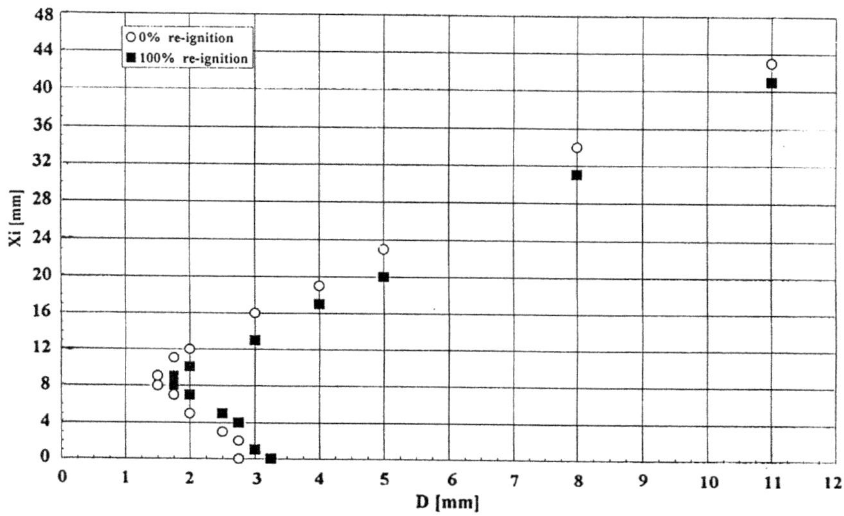

Figure 2.40 Results from experiments with a cylindrical 1 L primary chamber (Fig. 2.39). X_i is the distance from the electric spark ignition source on the chamber axis to the entrance of the axial cylindrical transmission hole of diameter D and length 12.5 mm. Explosive gas mixture: 4.2 vol.% propane in air.

From Larsen, Ø., Eckhoff, R.K., 2000. Critical dimensions of holes and slits for transmission of gas explosions. Some preliminary results for propane/air and cylindrical holes. Journal of Loss Prevention Process Industries 13, 341–347.

diameter for re-ignition increased far beyond the laminar flame transmission value. This is because, at large X_i values, the overpressures in the primary chamber reach high values at the moment of flame arrival at the hole entrance, and give rise to correspondingly high jet velocities. This in turn causes very fast mixing of the hot combustion products and the cold unburnt mixture, and the whole system cools down before the combustion chemistry gets under way.

Fig. 2.41 shows two high-speed video pictures at critical conditions for reignition, using a smaller primary chamber than that illustrated in Fig. 2.39. With $X_i = 8$ mm reignition occurred, with $X_i = 9$ mm there was no reignition. The characteristic difference between the two cases is the flame ball that appears in the downstream region of the jet in the case of reignition.

Results from some earlier similar studies by Wolfhard and Bruszak (1960) are given in Fig. 2.42.

In this case the gas mixtures in the primary and secondary chambers were different. Hence, the gas in the primary chamber was stoichiometric mixtures of methane, oxygen, and nitrogen in all the experiments. The ratio

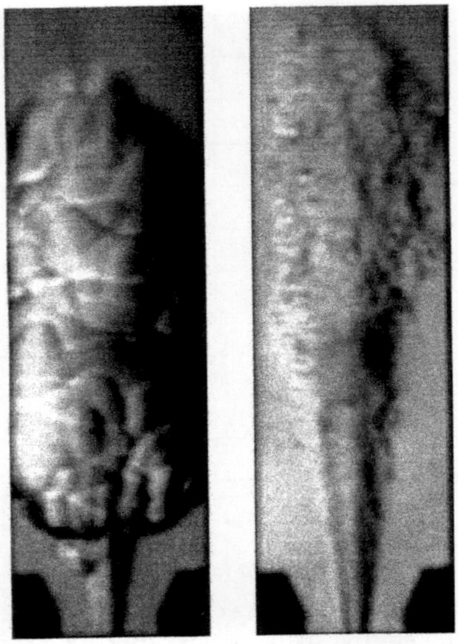

Figure 2.41 High-speed video Schlieren pictures of hot-gas jets expelled from a 0.021 L primary chamber into a larger secondary chamber via a straight cylindrical hole of diameter 4 mm and length 12.5 mm. 4.2 vol.% propane in air in both primary and secondary chamber. Distance from ignition point in primary chamber to hole entrance was 8 mm for reignition and 9 mm for no reignition. Pictures taken 6.25 ms after first sign of hot-gas jet out of hole.
From Larsen, Ø., 1998. Schlieren photo taken as part of work for the degree of cand.scient. (M.Sc.) of the University of Bergen.

of oxygen to nitrogen was varied and characterized by the "oxygen index" (OI), as defined by Eq. [2.11].

$$OI = (\text{vol. }\%O_2)/(\text{vol. }\%O_2 + \text{vol. }\%N_2) \qquad [2.11]$$

The purpose of the experiments was to determine the critical OI in the primary chamber for reignition in the secondary chamber. The three fuel gases used in the secondary chamber were methane, ethane, and a mixture of 90 vol.% CO and 10 vol.% H_2 respectively, and each gas was mixed with air to the respective stoichiometric composition. The gap for the electric spark used for igniting the gas in the primary chamber was located close to the entrance of the tube connecting the two chambers.

As Fig. 2.42 shows, the OI required for reignition increased systematically with increasing tube length for tube lengths >5–10 mm. The likely

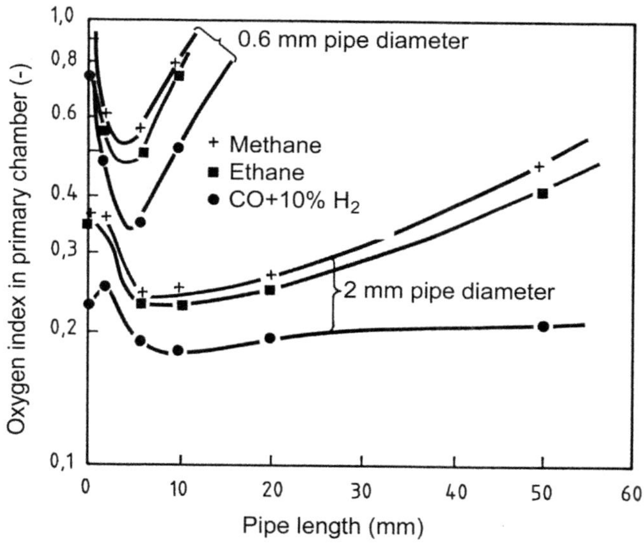

Figure 2.42 Critical oxygen index for ignition of an explosive gas mixture in a secondary chamber by a hot jet of combustion gases from a gas explosion in a primary chamber, as a function of the length of the cylindrical tube connecting the two chambers. Volume of primary chamber: 80 cm^3. Volume of secondary chamber: 3.8 L.
From Wolfhard, H.G., Bruszak, A.E., 1960. The passage of explosions through narrow cylindrical channels. Combustion and Flame 4, 149–159.

reason for this is cooling of the hot combustion gases by the tube wall, which will increase with increasing tube length. For very short tube lengths of a few mm, the critical OI started to rise again, probably due to changes in the jet characteristics. It is also seen that the critical OI for ignition increases with decreasing tube diameter, as would be expected, and with the type of fuel, with methane being the most difficult and $CO + H_2$ the easiest to reignite.

Various apparatuses have been used for determining standardized MESG values of combustible gases and vapors mixed with air (see data in Table 2.2). Normally, the primary chamber is spherical, whereas the secondary one is an annulus surrounding the sphere. The connection between the two chambers is a flange gap of length 25 mm and adjustable width. The aim of the test is to determine the critical maximum gap width that does not give reignition in the second chamber (MESG), for the actual combustible gas or vapor being tested. The highest values are normally obtained with the ignition source located fairly close to the gap entrance, rather than at the center of the primary chamber.

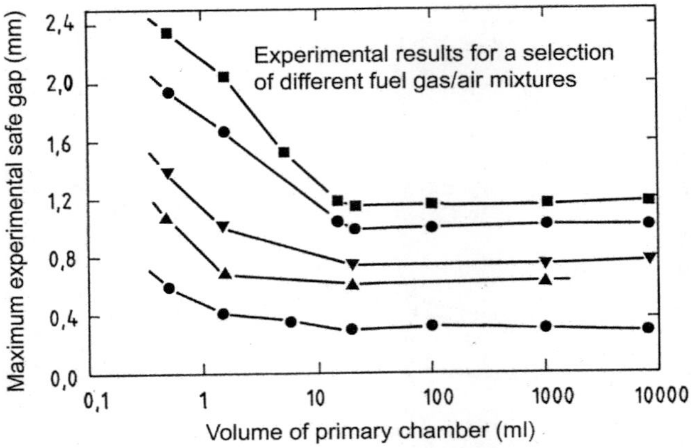

Figure 2.43 Influence of the volume of the primary chamber on MESG, determined experimentally by Physikalisch Technische Bundesanstalt (PTB), Germany.
From Phillips, H., 1987. A comparison of "standard" methods for the determination of maximum experimental safe gaps (MESG). In: Proceedings of the International Symposium on the Explosion Hazard Classification of Vapors, Gases and Dusts, pp. 83–108. Organized by the Committee on Studies of Hazardous Substances, National Materials Advisory Board, Publication NMAB-447, National Academy Press, 1987.

Experiments have confirmed that the volume of the primary chamber does not influence the measured MESG as long as this volume is at least 20 cm^3, and the volume of the secondary chamber is sufficiently large to prevent precompression of the gas there before the appearance of the hot jet. This is illustrated in Fig. 2.43.

As can be seen, no significant effect of the volume of the primary chamber was found over the range 0.020–8.0 L. For this reason it was decided to adopt a primary chamber of 0.020 L volume in the standard IEC (1971) test method. Fig. 2.44 shows the standard apparatus developed by IEC (1975b). Prior to an experiment, the width of the 25-mm long flanged gap is adjusted to the desired value by means of a spring-loaded micro-meter screw. The ignition source in the internal chamber is a 3-mm spark gap located perpendicularly to the gap plane, 14 mm away from the gap entrance. The chambers are flushed with the gas/air mixture to be tested until homogeneous composition is obtained throughout. Whether re-ignition occurs is observed visually through the glass window in the wall of the external chamber. The MESG data in Table 2.2 (Section 2.1.3.1) were determined by this method.

Fig. 2.45 gives some experimental results from Maskow (1950) showing the effect on MESG of varying both the length of the gap from the standard 25 mm and downward, and the fuel/air ratio.

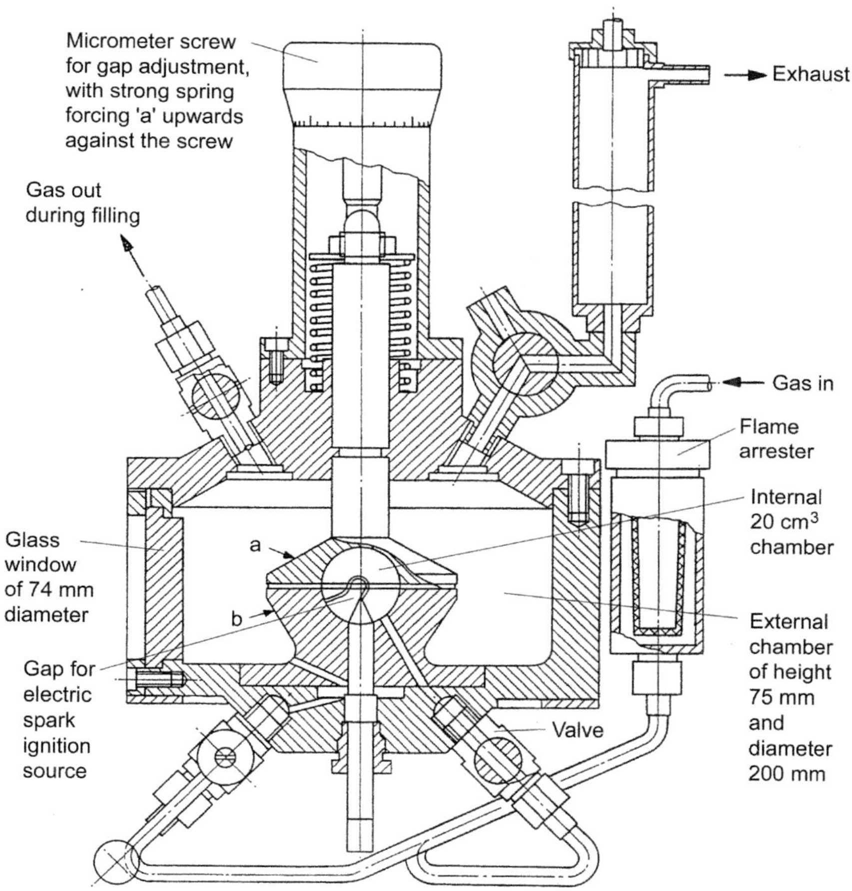

Figure 2.44 Illustration of the standard IEC apparatus used for determination of MESG for combustible gases and vapors mixed with air. Volume of primary spherical chamber is 20 cm^3. Standard gap length 25 mm.
From IEC, 1975b. First supplement IEC 79-1A to IEC 79-1 (1971). Electrical Apparatus for Explosive Gas Atmospheres. Part 1: Construction and Testing of Flameproof Enclosures of Electrical Apparatus. Appendix D: Method of Test for Ascertainment of Maximum Experimental Safe Gap. Central Office of International Electrotechnical Commission, Geneva, Switzerland.

This figure shows that the critical gap width for flame transmission (reignition downstream of gap) increased systematically with the gap length L, apart from a sharp edge (L = 0) being slightly less efficient than a short finite gap of length 1 mm. It is also seen that the fuel/air ratios that gave flame transmission most readily were close to the stoichiometic ratio.

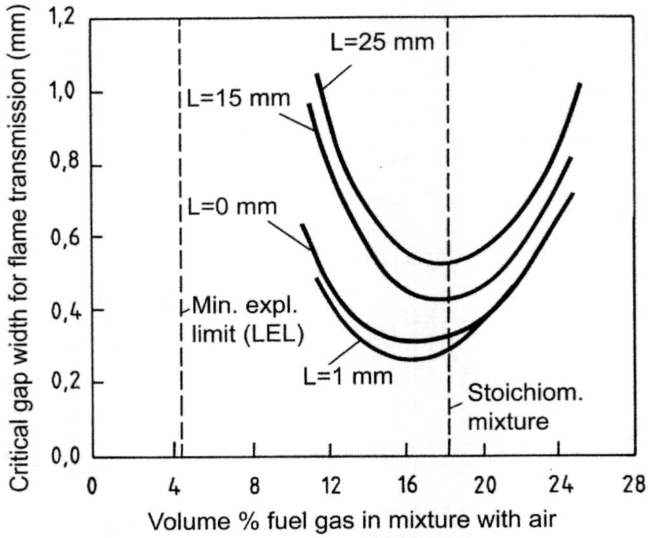

Figure 2.45 Maximum experimental safe gap, determined in a standard MESG test apparatus, for different gap lengths L and different fuel/air ratios. Fuel: 30 vol.% methane +70 vol.% hydrogen.
From Maskow, H., 1950. VDI-Fachberichte, vol.14, pp. 144–147, Verein deutscher Ingenieure, Düsseldorf, Germany.

2.2.8 Ignition by Rapid Adiabatic Compression

In the case of rapid adiabatic compression, much work has been performed in the context of internal combustion and diesel engines. Some useful information is given in Freytag (1965). If an explosive gas mixture is compressed rapidly adiabatically, or in a shock wave, the temperature will rise rapidly, and the cloud may ignite. The temperature increase mainly depends on the ratio between the pressure after compression and the initial pressure. Shock waves may be generated during sudden relief of high-pressure gases into pipelines. The shock wave then propagates into regions of lower pressure faster than the speed of sound. Very high peak pressures and temperatures can occur if the wave is diffracted or reflected by pipe bends, constrictions, connection flanges, closed valves, etc.

Fig. 2.46 shows pressure as a function of time during rapid compression of a mixture of hydrogen and air in a cylinder with a piston. The induction time between completion of compression and onset of ignition is a characteristic feature of this kind of process.

Figs. 2.47 and 2.48 show some experimental results for some hydrocarbon/air mixtures. As can be seen from these figures the induction time

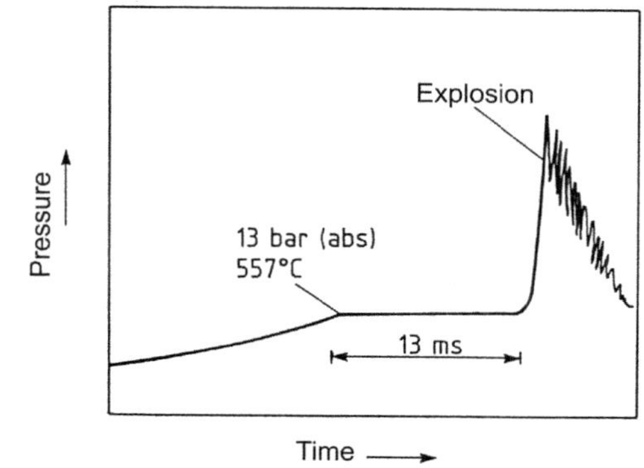

Figure 2.46 Pressure as a function of time during rapid compression of a mixture of hydrogen and air in a cylinder with piston. State of gas mixture at end of compression: 13 bar (abs) and 557°C. Ignition delay 13 ms.
From Freytag, H.H., 1965. Handbuch der Raumexplosionen. Verlag Chemie, Weinheim, Germany.

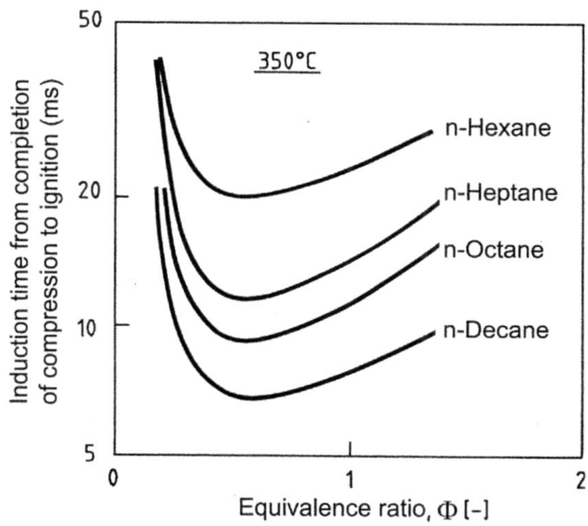

Figure 2.47 Induction time as a function of fuel/air ratio for some alkane/air mixtures compressed adiabatically to 350°C. ($\Phi = 1$ corresponds to the stoichiometric ratio.)
From Freytag, H.H., 1965. Handbuch der Raumexplosionen. Verlag Chemie, Weinheim, Germany.

depends on both the fuel type and the fuel/air ratio, and on the degree of compression (temperature of gas mixture just after compression has been completed). Fuel/air ratios of about half the stoichiometric composition seem to ignite most easily.

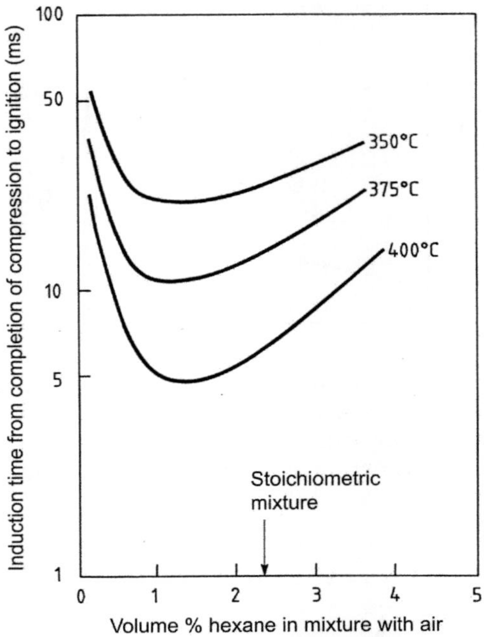

Figure 2.48 Induction time as a function of fuel/air ratio for hexane/air for various degrees of adiabatic compression.
From Freytag, H.H., 1965. Handbuch der Raumexplosionen. Verlag Chemie, Weinheim, Germany.

A more recent, detailed study of ignition following rapid adiabatic compression of heptane/air mixtures was undertaken by Minetti et al. (1995).

2.2.9 Ignition by Light Radiation

Some informative investigations were conducted by Welzel et al. (2000) and Thowle (2000). Welzel et al. studied ignition of a wide range of explosive gases and vapors mixed with air, by continuous optical radiation conveyed through an optical fiber, and absorbed by a solid black iron/manganese oxide target mounted at the end of the fiber. The combustible gases and vapors covered a wide range of autoignition temperatures. The experimental apparatus used is illustrated in Fig. 2.49.

The ignition chamber containing the explosive gas mixture was a vertical glass tube of length 50 cm and diameter 15 cm, ie, a volume of about 9 L. The oxide covered end of the optical fiber was positioned in the lower part of the chamber. The light source was a laser of wavelength 1064 nm. Fig. 2.50 shows a selection of experimental results.

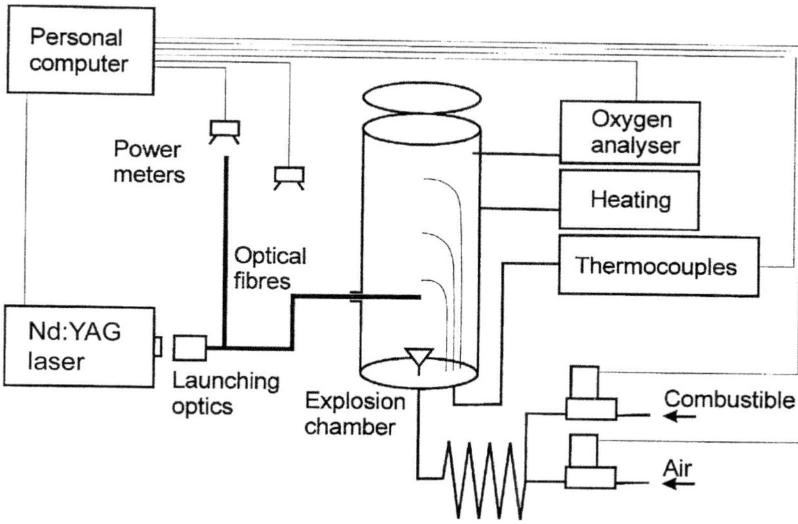

Figure 2.49 Experimental apparatus used for determining critical data for ignition of a wide range of explosive gases and vapors mixed with air by optical radiation absorbed by a solid iron/manganese oxide target attached to the end of an optical fiber.
From Welzel, M.M., Schenk, S., Hau, M., Cammenga, H.K., Bothe, H., 2000. Ignition of combustible/air mixtures by small radiatively heated surfaces. Journal of Hazardous Materials A72, pp. 1–9.

The figure also indicates conservative limits embracing all the data, viz. 35 mW for target areas smaller than 4 mm^2, and 5 mW/mm^2 for larger target areas.

Fig. 2.51 illustrates how the minimum radiant power required for ignition drops with increasing temperature of the explosive mixture.

Fig. 2.52, from Thowle (2000), summarizes data from several independent investigations, including that of Welzel et al. (2000), and the data confirm the validity of the conservative asymptotes suggested by Welzel et al. (2000). Note that the very low values in the lower right corner of Fig. 2.52 were obtained by targets heated electrically, not by light radiation.

Correspondence of the absorption bands of the combustible gas with the laser or radioactive radiation wavelength may cause ignition by direct excitation and ionization of the gas molecules. Radioactive radiation is generated by X-ray tubes and radioactive substances. Radioactive radiation can heat up a solid surface, owing to internal absorption of radiation energy, to such an extent that the minimum ignition temperature of the surrounding explosive gas mixture is exceeded. Such radiation can also cause direct chemical decomposition of the gas or other chemical reactions which can generate

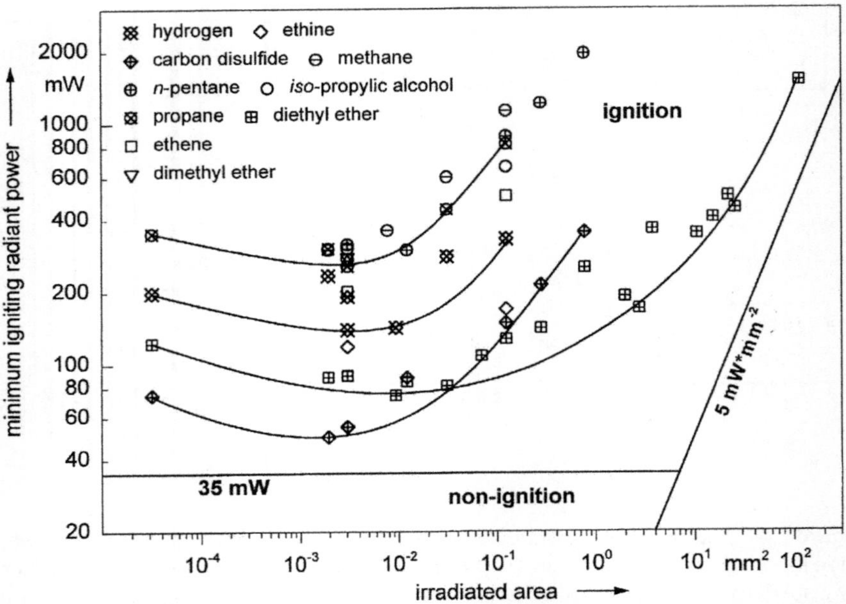

Figure 2.50 Minimum radiant power (mW) for ignition as a function of the surface area of the iron/manganese oxide target (mm^2) for a range of explosive mixtures of combustible gases/vapors and air.

From Welzel, M.M., Schenk, S., Hau, M., Cammenga, H.K., Bothe, H., 2000. Ignition of combustible/air mixtures by small radiatively heated surfaces. Journal of Hazardous Materials A72, pp. 1–9.

highly reactive radicals or unstable chemical compounds which can lead to ignition.

2.2.10 Concluding Remark

Critical ignition parameters, eg, the minimum ignition temperature and the minimum ignition energy, of a given explosive gas mixture can vary substantially with the actual ignition source characteristics, and the dynamics, pressure, and temperature of the gas mixture. This means that there is a need for methods that permit differentiation when specifying parameters for various practical situations. For example, it is essential that the minimum ignition temperature be specified as closely as possible for the particular industrial situation of concern. The application of just one value for a given explosive gas, based on a highly conservative laboratory test, irrespective of the actual industrial situation, may in some cases put industry to considerable unnecessary expense.

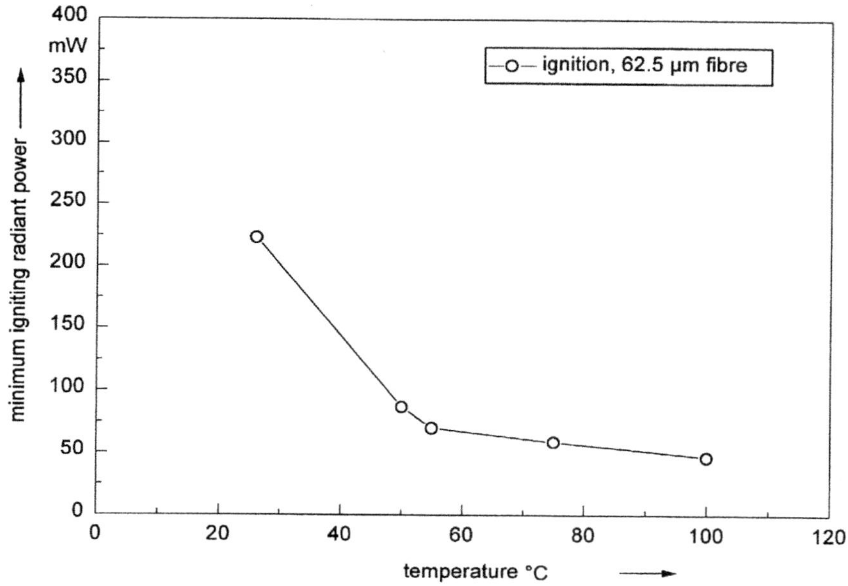

Figure 2.51 Minimum radiant power required for igniting a mixture of 12 vol.% diethyl ether in air, as a function of the temperature of the explosive mixture. The heated target was a thin layer of black iron/manganese oxide attached to an optical fiber of diameter 62.5 μm.

From Welzel, M.M., Schenk, S., Hau, M., Cammenga, H.K., Bothe, H., 2000. Ignition of combustible/air mixtures by small radiatively heated surfaces. Journal of Hazardous Materials A72, pp. 1–9.

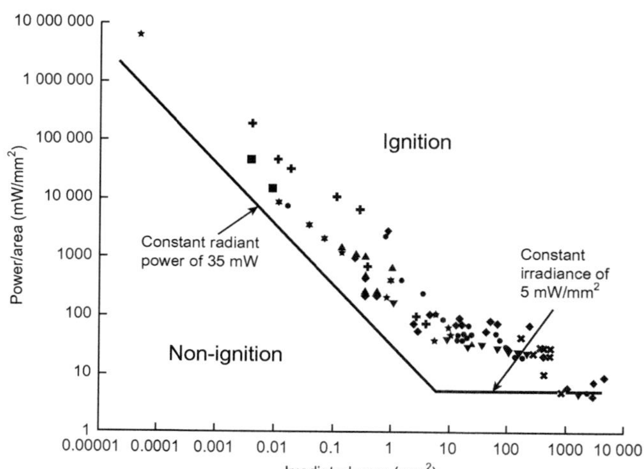

Figure 2.52 Summary of experimental data of minimum radiated power per unit target area (mW/mm^2) for igniting a range of explosive mixtures of combustible gases/vapors and air, as a function of the irradiated area (mm^2).

From Thowle, C., October 2000. Optical sources of ignition in hazardous areas. Power Engineering Journal, 238–242.

2.3 CASE HISTORIES OF ACCIDENTAL GAS/VAPOR CLOUD EXPLOSIONS

2.3.1 Motivation

Experience has shown that "learning by doing" is an effective way of acquiring new knowledge. Unfortunately, this also applies to learning about explosions, which can give rise to much human suffering and grief, as well as material damage and loss of profit. Those who have experienced an explosion accident, whether as workers or management in industrial plants, or elsewhere, have a profound appreciation of the realism of this hazard, beyond the reach of those who have only heard or read about accidental explosions in general terms. Real understanding in turn produces the proper motivation for minimizing the probability of occurrence of such events in the future. Clearly, accidental explosions are highly undesirable in any situation, and one therefore seeks less dramatic means of transferring knowledge and motivation. One way is the use of case histories, ie, fairly detailed accounts of explosions that have actually occurred.

The present section describes a number of gas explosion accidents in different types of industry and equipment, with emphasis on ignition sources, propagation mechanisms, and destructive effects. Much of the material is taken from Eckhoff (1996a).

2.3.2 Historical Perspective: Methane Explosions in Coal Mines

Natural gas is probably the first substance that exposed the human being to the hazard of gas explosions. Natural gas has been produced by nature itself as long as there has been life on our earth. It is produced whenever natural organic material is being decomposed in the absence of oxygen. Marsh gas is a typical example. It appears that natural gas was used as an energy source in China as early as 200–300 AD, presumably for cooking and illumination. At about 1600–1700 AD, natural gas was used for the same purpose in Italy. In 1778, the Italian scientist Volta discovered and was able to isolate methane as the main chemical component of natural gas. The first drilled well for industrial production of natural gas in USA was opened about 1820, and during the period 1850–1900 pipelines for distribution of natural gas became common throughout the eastern part of the country.

As coal mining developed in Europe during the 17th and 18th centuries, severe mine explosions became common. It was soon discovered that most often the origin of the explosions was the ignition of mixtures of flammable gas ("firedamp") and air, which accumulated in the mines. "Firedamp" is essentially methane, liberated from coal when pressure is released. In early coal mining in Europe, testing of mines for possible explosive gas was undertaken by volunteers creeping into the mine galleries wrapped in wet blankets and carrying an open flame on a stick. This is illustrated in Fig. 2.53.

In 1815, the "Sunderland Society for Preventing Accidents in Coal Mines," in United Kingdom , asked Sir Humphrey Davy to carry out systematic research on the causes of ignition of, and flame propagation in "firedamp"/air mixtures. As Davy recorded:

> *The great object was to find a light, which at the same time that it enabled the miners to work with security in explosive atmospheres, should likewise consume the firedamp.*

As pointed out by McQuaid (1990), one of the striking features of this investigation was the speed with which Davy moved from receiving the

Figure 2.53 A firefighter igniting methane under the roof of a mine gallery, prior to the miners being admitted to the gallery.
From McQuaid, J., March 14, 1990. Know Your Enemy. The Science of Accident Prevention. The Vernon Clancey Memorial Lecture 1990, City University, London.

problem to producing a successful solution. However, when the task of finding a safe source of light in coal mines was first put to Davy, he was not optimistic. It appeared to him that there was very little hope of finding an effective solution. Earlier attempts at finding a safe source of light had not met with much success. Designs of safety lamps in existence relied either on complete confinement, and so lamps could burn only for a short time, or else the air supply was pumped through water seals, and so lamps were not practicable, and the flame could easily be extinguished.

Despite his initial pessimism, Davy embarked on his research. He first investigated the chemical composition of the firedamp and then carried out numerous experiments on the circumstances under which it combusts. He established the limits of flammability of the firedamp in air and also established that it was much less easy to burn than other flammable gases. Under the test conditions used, it could be ignited by neither red-hot charcoal nor red-hot iron. The heat produced by it in combustion was less than that of most other flammable gases.

Davy found that in glass tubes one-quarter of an inch in diameter and 1-ft long, more than a second was required before the flame had traveled from one end to the other. When the tubes were one-seventh of an inch in diameter, flammable mixtures could not propagate a flame to the open end. Metallic tubes prevented flame propagation better than glass tubes. He now reasoned that the effect of the surfaces of small tubes in preventing combustion depended on their cooling powers, that is, upon their lowering the temperature of the burning mixture so much that it was no longer sufficient for continued combustion. He proceeded to conclude that a metallic tissue or gauze would offer the perfect barrier to flame propagation. He experimented until he found the geometry of the gauze and its material which was safe in all possible varieties of flammable mixtures, not only of firedamp but also of coal gas. His final design was produced in January 1816 and was immediately adopted in the coal mines. Fig. 2.54 shows an early version of the Davy lamp.

Davy had achieved his objective less than six months after receiving the letter from the Sunderland Society for Preventing Accidents in Coal Mines. And he did this without any of our modem analytical facilities. However, it seems reasonable to assume that this highly efficient and successful research was not only due to Davy himself. His young and extraordinarily talented assistant was Michael Faraday, who also most certainly must have played a major role. Much later, in 1844, Faraday was the first researcher to realize that also coal dust played a crucial part in many of the severe coal mine

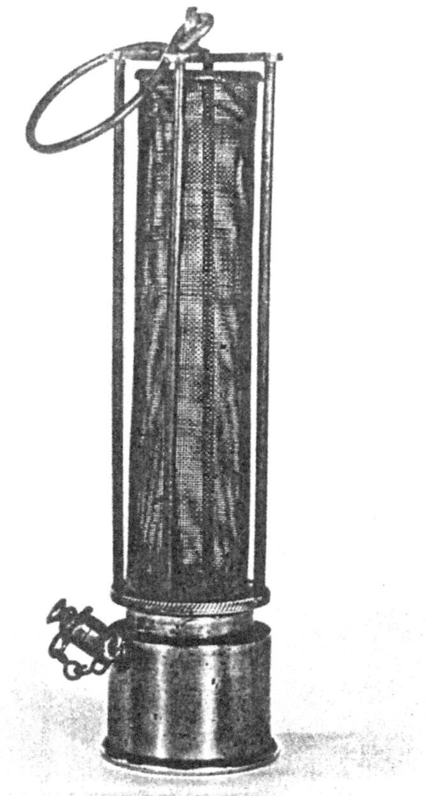

Figure 2.54 Early version of the coal mine lamp developed by Sir H. Davy about 1815.

explosions that Europe experienced in the growing coal mining industry at that time.

The excellent performance of the Davy lamp was confirmed in a letter from Mr. George Morris, Plas Issa, Wales, dated 27 January 1817, to his mine owner (from McQuaid, 1990):

> *Sir: You will be pleased to recollect that some time in the month of June last, I applied to you with a request you would send me immediately some of Sir Humphry[sic] Davy's safety lamps, in consequence of an explosion of the firedamp taking place in one of your coal mines, by which several of the men were dreadfully burnt and bruised. On the arrival of the safety lamps no accurate account of their use accompanied them. But I at length obtained (I think) the Edinburgh Review, in which was a detail of some experiments. This I read to the colliers, which gave them some confidence in the lamps, prior to which they secretly treated them with silent contempt; and I found, notwithstanding these interesting details, that a great doubt existed in their minds.*

The men had no sooner descended than the enemy (methane) was discovered (by ignition and combustion inside the Davy lamp), which they say very much alarmed them. They would have retreated if they could, but finding that impossible, took courage, and soon found they had destroyed the enemy so far; advancing a little farther, they found him again, and again destroyed him, and so on through the whole work. Thus, the first alarm was got over, when all the knowing men in the neighbourhood were got collected together to hear the result, all of which were astonished and amazed, that so simple looking an instrument should destroy and defy an enemy, heretofore unconquerable. The same precaution and use of the lamp, was gone through the second day, and when the firedamp was destroyed, we began working and continued to work in this way for some weeks.

However, despite this and other great achievements, methane and coal dust explosions continued to represent a severe hazard in coal mines. Cybulski (1975) summarized the major coal mine disasters that had occurred worldwide during 1900−1970. In 135 major explosions, more than 20,000 miners had lost their lives, ie, an average of about 150 in each of the disasters. Cybulski also described much of the extensive research that had been conducted in many countries up to about 1970, to reduce the explosion hazard in coal mines.

2.3.3 Some Older Published Reviews of Major Accidental Gas/Vapor Cloud Explosions

Gugan (1979) reviewed 100 major accidental gas/vapor cloud explosions in the process industries that had occurred worldwide up to the late 1970s. One of the main sources of information was the account by Davenport (1977). Gugan concluded that because of incomplete records, only 29 of the 100 explosion accidents could be analyzed in any depth. Although most of the 29 accidents produced significant blast effects, only 8 were documented sufficiently well to permit reasonably comprehensive analyses. Gugan (1979) gave a quite detailed discussion of five of the eight explosions.

In his book, Marshall (1987) presented 11 reviews and collections of case histories of major gas/vapor cloud explosions up to 1984, including the book by Gugan (1979) and the report by Davenport (1984). Marshall presented comprehensive case histories of the major dimethyl ether explosion at Ludwigshafen, Germany, in 1948; the propane explosion at Port Hudson, Missouri, in 1970; the cyclohexane explosion in Flixborough in 1974, and the isobutane explosion at Decatur, Illinois,in 1974.

Gow (1991) provided brief abstracts of published records of vapor/gas cloud explosions, including the explosions with propane (Sweden, 1981), vinyl chloride (United States, 1971), acetylenic alcohol (United States, 1967), ethanol (United States, 1972),1-pentol (United States, 1967), hydrogen-rich treat gas (United Kingdom, 1983), methanol synthesis gas (United States, 1983), natural gas (United States, 1986), hydrogen (United States, 1989), a mixture of ethane, propane, ethylene, and propylene (United States, 1967), ethylene oxide (United States, 1968), butadiene (United States, 1971), propylene (United States, 1976), propylene (United States, 1977), and hydrogen (Norway, 1989).

2.3.4 The Flixborough Disaster, United Kingdom (1974)

2.3.4.1 Summary

This is no doubt one of the most extensively investigated and quoted major accidental vapor cloud explosions in modern times. Some key data of this catastrophic event are given in Table 2.13.

The comprehensive public report by Parker et al. (1975) contains references to about 60 detailed reports on various aspects that were available by early spring of 1975. A further considerable number of articles, scientific papers, and reports were published after the public report had been issued. Examples are the papers by Warner (1975), Sadee et al. (1976/1977), Roberts and Pritchard (1982) and Berg (1985). Extensive reviews, and analyses of the Flixborough disaster, containing further references to papers and reports, were given by Gugan (1979, p. 12), Lees (1996, p. 20), and Marshall (1987, p. 20). Marshall includes an interesting analysis of blast damage of dwelling houses in the surroundings of the industrial plant, as a function of the assumed explosion epicenter. Roberts and Pritchard (1982) estimated the maximum explosion loads at various locations in the Flixborough explosion, on the basis of the degree of

Table 2.13 The Flixborough disaster. Some key data

Place:	Flixborough, Humberside, England
Date:	June 1, 1974
Fuel:	Cyclohexane
Type of plant:	Congested process area of cyclohexane oxidation section of caprolactam production plant (basic raw material for Nylon 6)
Fatalities:	28
Injuries:	89
Material loss:	Entire plant demolished; 1821 houses and 167 shops in nearby residential areas suffered various degrees of damage

deformation and damage on process equipment, lamp posts, and other mechanical structures.

The following summary of the various aspects of the Flixborough disaster is mainly based on the account by Lees (1996).

2.3.4.2 The Process and the Plant

The cyclohexane oxidation plant shown in Fig. 2.55 consisted of a train of six reactors in series in which cyclohexane was oxidized to cyclohexanone and cyclohexanol by air injection in the presence of a catalyst. The reactions are exothermic. The feed to the reactors was a mixture of fresh cyclohexane and recycled material. The product from the reactors still contained approximately 94% of cyclohexane. The liquid reactants flowed from one reactor to the next by gravity. In subsequent stages, the reaction product was distilled to separate the unreacted cyclohexane, which was recycled to the reactors, and the cyclohexanone and cyclohexanol, which were converted to caprolactam. The operating conditions in the reactor according to the design were 8.8 bar (g) and 155°C.

2.3.4.3 Events Prior to the Explosion

On the evening of March 27, 1974, it was discovered that Reactor No. 5 was leaking cyclohexane. The following morning the decision was taken to remove Reactor No. 5, and to install a bypass assembly to connect Reactors No. 4 and 6 so that the plant could continue in production. The diameter of the openings to be connected on these reactors were 28″, with bellows on the nozzle stubs, but the largest pipe available on site, which was considered to be suitable for making the bypass, had a diameter of 20″. Furthermore, the two flanges were at different heights so that the connection had to take the form of a dogleg of three lengths of 20″ pipe welded together with flanges at each end, which were bolted to the existing flanges on the stub pipes on the reactors. The bypass assembly is illustrated in Fig. 2.55. Calculations were performed to check that the pipe diameter was large enough for the required flow, and that it was capable of withstanding the same internal pressure as if it were a straight pipe.

However, apparently no proper calculations were made, which would have taken into account the forces arising from the dogleg shape of the pipe. No drawings of the bypass pipe were made other than a sketch in chalk on the workshop floor. The original stub pipes were connected to the reactors by bellows, as indicated in Fig. 2.55, but no calculations were done to check whether these bellows would withstand the particular forces caused

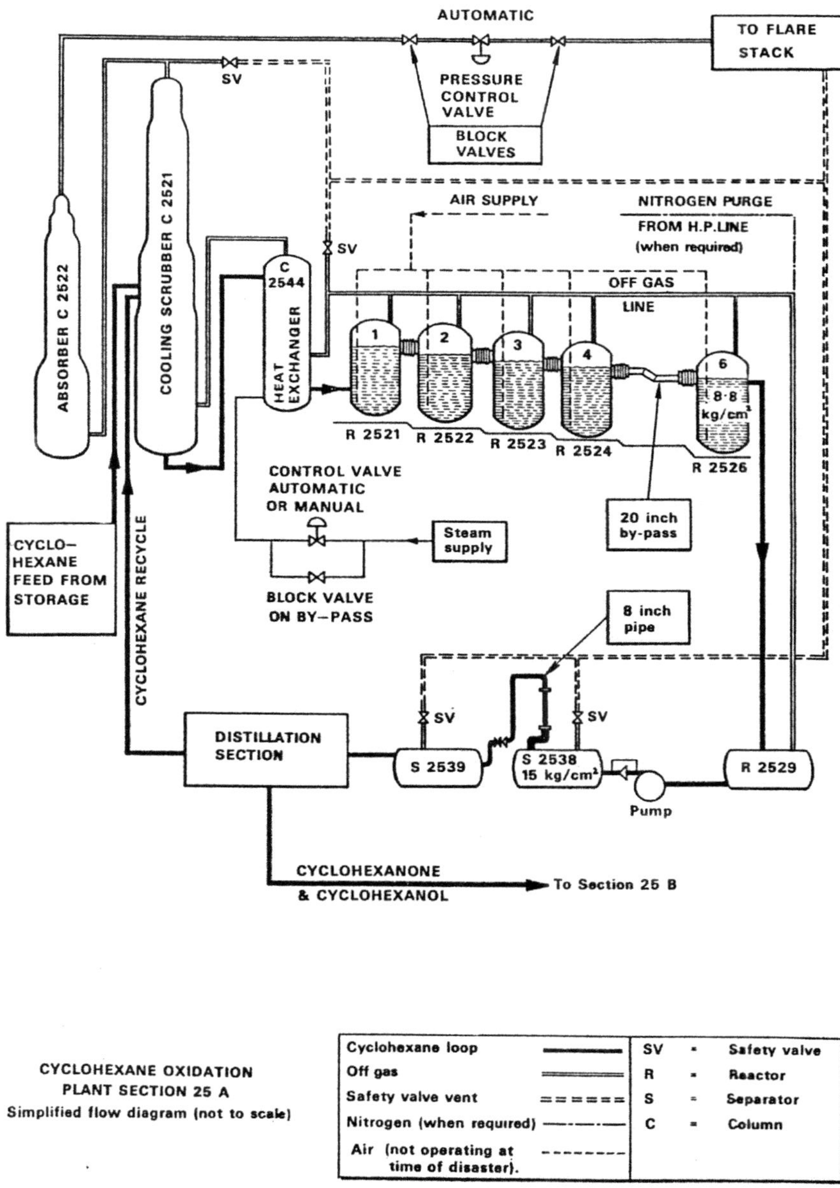

Figure 2.55 Simplified illustration of Section 25A of the cyclohexane oxidation plant of the caprolactam production plant at Flixborough.
From Parker, R.J., Pope, J.A., Davidson, J.F., Simpson, W.J., 1975. The Flixborough Disaster. Report of the Court of Inquiry to the Secretary of State for Employment. Her Majesty's Stationery Office, London, UK.

by the dogleg pipe. The bypass assembly was supported by a scaffolding structure, which was intended to support the pipe and to avoid straining of the bellows during construction of the bypass. It was not suitable as a permanent support for the bypass assembly during normal operation. No pressure testing was carried out on either the pipe or the complete assembly before it was fitted. A pressure test was performed on the plant, however, after installation of the bypass. The equipment was tested to a pressure of 9 bar (g), but not up to the safety valve pressure of 11 bar (g). The test was pneumatic, not hydraulic.

Following these modifications, the plant was started up again. The bypass assembly gave no trouble. There did appear, however, to be an unusually large usage of nitrogen on the plant, and this was being investigated at the time of the explosion. On May 29, the bottom isolating valve on a sight glass on one of the vessels was found to be leaking. It was decided to shut the plant down to repair the leak. On the morning of June 1 start-up began. The precise sequence of events is complex and uncertain. The crucial feature, however, is that the reactors were subjected to a pressure somewhat greater than the normal operating pressure of 8.8 bar (g). A sudden rise in pressure up to 8.5 bar (g) occurred early in the morning when the temperature in Reactor No. 1 was still only 110°C and that in the other reactors was less, whereas later in the morning, when the temperature in the reactors was closer to the normal operating value, the pressure reached 9.1–9.2 bar (g). The control of pressure in the reactors could normally be performed by venting the off-gas, but this procedure involved the loss of considerable quantities of nitrogen. Shortly after warm-up began, it was found that there was insufficient nitrogen to begin oxidation and that further supplies would not arrive until after midnight. Under these circumstances the need to conserve nitrogen would tend to inhibit reduction of pressure by venting.

2.3.4.4 The Explosion

During the late afternoon of the day of the explosion, the provisional 20″ bypass system failed, without or with a contribution from a fire on a nearby 8-in. pipe. This resulted in the escape of a very large quantity of cyclohexane. The cyclohexane formed a vapor cloud (and possibly also a liquid-droplet spray cloud, see Section 3.5.1), and the flammable part of the cloud found a source of ignition. At about 4.53 pm there was a huge, catastrophic vapor cloud explosion.

The explosion caused extensive damage and started numerous fires. The blast and the fires destroyed not only the cyclohexane plant but several other plants also. Many of the fires were in the tank farm. The blast of the explosion shattered the windows of the control room and caused the control room roof to collapse. Of the 28 people who died in the explosion, 18 were in the control room. Some of the bodies had suffered severe injuries from flying glass. Others were crushed by the roof. No one escaped from the control room. The main office block was also demolished by the blast of the explosion. Because the accident occurred on a Saturday afternoon, the offices were not occupied. If they had been, the death toll would have been much higher. The fires on the site burned for many days. Even after 10 days, the fires were hindering rescue work on the site.

2.3.4.5 The Investigation

Some of the possible causes of failure of the bypass assembly were outlined early in the inquiry. They included failure of the 20-in. diameter pipe due to a small pressure rise in the system. Several possible causes for a pressure rise in the 20-in. diameter pipe were suggested, viz. entry of high-pressure nitrogen into the system due to instrument malfunction, entry of water into the system, temperature rise in the system due to excessive heating by steam in reboiler of C2544, leakage of steam from a tube in C2544, explosion of peroxides formed in the process, and explosion due to air in the system.

The overall conclusion of the final inquiry was that the disaster was due to "ignition and rapid acceleration of the resulting deflagration, possibly to the point of detonation, of a massive vapor cloud formed by the escape of cyclohexane under a pressure of at least 8.8 bar (g) and a temperature of 155°C, and that the escape was from Section 25A of the cyclo hexane plant." There was no dispute that the main part of the cyclo hexane came from the 20-in. bypass assembly, but there was some uncertainly as to whether the mechanical failure of the assembly was the primary failure or whether it was a secondary failure caused by a preceding one.

2.3.5 The Beek Explosion, The Netherlands (1975)

2.3.5.1 Summary

Some central data of this disastrous event are given in Table 2.14.

Reviews of the accident were given by Gugan (1979) and Lees (1996). A comprehensive two-part report of investigation in the Dutch language and quoted by Anonym (1976) was issued by the Directorate of Labour of the Ministry of Social Affairs, in the Netherlands. This report contains quite detailed maps and photographs of the damaged plant, including a map indicating the contours of the explosive cloud prior to ignition.

Table 2.14 The Beek explosion. Some key data

Place:	Beek, The Netherlands
Date:	November 7, 1975
Fuel:	C3−C4 hydrocarbon fractions
Type of plant:	Congested process area of naphtha cracker installation
Fatalities:	14
Injuries:	104
Material loss:	Entire plant and a nearby tank farm destroyed. Window breakage in surroundings up to 2.5 km from explosion center.

2.3.5.2 Process and Plant

According to Wingerden et al. (1995) the explosion occurred in a naphtha cracker installation, covering an area of approximately 160 × 80 m. The cracking took place in furnaces and resulted in hydrogen, ethylene, and some other hydrocarbons. However, ethylene was the main product. Separation of the products was accomplished by distillation, compression, and cooling. The plant was subdivided into seven process sections, viz. the furnaces for cracking, various separation and distillation processes, a compression section, and cooling sections. Transport of raw materials and products between the various parts of equipment within each section, and between sections, occurred in pipes arranged in pipe-racks. Most process sections were densely packed with equipment, which was placed on either concrete floors or elevated grated floors.

2.3.5.3 Events Prior to the Explosion

In the morning on the day of the explosion, a leakage of a mixture of C_3−C_4 hydrocarbons occurred from one of the process sections. The most likely cause was low-temperature embrittlement of a weld in a feed drum. The prevailing wind caused the flammable vapor cloud to drift into the congested process area. After approximately 2 min from the start of the leak, the visible part of the cloud was approximately 2 m high. It was estimated that in total approximately 5500 kg of gas had escaped. At this point the cloud was ignited by one of the furnaces near one corner of the process area.

2.3.5.4 The Explosion

The resulting explosion caused severe damage inside the process area. Overpressures of up to 1 bar must have existed to account for the damage produced. Pipelines and beams were bent over, and several pipelines had ruptured, resulting in fires. Vessels had become dented and pushed away from their original locations. Part of a concrete platform had fallen down and another had tilted. Severe damage occurred to the concrete walls of the control room. The "engineers' room" was totally demolished. Fire broke out in a tank farm north of the process installation. Outside the plant area, damage was limited to window breakage up to a distance of 4.5 km (5 mbar). The 14 fatalities occurred in the control room and "engineers" room.

2.3.5.5 Computer Simulation

Twenty years after the accident, Wingerden et al. (1995) were able to perform a series of unique computer simulations of the Beek explosion disaster, using the comprehensive computational fluid dynamic (CFD)-based simulation code FLACS, developed in Norway since 1980. A comparison of local peak pressures estimated from the damage picture generated by the actual vapor cloud explosion in 1975, and the corresponding local peak pressures produced by the FLACS simulation, showed similar trends in the peak-pressure distribution across the process area. The simulations indicated that the explosion occurred with a fuel that was more reactive than pure propylene. Significant amounts of ethylene or butadiene must also have been present. The simulations also showed that for land-based process installations in the open, where partial confinement is mainly due to process equipment, the reactivity of the fuel is important with respect to the violence of an accidental explosion. The simulations further revealed that in an exploding heavy (flat) vapor cloud, a stoichiometric mixture of fuel and air is not necessarily the worst case. Overstoichiometric mixtures can be even more dangerous due to the rapid mixing of air above the cloud with the gas mixture in the cloud during the explosion process.

2.3.6 The Arendal Explosion, Gothenburg, Sweden (1981)

2.3.6.1 Summary

Some central data of this event are given in Table 2.15.

2.3.6.2 The Site of the Event

Details of this have been acquired from the Special Working Group (1981), Brandsjö (1988), and Nilsson (1981). The site of this explosion is not, strictly speaking, a process plant, but the event provides valuable

Table 2.15 The Arendal explosion. Some key data

Place:	Arendal, Gothenburg, Sweden
Date:	May 8, 1981
Fuel:	Propane
Type of plant:	Pipe track through urban area. Rupture of pipeline
Fatalities:	1
Injuries:	1
Material loss:	15 mill. Swedish Krona (SEK)

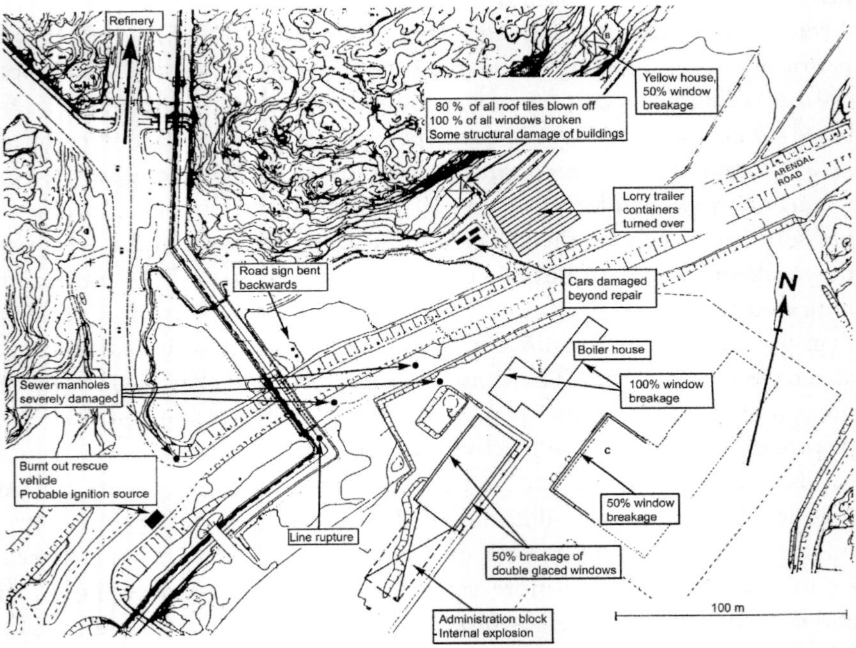

Figure 2.56 Plan of site and consequences of propane vapor cloud explosion at Arendal, Gothenburg, Sweden, in 1981.
From Nilsson, E., 1981. Gasmolnolyckan Göteborg 1981-05-08. Report from SAI (Swedish Explosives Inspectorate).

information of that is relevant even in the present context. Fig. 2.56 gives a view of the explosion site and indications of the local damage caused by the explosion. The site was a large obstacle field in the open, characterized by pipe tracks, bridges, buildings, cars, and trees, acting as turbulence-generating obstacles.

2.3.6.3 Leak and Vapor Cloud Formation

The gas leak giving rise to a large explosive gas cloud was probably caused by sabotage. A 5-km long 6-inch diameter pipeline, transporting liquid propane at 8—9 bar (g), became perforated, presumably by an explosive charge placed intentionally on the pipeline wall. About 95 m^3 of the liquid propane was flashed to the atmosphere through the resulting 90 cm^2 hole in the pipe wall. The gaseous propane mixed with the ambient air, and a major explosive cloud was generated in the surroundings. The report of the Special Working Group (1981) investigating the event, provides estimates of the leak rate of propane from the perforated pipeline, and of the total amount that escaped to the atmosphere. Visual observations of the size and shape of the vapor cloud prior to ignition are also included. However, Nilsson (1981) pointed out that the visible part of the cloud was due to water droplets generated by condensation due to cooling of the air by the propane evaporation, and that the extent of the real, considerably larger and explosive propane/air cloud could not be observed visually.

2.3.6.4 Ignition and Explosion

Ignition of the cloud occurred about 55 min after the pipeline had been perforated. The ignition source was probably a fire appliance that was driven into the invisible part of the explosive cloud. There was a series of distinguishable explosions, extending over about 30 s. The major fireball had a diameter of about 300 m. Maximum explosion overpressures at various locations, eg, in the sewer manholes shown in Fig. 2.56, were estimated from damage analyses.

2.3.7 Methane Explosion in a 17,000 m^3 Coal Silo at Elkford, British Columbia, Canada (1982)

2.3.7.1 Plant and Process

Storage of coal can present a gas explosion risk, due to spontaneous release of methane from some types of coal. An account of such an explosion in British Columbia, Canada, in 1982 was given by Stokes (1986).

The silo that exploded, height 48 m, diameter 21 m, and capacity of 15,000 tonnes, was used for storage and load-out of cleaned, dried metallurgical coal. Prior to the explosion, a methane detector had been installed in the roof of the silo. The detector activated a warning light in the silo control room when a methane concentration of 1% was detected, and an alarm light was activated when detecting 2% methane. A wet scrubber was located in the silo head house to remove dust from the dust-laden air in the silo during silo loading. A natural ventilation methane stack was also located in the silo roof to vent any buildup of methane gas from the silo.

The silo was full of coal 24 h prior to the explosion. However, during the evening before the explosion, 10,000 tonnes of coal were discharged, at the same time as conveying of deep-seam coal into the silo had been started. This continued until the explosion occurred. At the moment of the explosion, there were approximately 12,300 tonnes of coal in the silo, of which 7600 tonnes were deep-seam coal. Testing had shown that this quality of coal had a high methane emission rate and produced a low volatile coal dust. Clouds in air of this dust could not be ignited unless the air was mixed with methane.

2.3.7.2 The Explosion and Its Consequences

The explosion occurred early one morning in May. Witnesses stated that a flash was noticed in the vicinity of the head house, followed seconds later by an explosion displacing the silo top structures. This was followed by an orange-colored fireball that rolled down the silo walls, and extinguished prior to reaching the base of the silo. Fortunately, neither injury nor death resulted, and damage to surrounding structures was minimal, although large blocks of concrete, and reinforcing steel had been thrown several hundred meters from the silo. However, the plant itself had suffered substantial damage, including a devastated silo roof, head house, and conveyor handling system.

2.3.7.3 Possible Ignition Source

The ignition source was not identified, but the three possible sources were considered, viz. spontaneous combustion of the stored coal, an electrical or mechanical source, and hot coal from the thermal dryer. Spontaneous combustion had never presented a problem during 10 years of operation, with coal being stored in different environments for varying lengths of time. Consequently, spontaneous combustion was not considered a probable source of ignition. During demolition of the damaged silo, all electrical and mechanical components were recovered and inspected and did not show any evidence of having been the ignition source. Therefore, Stokes (1986) concluded that the remaining possibility, ie, hot coal from the thermal dryer, was the most probable source of ignition.

2.3.8 The "West Vanguard" Explosion in the North Sea (1985)

2.3.8.1 Summary

Some central data of this event are given in Table 2.16.

2.3.8.2 Site of Explosion

A detailed discussion of this accident is given in the report by the Public Commission of Investigation (1986). The explosion occurred on board the comparatively new, well-equipped and well-maintained mobile oil drilling platform "West Vanguard," while performing test drilling on the Halten Bank on the Norwegian continental shelf in the North Sea. The main platform structure is illustrated in Fig. 2.57.

2.3.8.3 Events Leading to the Explosion

An uncontrolled blowout of natural gas occurred during the early stages of a normal drilling operation. The blowout was a so-called "shallow-gas-kick" and happened before the blowout preventing valve had been mounted. Shallow gas pockets were expected in this area, but at greater depths than the 263 m below the seabed that had been reached when the blowout occurred. According to the "Public Commission of Investigation" (1986) a series of unfortunate circumstances led to the uncontrolled blowout, including inadequate reporting, partial lack of confidence in the process monitoring instruments, deficiencies in the well program, and reluctance to increase the mud weight. The gas pressure in the shallow pocket was on the order of 50 bar, and the velocity of the uncontrolled upward gas flow was very high. The gas diverter equipment was unable to withstand the heavy erosion by the sand and other solid particles in the flowing gas. Consequently, after a few minutes, the gas flow had perforated the equipment and started to flood the entire platform structure. The gas was effectively distributed to most parts of the platform via the ventilation system. Eventually, it reached an ignition source and got ignited.

Table 2.16 The "West Vanguard" explosion

Place:	Haltenbanken, Norwegian continental shelf
Date:	October 6, 1985
Fuel:	Natural gas
Type of plant:	Mobile drilling platform
Fatalities:	1
Injuries:	0
Material loss:	Several hundred mill. Norwegian Krone (NOK)

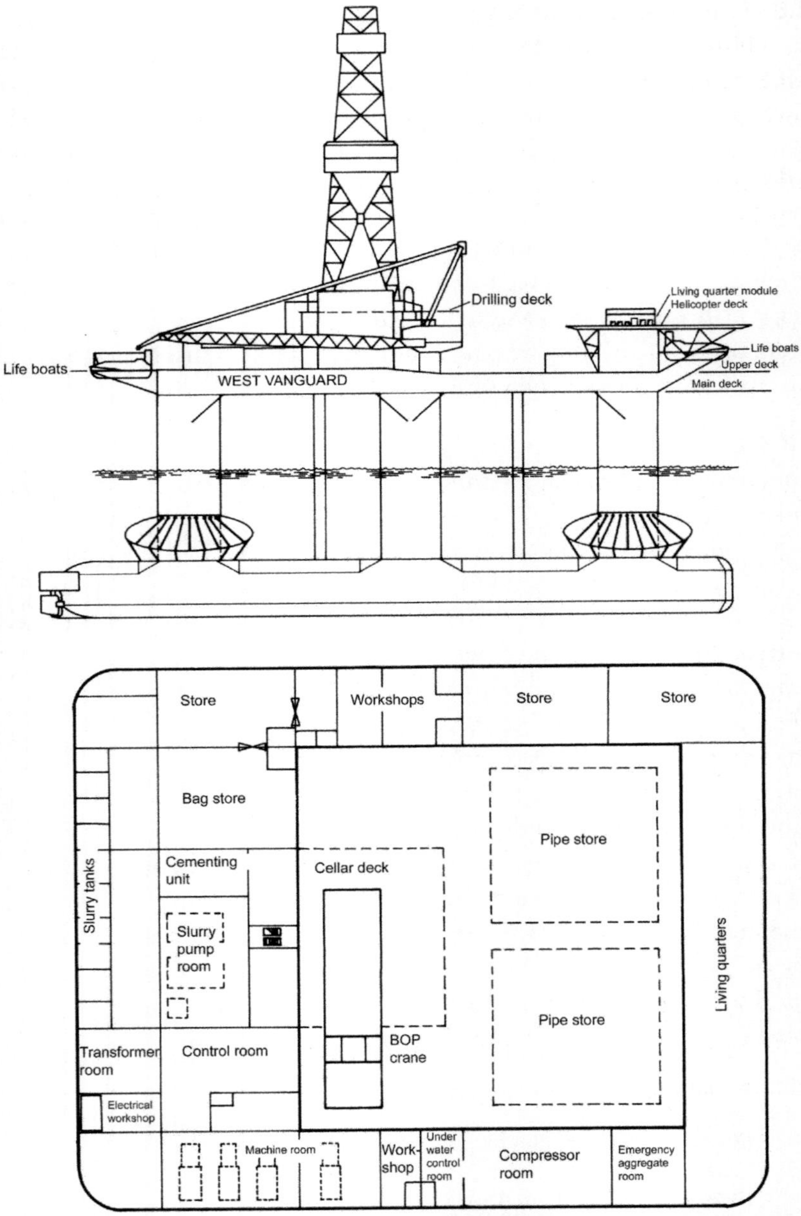

Figure 2.57 Illustration of the mobile drilling platform "West Vanguard," with a side view of the entire platform, and a top view of the main deck. The main deck is 85 m long and 75 m wide. The platform was kept floating by means of two 100-m long pontoons.

From Public Commission of Investigation, 1986. Ukontrollert utblåsning pa bore–platffiormen West Vanguard 6. oktober 1985. Norges Offentlige Utredninger, NOU 1986:16. Universitetsforlaget AlS, Oslo (1986). (In Norwegian.)

2.3.8.4 The Ignition Sources

The ignition sources were probably metal sparks or a hot surface generated in the cellar deck area during the intense erosion of metal during the blowout, and some electrical or other ignition source in the engine room of the platform. The former source was considered the most likely possibility for the first, strong explosion, because most witnesses claimed that this event occurred in the cellar deck area about 1 min before the explosion in the engine room.

2.3.8.5 The Explosions and Their Consequences

At least two distinct explosions were heard. The first and most powerful produced a large flame ball in the area of the drilling tower and across the pipe deck. This primary explosion probably occurred in the cellar deck. The second major explosion probably occurred in the engine room about 1 min after the first explosion. Fortunately, as soon as the blowout got out of control, and only a few minutes before the first explosion, the rescue operation had started, and 79 of the 80 persons on board were able to get into the lifeboats from which they were picked up by the standby ship. Large amounts of natural gas were bubbling through the sea to the sea surface, and from the standby ship it was observed that natural gas was burning across the sea surface in some areas round the platform, but fortunately not in the area from which the people were rescued. The platform suffered severe damage from the explosions and subsequent fires, mainly in the after part, including the area around the drilling tower and the adjacent areas and rooms. Roofs, bulkheads, and bolted steel doors were blown up or heavily deformed by the explosion pressures.

2.3.8.6 Computer Simulations of Gas Explosions

As part of the investigations after this accident, numerical simulations of explosions in various parts of the platform were performed by means of the FLACS code. This made it possible to compare the observed actual structural damage with that expected from various ignition scenarios. This work is discussed in more detail by Bjørkhaug et al. (1985).

2.3.9 Catastrophic Gas Explosion in Taegu, South Korea, April 1995

2.3.9.1 Overview

This gas explosion catastrophe did not occur in the process industry, but in the middle of the busy city of Taegu in South Korea. However, the lessons

Figure 2.58 Site of the liquefied petroleum gas (LPG) explosion disaster in Taegu, South Korea, April 28, 1995, which killed 100 people, and injured 117. Ten buildings were completely demolished, and 48 others partly damaged. Eighty vehicles were destroyed or damaged.
From Fire Investigation Report, 1995. Pages 101–107 were kindly made available to Eckhoff, R.K. by Mr. Jong-Kil Lee, manager of Korea Gas Safety Cooperation, Kyunggi-Do, South Korea.

to be learnt are also indeed relevant for the process industries. Fig. 2.58 shows the afflicted site after the event.

The following provisional summary is taken from the brief report by Spaeth (1995) published shortly after the accident:

> *At 07.50 h on April 28, 1995, an immense fireball tore through the heart of Taegu, South Korea's third-largest city, of a population of 2.3 million. Cars were hurled aloft, chunks of steel beams landed atop buildings, and 250-kg steel plates that had covered a subway excavation, rained down on vehicles, and pedestrians. A bus from the Youngnam Middle School pitched into the hole, killing all the students aboard.*

The description of the cause of the catastrophe in magazines and newspapers shortly after the accident were somewhat diverging. According to Spaeth (1995):

> *The explosion occurred when a drill operator working at a department store building site near the subway construction pit mistakenly drilled into a fuel pipeline. The liquid (LPG) poured out into the pit for about 20 minutes, vaporized, and the resulting vapor cloud was ignited, probably by a worker's blowtorch.*

Korea Times, a few days after the explosion, reported that

> *a 100-mm diameter gas pipeline was ruptured during digging near a department store adjacent to the explosion site. The pipeline was buried 1.3 m deep under the road. Gas leaking from the ruptured pipe flew into the subway construction site through a sewage line and got ignited by a spark from nearby welding work. Subsequent investigations disclosed a 7 cm hole in the gas pipe.*

However, the comprehensive Fire Investigation Report (1995) concludes that the exact cause of the explosion disaster could not be found. The following is a brief summary of the findings given in this report:

2.3.9.2 Circumstances Before the Explosion

The residents living near the construction site had, since the night of April 27, smelled gas several times and reported this to the City Gas Corporation and nearby police station, but the responsible persons had not responded to the reports. As late as 20 min before the explosion in the morning of April 28, workers on site again reported to the Corporation about the gas leakage, but no immediate action was taken. It later turned out that after receiving similar messages from the workers at the site about a gas leaks in both January and March, the City Gas Corporation had not taken any significant measures. Fortunately, in these two cases there were no explosions.

2.3.9.3 Possible Causes of Explosion

The exact causes of the explosion were not resolved, but three scenarios were suggested.

2.3.9.3.1 Scenario 1

Gas leaked from a loosened joint or a crack on one of the LPG pipelines passing through the construction site. The gas cloud produced got ignited and exploded. The construction company in charge insisted that they, before the explosion, had reported several times to the City Gas Corporation about smell of gas on the site, without any actions being taken. There were indications of poor quality of the work done when the pipelines were first laid.

2.3.9.3.2 Scenario 2

Generation of the explosive gas cloud was due to workers on the site having damaged the underground gas pipelines during their works. The City Gas Corporation received a report from a worker on the site saying that gas was leaking due to their mistakes during excavation work just before the explosion. In this scenario, the gas cloud was probably ignited by sparks/embers from welding operations. However, the company in charge of the whole subway construction strongly rebutted this hypothesis by claiming that the explosion had been initiated early in the morning before the work on the site had started. Furthermore, the gas pipelines were hanging about 15 m above the base of the working site, just below the perforated steel plates acting as a provisional road cover across the site. Therefore, it would not have been easy for the workers to accidentally touch the pipelines.

2.3.9.3.3 Scenario 3

Another possible cause of the gas leak might be formation of one or more cracks in the pipelines due to the continual vibrations induced by the heavy traffic across the perforated steel plates acting as the provisional road across the construction cavity.

2.3.9.4 Extent of Explosion and Its Consequences

Besides the violent noise of the explosion, a huge flame ball extending 50 m above the road level was observed. Numerous perforated steel plates used as provisional road cover for a road length of about 300 m, were flung into the air. One of these, weighing 280 kg (270 × 70 × 2 cm), was found hanging on a church cross on top of a three-story building near the explosion site. The explosion severely deformed and displaced the heavy-duty H-beams supporting the provisional road cover within 1 km of the explosion center. This caused numerous other steel plates to collapse under the load of all the vehicles, and 30 cars and buses were plunged into the 18-m deep construction cavity. Other cars soared to 10 m above road level. In all, 80 vehicles were damaged by the explosion. Ten buildings were completely demolished, and 48 others partly damaged. Gas pipelines, water pipes, and underground cable pits were also destroyed, the whole area was flooded, and the electrical power was cut off in thousands of homes. Ninety-seven people were found killed, 3 were missing, and 117 suffered injuries.

2.3.9.5 Rescue Operations

A total of 30 fire trucks, 10 cranes, and about 2500 rescue personnel were dispatched to the explosion site. However, because of the possibility of further explosions of remaining unburnt gas in the underground, rescue personnel could not be admitted to the site immediately after the explosion.

Flooding due to the broken water pipes delayed the rescue operations even more. Rescue operations could not start fully until about 8:00 pm of the same day, ie, about 12 h after the explosion.

2.3.9.6 Lessons Learnt

The overall cause of this catastrophe was an inadequate governmental bid system for construction work, including inadequate quality assurance of technical and managerial safety matters. The workers on the site did not have adequate understanding of the technical standards applying to the work they were doing, and the responsible authorities had not given sufficient guidance. The company in charge of the construction work had not checked the underground pipelines drawings prior to starting the excavation works. Furthermore, the construction site was not provided with any gas alarm system. The investigations also disclosed lack of updated maps of underground gas pipelines, electric cables, etc.

2.3.9.7 The"Human Factor"

The following story, also from Spaeth (1995), illustrates the important part played by human behavior, not only in the chain of events leading to accidents, but indeed also in the subsequent investigation to disclose what actually happened:

> *Five days after the explosion a local fire fighter told reporters that the gas leak had been reported to the local fire station by a street sweeper four hours before the explosion. According to the fire fighter, the complaint was entered in the station's log, but destroyed after the accident had occurred. When journalists found the sweeper, he said he reported a gas smell to a police box at 21.00 hours in the evening before the blast, and again at 04.00 hours the following morning. After the tragedy he was confronted by officials who threatened to have him fired from his job unless he repudiated his reports. He did so, and the retraction was videotaped. When his story was made public, however, the prosecutor general in Seoul ordered an investigation.*

2.3.10 Explosion at Buncefield Oil Storage Depot, United Kingdom in 2005

2.3.10.1 The Main Investigation

The present account is largely based on the BMIIB (2008) report (Buncefield Major Incident Investigation Board) and the paper by Johnson (2010). Early one Sunday morning in December 2005, a major vapor cloud explosion occurred at the Buncefield Oil Storage Depot, Hemel

Hempstead, Hertfordshire, United Kingdom. The explosion destroyed a number of fuel storage tanks on the site, and further explosions and a major fire resulted. According to Johnson (2010), the following sequence of events seems most likely:

- Overfilling of a gasoline storage tank with a subsequent 300 tonnes gasoline spillage into the surroundings for about 40 min, gave rise to the formation of a major gasoline vapor cloud. The overfilling of the gasoline storage tank occurred because the overfilling protection system did not function.
- The gasoline was reported to be a "winter mix" containing about 10% butane. The main components of the vapor cloud were probably butane and pentane.
- Due to the very low wind speeds prevailing at the time of vapor cloud formation, an explosive/flammable cloud was allowed to build up across an area of 120,000 m^2 with an estimated cloud height of at least 2 m over most of the area.
- Ignition of the cloud most probably occurred at an emergency water pump house. The collapse of the pump house was followed by a deflagration that started outside the house and propagated further into the undergrowth and trees in the surroundings.
- The flame accelerated grossly in the undergrowth and trees along a street through the tank farm area and reached flame velocities up to several hundred m/s, followed by a transition to detonation (DDT). The detonation then propagated through the remaining part of the vapor cloud.

Fig. 2.59 shows a cross section of the type of tank that was overfilled and from which the initial leakage of petrol occurred. Fig. 2.60 shows the damaged pump house. Fig. 2.61 shows the large fire engulfing a major part of the tank farm area. The fire persisted for several days. Fig. 2.62 shows the tank farm area after the fire had finally been extinguished.

Presumably, because of the time of the explosion (early Sunday morning), there were fortunately no fatalities, but 43 people were injured, and significant damage occurred to both commercial and residential properties in the vicinity of the tank farm. Total damages were on the order of US $1.5 billion.

In the investigation by BMIIB (2008), the reason for the extreme severity of the initial major explosion was one central issue requiring an explanation. Because of the very limited congestion provided by pipework, etc., along

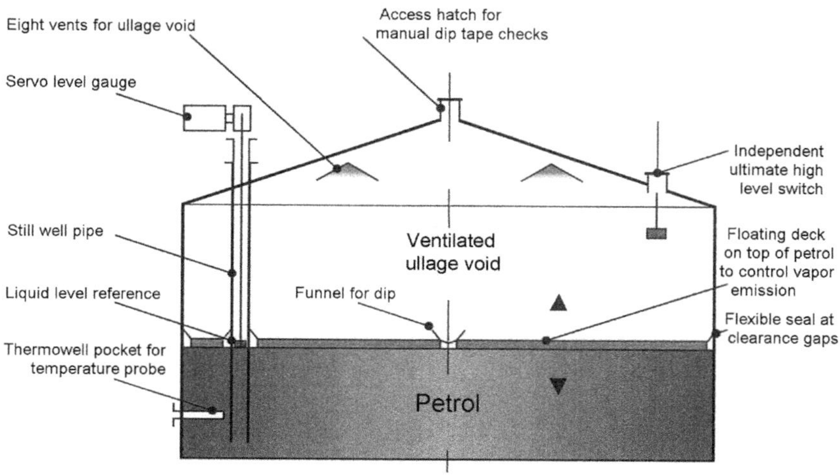

Figure 2.59 Cross section of the type of tank at Buncefield Oil Storage Depot, which was overfilled, and from which the initial leakage of petrol occurred.
From BMIIB, 2008. The Buncefield Incident 11 December 2005. The final report of the Major Incident Investigation Board, vol. 1. Crown Copyright. ISBN:9780717662708.

Figure 2.60 Remains of the fire pump house at Buncefield Oil Storage Depot where the first major vapor cloud explosion was initiated.
From BMIIB, 2008. The Buncefield Incident 11 December 2005. The final report of the Major Incident Investigation Board, vol. 1. Crown Copyright. ISBN:9780717662708.

Figure 2.61 The major fire at Buncefield Oil Storage Depot, which followed the vapor cloud explosion(s). Contains public sector information published by the Health and Safety Executive and licensed under the Open Government Licence.
From BMIIB, 2008. The Buncefield Incident 11 December 2005. The final report of the Major Incident Investigation Board, vol. 1. Crown Copyright. ISBN:9780717662708.

Figure 2.62 View of the Buncefield Oil Storage depot after all fires had been extinguished. The tank that was overfilled and gave rise to the initial major petrol vapor cloud is seen in the center of bund at the picture bottom. Contains public sector information published by the Health and Safety Executive and licensed under the Open Government Licence.
From BMIIB, 2008. The Buncefield Incident 11 December 2005. The final report of the Major Incident Investigation Board, vol. 1. Crown Copyright. ISBN:9780717662708.

the path of the explosion, the geometrical conditions required for producing the high explosion overpressures that actually occurred apparently did not exist. It was also thought that the extensive pressure damage to instrument boxes, drums, etc. located within the vapor cloud, could not be explained by flame acceleration due the bushes and trees. Most of this damage indicated that overpressures had been in excess of 2 bar, even in areas some distance from the congestion provided by the bushes and trees.

2.3.10.2 Were the High Overpressures in the Major Explosion Nevertheless Caused by Trees and Bushes?

An interesting numerical simulation study of the effect of trees and bushes on the development of gas explosions, with particular reference to the Buncefield tank farm explosion, was undertaken by Bakke et al. (2010). They used the CFD-based numerical simulation code FLACS as their simulation tool. Because the tank farm site itself was not much congested, it was not in itself expected to cause high overpressures in case of a hydrocarbon leak explosion. Therefore, focus was on the obstacles provided by bushes and trees alongside the roads surrounding the tank farm, and these were included in the FLACS simulation model. The extent of the actual obstacle field provided by bushes and trees was documented by video recordings made in the summer of 2006.

In the simulations, a large, shallow, heavier-than-air gas cloud was assumed to cover part of the site and surroundings. Following ignition an initial flame was first established in the gas cloud. The flame then propagated and accelerated through the trees and bushes along the roads surrounding the tank farm. The simulations showed that the bushes and trees gave rise to massive flame acceleration, which in turn resulted in high overpressures of several bar (g). Fig. 2.63 shows the scenario of tanks and vegetation adopted in the FLACs simulations.

The FLACS software had previously been extensively validated experimentally only with congestions consisting of typical industrial process equipment including piping. In the present investigation, therefore, a set of real physical explosion experiments, using propane/air as the explosive gas mixture and real bushes and trees as obstacles, were carried out to investigate the extent to which the enhanced combustion due to bushes and trees that appeared in the simulations would also occur in reality. The experimental facility consisted of a 20-m long plastic tent with a 4 m^2 semicircular cross section and, hence, a volume of approx. 80 m^3, as illustrated in Fig. 2.64.

Figure 2.63 Representation of fuel tanks and vegetation used in the FLACS simulations of the Buncefield tank farm explosion in 2005.
From Bakke, J.R., Wingerden, K., van Hoorelbeke, P., Brewerton, B., 2010. A study on the effect of trees on gas explosions. Journal of Loss Prevention in the Process Industries 23, 878–884.

Figure 2.64 Interior of 20-m long plastic tent used in the propane/air explosions experiments with real bushes and trees as explosion-enhancing, turbulence-generating obstacles.
From Bakke, J.R., Wingerden, K., van Hoorelbeke, P., Brewerton, B., 2010. A study on the effect of trees on gas explosions. Journal of Loss Prevention in the Process Industries 23, 878–884.

This provided sufficient space for using various selections of lanes of bushes and trees in the experiments. The experimental program involved different sizes and densities of trees and bushes in the plastic tent. The experiments were performed with stoichiometric propane—air mixtures, and they resulted in continuously accelerating flames over the full 20-m length of the tunnel for some of the scenarios investigated. The main conclusions drawn by Bakke et al. (2010) from these experiments were that trees and bushes can give rise to significant flame acceleration in explosive gas/air clouds, and that advanced CFD-based simulation models, such as FLACS, can be used to study such phenomena. However, they also concluded that more experiments would be needed before codes like FLACS would be able to predict flame acceleration in dense vegetation with acceptable accuracy.

2.3.11 The Tesoro Anacortes Refinery Explosion

2.3.11.1 Overview of Similar Accidents in the United States

The following account is based on the investigation report CSB (2014). In 2014, there were about 150 petroleum refineries in the United States. However, this number represents only a small fraction of the many thousands of chemical processing facilities in the country. Despite this, CSB found that during the decade 2004—2014, an appreciable number of significant and deadly incidents occurred just at petroleum refineries. In 2012 alone, CSB recorded 125 significant incidents just at US petroleum refineries.

2.3.11.2 Introduction to Anacortes Refinery Explosion

This accident occurred on April 2, 2010, at the petroleum refinery of Tesoro Refining and Marketing Company in Anacortes, Washington, United States. They experienced a severe explosion accident with a subsequent fire, causing the death of seven employees. At the time of the explosion, approximately 350 employees worked at the refinery, of which nearly 200 were operations and maintenance workers. The event fatally injured seven employees (one shift supervisor and six operators) working in the immediate vicinity of the heat exchanger at the time of the accident. Up to at least 2010, this was the largest fatal process accident at a US petroleum refinery since the British Petroleum (BP) Texas City catastrophe in 2005.

2.3.11.3 The Chain of Events

The initiating event was a catastrophic rupture of one of the heat exchangers in the *naphtha-hydro-treatment (NHT) unit*. The NHT unit contained two parallel banks of three heat exchangers each (A/B/C and D/E/F), as illustrated in Fig. 2.65. The purpose of this equipment was to preheat the process fluid before it entered a reactor. The heat exchanger (E) that ruptured was constructed of carbon steel. The same applies to heat exchanger B, and as Fig. 2.65 indicates, the investigations revealed that there was in fact a crack also in that unit, at about the same location as the rupture occurred in exchanger E.

At the time of the accident, the Tesoro workers were in the final stages of starting up the process of putting the bank of heat exchangers back in service, following a cleaning operation of the interior of the pipework of the

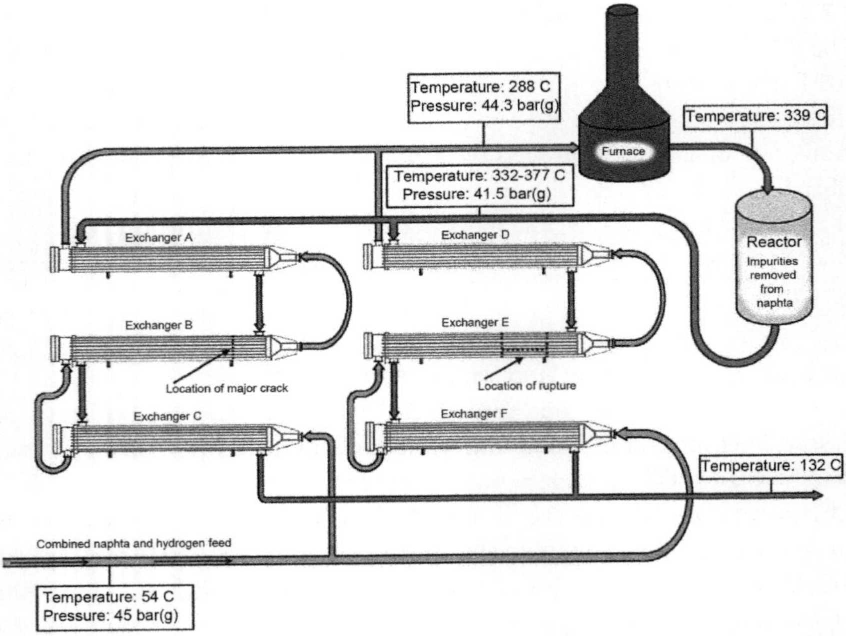

Figure 2.65 Illustration of the heat exchangers of the Tesoro Anacortes Refinery NHT Unit. There are two banks of three heat exchangers: A/B/C bank and D/E/F bank. The E heat exchanger was the one that ruptured catastrophically on April 2, 2010.
From CSB, 2014. Catastrophic rupture of heat exchangers in the Tesoro Anacortes Refinery. Report No. 2010-08-1-WA, May 2014, U.S. Chemical Safety and Hazard Investigation Board (CSB), Washington DC, USA.

exchangers. The refinery had previously experienced frequent leaks, and occasionally fires, during this kind of startup activity. On April 2, 2010, the catastrophic rupture of one of the heat exchangers occurred just while the operations staff was performing the startup operation. According to CSB (2014), the rupture occurred because of a *high-temperature-hydrogen-attack* (HTHA). A highly flammable mixture of hydrogen and naphtha at more than 260°C was released from the ruptured heat exchanger and got ignited. The result was a violent explosion followed by an intense fire that burned for more than 3 h.

HTHA can occur when steel at elevated temperatures is exposed to hydrogen, as can happen in refineries, petrochemical, and other chemical plants. HTHA is different from embrittlement or other forms of low-temperature damage of steels caused by hydrogen. In the case of HTHA, hydrogen molecules dissociate, atomic hydrogen diffuses into and penetrates the steel and reacts with the carbon dissolved inside the bulk of the material to form methane. If the methane then accumulates at grain boundaries and/or other interfaces instead of diffusing out of the steel, internal fissures and cracks that are typical of HTHA are generated.

2.3.11.4 Further Analysis of the HTHA at the E Heat Exchanger

HTHA tends occur in high-stress areas in carbon steel, such as near and in non-heat-treated welds. The welds of the carbon steel of the heat exchanger that ruptured had not been heat-treated after welding. The postaccident analysis by CSB (2014) disclosed that the HTHA damages were in the high-stress areas near the welds. The location of the catastrophic rupture was in fact also along these high-stress weld regions and was attributable to small cracks caused by HTHA.

2.3.11.5 Recommendations and Concerns Expressed by CSB

2.3.11.5.1 Technical Matters

The process hazard analyses (PHAs) conducted at the plant had not evaluated in an adequate way the effectiveness of the recommended measures for preventing HTHA in process equipment. Had the adequacy of the recommended safeguards been verified, proper safeguards for protecting against HTHA failures could have been recommended.

CSB did not consider inspection to detect HTHA at an early stage being sufficiently reliable to ensure mechanical integrity and prevent HTHA equipment damage. Such inspection is very difficult because the damage can be microscopic and may be present only in small-localized areas of the equipment. Successful identification of HTHA is highly dependent on the specific techniques employed and the skills of the inspectors.

CSB had been informed of a number of recent refinery accidents in which HTHA had occurred below accepted maximum safety limits for carbon steel. This confirmed a strong industry-wide need for better tools for reliable predictions of safe operation limits to prevent HTHA.

A main technical recommendation by CSB for preventing further HTHAs was to implement the concept of inherently safer design. One recommended measure was then to use high-chromium steels rather than carbon steel. High-chromium steels are significantly more resistant to HTHA than carbon steels. Therefore, the use of such materials in heat exchangers will significantly reduce the likelihood that HTHA will occur.

2.3.11.5.2 Organizational Concerns

The PHAs conducted at the plant had not evaluated in an adequate way the effectiveness of the recommended measures for preventing HTHA in process equipment. Had the adequacy of the recommended safeguards been verified, proper safeguards for protecting against HTHA failures could have been recommended.

The start-up of the NHT heat exchanger banks was hazardous nonroutine work. Leaks often developed that presented hazards to workers conducting the start-up activities. Process hazard analyses conducted at the refinery did not ensure that these hazards were controlled and that the number of workers exposed to them was minimized.

The existing procedures did not prohibit or effectively limit the use of additional personnel during the nonroutine hazardous start-up of the NHT heat exchangers. The heat exchanger start-up procedure specifies the use of only one outside operator to perform start-up operations of the NHT heat exchanger banks. However, on the day of the incident, a supervisor requested five additional operators to assist with the start-up of the heat exchanger banks.

The NHT heat exchangers frequently leaked flammable hydrocarbons during start-up, sometimes resulting in fires. Tesoro management had been complacent about these hazardous leaks and did not always investigate the cause of the leaks. Tesoro did take some actions to prevent the leaks, but these actions did not effectively prevent the leaks before the April 2010 incident. Additional operators, such as those present during the April 2010 heat exchanger start-up, were frequently needed during start-up of the NHT heat exchanger banks to respond to potential hydrocarbon leaks or fires.

This past practice contributed to the presence of the six additional workers in the unit during the April 2010 incident.

The CSB found several indications of process safety culture deficiencies at the Tesoro Anacortes Refinery. Refinery management had normalized the occurrences of hazardous conditions, including frequent leaks from the NHT heat exchangers, by using steam to mitigate leaks, ineffectively identifying methods to prevent leaks from the heat exchanger flanges and gaskets, commonly requiring additional operators during NHT heat exchanger start-ups and exceeding the staffing levels that procedures specified.

The process safety culture of the refinery required proof of danger rather than proof of effective safety implementation. For years, technical experts used design parameters to evaluate the E heat exchanger for HTHA susceptibility. Data for actual operating conditions were not readily available, and in drawing their conclusion that the E heat exchanger was not susceptible to HTHA damage, the technical experts were not required to prove the effectiveness of the safety measures recommended.

2.4 MEANS OF PREVENTING AND MITIGATING/CONTROLLING GAS/VAPOR EXPLOSIONS IN THE PROCESS INDUSTRIES

CEN (1997) issued a standard describing basic concepts and methodology in explosion prevention and mitigation in the process industries. Explosions and fires involving combustible gases and vapors constitute a major hazard in process industries and other environments in which such materials are produced, used, and handled. Therefore, the efforts to minimize the risk of explosions and fires in these industries continue nationally as well as internationally, and much effort is spent on preventing and mitigating accidental gas and vapor cloud explosions on offshore process installations.

Explosion risk is most often defined as the product of the probability of an explosion and its expected consequence. Therefore, the basic principle of gas explosion risk management is to minimize the explosion probability as well as the expected explosion loads that are related to the explosion consequences. Reduced gas explosion consequences can be obtained by active as well as passive measures. The following overall aims have high priority in gas explosion risk management:

- Prevention and control of potential leaks of combustible gases
- Good ventilation to minimize extent and duration of explosive atmosphere due to gas releases

- Prevention and control of possible ignition sources
- Minimizing high equipment congestion/blockage to ease explosion venting, and to minimize flow-induced turbulence during gas cloud formation and explosion
- Installation of blast and fire barriers

The most important means of prevention and mitigation/control are summarized in Fig. 2.66.

2.4.1 Application of the Concept of "Inherently Safer Plant Design" to Prevention and Mitigation/Control of Accidental Gas/Vapor Explosions

2.4.1.1 Outline of Basic Concept

According to Kletz (2003), Khan and Amyotte (2003), Hendershot (2006), and Amyotte et al. (2007), inherently safer plant design is based on the four main principles of *minimalization*, *substitution*, *moderation*, and *simplification*. *Minimalization* implies using so little of the hazardous material that there is no significant risk of hazardous effects. *Substitution* implies using a material or a process, which is less likely to give rise to hazardous situations. *Moderation* implies using the hazardous material in the least hazardous form. *Simplification* implies elimination of unnecessary complexity, and design of "user friendly" plants.

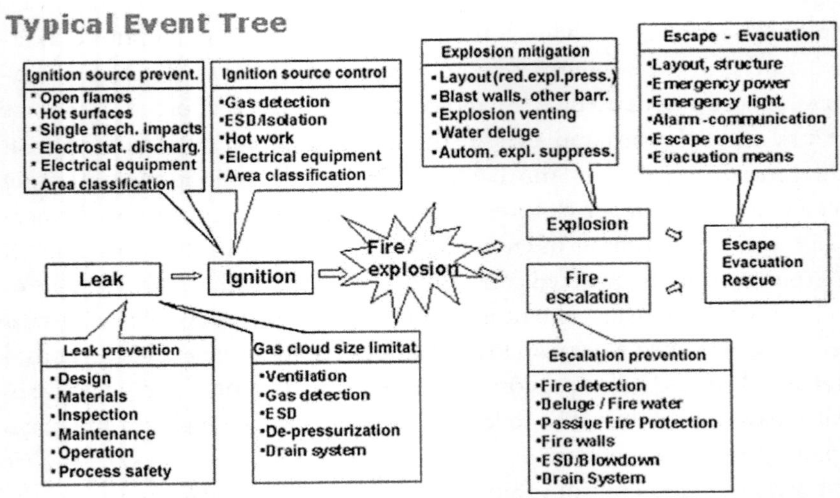

Figure 2.66 Event tree illustrating the various means for preventing and mitigating/controlling gas cloud explosions in the process industries.

Measures for improving inherent safety may be introduced at any stage of a process development project, but the maximum effect is obtained if the measures are introduced as early as possible in the process development. As a process evolves through its life cycle, changes in basic technology become gradually more difficult and expensive. In addition, in a late phase of process development it may not be possible to invest the time and money required to develop an alternative production scheme.

2.4.1.2 Examples

The following examples are taken from Khan and Amyotte (2003).

2.4.1.2.1 Minimalization

Minimalization means to use as small quantities of hazardous substances as practically possible. This may, eg, be achieved by using efficient continuous reactors (stirred tanks, loop reactors, or tubular reactors) in place of large batch reactors.

Although the *reaction of nitric acid and glycerol* is quite rapid, early industrial implementation of it used large batch reactors. In current processes, however, one has obtained very efficient mixing of the reactants in continuous reactors that are orders of magnitude smaller than the older batch reactors. This greatly reduces the potential hazard from this extremely hazardous reaction.

A similar situation exists for production of the very toxic and irritant phosgene. This is can now be done in continuous reactors on demand, thus eliminating the need for storage of liquid phosgene.

2.4.1.2.2 Substitution

Substitution means that one material is being replaced by a less hazardous substance. One familiar example is replacement of *organic-solvent-based paints* and coatings by corresponding *water-based products*. The same applies to many cleaning and de-greasing operations, in which the organic solvents have been replaced by water-based systems.

A more specific example is the *manufacture of acrylic esters* by oxidation of propylene to produce acrylic acid, followed by esterification to manufacture the various esters. This is an inherently safer procedure than the older Reppe process, which used acetylene, carbon monoxide, and nickel carbonyl.

Another specific example is the *production of di-sodium iminodiacetate (DSIDA)*, an intermediate used for agricultural chemicals. This material was traditionally manufactured using a process requiring ammonia, formaldehyde, hydrogen cyanide, and hydrogen chloride. The Americal company

Monsanto developed a new process for DSIDA production, which eliminated the use of hydrogen cyanide and formaldehyde, and which both had a higher yield and were simpler, and still produced a product of sufficiently high purity.

A polymerization process was conducted using a gradual-addition batch process, with a large quantity of organic solvent being required to keep the system viscosity low enough for effective mixing and heat transfer. In the event of an uncontrolled runaway polymerization, a large quantity of flammable and toxic material would be ejected through the reactor rupture disk. Instead of relying on an expensive and elaborate emergency relief discharge system to control this serious potential hazard, the entire basic process philosophy was changed. It was found that it was possible to produce the polymer using a suspension polymerization process in water. Because of the significantly lower amount of solvent required in the new process, the material released in a possible runaway process would be mainly water, mixed with just a small amount of solvent and unreacted monomer. Runaway was also less likely because of the higher heat capacity of water, allowing it to absorb more of the heat of reaction during a runaway.

2.4.1.2.3 Moderation

Moderation means to use less hazardous conditions, a less hazardous form of a material, or production facilities that minimize the impact of a release of hazardous material or energy.

In the 1930s, ammonia plants typically operated at very high pressures of up to 600 bar. Over the years, improved understanding of the chemistry has resulted in a gradual reduction of the operating pressures in ammonia plants, and in the 1980s plants operating in the range of 100−150 bar had become common. These newer, low pressure plants are inherently safer, cheaper, and more efficient than their high-pressure counterparts.

2.4.1.2.4 Simplification

Simplification means to design facilities in which unnecessary complexity is eliminated, and operating errors are made less likely.

An adequate understanding of the basic process chemistry allows process chemists, and engineers to design simpler reaction systems, and facilities. *One-pot*, and in situ processes allow hazardous intermediate reaction products to be generated in the same vessel in which they will be further chemically converted. This eliminates the need for intermediate hazardous storage of such intermediates, or the need for conveying them in piping systems in the plant. The required inventory of hazardous material is thus limited to a maximum of one batch.

This example illustrates the beneficial overlap that can be obtained between different inherent safety principles (in this case, simplification and minimization). On the other hand, simplification sometimes involves a trade-off between the complexity of an overall plant and the complexity within one particular piece of equipment. For example, a modified reactive distillation process for producing methyl acetate required only three columns and associated support equipment, whereas the older process required a reactor, an extractor, and eight columns, and associated support equipment. The new process was simpler, safer, and more economical. On the other hand, the operation of the reactive distillation component of the new simplified system was more complex and knowledge intensive than operating the distillation system of the original more-diversified process.

2.4.2 Preventing, and Limiting Size of, Explosive Gas/Vapor Clouds

2.4.2.1 Preventing Gas Leaks from Process Equipment

The following is based on information from Statoil (2004).

2.4.2.1.1 Piping and Vessels

Piping, pipelines, tanks, heaters, mechanical equipment and vessels design integrity and related inspection, maintenance, and operation activities shall ensure that leaks of hydrocarbon fluids (and chemicals and toxic gases) do not occur. This is accomplished by the following measures:

- The inspection system shall verify that the chosen monitoring parameters are within specified limits
- A tailored condition monitoring strategy shall be established for any given plant
- A working process for establishing inspection programs shall be available
- Inspection acceptance criteria shall be defined
- Systems for reporting and follow-up of findings shall be established
- Blowdown systems are to be designed with a particular focus on pressure transients, surge, and vibration

2.4.2.1.2 Flanges and Connections

The design, activities of operation, inspection, and maintenance of flanges and other mechanical connections shall ensure that hydrocarbon leaks do not occur. The amount of flanged connections in systems containing hydrocarbon fluids shall be kept at a minimum by observing the following:

- Prior to undertaking constructive modifications of the plant the possibility of using welded connections as opposed to flanges shall be investigated, based on a coherent safety philosophy. For example, any mechanical connections (eg, flanges) on the pipeline side of risers on offshore installations should be avoided. If such connections are chosen, appropriate protection must be applied.
- Mechanical connections shall be designed to be able to sustain mechanical vibrations.
- Adequate procedures for assembling flanges and other connections, including bolt tensioning, must be developed and followed.
- Compact type flanges are preferred to the American National Standards Insitute (ANSI) type due to a lower predicted leak frequency.
- Depressurization (blowdown) pipes should normally be designed without flanges.

2.4.2.1.3 Instrumentation, Valves, and Rotating Machinery

The design, activities of operation, inspection, and maintenance of in-stream instrumentation, valves, and rotating machinery (eg, pumps and compressors) shall ensure that hydrocarbon leaks do not occur. A maintenance program shall cover all equipment that could be a potential gas/vapor leak source.

2.4.2.2 Minimizing Size of Gas Cloud in Case of Accidental Leaks by Early Leak Detection and Effective Shutdown

The following is based on Statoil (2004).

2.4.2.2.1 Overall Objective and Performance Criteria of Gas Detection Systems

Detectors for flammable gas shall be provided in areas (risk volumes) where explosion loads from ignited gas clouds could cause unacceptable damage, ie, loss of safety functions. In these areas, the smallest gas cloud that has the potential of causing such damage shall be specified as the detection target. The gas detection function shall provide reliable and fast detection of leaks before any accidentally released gas cloud reaches a concentration and size that could cause significant risk to personnel and plant. The detector coverage (sensor number and distribution in the process area of concern) shall be based on realistic worst-case gas leakage scenarios (source locations and leakage rates) within the process area of concern by considering the dispersion mechanism transforming the leak jet into a gas cloud. Due to the complex influence of the geometry and arrangement of process equipment, environmental conditions such as ventilation, numerical gas dispersion simulations, using eg, the FLACS code, may have to be performed for optimization of location of detectors within specific areas.

2.4.2.2.2 Detector Types

For hydrocarbon gases, infrared detectors shall be the preferred whenever suited for the gas to be detected. Optical beam detectors with long range shall be used to increase detection probability of small leaks. Catalytic detectors should only be used if proper detection performance by other types cannot be achieved. Maximum response time of the gas detection function shall be defined to ensure that total reaction time for each safety function can be fulfilled. Typical response times are as follows:

- Infrared detectors: <5 s for general area applications
- Catalytic detectors: <10 s from low alarm (20% LEL) to high alarm (30% LEL)
- Acoustic detectors: <30 s including delays to improve false-alarm immunity

2.4.2.2.3 Coverage and Location of Gas Detectors

The following overall principles shall be applied with respect to location of detectors:

- Detectors shall be spread out evenly, but not symmetrically, to ensure adequate coverage of long "corridors" in the main ventilation flow direction. Natural flow "corridors," eg, walkways along expected flow direction, should be covered.
- Detectors shall not be located in zones with no or little ventilation (dead zones) as gas concentration rises slower in these areas as compared to areas with higher ventilation rates. Examples of dead zones are corners, between large roof beams, and congested areas without leak sources.
- In process areas of considerable height, eg, modules on offshore platforms, detectors should be positioned in a minimum of two levels

Crucial gas detector locations include:

- Zone 1 and zone 2 areas
- Inlets of heat, ventilation, and air conditioning systems (HVAC)
- Ventilation outlets from classified areas
- Combustion air intakes and turbine hoods

- Rooms that contain potential ignition sources and are equipped with doors, trapdoors, hatches, etc. toward classified areas (see Chapter 10)
- Enclosed areas where gas clouds can enter or be formed
- Rooms containing potential ignition sources (continuous monitoring in emergencies)
- Battery rooms if hydrogen can be formed in dangerous concentrations
- Gas turbine hoods

Gas detectors in ventilation intakes shall ensure that a possible gas leak be detected and dampers closed before dangerous concentrations can reach ventilated areas. In the case of hydrocarbons, use of infrared-type detectors located as closely as possible to the ventilation intake allows actions to be initiated within a short period, typically a few seconds.

2.4.2.2.4 Gas Detection Alarms

The gas detection system shall automatically activate predetermined actions in accordance with the actual plant-safety-strategy-and-philosophy documentation. Each individual gas detector shall provide alarm upon gas exposure. The explosive gas concentrations at which alarms are activated are typically in the range 20–30% of LEL. Alarms shall be given as quickly as possible to warn and guide personnel upon detection of gas. Detection of gas, failure to execute actions upon demand, and system defects shall be presented as alarms in central control room.

2.4.2.2.5 Other Issues of Concern Related to Gas Detection Systems Include

- Systems independence
- Systems operation integrity
- Systems survivability
- Systems inspection, maintenance, and modification
- Man–machine interface
- Documentation

2.4.2.2.6 Mist/Spray Detectors

In the case of release of droplets of flammable liquids that evaporate slowly after being released to the atmosphere, the detectors shall be located close to the floor of the process area, eg, 0.5 m above the deck on offshore platforms.

Oil mist detectors (optical obscuration) should be installed in machinery spaces and turbine exhaust channels, indicating the release of high-pressure flammable liquids such as fuel or lubrication oil that may lead to aggressive and damaging fires.

2.4.2.3 Emergency Shutdown Systems (ESD)

The following is based on Statoil (2004).

2.4.2.3.1 Purpose of ESD Systems

The purpose of emergency shutdown systems at large is to prevent escalation of abnormal conditions into major hazardous events, and to limit the extent and duration of any such conditions. In the present context this means to limit, to the extent possible, any accidental releases of explosive gas, vapor, and mist from process equipment, and to eliminate possible ignition sources, to avoid hazardous explosions and fires.

2.4.2.3.2 Manual ESD Activation

Manual activation of the ESD system shall be possible from strategic positions along escape routes and at stations where accessibility and manning in a hazard situation is adequate. Manual activation stations shall be provided in the central control room, and in strategic locations in the process areas subject to assessment of plant arrangement and applied safety strategy. Manual activation buttons shall be protected against inadvertent activation, eg, protective covers or rotary switch. Each post shall be clearly marked, and a consistent color coding shall be applied for the activation buttons, eg, yellow.

2.4.2.3.3 ESD Actions and Documentation

Once initiated, all required ESD actions shall be automatically activated. The ESD functions shall be arranged in a tree-structured level hierarchy according to extent of actions and plant process facilities. All ESD levels and actions shall be described in detail within the documentation for operation of the plant. A superior ESD level will initiate lower levels. The following automatic actions shall be executed:

- Shutdown of wells (platform and subsea) and import/export pipelines including stop of export from upstream installations
- Shutdown and sectioning of the hydrocarbon process facilities
- Initiation of depressurization of pressurized process plant ("blow-down") at higher ESD levels
- Ignition source isolation/deactivation (see Section 2.4.4)
- Shutdown of main power generation

- Start/stop of emergency power generator
- Shutdown of drilling and work-over equipment not required for oil/gas well control.

2.4.2.3.4 ESD Alarms

Effective alarms shall be given as quickly as possible to warn personnel upon ESD level initiation. Activation of main ESD levels, failure to execute actions upon demand and system defects shall be presented as alarms in the central control room. An ESD shutdown action shall, in addition to general plant alarm, also be announced at other strategic locations, eg, by visual display of ESD alarms in local control rooms. The ESD system status shall be continuously available in the central control room, and the system shall raise alarms there for operator awareness or action.

2.4.2.3.5 ESD Response Time/Rate

The required response characteristics of the various elements of the ESD function loop shall be quantified to ensure that the total reaction time for each safety function is sufficiently short. The actual response times/rates of any specific ESD action to achieve safe state, shall be clearly identified in relevant documentation. An example is that the time from activation to start of deenergizing of a solenoid valve shall be <2 s.

2.4.2.3.6 ESD Independence, Operating Integrity, Reliability, and Survivability

The following requirements shall be satisfied:

- The ESD system shall operate as an independent system.
- The classified ESD safety instrumented functions shall be functionally and physically segregated from other systems/functions.
- ESD functions shall be fully operational whenever the plant is "live" with combustible process liquids or gases, and appropriate testing of the system shall be possible without interrupting operation.
- Reliability and safety unavailability criteria for ESD functions shall be documented and maintained throughout the life cycle of the ESD system.
- The 1ogic solver and essential utilities shall be located in a protected area.
- Equipment that is critical for the effectuation of system actions shall be protected against accidental loads such as explosion, fire, and falling loads.

2.4.2.3.7 ESD Fail-to-Safe Principle

In the event of malfunction, the entire ESD system or affected parts of it shall automatically go to a predefined safe state. If this principle ("fail-to-safe") is not applied, an equivalent level of safety shall be achieved by redundancy, diagnostics, and alarms to control room.

2.4.2.4 ESD Man—Machine Interface

The man—machine interface in the central control room shall provide the means for operator awareness and actions and be suitable during emergencies. The ESD status overview should be clearly displayed in the central control room at all times. All man—machine-interface facilities shall be based upon an established philosophy, and the presentation of the system shall be clear and concise. Shutdown alarms and critical system failures shall be presented in proper sequence/priority. The hard-wired safety panel shall provide the means for manual activation, and status observation of relevant actions, eg,

- Plant alarm
- Depressurization
- Ignition source isolation/deactivation at relevant levels
- Fire water-pump start
- Fire extinguishing systems status

Regarding documentation and modification, the following points must be observed:

- Documentation describing the ESD system shall be available in a formal updated revision that reflects current design and operation.
- Relevant documentation for operation and maintenance shall be available and updated.
- Modifications and changes to the ESD system shall be made in accordance with appropriate written procedures for initiation, documentation, review, and approval.

2.4.2.5 Fast Dilution with Air of Gas Leaks by Natural Ventilation and Forced Heating, Ventilation, and Air Conditioning Systems (HVAC)

The following is based on Statoil (2004).

2.4.2.5.1 Overview

The purpose of natural ventilation and HVAC systems is to prevent escalation of accidental leaks into major hazardous gas clouds by limiting the extent and duration of any such clouds that may occur. Natural ventilation shall:

- Dilute gas concentrations and reduce the size of flammable gas clouds
- Dilute harmful concentrations of smoke or toxic gases
- Ensure acceptable working and equipment environment

HVAC shall:

- Dilute gas concentrations and reduce the size of flammable gas clouds
- Provide pressurization of rooms to prevent ingress of smoke or gas
- Provide smoke-extract ventilation for internal fire conditions
- Ensure acceptable working and equipment environment

2.4.2.5.2 Natural Ventilation in Classified Areas

For definitions of classified areas, see Section 10.1 in Chapter 10. The natural ventilation system shall for 95% of the time provide a minimum-specified average ventilation rate throughout the area to avoid stagnant zones. The following points should be considered:

- Location and sizes for ventilation louvers (eg, in walls of offshore platform modules) shall be optimized to give required minimum ventilation air change rates for dilution of flammable gases but shall, in addition, comply with safety risk analysis (gas dispersion, explosion loads, etc.) and maintain acceptable working environment/weather protection for personnel.
- Natural ventilation shall be documented by calculations and/or model testing. Potential stagnant zones shall be evaluated and precautions taken when considered necessary. During operation, compliance with design documentation shall be verified, ie, has additional equipment been installed, have louvers been changed or blocked, etc.

2.4.2.5.3 Forced/Mechanical Ventilation in Classified Areas

The following points must be observed:

- A ventilation rate to satisfy specified minimum requirements shall be provided throughout the area to avoid stagnant zones.
- Alarm shall be given upon too low airflow.
- In the event of an internal gas leak in special systems such as turbine enclosures, battery rooms, gas analyzer houses, etc., all available ventilation capacity should be used to maintain the "unclassified" status.

- There shall be a reliable overpressure of minimum 5×10^{-4} bar in unclassified rooms relative to surrounding classified areas. Alarm shall be given upon too low overpressure.
- It must be possible to perform a manual disconnection of ignition sources in rooms with self-closing doors toward classified areas.
- Rooms without such doors, but which are next to classified areas shall be treated as rooms with doors. Access from nonclassified to classified areas should be avoided.
- Fire/smoke/gas "dampers" shall provide quick, reliable, and effective prevention of spread of gas, fires, and smoke into other areas via the HVAC system. Gas-tight fire dampers shall be installed in HVAC inlets. Dampers shall close automatically within 6 s upon gas and/or smoke detection in the air intake. Dampers and fans shall be interlocked to avoid abnormal pressure configurations. Damper position shall be monitored from the central control room.
- Ventilation functions and equipment shall be fully operational whenever the plant is live with flammable gases/liquids, and appropriate testing of the function shall be possible without interrupting operations.
- Reliability and safety unavailability criteria for the HVAC system shall be documented and maintained through the life cycle of the system. In case of malfunction, the HVAC system or affected parts of it shall go to a predefined safe state. The HVAC system and its components shall be designed and protected to ensure that it will remain operative during incidents in which the system has a role as or forms a safety barrier.

2.4.2.5.4 Documentation and Modification

The following points must be observed:

- Documentation describing the ventilation system shall be available in a formal updated revision that reflects current design and operation.
- Relevant documentation for operation and maintenance shall be available and updated.
- Safety requirements and philosophies shall be adhered to, and safety assessment shall be performed prior to changes affecting system safety functions.

2.4.3 Preventing Ignition Sources

2.4.3.1 Introduction

The various types of potential ignition sources and their properties are discussed in Section 2.2. Chapter 10 discusses requirements to ensure that *electrical* equipment does not become an ignition source if making contact with explosive gas. Chapter 10 also discusses the concept of "hazardous area classification," which implies that the various regions/areas of the industrial plant are grouped in three categories depending on the likelihood that explosive gas may exist there. Regions where this likelihood is regarded as zero in comparatively normal operation, are denoted "unclassified" or "nonclassified."

Electrical as well as nonelectrical equipment installed and used in locations where an explosive gas atmosphere may occur shall comply with the requirements corresponding to the hazardous area class of the area in which the equipment is to be used. Equipment, both electrical and nonelectrical, used in hazardous areas shall provide a minimum level of explosion protection as follows:

- Zone 0: Certified equipment of category 1
- Zone 1: Certified equipment of category 2
- Zone 2: Certified equipment, or equipment suitable for use, of category 3

In general specifications for electrical equipment to be used in areas where an explosive gas atmosphere will or may exist are to be in accordance with recognized international/European (IEC/ European Standards (EN)) or other relevant standards/norms. In addition, for nonelectrical equipment, specifications should be based on recognized European or other relevant standards/norms.

2.4.3.2 Open Flames

The use of open flames, for operational or any other purpose, should be prohibited in all areas where explosive gas atmospheres may occur. Certain devices containing flames may be permitted if the flames are safely enclosed, and if the temperature of the external enclosure surface does not exceed the prescribed maximum value for the gas in question. Furthermore, it should be ensured that the enclosure is adequately resistant to the effect of the flame, and that flame propagation into classified areas cannot occur.

Hot exhaust gases may be introduced into areas that may contain explosive gas atmospheres if suitable devices are available to guarantee that the temperature of the exhaust gases at the inlet to the area is sufficiently low for ignition of the explosive atmosphere to be impossible.

2.4.3.3 Hot Surfaces

The standard rule is that the temperature of hot surfaces such as exhaust pipes and channels shall not exceed the autoignition temperatures (AIT) of flammable fluids that can be present due to accidental leaks. As discussed in Section 2.2.4, this may be a very conservative requirement in some cases. Thermal insulation materials shall cover the entire hot surface, and necessary means of protection shall be provided to prevent flammable fluids from penetrating into the insulating material.

In addition to easily recognizable hot surfaces such as heating coils, excessively heated surfaces can also be generated by mechanical processes/equipment converting mechanical energy into heat, ie, friction clutches and mechanically operating brakes. Furthermore, all moving parts in bearings, shaft passages, glands, etc. can become sources of ignition if they are not adequately maintained and lubricated. In tight housings of moving parts, the ingress of foreign bodies or shifting of the axis can also lead to friction which in turn can lead to high surface temperatures, in some cases quite rapidly.

The temperature of any surface in an area where an explosive atmosphere is likely to be present during normal operation should never exceed 80% of the minimum ignition temperature in °C (see Table 2.2) of a combustible gas or liquid, not even in the case of rare equipment malfunctions. In areas where explosive atmospheres are likely to occur only in abnormal situations, the temperatures of all apparatus surfaces that can come into contact with the explosive gas mixture should be lower than the minimum ignition temperature.

2.4.3.4 Burning Metal Particles, "Thermite" Reactions, and Transient "Hot Spots"

Grinding and cutting operations must be prohibited in areas where explosive atmospheres may occur. Hot spots and spark showers from such operations may ignite explosive gas atmospheres, including natural gas/air.

Impacts involving rust and light metals (eg, aluminum and magnesium) and their alloys can generate thermite flashes that can cause ignition of a wide range of mixtures of combustible gas or vapor and air, eg, natural gas/air. Titanium and zirconium can form incendiary metal particle sparks by impact or friction against any sufficiently hard material, eg, concrete.

Hot metal beads that are produced during welding or cutting operations may be regarded as metal sparks with large surface areas. They are effective ignition sources, eg, for natural gas.

In areas where an explosive atmosphere is likely to occur under normal operation, apparatus and equipment which, even in the case of rare operational malfunctions, can give rise to incendiary friction, impact, or abrasion sparks should be excluded. The possibility of impacts between aluminum (excluding alloys with less than 10% aluminum and paints and coatings with less than 25% aluminum by weight) and steel (except stainless steel, if the presence of rust particles can be excluded) must be prevented. Friction and impact between titanium or zirconium and any hard material must be avoided. It is possible in some cases to protect light metals from mechanical contact with rust by applying a coating. However, when using coatings of electrically nonconductive materials such as plastics, the risk of generation of incendiary electrostatic discharges must be considered (see Section 2.4.3.5).

The use of various metal tools for maintenance and repair work may generate metal particle sparks, and the possibility of these as potential ignition sources must be considered. In the case of natural gas, sparks from conventional steel hand tools are unlikely to cause ignition (see Table 2.9).

2.4.3.5 Ignition by Electric Sparks, Arcs, and Electrostatic Discharges

All *electrical apparatuses* to be used in areas were explosive gas atmospheres may occur, must be manufactured, installed, and operated in such a way that the likelihood of such apparatuses igniting an explosive gas atmosphere is acceptably low. Electrical apparatuses intended to prevent ignition of explosive atmospheres by electric sparks, arcs, and electrostatic discharges arising from such apparatuses, have to satisfy standardized requirements, as discussed in Chapter 10.

Lüttgens and Glor (1989) and Lüttgens and Wilson (1997) gave extensive information on both fundamental and practical aspects of preventing ignition by *electrostatic discharges*.

Objects of electrically conducting materials that may become electrostatically charged shall be connected to earth or be bonded according to prevailing standards and guidelines. Objects shall be protected against triboelectric charging (due to physical contact followed by separation of electrically different materials). Such charging can also take place during rapid flow in pipes of electrically nonconducting liquids or gases containing electrically nonconducting liquid droplets or solid particles. But charge can also be produced in high-velocity water jets. Particular precautions will be required if nonconductive construction materials are applied, in which apparent connections to earth may not be functioning. Electrostatic charging and generation of incendiary electrical discharges may be a problem in situations including:

- Fueling operations
- Filling of containers, tanks and pressure vessels
- High fluid velocities (high water sprays, gas jets),
- Shot blasting
- Steam cleaning
- Loading operations

Stray currents flowing in electrically conductive systems or parts of systems can give rise to spark or arcs. Such currents can appear:

- as return currents in power-generating systems, eg, in the vicinity of large welding systems, for example when conductive electrical system components such as pipes and cable sheathing laid underground lower the resistance of the return current path,
- Because of any short-circuit, or of a short-circuit to earth, owing to faults in the electrical installations,
- Because of lightning (see separate discussion that follows),
- Because of magnetic induction, eg, near electrical installations operating at high currents or high frequencies (see separate discussion in the following).

If parts of a system able to carry stray currents are disconnected or connected/bridged, even in the case of slight potential differences, electric sparks or arcs can be formed, which may be able to ignite an explosive gas cloud. Moreover, ignition can also result from the heating of the current paths.

When impressed-current *cathodic corrosion protection* is used, this ignition risk has to be expected, and special protective measures should be provided. However, if sacrificial anodes are used, ignition risks due to electric sparks do not normally have to be accounted for.

The possibility of compensating for the electric potential should be considered for all conductive parts of the installation, including those that are not adjacent to electrical equipment. It is permissible to deviate from this requirement within areas enclosed by conductive walls that are included in a potential compensation system. Before introducing conductive parts, eg, ventilation and suction pipes in tanks, such parts should be included in a potential compensation system.

For conductive parts of the system which are not adjacent to electric installations, it is permissible to dispense with special measures to compensate the potential difference when a compensation system already exists, formed by interconnected parts of the electrically conductive systems, eg, pipe networks or *extensive earthing systems.*

Before the connections of conductive parts of the system are opened or closed, eg, during dismantling of fittings and parts of pipes, bridges comprising connection lines with an adequate cross section should be provided if it is possible that the adequacy of the interconnection will be impaired.

Ignition will always occur if *lightning* strikes an explosive gas cloud. Moreover, there is also a possibility of ignition due to the high temperature reached by lightning conductors. Large currents flow from where the lightning strikes and these can produce sparks in the vicinity of the point of impact. Even in the absence of lightning, *thunderstorms can cause high induced voltages* in plants and equipment, which in turn can give rise to sparks and arcs. Installations should be furnished with appropriate lightning protection measures, as specified in installation guidelines for electrical systems and for cathodic corrosion protection systems.

As discussed by Harrold (1992), *high-frequency electromagnetic waves* are emitted by all systems that generate high-frequency (HF) electrical energy (10^4–10^{11} Hz), eg, radio transmitters or industrial high-frequency generators for heating, drying, hardening, welding, cutting, etc. All conductive parts located in the radiation field function as receiving aerials. Ignition sources may be generated if the field is powerful enough and if the receiving aerial is sufficiently large. The primary process is generation of electric currents via electromagnetic induction. The currents induced may make thin wires glow or may generate sparks during the contact or interruption of conductive parts. If induced currents and inductances of conducting parts can be estimated, Fig. 2.31 can give guidance as to whether such sparks can ignite the explosive mixture of concern. In the case of particularly strong HF fields, for example in the immediate vicinity of high-power HF generators, nonconductive materials can also become heated due to resonance absorption, and can thus become a source of ignition. The energy picked up by the receiving aerial, and the received HF power at any particular wavelength, depend mainly on the distance between the transmitter and the aerial and the dimensions of the aerial.

2.4.3.6 Ignition by a Jet of Hot Combustion Products

The basic knowledge about ignition of explosive gaseous atmospheres by jets of hot combustion products is applied in practice in the design of so-called "flameproof" enclosures. These are enclosures for electrical components and apparatuses designed to satisfy two essential requirements. First, the enclosure must be sufficiently strong to withstand the full explosion pressure if a gas explosion occurs inside the enclosure. Secondly, any slits or holes in the enclosure wall must be sufficiently narrow to eliminate the hazard of ignition of a possible explosive atmosphere on the outside by the jet of hot combustion products emerging from the slit. International standards specify requirements for design of flameproof enclosures for industrial use. This kind of electrical apparatus, named"flame-proof," is discussed in Section 10.2.1.2 in Chapter 10.

2.4.3.7 Ignition by Light Radiation

Focused radiation in the optical spectral range ($10^{11}-10^{15}$ Hz) can become a source of ignition through absorption. For example, sunlight can cause ignition if focused, eg, by bottles acting as convex lenses or by concave reflectors. Focused light beams may occur as short pulses or as continuous radiation. A small solid object in the explosive gas mixture, hit by a hot focal point created by such radiation, can resemble either an electric spark or arc. Depending on the temperatures generated on the solid object, it may become an effective ignition source.

With laser radiation, eg, in communication systems, distance-measuring devices, surveying work, and visual-range meters, the power density of even an unfocused beam, even at great distances, can be so great that ignition is possible. Here, the heating process occurs mainly when the laser beam is absorbed by a solid body surface or dirty transparent parts (see Section 2.2.9).

2.4.4 Controlling Ignition Sources

This section is based on Statoil (2004).

2.4.4.1 Introduction

The concept of ignition source *control*, as opposed to ignition source *prevention*, implies that potential ignition sources may exist in an area as long as there is no significant concentration of combustible gas or vapor in the atmosphere. However, as soon as significant concentrations of combustible gas/vapor are detected, these potential ignition sources shall be removed from the area, deenergized, or in some other relevant way be brought in a state/situation in which they are unable to initiate an explosion.

2.4.4.2 Equipment in Naturally Ventilated, Normally Nonhazardous, Areas

Electrical and nonelectrical equipment installed in nonhazardous, naturally ventilated areas shall at least comply with the requirements for equipment to be used in zone 2 (see definition of zones in Section 10.1.2.7 in the present book). The equipment must not act as an ignition source whenever explosive gas enters a normally nonhazardous area. Use of equipment that is not specifically protected with regard to becoming an ignition source shall be based on a specific safety assessment considering realistic worst-case scenarios of explosive gas cloud formation. In the offshore industries, dispensations for use of noncompliant equipment shall be based upon a specific safety assessment as well as measures for avoidance of effective ignition sources, eg, isolation (eg, deenergizing) upon gas detection.

Safety-related equipment that should remain operative after completion of an extensive evacuation of, eg, an offshore oil/gas installation, shall as a minimum comply with hazardous area zone 2 requirements. Such equipment includes:

- Emergency lighting
- Communication equipment
- Navigation aid flooding lights for lifeboat area and helicopter deck
- Fire pumps

2.4.4.3 Isolation of Electrical Ignition Sources

Potential electrical ignition sources, equipment, and activities shall be identified, and a philosophy/strategy shall be implemented for disconnection/isolation in case of an accidental gas leak. Explosion-unprotected equipment is to be categorized in one of three groups, depending on the need for the equipment to be in operation in the case of a gas leak.

2.4.4.3.1 Group 1

This group contains electrical equipment that does not affect production availability or safety integrity. Such equipment should be disconnected/isolated on one single "low" gas alarm anywhere on the installation, or by the signal "confirmed gas in hazardous area." Group-1 equipment includes:

- welding sockets and sockets for hand tools
- air-driven tools
- temporary electrical equipment
- electrical heating elements

- hot-work activities

Explosion-unprotected electrical equipment including electrical heating elements located in nonhazardous mechanically ventilated areas shall be isolated upon a single low-gas detection alarm in HVAC intakes. Electrical equipment, except heating elements, located inside rooms may be kept alive until detection of confirmed gas in HVAC intake, provided a safety analysis has documented sufficiently low ignition probability.

2.4.4.3.2 Group 2

This group contains electrical equipment that must be kept powered to maintain production, drilling operations, etc. Such equipment should be disconnected/isolated on the signal "confirmed gas in nonhazardous areas." Group-2 equipment includes:

- main power generators
- main electrical distribution panels
- diesel engines
- heaters
- boilers
- ventilation systems
- emergency generators, unless defined as a safety critical item.

Explosion-unprotected equipment located inside mechanically ventilated rooms shall be subject to instantaneous automatic isolation upon confirmed gas detection in HVAC inlets. Explosion-unprotected equipment located inside rooms with doors/air locks facing classified areas or located in the close vicinity of classified areas shall be subject to instantaneous automatic isolation; either upon confirmed gas detection in the room due to ingress via doors/air locks or by coincident loss of ventilation flow and single low-gas alarm in adjacent areas.

2.4.4.3.3 Group 3

This group contains safety critical electrical equipment that must be in operation to ensure safe escape and evacuation, and/or to prevent accident escalation. In the case of a major emergency such equipment shall be disconnected/isolated with a time delay. Typical Group-3 equipment constitute parts of larger systems like:

- uninterruptable power supplies (from batteries) (UPS)
- fire and gas detection systems (F&G)
- emergency shutdown systems (ESD)

- process shutdown systems (PSD)
- public address systems (PA)
- emergency communication systems (EC)

On offshore platforms, equipment required to secure oil or gas wells in an emergency situation shall be defined in an emergency shutdown procedure. A backup procedure for manual shutdown shall also be established.

On offshore installations, safety control units, uninterruptable power supplies, and emergency switch gear shall be located in protected rooms inside the living quarters or in the vicinity of the central control room, based upon safety assessments. If such equipment is not properly located, the minimum requirement is that it shall be suitable for use in zone 2 areas (see definition in Chapter 10). This will normally apply to all active field equipment such as detectors and loudspeakers, including those located inside nonhazardous rooms and living quarters.

2.4.4.4 Isolation of Nonelectrical Ignition Sources

Diesel- or gas-fueled equipment such as prime movers and heaters shall be stopped immediately upon confirmed gas detection in ventilation air intake to the enclosure or combustion chamber, and the equipment (nonexplosion protected) shall be shut down in accordance with the assigned ignition source isolation groups and equipment location.

2.4.4.4.1 Group 1

Group-1 equipment comprises nonessential, nonelectrical equipment. Examples are rotating non-Ex equipment, mostly internal combustion engines, which power electricity generators and various tools, eg, high-pressure water washing systems. For such equipment the following points should be observed:

- This type of equipment shall normally be located in enclosed areas/compartments.
- Explosion-unprotected equipment being accepted for location in naturally ventilated areas shall be shut down instantaneously by single low-gas alarm anywhere on the installation. Explosion-unprotected equipment located in a nonclassified mechanically ventilated area shall be isolated/deenergized upon specified gas alarms or confirmed gas.
- Exhaust pipes and hoses shall be fitted with spark arrestors.

2.4.4.4.2 Group 2

This group comprises essential nonelectrical equipment, which shall be located in enclosed areas/compartments that are either safe by pressurization, or the probability of gas ingress is found to be sufficiently low to ensure an acceptably low risk of explosion. This group comprises rotating non-Ex equipment, mostly internal combustion engines, which power equipment necessary for maintaining production. Normally such equipment is shut down because of confirmed combustible gas or fire in classified areas. For such equipment, the following points should be observed:

- Explosion-unprotected equipment located inside mechanically ventilated rooms shall be subject to instantaneous automatic isolation by confirmed gas detection in the HVAC inlet of the room or in combustion inlets, or by overspeed of rotating machinery.
- Explosion unprotected equipment located inside rooms with doors/air locks facing classified areas, or located in the close vicinity of classified areas, shall also be isolated/deenergized automatically upon either confirmed gas detection in the room due to ingress via doors/air locks, or coincident loss of ventilation flow, or gas alarm in adjacent areas.

2.4.4.4.3 Group 3

This group comprises internal combustion engines/rotating equipment that are safety-critical eg, emergency electricity generators, water pumps for firefighting, and kill pumps for wells. This group of equipment shall be deenergized upon confirmed gas detection in air inlets, and/or in the case of a major emergency, after a specified time delay.

2.4.4.5 Cranes

To reduce the ignition probability all crane operations shall be stopped and cranes shut down, once a leakage of flammable gas has been detected.

If the crane is in operation in a situation of a single "low gas" alarm in any location on the installation, the crane operator shall immediately take the crane into a safe position, secure the load, and perform a manual shutdown. Upon confirmed gas detection in crane ventilation/combustion air intake, an automatic shutdown of the crane, without time delay, shall be initiated. Adequate means for crane operator warning, and consistent operating procedures shall be available.

When the crane is not in operation, potential ignition sources in the crane shall be isolated/deenergized automatically upon a single "low gas" alarm anywhere on the installation. In the case of cranes located in nonhazardous areas, external equipment shall generally be certified for use in hazardous area zone 2, as a minimum. The same applies to internal/local equipment, if operative after main crane shutdown, and isolation/deenergizing.

2.4.4.6 Human–Machine Interface (HMI) in Central Control Room

Human-machine-interface (HMI) in the central control room shall present extensive system information and means for operator interactions. Therefore, the HMI shall include the operator stations and a simplified hardwired safety matrix panel allowing manual activation of major safety barrier functions from the central control room. This also includes operator monitoring and control of ignition sources in an acceptable manner.

2.4.4.7 Operation, Inspection and Maintenance

All equipment that are potential ignition sources shall be operated, inspected, and maintained to ensure that any such equipment is in a proper condition at all times. A maintenance program including descriptions and instructions for testing, inspection, and repair shall be available. Safety-critical functions and components shall be tested regularly according to accepted standards to verify that the specified safety integrity level and failure rates are fulfilled.

2.4.5 Mitigating/Controlling Gas/Vapor Cloud Explosions Once Initiated Despite Preventive Measures

2.4.5.1 Control and Mitigation—Two Different Concepts?

Tam (2002) suggested that the terms *control* and *mitigation* of accidental explosions should have different specific meanings. He defined *mitigation* as an action causing a reduction of the consequences of an explosion event, without there being any reduction in the explosion severity at the explosion source. *Control*, on the other hand, would be when a device or technique has a direct impact on the severity of an explosion at its source. Automatic explosion suppression and partial inerting would then be defined as controlling techniques, whereas explosion venting would be a mitigatory technique.

Tam's distinction between control and mitigation is intriguing, but the distinction may not always be entirely clear in practice. In the present text, the term mitigation is used to cover both aspects.

2.4.5.2 *Explosion Isolation: Minimizing Explosion Propagation* Inside *Process Equipment Using Shut-off Valves and Other Physical Barriers*

2.4.5.2.1 Reasons for Using Explosion Isolation

CEN (2004b) issued a standard for design of explosion isolation systems. The term "isolation" is somewhat ambiguous. In some cases, it implies closure of pipelines and other connections between sections of a process plant. In the present context, the concept "explosion isolation" comprises means for preventing gas explosions from spreading from one process volume to another via pipes and ducts. There are three main reasons for adopting this concept. Firstly, there is always a desire to *limit the extent of the explosion* as far as possible.

Secondly, a gas flame propagating in a duct between two process units tends to accelerate due to flow-induced turbulence in the gas ahead of the flame. For a sufficiently long duct, this may result in a *vigorous flame jet* entering the process unit at the downstream end of the duct. The resulting extreme combustion rates can generate very high explosion pressures even if the process unit is generously vented. This effect was demonstrated in a dramatic way for flame jet-initiated explosions of propane/air in a generously vented 50-m^3 vessel, by Eckhoff et al. (1984).

The third main reason for preventing flame propagation between process volumes is *pressure piling*. This implies that the pressure in the unburnt gas mixture in the downstream process unit(s) increases above atmospheric pressure due to compression caused by the expansion of the hot combustion gases in the unit in which the explosion starts, and in the connecting duct(s). Experiments have shown that the final explosion pressure in a closed vessel is proportional to the initial pressure of the explosive mixture. Therefore, in a coupled system, higher explosion pressures than would be expected from atmospheric initial pressure can occur transiently due to pressure piling.

This was demonstrated in a laboratory-scale gas explosion experiment by Heinrich (1989) as shown in Fig. 2.67.

Despite the marked cooling by the walls in this comparatively small experiment, the transient peak pressure in V_2 significantly exceeded the adiabatic constant volume pressure of about 7.5 bar (g) for atmospheric initial pressure. Extremely serious situations can arise if flame jet ignition and pressure piling occurs simultaneously. It is important, therefore, to take effective measures to prevent propagation of gas and vapor explosions from one process item to another through connecting pipes and ducts.

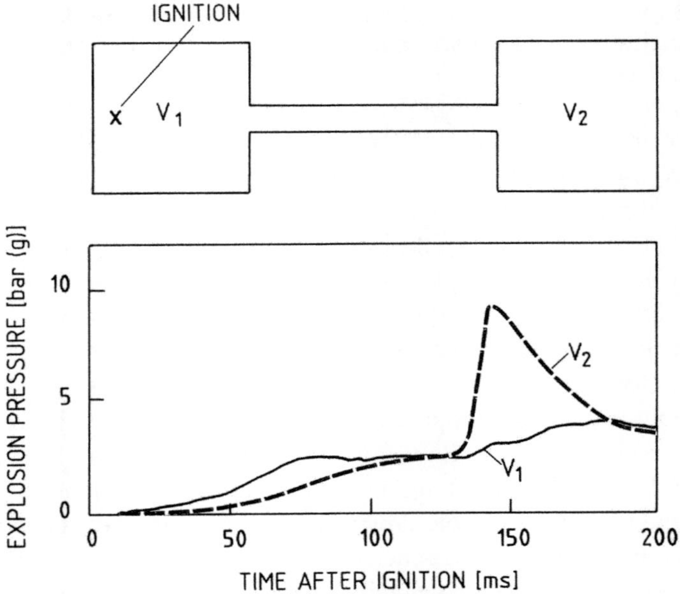

Figure 2.67 Pressure development in two closed vessels of 12 L each, filled with 10 vol.% methane in air at atmospheric initial pressure and normal temperature, and connected with a 0.5-m long duct, following ignition at the point indicated in the figure.

From Heinrich, H.-J., 1989. Ablauf von Gas-und Staubexplosionen — Gemeinsamkeiten und Unterschiede, VDI-Berichte, vol. 709. VDI-Verlag GmbH, Düsseldorf, Germany, pp. 93—112.

2.4.5.2.2 Mechanical Valves

One means to stop explosion propagation is to install various types of mechanical valves in the connecting pipes/ducts. Such valves can be either passive or active. One passive device for interrupting gas and vapor explosions in ducts is the *Ventex valve* described by Rickenbach (1983) and illustrated in Fig. 2.68.

In normal operation, the dust cloud being conveyed in the duct flows around the valve poppet without causing any significant offset as long as the flow velocity is less than about 20 m/s. However, in case of an explosion in the duct, the preceding blast pushes the valve poppet in the axial direction until it hits the neoprene gasket, in which it is held in position by a mechanical catch lock, which can be released from the outside.

Active Ventex valves are also being used. In this case, a remote pressure or flame sensor activates a separately powered system that closes the valve in the desired direction prior to arrival of the flame. Other active mechanical

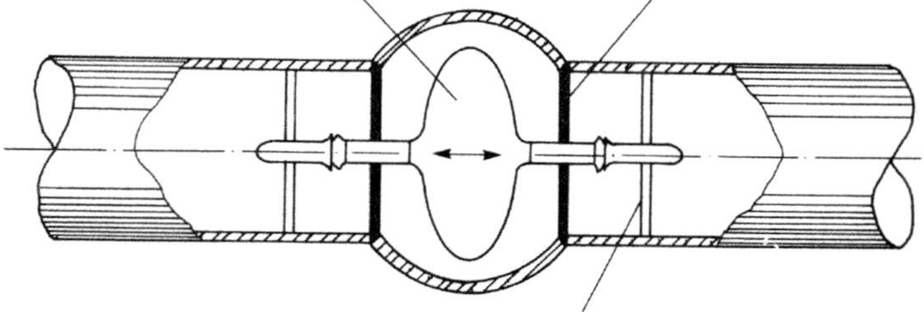

Figure 2.68 Illustration of the Ventex valve described by Rickenbach (1983). *From Eckhoff, R.K. (2003): Dust Explosions in the Process Industries. 3rd edition, Gulf Professional Publishing/Elsevier. (1st edition published in 1991, 2nd edition in 1997), ISBN 0-7506-7602-7.*

valves include various types of fast-closing slide valves. The closure is triggered by the signal from a flame or pressure-rise probe positioned upstream of the valve. The required closing times of such valves depend on the distance between the remote pressure or flame sensor and the valve, and on the type of explosive gas. Often, closing times as short as 50 ms, or even shorter, are required. This most often is obtained by using an electrically triggered explosive charge for releasing compressed air or nitrogen operating the valve. The valves must be sufficiently strong to resist the high pressures that can occur on the explosion side after valve closure. In the case of pressure piling effects and/or detonation, the transient pressure peaks can be very high.

2.4.5.2.3 Flame Arresters

Flame arresters are a well-proven means of effectively stopping flame/explosion propagation in pipelines, as well as out of or into process enclosures. The following account is based on Halstrick (1995).

Depending on the individual operating conditions, different combustion processes such as deflagrations, detonations, and endurance burning are possible. This calls for the application of flame-arresting devices that are suitable for the actual situation, ie, deflagration-proof, detonation-proof, or endurance-burning-proof devices. Except for very small devices for special applications, all-purpose flame-arresting devices, ie, devices that stop flame propagation in all three combustion modes, do not presently meet all standard requirements for being installed. The reason for this is that the design/construction of flame arresters for protection against deflagrations, detonations, and endurance burning differ considerably.

One of the most important components of a flame-arresting device is the *flame trap*, the purpose of which is to stop the flame propagation, whether a deflagration, a detonations, or endurance burning. The range of different design concepts in use can be categorized as either *static* or *dynamic* flame traps, of which the former if further subdivided into *dry* and *liquid* types.

Dry static flame traps are available as close-meshed sieves, plate arresters, sintered metals, and crimped ribbon-type traps. It is crucial that the width of gaps in the trap can be manufactured to exact, reproducible tolerances. This is not possible for flame traps of metal—wool, metal—foam, or knitted-metal mesh. Simple wire gauze ("Davy screens") is also inadequate because it is easily destroyed even when exposed to quite harmless low over-pressure explosions. The most common and reliable design of dry static flame traps is the crimped-ribbon-type trap illustrated in Fig. 2.69. Such traps can be

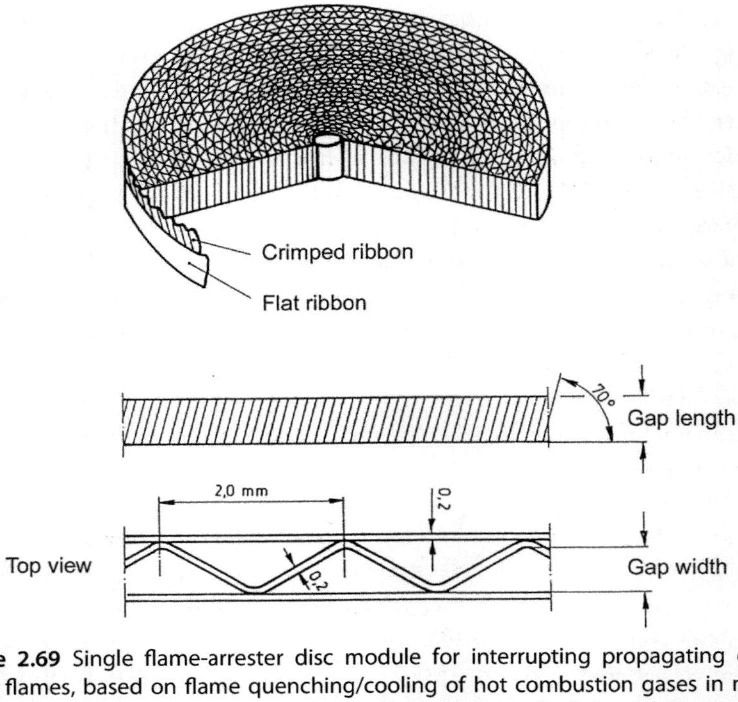

Figure 2.69 Single flame-arrester disc module for interrupting propagating gas or vapor flames, based on flame quenching/cooling of hot combustion gases in narrow channels made from corrugated metal strip. Typical standard gap widths are 0.15, 0.2, 0.3, 0.5, 0.7, and 0.9 mm.
From Halstrick, V., March 3, 1995. Protego fundamentals, Protego report no. 04. Published by Braunschweiger Flammenfilter GmbH, Germany.

manufactured with high reproducibility to the desired gap width based on the MESG (see Section 2.2.7 and Table 2.2 in Section 2.1.3.1) of the combustible gas or vapor in question.

Depending on the application of the flame trap, whether deflagration-proof, detonation-proof, or endurance-burning-proof, two, three, or more disc modules are mounted in series in a specially designed enclosure that satisfies the mechanical strength requirements for the actual application and provides the appropriate gap length for the required heat dissipation.

Liquid static flame traps are used in process systems in which the combustible fluid is partly or fully in the liquid phase. This liquid is then used as the sealing medium in the flame trap. This trap design is always adopted for flame arresters to be used in liquid-filled pipelines that may drain empty. An example is petrol-filling and -emptying lines, in which the petrol is used as the sealing liquid operating as the flame trap. The absence of oxygen after a deflagration/detonation prevents ignition of the flammable sealing liquid. Static liquid flame traps can only be used for stopping deflagrations, and detonations, and not in process systems in which endurance burning can be expected. Fig. 2.70 illustrates the principle of a liquid static flame trap.

A dynamic flame trap is one which prevents a flame on the low-pressure side of process equipment to propagate into the high-pressure side. The basic principle of operation is that the flow speed of the combustible medium from high to low pressure is higher than the flame speed in the opposite direction. Because the flow speed is low close to the outlet wall, special precautions are required to prevent the flame from entering the high-pressure side there. One application of dynamic flame traps is as safety-relief valves that open at quite high pressures, which result in high flow speeds. Examples are incinerator inlets, and flare systems in which the flow speeds are maintained well above the flame propagation speed in the respective mixture. *Explosion proof end-of-line flame arresters* must be designed to prevent flame flashback into the pipeline in the case of an external atmospheric vapor cloud explosion.

Volume explosion flame arresters are designed for specified volumes. This means that these arresters provide safety only for conditions that are almost equivalent to the original test conditions (size and type of explosion volume and size of opening).

Explosion proof in-line flame arresters consist of an explosion-pressure-tight body with the flame-arresting element which in turn is composed of the actual flame filter disc of two or more layers and the enclosing cage.

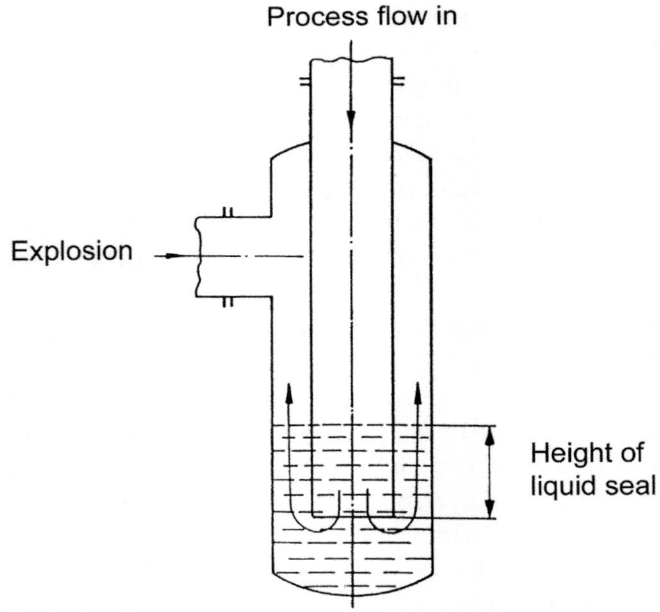

Figure 2.70 Illustration of a liquid static flame trap for stopping deflagrations, and detonations in pipelines.
From Halstrick, V., March 3, 1995. Protego fundamentals, Protego report no. 04. Published by Braunschweiger Flammenfilter GmbH, Germany.

The gap width of the flame filter disc has to be adjusted to the appropriate MESG to prevent flame transmission. It is important that this type of arrester is not exposed to other flame propagation speeds and explosion pressures than the range for which the devices have been tested and approved.

Dry detonation arresters with a static flame trap consists of a detonation shock absorber in front of a flame-arresting element, as illustrated in Fig. 2.71. The purpose of the shock absorber is to convert a detonation into a deflagration before the flame arrester is reached. It is crucial that the detonation shock absorber points toward the possible point of ignition. If pointing in the opposite direction only deflagrations will be arrested. In some situations, approach of detonations has to be expected from both sides, and bi-directional detonation arresters are required. It is important to note that in-line flame arresters in general, and detonation arresters in particular, do not prevent flashback in endurance-burning situations, and to handle such situations additional measures have to be taken.

Filling and emptying lines of storage tanks for flammable liquids have to be protected by detonation arresters unless, during operation, the lines are

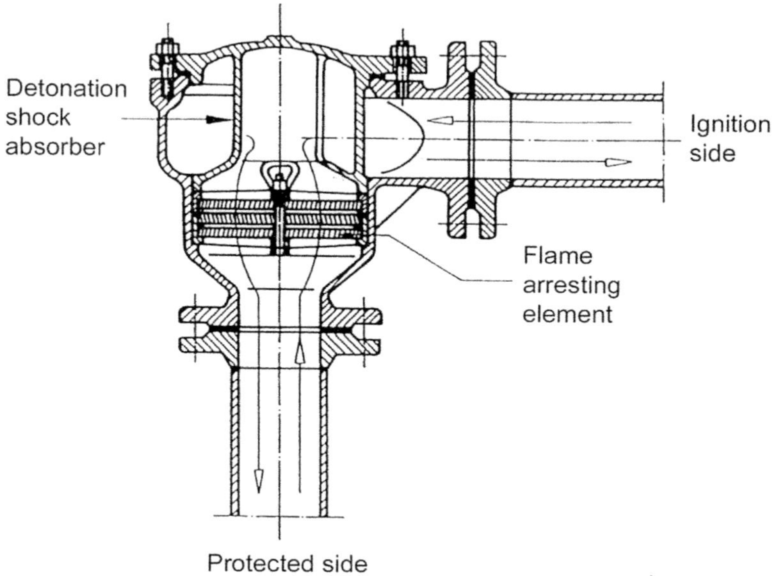

Figure 2.71 Dry detonation arrester with static flame trap.
From Halstrick, V., March 3, 1995. Protego fundamentals, Protego report no. 04. Published by Braunschweiger Flammenfilter GmbH, Germany.

permanently filled with liquid. If this is not the case, there is a possibility of explosive vapor/air mixtures being developed in the system, and *liquid-type detonation arresters* may be installed. In this case the liquid itself forms the barrier absorbing the detonation shock. Such devices constitute a complete protective system and shall therefore ensure that the liquid barrier is maintained under all circumstances. For example, if the line connected to the vessel is to be emptied by a suction pump, a static flame trap is installed as a bypass to prevent the suction operation from emptying the liquid barrier.

Special *flashback-proof detonation foot valves* are also used for preventing flame transmission between vessels and connecting lines, while emptying the line. These devices are also based on the use of liquid barriers.

To prevent flame flashback into a line in the case of *endurance burning* outside the line end, both dry static and dynamic end-of-line flame traps are used. When explosive atmospheres may be expected in storage tanks for flammable liquids, endurance-burning-proof flame arresters are used in the direct venting system to the atmosphere to prevent flames in the lines from entering the tanks. Vent caps are used if loss of flammable liquid by evaporation is of minor importance, whereas pressure/vacuum relief valves are applied whenever evaporation losses have to be avoided for environmental or economic reasons, or both.

2.4.5.2.4 Flame Interruption by Automatic Injection of Suppressant

A third active-explosion isolation method is interruption by fast automatic injection of extinguishing chemicals ahead of the flame. This is a special application of the automatic explosion suppression technique described in Section 2.4.5.8. Important design parameters are type of combustible gas/vapor; size of the enclosure in which the explosion starts; ignition location; duct diameter; distance along the duct from the primary explosion enclosure to the point of suppressant injection; method used for detecting onset of primary explosion,; and type, quantity, and rate of release of suppressant.

2.4.5.3 Minimizing Gas Cloud Size and Controlling Gas Cloud Location Outside Process Equipment Using Physical Barriers

The following account is in part based on the paper by Tam (2002).

2.4.5.3.1 The Hazardous Situation in a Brief Historical Perspective

In the chemical and process industries, controlling/mitigating potential gas/vapor explosions outside process equipment, due to leaks from such equipment, has been and still is a major concern. In the offshore oil and gas industries, the main emphasis has been on controlling explosions in large, fully, or partially enclosed areas containing process equipment ("modules"). The hazardous explosive gas clouds are then generated by leaks from process equipment and pipe works run at pressures considerably higher than atmospheric. In the past, halogenated hydrocarbons ("Halons") were used for controlling the development of gas and vapor explosions in confined enclosures of smaller volumes, eg, turbine housings. However, due to their negative effects on the environment the "Halons" are no longer used. In smaller confined volumes such as turbine housings, it has largely been possible to replace "Halons" by water mist. In more-open larger volumes (modules), the explosion-controlling methods used include optimization of equipment and pipework layout to reduce flow-generated turbulence, and hence reduce the explosion rate. Other controlling methods include the use of blast walls and total water deluge.

2.4.5.3.2 Definition of a Physical "Barrier"

As the name indicates, a physical "barrier" is a physical arrangement that controls the severity of a gas/vapor explosion. Such an arrangement can, for example, be a wall. Total area water deluge would not be classified as a barrier technique, but as a total flooding technique. Tam (2002) distinguished between two main categories of barrier methods, viz. *cloud size and location control*, and *explosion development control.*

2.4.5.3.3 Cloud Size and Location Control Barriers

The main purpose of this class of barrier is to minimize the volume of flammable gas that a gas explosion is allowed to consume, and to control the location of this volume. Generally, the overpressure load on structural surfaces is expected to increase with increasing size of an explosive gas cloud. It is, therefore, important to limit the size of any potential accidental explosive cloud as far as possible. The most commonly used barrier type is the *blast wall.* In some cases a main function of the blast wall is to limit the size of the explosive cloud; in other cases, its purpose is mainly to reduce the impact of a potential explosion, by shielding the structures outside the wall from excessive overpressure loading. As pointed out by Tam (2002), the concept of using a wall to restrict the size of the explosive gas clouds also includes *weaker walls* than blast walls, and even deliberately weakened walls. The feasibility of the concept of weak-wall explosion barriers was investigated by means of demonstration calculations using a CFD gas explosion code. The process area was divided into two halves using various arrangements of weak walls. It was shown that maximum overpressures developed on the structural surfaces could be reduced by 50—75% depending upon the arrangement of the weak barriers. *Soft barriers* are very weak and light physical barrier walls. This type of barrier offers the possibility to separate an area into a number of separate cells, allowing control of the size and location of the accidental gas cloud within a hazardous area. The soft-barrier control method has been used on oil and gas platforms.

Soft triggered barriers are either *nonsuppressive* or *suppressive.* The most commonly used nonsuppressive *soft* barrier is a water curtain that has no specific suppressive effect. According to Tam (2002), such curtains have been used successfully in the process industries onshore to contain hazardous gas clouds within a limited area, or to divert such clouds to safer areas. A number of extensive experimental research programs comprising various scales up to full size, have demonstrated that water-spray barriers can be effective in containing hazardous gas clouds. However, some types of water curtains are also categorized as *suppressive barriers.* Experiments have shown that this type of curtain, in addition to limiting the size of a dispersing explosive gas cloud, also causes the flame propagation to decelerate as the flame passed through the curtain. The effect is a significant reduction of the maximum explosion overpressure. *Automatic explosion suppression* (see Section 2.4.5.8) can also be used for establishing soft-triggered suppressive barriers.

2.4.5.4 Use of Physical Barriers for Controlling Explosion Violence

Tam (2002) emphasized that two effects contribute to the increased hazard associated with gas explosions in congested areas, viz.

- precompression of the explosive gas cloud in the congested area by an explosion developing in adjacent areas (pressure piling), and
- flame propagation in the congested, precompressed volumes

2.4.5.4.1 Controlling Precompression (Pressure Piling)

In a process area, one may find regions where both the degree of confinement and the local aspect ratio (length-to-width/diameter) are high. If a gas explosion occurs in such regions, one can expect high explosion overpressures. High overpressures can develop in such areas even if the ignition source is located outside the area, because the explosion can propagate into the area. It is important, therefore, to prevent gas explosions in such areas.

As shown in Section 2.4.5.2, precompression has a multiplicative and not an additive effect on the maximum explosion pressure. Barriers that isolate the confined volumes from the rest of the area could be an effective means of controlling the maximum explosion pressure that can occur, by preventing the explosive gas from entering these confined volumes. Hard barriers that can withstand the explosion force will effectively eliminate the problem, but soft barriers can also have a significant controlling effect.

2.4.5.4.2 Controlling Turbulence Generation

It is well known that severe gas explosions can develop in regions of high congestion, due to the enhanced combustion rate caused by the turbulence generated in the still unburnt explosive cloud ahead of the flame when it is pushed past equipment and pipe work. Experiments of the type illustrated in Fig. 2.10, have demonstrated the dramatic increase of the explosion violence that can result with certain obstacle configurations. Tam presented results from small- and medium-scale experiments demonstrating the effect of soft barriers in such geometries. The experimental arrangement is illustrated schematically in Fig. 2.72.

The tests comprised two arrangements, one without a soft barrier covering the obstacles (upper part of figure) and one with a soft barrier

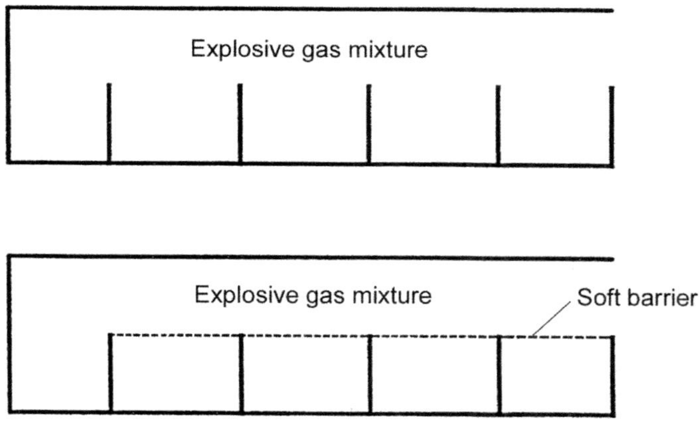

Figure 2.72 Schematic illustration (side elevation) of small- and medium-scale experiments conducted by GexCon, Norway (previously Christian Michelsen Research (CMR)), to investigate the effect of soft barriers on the violence of methane/air explosions in confined spaces containing turbulence-generating obstacles.
From Tam, V.H.Y., November 12–15, 2002. Barrier Methods for Gas Explosion Control, Lecture given at "Course on explosion prediction and mitgation", at University of Leeds, Energy and Resources Research Institute, School of Process, Environmental and Materials Engineering.

(lower part of figure). With the barrier, the explosive gas occupied only the upper part of the volume of the test structure prior to ignition. In the small-scale experiments, the maximum explosion overpressures were >250 mbar without the barrier, and only about 5 mbar with the barrier. In the intermediate-scale experiments, the effect of the barrier was also significant, viz. 600 mbar without barrier, and 150 mbar with barrier.

Other experiments, as pointed out by Tam (2002), have shown that for a given volume of obstructions, the combustion enhancing effect is greater with a large number of small obstacles than with a small number of large obstacles. One potential technique for controlling the explosion violence could then be to enclose a large number of small obstacles (eg, pipe work) in a larger hard or soft shield into which the explosion would not propagate. Tam discussed a number of potential solutions. In some process areas, precompression and turbulence generation can mutually enhance each other's effects. It is then important to limit the volumes of such areas as far as possible.

2.4.5.4.3 Suitable Materials for Soft Barriers

Tam (2002) pointed out that a soft barrier does not have to sustain the total gas explosion. It only needs to survive long enough to prevent substantial amounts of flammable gas from entering the fuel-free zone during the passage of the flame. For this purpose, low-pressure and low-tension membranes such as meteorological balloons could be used. Such balloons, in effect, divide the fuel-free zone into many fuel-free cells. Furthermore, the balloons can be filled with oxygen-free gases or mixtures of air and inert gas. Low-tension and low-pressure balloons can "mold" themselves to the shapes of hard surfaces and overcome the problem of flammable gas being trapped in the volumes between conventional spherical balloons (high pressure, high tension). Tam (2002) refers to large-scale experiments that have demonstrated that low-pressure low-tension balloons could last long enough for the explosion flame to pass the area of the balloon before the balloon collapses. However, so far these methods are not widely used in practice.

2.4.5.4.4 Additional Hazards to Be Considered When Using Soft Physical Barriers Other Than Water

Focusing only on the reduced maximum explosion pressure, when using soft physical barriers, is not sufficient. Some other important safety issues must also be resolved, viz.;

- possible emission of missiles or fragments resulting from collapsing barriers
- creation of small volumes/pockets of explosive gas that cannot be reached by the ventilation systems, because of the barriers
- unacceptable reduction of operational efficiency due to the presence of the barriers
- more complicated maintenance procedures due to the presence of the barriers

For these and other reasons, the use of soft barriers (other than water) has only gained limited terrain so far.

2.4.5.5 Design of Buildings to Prevent Damage by Gas Explosions

It is important to distinguish between ideal requirements and realistic possibilities. In all circumstances it is strongly recommended that the explosion problem be taken into account as early as possible in the planning process, whether a completely new plant is to be constructed, or an existing plant rebuilt.

Ideally, any plant in which gas explosions may occur should be located at a safe distance from other activities, buildings, etc. Furthermore, the various parts of the plant should be separate to enable effective isolation of the explosion to within the section of the factory in which it starts. Buildings should be one-story whenever otherwise suitable. If multistory buildings have to be used, the parts of the plant representing the greatest explosion hazard should be located as high up as possible, preferably on the roof.

In the past, floors and roofs of factory buildings were often supported by recesses in comparatively weak walls with no reinforcement as illustrated in Fig. 2.73a. In the case of an explosion, the walls were displaced outward even at very modest over-pressures, and the floors and roof fell down into the building as illustrated in Fig. 2.73b. Clearly, under such circumstances the consequences of even minor dust explosions in the building could be catastrophic. A much better solution is that the roof and intermediate floors be supported by a strong frame structure. The walls are then low-mass panel sections that function as vent covers, should an explosion occur. If required, the panels must be anchored to the frame structure.

In some cases, the hazardous plant items can be placed in special, isolated, well-vented niches as illustrated in Fig. 2.74. Depending on the location, the floor and roof of the niche may also have to be explosion proof. However, if the building is constructed of reinforced concrete, it can be made sufficiently strong to enable the windows to serve as vents

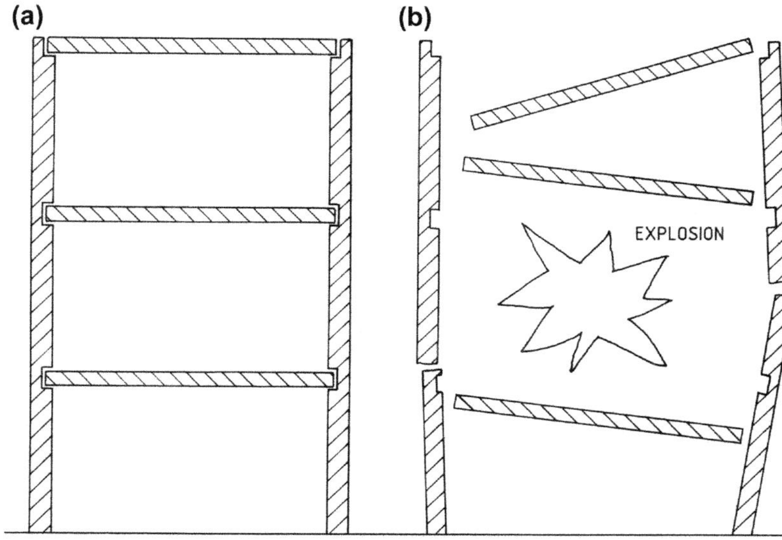

Figure 2.73 Inappropriate construction of multistory building with regard to preventing hazardous collapse of the floors in case of an explosion inside the building.
From Eckhoff, R.K. (2003): Dust Explosions in the Process Industries. 3rd edition, Gulf Professional Publishing/Elsevier. (1st edition published in 1991, 2nd edition in 1997), ISBN 0-7506-7602-7.

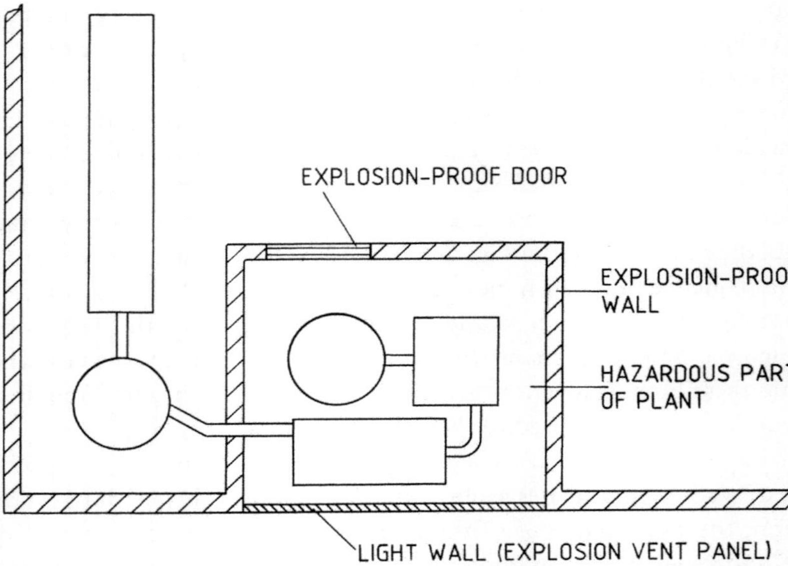

Figure 2.74 Illustration (top view) of arrangement in which the hazardous part of the plant is located in an isolated well-vented niche inside the building.
From Eckhoff, R.K. (2003): Dust Explosions in the Process Industries. 3rd edition, Gulf Professional Publishing/Elsevier. (1st edition published in 1991, 2nd edition in 1997), ISBN 0-7506-7602-7.

It is then important to ensure that flying pieces of glass do not present a hazard to humans. To avoid this hazard, it may be necessary to replace glass panes by anchored, transparent plastic panes.

Some final points to be taken into account when planning layout and construction of factory buildings with the aim to reduce the explosion and fire hazard include:

- Safe escape routes in case of explosion and fire
- Fire-resistant construction materials
- Fire-resistant doors
- Electrical installations according to latest regulations/recommendations.

2.4.5.6 Explosion Venting

2.4.5.6.1 What Is Explosion Venting?

The basic principle is illustrated in Fig. 2.75.

The maximum explosion pressure in the vented explosion, P_{red}, is a result of two competing processes:

- Burning of the explosive cloud, developing heat, and hence increasing the pressure.
- Flow of unburnt, burning, and burnt cloud through the vent, relieving the pressure.

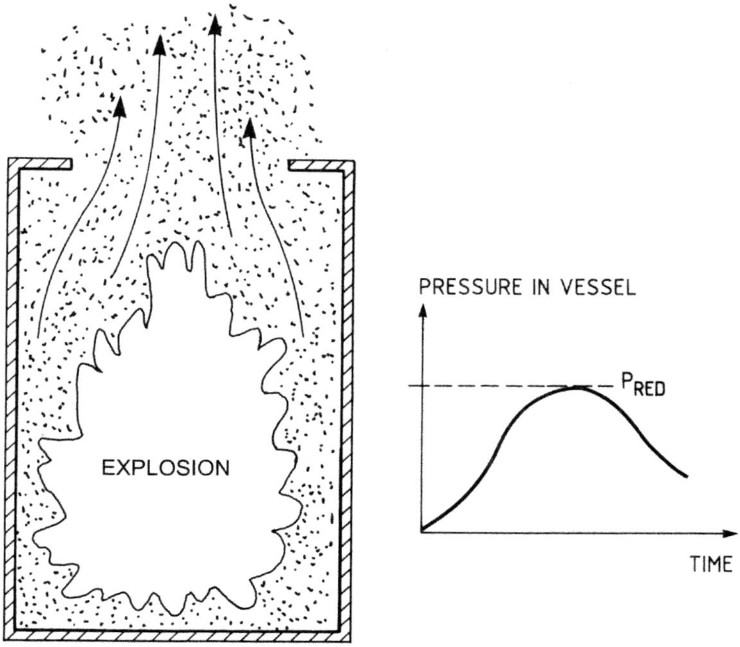

Figure 2.75 Illustration of the principle of explosion venting.

The two processes can be coupled via flow-induced turbulence that can increase the burning rate. The maximum permissible pressure, P_{red}, depends on the construction of the enclosure, and on whether plastic deformation or only elastic is acceptable in the case of an explosion.

Explosion venting of simple process enclosures is less common in the case of gases and vapors than in the case of explosive dust clouds (see Section 7.5.4.4). Gas explosion venting is more often applied to rooms of onshore facilities and processing modules on offshore gas and oil installations. This reflects the fact that accidental gas explosions most often occur in explosive clouds generated outside process equipment by accidental leaks, whereas dust explosions are practically always initiated in dust clouds generated inside process equipment as part of the process.

2.4.5.6.2 Vent Covers

CEN (2003) produced a draft standard for design of specific explosion venting devices. Vent covers for process enclosures are discussed in Section 7.5.4.4. For rooms, eg, offshore process modules, as pointed out by Bjerketvedt et al. (1997), typical vent covers include:

- Louvered walls
- Solid wall cladding

- Relief walls (also called wind walls or weather cladding)
- Glass windows (not recommended)

2.4.5.6.3 Potential Hazards Caused by Venting

See Section 7.5.4.4 (dusts).

2.4.5.6.4 Use of Vent Ducts

See Section 7.5.4.4 (dusts).

2.4.5.6.5 Reaction Forces Caused by Venting

See Section 7.5.4.4 (dusts).

2.4.5.7 Use of Numerical Computational-Fluid-Dynamics-Based (CFD) Gas Explosion Simulations in Design of Effective Measures for Controlling/Mitigating Accidental Gas Explosions in Complex Geometries

2.4.5.7.1 Background

In some industries, eg, petrochemical plants and oil refineries, the process areas are highly congested, as illustrated in Fig. 2.76. In such cases the inherent positive feedback loop of the explosion development can generate very high burning rates and pressure piling effects, and hence very high, destructive explosion pressures. Today, one can use comprehensive numerical computer simulation models (eg, FLACS) to predict the development and strength of gas explosions even in such extreme situations. This makes it possible to take appropriate precautions, eg, in terms of adequate safety distances to neighboring activities.

The following summary is based on the text of Czujko (2001).

2.4.5.7.2 CFD-Models in General

In the general CFD approach, one aims resolve the physics of flow development numerically by dividing the actual space into small boxes (control volumes) and implementing models for the various physical phenomena like fluid flow and turbulence. In each cell, all variables are assumed constant in one time step, and based on the flow balance and fluxes, as well as the physics taking place inside the cell in the next time step, the variables may change. Furthermore, in CFD explosion simulation not only physical flow but turbulence must be modeled. It is also essential to model combustion/flame development. Thus, the conservation equations for

- masses
- impulses

Figure 2.76 Highly congested process plant in which very high explosion pressures can be generated in case of a gas explosion.

- turbulence fuel transport and mixture fraction, and
- enthalpy

are solved for each time step and control volume.

Some CFD models are designed especially for simulating gas explosions, and they have been extensively validated by experiments at various scales. The most well-known models of this type are FLACS, EXSIM, and AutoReaGas. During recent years, the development of computers has substantially increased the possibility of applying CFD for simulating explosion development in complex large-scale systems. There are nevertheless still strong limitations on the spatial and temporal resolution that can be obtained in realistic industrial scenarios. For this reason, one incorporates so-called subgrid models to resolve the phenomena that occur on a smaller scale than typical numerical cell sizes (0.5—1 m). Such models take care of, e.g.:

- turbulence
- flame folding
- effect of water deluge

The need for subgrid models is one reason why development of comprehensive CFD codes has to be based on extensive validation by experiments.

2.4.5.7.3 The FLACS Model and Its Applicability

FLACS has been developed at GexCon, Bergen, Norway (previously Christian Michelsen Research (CMR), and initially Chr. Michelsen Institute (CMI)) since 1980. Björn H. Hjertager was the pioneering researcher and the inventor of FLACS. His first international presentation of the very first version of FLACS was Hjertager (1981). The development work was supported by major oil and gas companies for more than a decade. The code is currently used worldwide and is being improved continuously.

FLACS uses a 3D Cartesian-implicit Navier—Stokes solver based on the Simple method by Patankar with a compressible extension. Implicit/explicit second-order terms are implemented for increased accuracy. Flow equations are closed by means of a k-ε model for turbulence. Models for wall function and turbulence generation from subgrid geometry are required and implemented. A distributed porosity concept is used for mapping the complicated geometries onto the simulation mesh. In developing such a mapping model, there are several pitfalls, and the "perfect" solution seldom exists. Further, the two equations for handling fuel transport and consumption are solved. To model the combustion process properly, a range of subgrid models is needed. A flame model describes the extent of the reaction zone, whereas burning velocity models decide the flame-front velocity. This velocity, in turn, is a function of fuel composition and concentration, oxygen concentration, temperature, pressure, turbulence level, various flame instabilities, and presence of nonflammable gas dilution or water mitigation. Corrections for increased flame area due to flame folding around subgrid obstacles are necessary.

Special FLACS models simulate gas leaks, gas flow across boundaries, effect of water spray on explosion development, effect of pressure-relief panels, and heat, ventilation, and air conditioning systems (HVAC) (see Section 2.4.2.5). A multiblock solver for propagation of blast waves in the far field was developed in 1994. Some versions of FLACS also have models for heat transfer through walls and for thermal radiation. Further models for implementation in FLACS have been developed recently, are under development, or are being planned, including:

- Models for predicting deflagration-to-detonation transition (DDT) and for simulating detonations

- Models for predicting autoignition
- Models for simulating various types of flame instability models (important for low-confinement/low-congestion scenarios)

In addition, work is in progress to develop:

- Models for oil mist leaks, evaporation, deposition, dispersion, and explosion (Chapter 5)
- Models for dust cloud generation and explosion (Chapter 7)

2.4.5.7.4 Adequate Representation of Geometry Essential

Because the geometry of the system in which the explosion takes place can have a decisive influence on the explosion development, and hence on the maximum explosion pressures produced, it is essential that CFD models for gas explosion simulation represent this geometry properly. Geometries can be either defined manually or imported from CAD systems. In FLACS simulations a porosity-distributed flow-resistance concept is used for automatic geometry representation and contains no subjective estimates of blockage, which is often encountered in other CFD simulators.

2.4.5.7.5 Ventilation and Dispersion Simulations

Transient leaks of combustible gas, sonic as well as low momentum, in combination with natural and forced ventilation, can be simulated by means of the FLACS code. Liquid leaks that are converted into equivalent gas leaks through evaporation can also be simulated. The numeric representation of the geometry of the explosion environment includes even minute details, which are important for ventilation calculations, and can be very important for gas dispersion calculations.

2.4.5.7.6 Gas Explosion Simulations

The size range of systems that may be simulated span from less than 1 L to thousands of m^3. The gas cloud can be either predefined or calculated from a dispersion simulation preceding the simulation of the explosion. The gas clouds that can be handled by FLACS comprise any combination of methane, ethane, propane, butane, propylene, ethylene, acetylene, hydrogen, CO, and H_2S, mixed with either air or oxygen-enriched air. Initial pressure, temperature, and turbulence can be varied. Systematic simulation studies can be performed to quantify the effects on the maximum explosion pressure of relocation and rearrangement of process equipment, altering the degree of confinement of the area studied, use of various types of relief panels, use of inert gas dilution (N_2 or CO_2), and water deluge, etc.

2.4.5.7.7 Far-Field Blast Calculations

Using a multiblock concept, shock/pressure waves from the explosion that propagates into the far field and interacts with objects there, can also be simulated. (A discussion of blast waves from accidental explosions and their destructive effects is given in Chapter 9.)

2.4.5.8 Automatic Explosion Suppression

2.4.5.8.1 General Concept

CEN (2002b) drafted a standard for explosion suppression systems. The draft does not, however, give any quantitative guidelines for design of such systems. The first patent for a fast automatic fire-suppression system was allotted to a German company as early as 1912. The Second World War accelerated the development. The British Royal Air Force found that 80% of the total losses of aircraft in combat were due to fire. Based on this evidence, a military requirement was issued specifying a lightweight high-efficiency automatic fire-extinguishing system for protecting aircraft engines and their fuel systems. A similar situation arose in Germany. As a result, new, fast-acting fire extinguishers were developed based on three main principles:

- Extinguishing agent permanently pressurized
- Large-diameter discharge orifice
- Very fast automatic opening of valve for immediate release of extinguishing agent by means of an explosive charge

These principles, combined with a fast-response optical or pressure-rise flame-detection system, form the basis of today's automatic explosion suppression systems, as illustrated in Fig. 2.77.

The suppressor contains a suitable extinguishing agent (suppressant) and a driving gas, normally nitrogen at 60—120 bar. If ignition of the explosive atmosphere should occur, the onset of pressure rise in the vessel due to the growing flame is detected automatically by the pressure sensor, which produces an electric signal that triggers an explosive charge opening the suppressor valve almost instantaneously. A special nozzle design ensures that the suppressant is distributed as evenly throughout the vessel volume as possible. In principle, the pressure sensor can be made sensitive enough to detect even a very small initial flame. However, if the minimum pressure rise for triggering the opening of the suppressor valve is chosen too small, minor pressure variations that may occur due to normal plant operation,

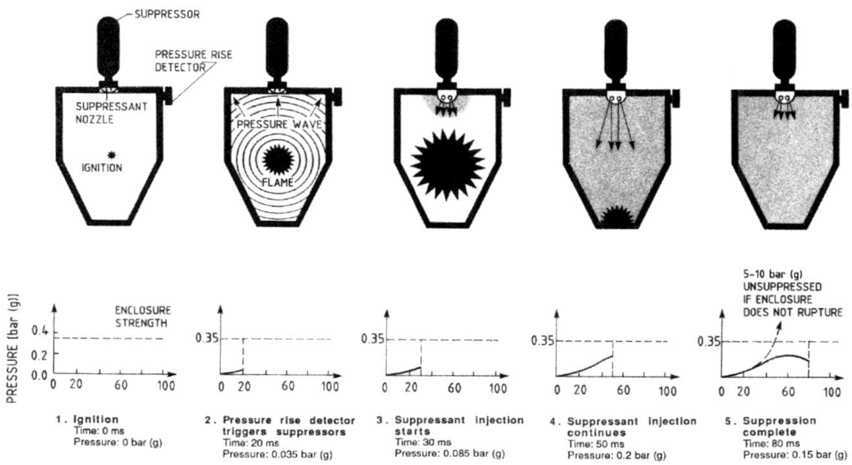

Figure 2.77 Illustration of system for automatic suppression of accidental gas, mist/spray, and dust explosions.
From Eckhoff, R.K. (2003): Dust Explosions in the Process Industries. 3rd edition, Gulf Professional Publishing/Elsevier. (1st edition published in 1991, 2nd edition in 1997), ISBN 0-7506-7602-7.

can cause false activation of the suppression system. The use of two pressure detectors oriented at 90 degrees to each other can make it easier to discriminate between pressure rise due to explosions and other disturbances. Also, natural pressure variations can be filtered out electronically if the rate of rise of such normal variations is significantly lower than the initial rate of rise of the explosion pressure.

In the case of gas and vapor explosions, ultraviolet (UV) or infrared optical flame sensors can be used instead of pressure sensors for detecting the initial flame. A problem with the ultraviolet detectors is that they do not discriminate between hydrocarbon flames and some other light sources, for example welding arcs. Therefore, false alarms may become a problem. As a result, ultraviolet detection systems are often switched off during maintenance and repair work requiring welding, despite the fact that the risk of gas leaks may be at a maximum during just such periods of irregular operation. In some cases this problem has been solved by using a combined system of UV radiation and gas detectors, requiring that both types of detectors trigger before the suppressant is released. Then radiation from welding will be detected, but without this leading to suppressant injection unless a significant concentration of hydrocarbon gas is detected simultaneously by the independent gas detectors.

As mentioned in Eckhoff (1983), the infrared pyrometer technique offers an alternative solution to this problem. The basic principle is to measure the relative intensities of the radiation emitted by the actual source at two or more discrete wavelengths. From this information, using an integrated computer, the Planck radiation distribution curve that has the same intensity ratios for the same wavelengths can be determined. Any such set of ratios can be associated with a specific light-source temperature, and the detector system is thus able to indicate whether the actual temperature indicated is typical of say a hydrocarbon flame, or whether it is significantly higher, eg, on the order of welding arc temperatures.

2.4.5.8.2 Basic Approach to Design of Explosion Suppression Systems

Moore (1981) presented a basic approach to design of industrial suppression systems. He introduced the concept of critical mass M_t of suppressant that is just sufficient for suppressing the flame when being evenly distributed throughout the flame volume. He assumed that there exists a critical minimum mass concentration of any given suppressant for suppressing a flame of a given explosive mixture, and that a suppressant cloud of this concentration or higher must occupy at least the flame volume for successful suppression. It then follows that the critical mass M_t increases with time because the flame volume increases with time. A similar line of thought was applied to the mass of suppressant actually delivered at any time after onset of flame development. Successful suppression would result if

$$M_{t,\text{delivered}} > M_{t,\text{required}} \qquad [2.12]$$

Eckhoff (2003) gives a somewhat more detailed discussion of Moore's model. It is customary to distinguish between three main overall suppression strategies:

- *Advance inerting*: Detect initial explosion, identify its location, activate appropriate suppressors, and establish suppressant barriers to prevent explosion spread to other process units
- *Local suppression*: Detect initial explosion, identify its location, activate appropriate suppressors for ensuring no flame propagation beyond explosion kernel
- *Total Suppression*: Detect explosion, and deluge entire system with suppressant to ensure that explosion is totally suppressed.

The design of any particular industrial suppression system depends on the suppression strategy chosen, the type of suppressant, the chemical, and explosion properties of the explosive mixture, the nature of the process/enclosure to be protected, the volume and shape of the enclosure, and on other actions taken to prevent or mitigate accidental explosions in the plant.

In large process volumes it can be difficult to obtain adequate distribution of the injected suppressant if the gas leak to the atmosphere is in the form of a strong gas jet. Depending on the velocity and size of the leak jet the injected suppressant may or may not be able to penetrate the jet region.

2.4.5.8.3 Sophisticated Optical Detection Systems May Be Required in Large Process Volumes

If the volume of the enclosure or area to be covered by the suppression system is very large, eg, of the size of typical offshore oil and gas platform modules, very large flames can develop before a detectable rise in pressure has been generated. In such cases, as pointed out by Eckhoff (1983), pressure rise detection of the onset of explosion is inadequate, and optical flame detectors, either UV or infrared (see Section 2.4.5.8.1) have to be used. Ultraviolet detectors are able to detect a fireball of 1-m diameter at a distance of 10–20 m within about 15 ms, during which an initial flame will only have propagated a distance of a few cm. In principle, therefore, such detectors should be able to detect any initial fireball in modules typical of offshore platforms at a sufficiently early stage to ensure effective local suppression. However, an important condition is that the light path between the initial fireball, and the UV detector is not obstructed by walls or process equipment, etc. Furthermore, the suppressant containers have to be located close to the origin of the explosion, and be able to throw the suppressant in the appropriate direction. It is clear that for such applications a quite advanced detection system, supported by computers, will be required. First, the precise location of the local flame must be identified, and then the adequate quantity of suppressant must be released locally in the area where the flame has been detected.

To cover large process volumes adequately, a number of such detectors may have to be located in an optimal matrix covering the entire volume, with the aim to localize the initial flame with the required accuracy. This requirement is rather severe if it is necessary to identify the precise location of flames as small as 1 m^3 in volumes of several thousand m^3 or more.

It should finally be mentioned that violent injection of suppressant into a fairly quiescent explosive gas cloud will inevitably increase the turbulence in the cloud, and hence also its burning rate, which is, of course, undesirable.

CHAPTER THREE

Boiling Liquid Expanding Vapor Explosions (BLEVEs)

3.1 DEFINITIONS AND CHARACTERISTICS OF BLEVEs

This chapter is essentially identical to the paper by Eckhoff (2014). According to the extensive review by Abbasi and Abbasi (2007), the Boiling Liquid Expanding Vapor Explosion (BLEVE) was probably adopted for the first time in 1957 by researchers at Factory Mutual at Factory Mutual Research Corp., United States. They had analyzed the failure of a vessel that contained an overheated mixture of formalin and phenol and suggested that the container suffered a BLEVE. Later, Walls (1978, 1979) defined a BLEVE as "a failure of a major liquid-filled container into two or more pieces at a moment when the temperature of the contained liquid is well above its boiling point at normal atmospheric pressure."

Birk and Cunningham (1994) defined a BLEVE as "the explosive release of expanding vapour and boiling liquid when a container holding a pressure-liquefied gas fails catastrophically." Catastrophic failure was defined as "the sudden opening of a tank/container to release its contents nearly instantaneously."

The definition of a BLEVE presented by CCPS (1999) (Center for Chemical Process Safety) is "a sudden release of a large mass of pressurized superheated liquid to the atmosphere." The sudden release is due to a sudden containment failure caused by fire, a missile, corrosion, a manufacturing defect, internal overheating etc.

Berg et al. (2006) discussed the discrepancies between various BLEVE definitions. Some confusion results from the fact that catastrophic failure of vessels containing liquefied gases/vapors is often a result of a fire that has heated up the vessel for some time. If the liquid is flammable, a sudden rupture of the vessel wall and an explosive ejection of the superheated liquid into the surroundings will in most cases immediately set the ejected liquid on fire, and a large fireball will result. Consequently, the entire chain of events, including the formation of the fireball, is sometimes defined as a BLEVE. This is the philosophy behind the illustration given in Fig. 3.1.

Explosion Hazards in the Process Industries
ISBN: 9780128032732
http://dx.doi.org/10.1016/B978-0-12-803273-2.00003-7

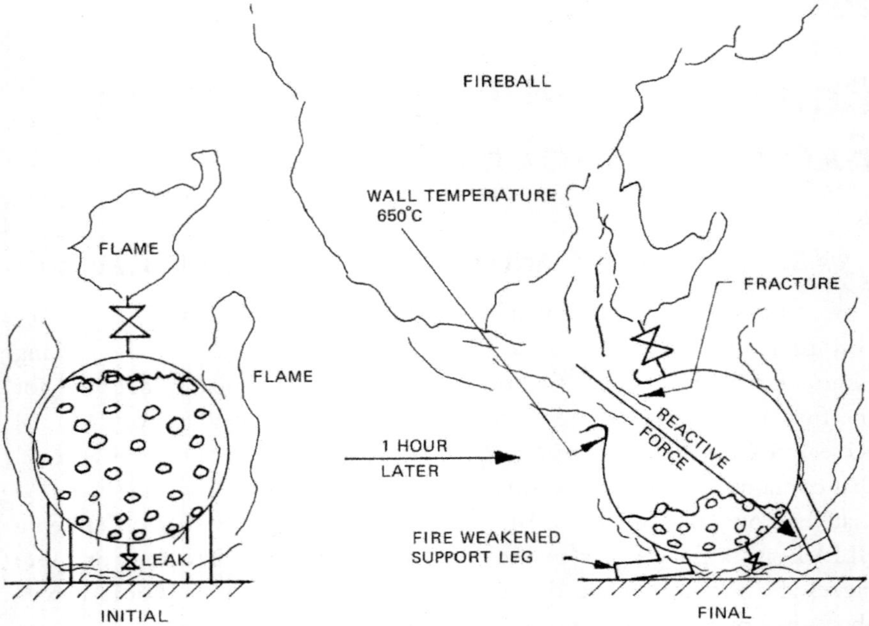

Figure 3.1 Illustration of a BLEVE, including the major secondary fireball following ignition of the ejected vapor/liquid.
From Marshall, V.C., 1987. Major Chemical Hazards, Elis Horwood Series in Chemical Engineering. Ellis Horwood Ltd/Halsted Press, A Division of John Wiley & Sons, Chichester/New York, ISBN:0-470-20813-9.

Berg et al., however, suggest that the definition of a BLEVE should not include a possible subsequent generation of a fireball, but comprise only the explosive rupture of the pressure vessel and the subsequent flash evaporation of its superheated liquid content. Some definitions of BLEVEs require that the liquid initially confined in the vessel has to be heated above its superheat limit temperature (SLT) at atmospheric pressure (see Section 3.4.2 in the following) before vessel rupture takes place. Other, less strict, definitions only require that the liquid temperature at vessel rupture significantly exceeds the liquid boiling point at atmospheric pressure. In any case, the higher the temperature of the liquid is at the moment of tank rupture, the more severe will be the accident.

3.2 CHAIN OF EVENTS LEADING TO BLEVEs AND THEIR CONSEQUENCES

The following summary to a significant extent follows the same main lines as in the very extensive review by Abbasi and Abbasi (2007).

3.2.1 A Vessel Containing Pressurized Liquefied Gas Is Accidentally Exposed to Heat (Fire)

A pressure vessel can fail even at normal ambient temperature by missile impact, fatigue, corrosion, or an accidental excessive rise of the process pressure. However, in the present context the main focus will be on excessive pressurization due to accidental heating of the vessel from the outside. It is then important to keep in mind that a pressure vessel is designed to withstand the relief valve set pressure only at the design temperature conditions, which may be just ambient atmospheric temperature.

Pressurized liquefied gas (PLG) is a substance that is in the gaseous state at normal ambient temperature and pressure, but which has been liquefied by compression and is kept as a liquid at normal ambient temperature in a pressure vessel. Propane and butane are common examples. In the case of accidental heating of a vessel containing PLG, eg, by the heat from a fire, as illustrated in Fig. 3.1, the already elevated vapor pressure inside the vessel will rise further. When the pressure reaches the set pressure of the pressure relief valve of the vessel, vapor from the liquid in the vessel is expelled into the open atmosphere, and the liquid level in the vessel will drop as further liquid continues to evaporate. As long as most of the vessel volume is occupied by the liquid, most of the vessel wall will be effectively cooled by the liquid. However, as the liquid continues to evaporate and escape via the pressure relief valve, the proportion of the vessel wall that is effectively cooled decreases. Eventually the portion of the vessel wall that is not in contact with the liquid weakens due to the temperature rise caused by the external fire load, and may fail. Schulz-Forberg et al. (1984) investigated the failure mechanisms of propane storage tanks exposed to thermal stress, including fires.

3.2.2 Sudden Depressurization and Evaporation of the Hot Liquid

When the vessel fails, the liquid pressure suddenly drops to atmospheric pressure. Because the liquid temperature is well above the atmospheric pressure boiling point, the liquid will evaporate abruptly and violently and expand to a cloud of a volume between several hundred and 2000 times that of the original liquid volume. If the liquid temperature is significantly higher than its SLT (see Section 3.4 in the following) at atmospheric pressure, most of the liquid volume can undergo extremely fast and homogeneous nucleation and evaporation. In that case, the vaporization/expansion process will be extremely fast and violent, and typically occur within 1 ms after the abrupt vessel depressurization caused by the sudden vessel failure.

The nucleation and vaporization processes require heat, which is taken from the liquid itself. Therefore, the liquid temperature will drop as the volumetric evaporation progresses. As soon as the liquid temperature drops below the SLT, the very rapid volumetric nucleation and vaporization terminates and normal boiling at the hot surfaces becomes the dominating process of vapor production. Hence, the more the liquid temperature at vessel failure exceeds the SLT of the liquid, the greater will be the fraction of the liquid that can flash almost instantaneously to vapor.

It is important to emphasize that a BLEVE can occur even when the temperature of the suddenly depressurized liquid is below the SLT of the liquid at atmospheric pressure, but still significantly higher than the atmospheric boiling point. However, clearly the intensity of the evaporation will then be lower than if the initial liquid temperature exceeds the SLT at atmospheric pressure.

3.2.3 Blast Wave Emission From BLEVEs

Berg et al. (2006) discussed various models for calculating the strength of blast waves emitted from BLEVEs.

Acoustic modeling of blast wave emission requires full knowledge of the source overpressure as a function of time. This, in turn, requires full knowledge of the physical characteristics of the liquid release as a function of time, from the onset of the crack development in the pressurized vessel. Crack development is a complex process that depends strongly on the design and structure of the actual vessel. Therefore, simplifying assumptions are required. One such assumption is that the liquid release rate grows linearly with time. Using this assumption, Berg et al. considered as an example a BLEVE from a 50 m^3 propane tank in the open. They then calculated the maximum overpressure at a distance 10 m away from the tank, as a function of total release time Δt. With Δt equal to 0.05, 0.2, and 1.0 s, respectively, the corresponding maximum overpressures Δp were 1.06, 0.07, and 0.0003 bar.

Gas dynamic modeling gives results that are more accurate provided sufficient input information is available for defining the actual scenario. In addition, good experimental data are required for adequate validation of the models. As an example, Berg et al. discussed the most valuable and controlled BLEVE experiments by Giesbrecht et al. (1981), who used vessels of volumes ranging from 0.226 to 1000 L, containing liquefied propylene. The vessels, located at ground level, were heated to 50–80°C ie, to 40-bars overpressure and deliberately ruptured more or less instantaneously by means of a small high-explosive charge. Fig. 3.2 shows the experimentally determined correlation between the maximum overpressure in the blast wave emitted into the surroundings from a BLEVE and the scaled distance from the BLEVE, together with the correlation obtained by Berg et al. (2006) by numerical simulation.

In numerical modeling of various aspects of BLEVEs, the report by Xie (2007) may be a useful source of information.

3.2.4 Shattering of Secondary Vessels due to Missiles From the Shattering of the First Vessel

The shattering of the first vessel in a series of BLEVEs is normally due to heating by fire exposure. However, missiles from shattering vessels present a much greater danger of "domino" effects than fireballs or blast waves. The bursting of the first vessel causes vessel fragments to be ejected into the surroundings. These missiles often damage other vessels that store PLG, causing these vessels to undergo BLEVEs as well. This "domino" effect actually took place in the Mexico City disaster in 1984 (see Section 3.6.1.2), causing the largest number of lives lost ever recorded in one single

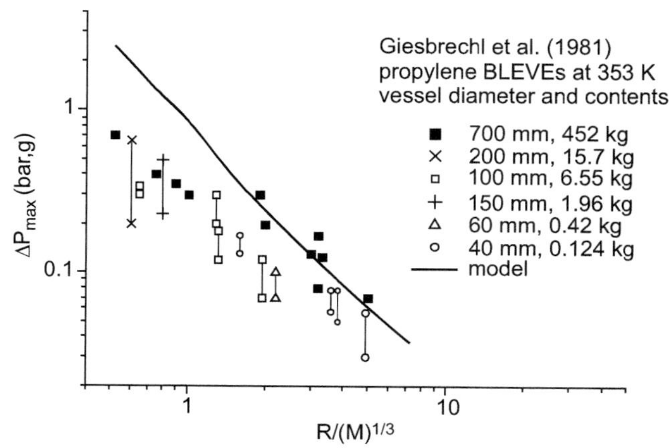

Figure 3.2 Experimental and computed blast overpressures originating from exploding vessels of various sizes containing propylene.
From Berg, A.C., van den Voort, M.M., van der Weerheijm, J., Versloot, N.H.A., 2006. BLEVE blast by expansion-controlled evaporation. Process Safety Progress, 25, 44–51.

explosion/fire accident in the process industry. Large parts of vessels can fly as projectiles over long distances. In one accident (Port Newark), a large part of a spherical vessel traveled more than 800 m before hitting and destroying a petrol bunker.

According to Abbasi and Abbasi (2007), the likely consequence of a BLEVE series in terms of the duration and propagation of missiles depends on the following factors:

- likely number and mass distribution of missiles
- velocity and range distributions of missiles
- likely directions of propagation of missiles
- penetrability and destructive potential of missiles

3.2.5 Generation and Development of Fireballs

If the liquid involved in the BLEVE is neither combustible nor toxic, as is the case with water, the pressure wave and the missiles from the vessel shattering are the only hazardous effects of the explosion.

However, if the liquid is flammable, as is often the case, the mixture of liquid/gas released by the explosion catches fire generating a fireball. Analyses of actual BLEVEs have shown that more than two-thirds involved flammable chemicals.

The shape and size of the fireball, and the heat load produced by it, depend on several factors. At the outset, the expelled mass of fuel can burn only at its periphery because there is not sufficient air inside the mass, ie, the fuel/air mixture ratio is above the upper flammability limit. Furthermore, not all the fuel initially contained in the tank may become involved in the primary fire. Some of the liquid may leak to the surroundings before the explosion via a crack or other opening in the vessel. Alternatively, some of the liquid may be entrained in the wake formed by flying fragments. In the Mexico City disaster in 1984 (see Section 3.6.1.2), fragments of shattered vessels carried with them parts of the flammable liquid, which in themselves caused the fires to spread.

As the main fireball from a BLEVE grows, the turbulence of the flame entrains air into the fireball. At the same time thermal radiation from the flame vaporizes liquid droplets and heats the combustible cloud. Because of all these processes, the entire cloud increases in volume, rises, and attains an approximately spherical shape. Depending on the amount of combustible material involved, such fireballs can become very large and cause very intense thermal radiation. The size, lifetime, and radiation intensity of a fireball may also depend on the initial temperature of the liquid originally contained in the vessel. Whether the loss of confinement of the liquid occurred while the pressure inside the vessel was still rising is a further factor. Although BLEVE fireballs are spherical when fully developed, they acquire a mushroom-like shape during lift-off.

Fireballs resulting from two-step BLEVEs (see definition later in this section) may be approximately ellipsoidal in shape.

According to Lees (1996), it may happen that BLEVEs involving combustible materials are not ignited at the release point, but at a later stage. In such cases the resulting event may either be a major flash fire or a vapor cloud explosion.

3.2.6 Pool Fires

Some of the liquid expelled from the shattered vessel may be splashed and hit the ground nearby forming short-lived pools before vaporizing. In the case of flammable liquids this may give rise to pool fires.

3.2.7 Emission of Toxic Gases/vapors

BLEVEs giving rise to emission of toxic substances have occurred with a number of toxic compounds, such as ammonia, chlorine, chlorobutadiene, and phosgene. Of the one-third of recorded BLEVEs that did not involve flammable liquids, the majority involved toxic gases. Of these, chlorine (14%), ammonia (10%), and phosgene (2%) account for 76% of all BLEVEs involving nonflammables. With such chemicals, the fatalities were caused by the toxic material expelled with the blast wave, and also by missiles. According to Abbasi and Abbasi (2007), chlorine accounted for most of the fatalities in the major BLEVEs with noncombustible but toxic materials during 1926—1981, followed by ammonia.

3.2.8 Emission of "Cold" Nontoxic Gases

BLEVEs have occurred with tanks containing carbon dioxide and water, even in the absence of a fire. Abbasi and Abbasi (2007) suggested that many boiler explosions, which are far more common than explosions involving other chemicals, are in fact BLEVEs, without being commonly acknowledged and classified as such. Indeed, if all boiler explosions that occur with superheated water were to be included in the BLEVE statistics, it might well be that the most frequently involved liquid in BLEVEs turns out to be water.

3.2.9 Potential Consequences of BLEVEs

The energy of explosion, or "burst energy," determines the severity of the blast wave generated by the BLEVE and the velocity (hence the range and the penetration) of the shattered vessel fragments. The mode of release of the PLG in the shattered vessel determines the size, duration, and heat flux of the fireball if the PLG is flammable or it determines the pattern of atmospheric dispersion if the PLG is toxic. There is a much greater degree of uncertainty associated with predicting the consequences of BLEVEs than with predicting consequences of gas/vapor cloud explosions. This is because of the central roles played by superheated liquids and pressurized gases.

3.3 INDUCTION TIME PRECEDING BLEVEs: LONG-DURATION BLEVEs

Experience from accident histories has shown that some vessels exploded within a few minutes of fire engulfment or missile hit, whereas others did not explode until after several hours. In some cases, the induction time from fire engulfment or missile hit to vessel explosion was up to 24 h. In the Feyzin accident (see Section 3.6), the induction time was about 90 min. In the Mexico City catastrophe (see Section 3.6), the induction times varied between 3 and 30 min.

Clearly, having a good estimate of induction times in potential BLEVEs is important for optimization of damage control. Abbasi and Abbasi (2007) discussed a number of investigations aimed at quantifying induction times. For example, Blything and Reeves (1988) analyzed BLEVEs from horizontal cylinders filled to 75% of their capacity with liquid butane. The fire load was either by partial fire engulfment or by jet flame impingement. From their analysis they concluded that a BLEVE would occur with induction times between 4 and 48 min. In another investigation considering a full 2000 tons PLG storage sphere being exposed to flames, Selway (1988) found that 7−11 min, 25−38 min, and 6−7 min were likely induction times in the cases of total fire engulfment, partial fire engulfment, and jet flame impingement, respectively. According to Selway, these times would be shorter if the vessel is not completely filled with liquid.

In both fire-induced and missile collision-induced initial vessel damage, the first tear or crack may or may not propagate straight away to a sufficient size to instantaneously cause a BLEVE. In some cases, an initial tear or crack that is too small to directly initiate a BLEVE, may restart propagation again after some time and increase to the size required to cause a BLEVE. Typical crack propagation speeds are then >200 m/s. About 20% of the BLEVEs observed by Birk and Cunningham (1994) were of this long-duration type. This two-stage process is illustrated diagrammatically in Fig. 3.3. Such events may occur with tanks that are almost strong enough to resist a total loss of containment (TLOC), but are "pushed over the edge" by the violent boiling

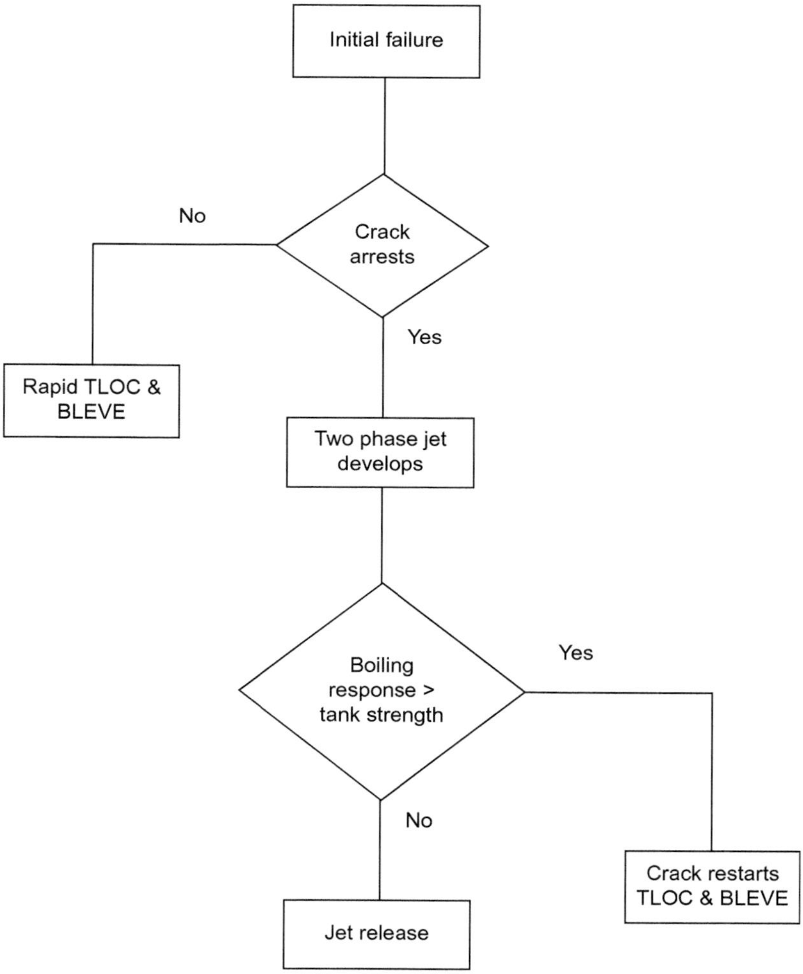

Figure 3.3 Chain of events in a two-step BLEVE.
From Abbasi, T., Abbasi, S.A., 2007. The boiling liquid expanding vapor explosion (BLEVE): mechanism, consequence assessment, management. Journal of Hazardous Materials, 141, 489–519. based on Birk, A.M., Cunningham, M.H., 1994. The boiling liquid expanding vapor explosion. Journal of Loss Prevention in the Process Industries, 7, 474–480 and Birk, A.M., van der Steen, J.D.J., 2002. On the transition from no-BLEVE to BLEVE for a 1.8 m³ propane tank. The American Society of Mechanical Engineers Pressure Vessels and Piping Division PVP, 446, 143–152.

of the depressurization-induced superheated liquid. The occurrence of long-duration BLEVEs are strongly related to the setting of the pressure relief valve (PRV), because this ultimately determines the liquid temperature.

Birk and Cunningham (1994, 1996) conducted a series of experiments with 400 L propane tanks, subjecting the tanks to fire engulfment and studying the pattern and the duration of vessel failure. The tanks were equipped with PRVs. Table 3.1 summarizes some results obtained by Birk and Cunningham (1994).

In general, it is not possible to forecast how much time a vessel exposed to a fire may take before undergoing a BLEVE. This can vary from a few seconds to several hours. This fact makes it very dangerous for firefighters to operate close to a fire-engulfed vessel containing a PLG/vapor. There have been many instances when a vessel has exploded even after the

Table 3.1 Summary of experimental results obtained by Birk and Cunningham (1994)

Test	Fire conditions	No. of vents	Outcome	Comments
Vent to empty	Pool only Pool and torch	5 2	Entire tank contents lost through the PRV	Fire conditions were not severe enough to initiate a local failure.
Partial failure	Pool and torch Torch only	7 4	Partial failure occurred in the tank causing a two-phase liquid/vapour mixture to be released from the tank. In two cases the jet was angled such that it propelled the tank up to 30 m	Local thermal weakening was such that pressure stresses in the wall exceeded local tank wall strength, causing a failure (crack). The tank wall was strong enough to arrest the crack
TLOC and BLEVE	Pool and torch Torch only	9 2	The tank failed catastrophically releasing the entire contents as expanding vapour, boiling liquid and dispersed droplets. Failures resulted in blasts and fireballs	Not all BLEVEs were the same The characteristics of BLEVEs varied with tank properties and lading conditions at the time of failure

PRVs have been venting the vessel for several minutes. Of all the harmful effects of BLEVEs, the one with the greatest impact is rocketing fragments. Often deaths of people quite a distance away from the site of the BLEVE and secondary accidents in other process units, are caused by such missiles.

3.4 THE ROLE OF SUPERHEAT LIMIT TEMPERATURE OF THE LIQUID IN BLEVE DEVELOPMENT

3.4.1 Boiling and Boiling Points of Liquids

In the context of BLEVEs it is essential to distinguish between the two different boiling points of a liquid, viz. its normal boiling point (BP) and its SLT. The SLT of a liquid is quite different from its BP. The basic definition of BP, as experienced in normal cooking, is the temperature at which the liquid's vapor pressure equals the ambient pressure above the liquid surface, plus the liquid head above the point in the liquid being considered. Therefore, a given liquid in an evacuated environment has a lower boiling point than when the same liquid is at atmospheric pressure. If the liquid is in an environment of pressure above atmospheric, it will have a higher boiling point than at atmospheric pressure. For vapor bubbles to form and expand in normal cooking, the presence of "active points" at the bottom and/or wall of the cooking device or on impurities in the liquid is essential. Such points can be sharp edges, scratches or pores in the surface, or small solid particles adhering to the surface. The function of an "active point" is to allow a vapor bubble to form without being fully embraced by a liquid/vapor interface.

A pure liquid, with no "active points," can be heated to a temperature considerably higher than its boiling point at the prevailing pressure before significant boiling starts. This temperature is called the "superheat limit temperature" (SLT), which is synonymous with the alternative term "homogeneous nucleation temperature." In the absence of "active points," fully enclosed vapor bubbles must become formed within the pure liquid. In this case the surface tension of the liquid at the liquid/bubble interface will suppress the transfer of liquid molecules into vapor phase molecules. Therefore, a pure enclosed liquid that is free from "active points" will have to get heated to above its normal boiling point before the vapor pressure becomes sufficiently high for onset of boiling. The excess temperature and hence liquid vapor pressure required to overcome the surface tension of a bubble wall in a liquid is inversely proportional to the bubble diameter. In the absence of "active points," therefore, a higher pressure is required to initiate formation of the initial micro-bubble than to continue its growth. Consequently, once the initial small bubbles are formed, they expand rapidly and violently at an accelerating pace.

This is in a way similar to the maximum initial resistance experienced just when one starts to blow up a rubber balloon. The pressure of the air blown into the balloon must not only exceed the ambient pressure, but also the additional overpressure required for exceeding the maximum initial resistance of the balloon wall against being stretched out.

3.4.2 The Superheat Limit Temperature in Relation to BLEVEs

When a vessel containing a PLG gets excessively heated, eg, by a fire, the temperature and pressure inside the vessel will increase accordingly. If the temperature of the pressurized liquid in the vessel ultimately exceeds the SLT at ambient atmospheric pressure, and the vessel suddenly ruptures, as it will do in a BLEVE, the liquid momentarily becomes excessively superheated. In such a situation a fraction of the liquid will evaporate practically instantaneously, carrying with it liquid droplets. This occurs extremely fast, within a few milliseconds. The increase in volume caused by this instantaneous vaporization is enormous, and together with the rapid expansion of the compressed vapor existing in the vessel even before the rupture, this generates a strong pressure wave with a significant damage potential.

Clearly, if the liquid is combustible and its vapor and spray gets ignited immediately after a BLEVE has taken place, the accident potential of the entire event will escalate substantially.

From what has been said previously, it is clear that the difference between the actual temperature of the superheated liquid at the moment of vessel burst, and the SLT of the liquid at atmospheric pressure is of prime importance with regard to the extent to which explosive evaporation will take place. The SLT provides a basis for developing means of predicting both likelihood of possible BLEVEs and the severity of their consequences. A logical safety strategy that makes direct use of the SLT is to operate PRVs of PLG vessels in such a manner that the vessels are depressurized before the vessel contents reaches the SLT.

3.4.3 Experimental Determination of Superheat Limit Temperature

The droplet explosion method is the most common and most reliable experimental means of determining SLT. The apparatus is illustrated in Fig. 3.4.

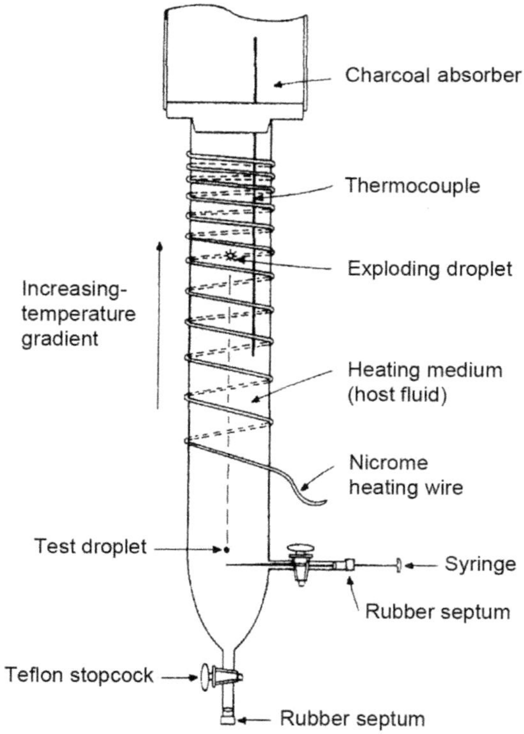

Figure 3.4 Droplet explosion apparatus for determining superheat limit temperatures (SLTs) of liquids.
From Patrick-Yeboah, J.R., Reid, R.C., 1981. Superheat-limit temperatures of polar liquids. Industrial & Engineering Chemistry Fundamentals, 20, 315–317.

A small droplet of the test liquid is introduced at the bottom of a vertical tube that is filled with a host liquid of a higher density than the liquid to be tested. This will ensure that the droplet will rise in the host liquid. In addition, the host liquid must be immiscible with the test liquid and should have a BP well above the SLT of the liquid to be tested.

By means of a specially designed heating system, a vertical temperature gradient is established in the host liquid column, the lowest temperature being at the column bottom and the highest at the top. Therefore, as the test droplet rises through the column its temperature will increase gradually. This continues even beyond the BP, because the liquid in the droplet is not in touch with any active points (eg, solid surface) and no boiling will take place. However, as soon as the temperature reaches the SLT the droplet will evaporate explosively. This temperature is known from the position of the droplets at the moment of explosive evaporation. The temperature in the column as a function of vertical position is monitored by means of a thermocouple that can be positioned at any desired point in the column.

One problem inherent in the droplet explosion technique is that the initial unstable bubble nuclei being formed when the temperature approaches SLT have sizes on the molecular scale. It is likely, therefore, that even in the most careful experiments, microscopic temperature gradients in the liquid could initiate formation of bubbles at a global liquid temperature significantly below the superheat limit. Another concern is whether the global droplet temperature is actually equal to that of the surrounding liquid at any time during the rise of the droplet. The temperature difference will depend on the droplet size, the droplet rise velocity, the steepness of the vertical temperature gradient in the host liquid column, and the heat conductivity of the droplet liquid. Nevertheless, the droplet explosion technique is the best method currently available for experimental determination of SLT.

Another experimental method for determining SLT is based on immersion of very thin and rapidly electrically heated wires in a very clean test liquid. The temperature of the wire is recorded continuously as the wire is heated up. As soon as the wire temperature has reached the SLT of the liquid, the liquid layer adjacent to the wire will vaporize explosively. This approach requires that the dynamics of the thermocouple/temperature measuring device must be faster than the dynamics of the heating of the wire. The nature of the wire surface also influences nucleation. This method also gives fairly reproducible results, but it is considered less accurate than the rising-bubble method.

In Table 3.2, from Salla et al. (2006), the boiling points, the SLTs, and the critical temperatures are given for a selection of substances.

Table 3.3 is based on data from Wakeshima and Takata (1958). The theoretical SLT values were calculated using the classical Döring theory.

3.4.4 The Role of Superheat Limit Temperature in BLEVE Theories

3.4.4.1 Overview

As pointed out by Abbasi and Abbasi (2007), several complex questions related to BLEVEs call for theories that can predict both the development and consequences of such events in given practical situations.

In most theories on the development of BLEVEs, the SLT plays a central role. Clearly, this is because the explosive volumetric boiling of liquids at temperatures above the SLT at atmospheric pressure is so much more violent than boiling below SLT.

Table 3.2 Comparison of experimental boiling points and critical temperatures, and various calculated SLTs for some substances, based on a table in Salla et al. (2006)

Substance	Boiling point (K)	SLT (K)	Crit. Temp. (K)
Methane	112	160−177	191
Ethane	185	259−279	305
Ethylene	169	239−261	282
Propane	231	313−332	370
Propylene	226	309−328	365
n-Butane	273	360−382	425
n-Pentane	309	379−422	470
n-Hexane	342	407−456	508
n-Heptane	372	430−484	540
n-Octane	399	467−510	569
Cyclohexane	354	429−497	554
Benzene	353	453−504	563
Toluene	384	501−531	592
Water	373	547−607	647
Ammonia	240	343−377	406
Chlorine	239	353−378	417
Carbon dioxide	195	257−294	304
Hydrogen	20	29−33	33
Nitrogen	77	107−119	126
Oxygen	90	131−145	155

Table 3.3 Experimental boiling points and experimental and theoretical SLT values based on a table in Wakeshima and Takata (1958)

Substance	Boiling point (K)	$SLT_{eksper.}$ (K)	$SLT_{theor.}$ (K)
n-Pentane	309	383	383
n-Hexane	342	386	385
n-Heptane	371	386	387
Cyclo-hexane	354	408	411

According to Nesis (1966), the first major contributions to a basic, theoretical understanding of superheating was provided by the pioneer J.W. Gibbs in the last part of the 19th century. Using thermodynamic concepts he explained why, in the absence of the "active points" required for boiling at the BP, it is possible to heat a liquid up to a temperature above its BP before boiling starts. The reason is that the liquid has to cross over a certain thermodynamic potential energy barrier ("activation energy") before it can transit into a stable vapor state. When a microscopic vapor bubble is first produced inside a liquid at a temperature above its BP, the thermodynamic potential energy of the system will change with time as a result of two opposing simultaneous processes. The potential energy will increase because surface energy is accumulated at the interface between the liquid and the vapor bubble. However, when the unstable molecules in the liquid state transit to the more stable vapor state, the potential energy will decrease. For very small, microscopic bubbles the first effect is dominating, whereas the latter effect takes over as the bubbles grow. Therefore, there is a given critical microscopic bubble size at which the thermodynamic potential of the system reaches a maximum. As soon as the bubble has grown beyond this critical size, it will grow further spontaneously. The minimum liquid temperature that allows this critical bubble size to be reached is per definition the SLT of the liquid (cfr. The rubber balloon analogy at the end of Section 3.4.1). Just as with the BP, the SLT also varies with the pressure of the environment in which the liquid is kept. However, in the context of BLEVEs, the SLT for ambient atmospheric pressure is of prime importance and significance. Hence, when the acronym SLT is used in the following, it generally means SLT at ambient atmospheric pressure.

3.4.4.2 SLT Is Not an Absolute Lower Temperature Limit for BLEVEs to Occur

Comprehensive research conducted by Birk and coworkers has revealed that the SLT is not an absolute lower temperature limit for a BLEVE to occur. Depending on the circumstances leading to vessel rupture, BLEVEs can occur even at temperatures significantly below the SLT. It is clear, however, that the blast overpressures generated by such BLEVEs will be considerably lower than those produced by BLEVEs occurring above the SLT.

3.4.4.3 Flammable Liquids Giving Rise to Fireballs

The size, shape, and radiation intensity of the fireballs which are formed when flammable liquids undergo BLEVEs do not appear directly related to the extent to which the liquid is superheated.

3.4.4.4 Theory of Boiling-Liquid-Collapsed-Bubble Explosion

According to Abbasi and Abbasi (2007), the understanding of the BLEVE process has been further refined by careful analysis of some more recent BLEVE accidents, followed up by controlled experiments. One conclusion is that all BLEVEs are in fact the result of a more or less universal two-stage process. The first step is the formation of a limited crack in the vessel wall, giving rise to a moderate initial leak ("leak before break"). In the second step, there will be waves of repeated depressurization and repressurization caused by further crack propagation and evaporation and ejection of fluid, leading to the final major failure. The delay time between crack initiation and catastrophic failure depends on how much of the vessel volume is initially filled with liquid. A typical delay time for 20% filling is 40 s and for 85% filling 1.4 s.

3.5 PROPERTIES AND EFFECTS OF FIREBALLS FROM BLEVEs

Bosch and Weterings (1997) defined a fireball as a fire that burns with sufficient rapidity for the burning mass to raise into the air as a cloud or ball. In all BLEVEs involving combustible liquids there is an almost instantaneous two-phase release of most of the liquid which immediately auto-ignites to form the fireball. Therefore, the fireball is more or less an inevitable consequence whenever a vessel containing a flammable liquid suffers a BLEVE. For this reason, the fireball is often considered an inherent part of any BLEVE. However, about one-fifth of all BLEVEs occur with nonflammable liquids (including fire suppressants like nitrogen, carbon dioxide, and water), and obviously no fireball is generated in such cases.

Although fireballs from BLEVEs are predominantly influenced by momentum forces, fireballs generated when more stagnant vapor clouds are ignited are mostly moved by buoyancy forces.

In some rare cases a vessel containing a flammable PLG may first release a sufficient mass of vapor to generate a vapor cloud that gets ignited before the vessel fails in a BLEVE that generates a much bigger fireball. According to Makhviladze, Roberts, and Yakush (1999), the fireball from a BLEVE releasing 100 tons of flammable liquid hydrocarbon develops about 5×10^{12} J of thermal energy within 10—20 s. A significant part of this energy appears as thermal radiation, which is powerful enough to scorch people, damage property, and trigger secondary fires.

Frame-by-frame analysis of full-scale fireball records conducted by Crawley (1982) show that the fireball passes through three phases, viz. growth, steady burning, and burn-out. The growth phase comprises two time intervals, each of about 1 s. During the first interval, the fireball grows to about half its final diameter, and the fireball boundary is bright yellowish-white, indicating a flame temperature of about 1300°C. In the second time span of the first phase, the fireball attains its maximum volume, but about 10% of the surface is now dark and sooty, whereas the major 90% is white, yellowish-orange or light red, indicating flame temperatures in the range 900–1300°C, with an estimated effective average flame temperature of 1100–1200°C.

In the second steady-burning phase, which lasts for some 10 s, the fireball, which is now roughly spherical, is no longer growing. At the start of this phase it begins to lift off, rises, and attains the characteristic mushroom shape. The estimated average effective flame temperature remains at 1100–1200°C.

In the third burn-out phase, which typically lasts for about 5 s, the fireball size remains constant, but the flame becomes less sooty and more translucent.

According to Abbasi and Abbasi (2007), the following issues must be addressed to enable prediction of the size and duration of a fireball, and the thermal radiation from it:

- mass of flammable substance released in the BLEVE
- mass of flammable substance consumed in the fireball
- fireball development as a function of time
- fireball full size and duration
- radiation heat load generated
- view factor

Bosch and Weterings (1997) presented the following list of items that need to be quantified to enable estimation of the thermal radiation impact of a BLEVE fire-ball:

- amount of flammable material released in BLEVE
- heat generated by the fireball
- fireball radius
- fireball duration
- fireball lift-off height
- radiated heat flux
- net available heat for thermal radiation
- radiation absorption by water vapor between fireball and point of impact
- radiation absorption by carbon dioxide between fireball and point of impact
- atmospheric transmissivity
- distance from center of fireball to actual point of impact
- "view factor" at point of impact

Bosch and Weterings (1997) defined the "view factor" as the ratio between the received and the emitted radiation energy per unit area, or the fraction of the fireball that is "seen" by the target at the point of impact.

The view factor incorporates the orientation of the object relative to the fireball, as well as its distance from the fireball center.

As is the case with predicting the other BLEVE impact parameters, considerable uncertainty is also to be expected in the estimation of the thermal parameters of fireballs. These problems appear right from the initial assessment of the amount of combustible material (the "flashing fraction") that is released instantaneously on vessel failure. Bosch and Weterings (1997) assume that the entire vessel content will flash over and contribute to the fireball, whereas Marshall (1987) and others assume that the fraction of the fuel that participates in the fireball is only about one-third of the total amount released in the BLEVE.

3.6 BLEVE CASE HISTORIES

This section is based on the extensive review by Abbasi and Abbasi (2007).

3.6.1 BLEVEs in Storage Tank Facilities

3.6.1.1 Tank Storage Farm in Feyzin, France, in January 1966

This BLEVE catastrophe is one of the worst accidents involving PLG that has ever occurred. The scene is illustrated in Fig. 3.5.

The present summary is based on the detailed account by Török et al. (2011) and the review by Abbasi and Abbasi (2007). The series of events leading to the catastrophe was as follows:

- Three workers started to drain water from the bottom of tank No. 4 in Fig. 3.5. This was a 1200 m^3 spherical propane storage tank, which was nearly full of liquid propane. There was a system of three draining valves underneath the tank. The first valve was closest to the tank bottom and its outlet pipe was split by a Y-connection leading to two parallel valves further downstream.
- To perform the draining, the valve closest to the tank bottom was first opened, followed by the opening of one of the two lower parallel valves. After some time the appearance of traces of oil in the drained water indicated that the draining was nearly complete. Then the first valve was closed, and subsequently partly reopened to complete the draining. In the meantime, the line had become blocked, presumably by hydrate, and no liquid flow appeared. The first valve was then fully reopened. The choke then suddenly cleared, and the operator and the two other men were splashed with liquid propane. The handle came off

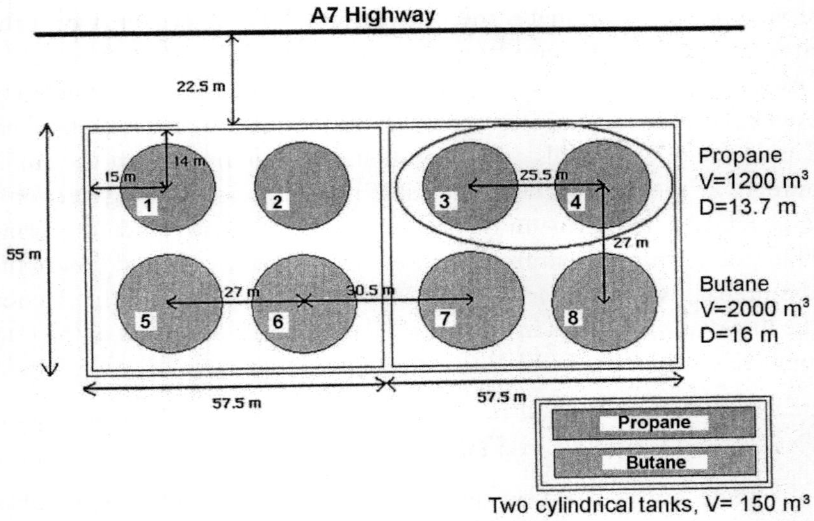

Figure 3.5 Tank farm at Feyzin, France, consisting of four spherical tanks of capacity 1200 m^3 each, four spherical tanks of capacity 2000 m^3 each, and two horizontal cylindrical tanks of capacity 150 m^3 each.
From Török, Z., Ajtai, N., Turcu, A.-T., Ozunu, A., 2011. Comparative consequence analysis of the BLEVE phenomena in the context of land use planning. Case study: The Freyzin accident. Process Safety and Environmental Protection, 89, 1–7.

the valve and it was impossible to get it back in place. The lower valve had become frozen and could not be closed. Access to the valve area was difficult because the valves were immediately below the tank bottom, which was only 1.4 m above the ground.

- The propane leak was now completely out of control and soon formed a large visible 1 m thick vapor/mist cloud, which spread out across the ground as a layer of radius of 150 m.
- Twenty-five minutes after the leak had started the large vapor cloud was ignited by an automobile that had stopped on a nearby road. The fire flashed back to the storage tank and continued to burn there, but there was no immediate explosion.
- The tank was equipped with a water deluge system but the water supply was inadequate for cooling the tank. Furthermore, as soon as the fire brigade began to use their hoses, the water supply to the deluge system ran dry. Apparently the firemen had used up the available water for cooling the neighboring tanks to prevent the fire from spreading to them. They may have assumed that the tank on fire was protected sufficiently by the opened relief valve.
- Ninety minutes after the fire started, the tank BLEVEed.

Ten of the 12 firemen operating within 50 m from the tank were killed. People 140 m away were badly burned by a wave of propane which passed over the compound wall. Altogether, 15–18 men were killed and about 80 injured. Flying debris broke the legs of an adjacent spherical tank which fell over. Its relief valve discharged combustible liquid which added to the fire, and 45 min later even this tank BLEVEed. Altogether, five spherical and two other pressure vessels ruptured catastrophically, whereas three suffered less severe damage. The fire also spread to gasoline and fuel oil tanks.

3.6.1.2 PEMEX PLG Terminal Catastrophe in Mexico City, in November 1984

The large Petroleos Mexicanos (PEMEX) PLG terminal in San Juan Ixhuatepec, Mexico City, received supplies from three different gas refineries every day. In the morning of the accident, the vessels at the terminal were being filled with PLG arriving in a pipeline from a refinery 400 km away. Then a drop in pipeline pressure occurred due to the rupture of an 8-in. pipe that connected one of the storage spheres to a series of cylindrical tanks. However, the operators did not imagine such a possibility and the release of the PLG from the leaking pipeline continued for 5–10 min. This led to the major catastrophe illustrated in Fig. 3.6.

The escaping gas formed a 2-m high cloud covering an area of 200 m × 150 m, which drifted toward a flare tower where it caught fire and prepared the ground for the first BLEVE in a series of several.

Four spherical tanks, each containing 1500 m^3 of PLG, and several smaller cylindrical tanks containing between 45 and 270 m^3 PLG each suffered BLEVEs. All the BLEVEs generated fireballs, which raged through the streets of Ixhuatepec for about 90 min. A block of perhaps 200 houses built mostly of wood, cardboard, and metal sheets was demolished by these fireballs. Masses of fragments of tanks and pipes weighing up to 40 tons, were blown into the air and landed as far away as 1200 m. The BLEVEs ejected fragments of four of the six larger spherical tanks. The fragments were wrapped in burning PLG and rocketed in all directions. Some of the fragments hit other vessels which were also damaged. Other fragments initiated local fires which engulfed other vessels. Fifteen of the 48 smaller cylindrical vessels were ejected as "bullets" into the surroundings. Fig. 3.6 indicates some vessel fragments and "bullets" retrieved after the accident.

The disaster caused about 650 deaths and more than 6400 injuries. The entire PEMEX terminal was devastated. The total damage was estimated at more than 30 million US$.

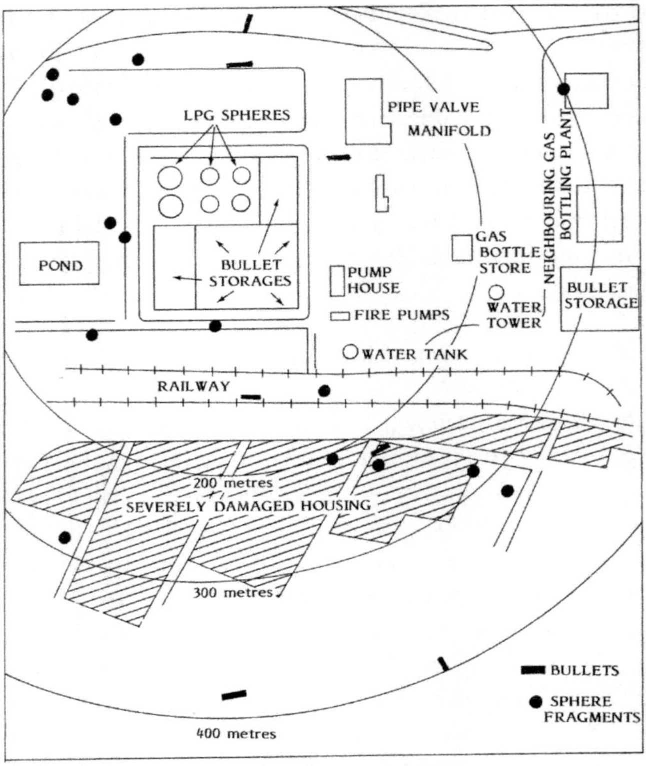

Figure 3.6 Site of the Mexico City BLEVE catastrophe.
From Marshall, V.C., 1987. Major Chemical Hazards, Elis Horwood Series in Chemical Engineering, Ellis Horwood Ltd/Halsted Press, A Division of John Wiley & Sons, Chichester/New York, ISBN:0-470-20813-9.

3.6.1.3 Boral PLG Distribution Depot in Sydney, Australia, in April 1990

In the evening of April 1, 1990, a BLEVE occurred in the Boral PLG storage/distribution terminal outside Sydney. This led to a series of further BLEVEs throughout the night.

The first event was an explosion of a small gas tank. The resulting fire then spread along ruptured gas pipes to the four main 100-tons steel PLG storage tanks containing at total of at least 40,000 L of PLG. The fire heated up the tanks until the 15-cm thick steel walls failed and gave rise to very powerful BLEVEs. The resulting fireballs and gas flares extended hundreds of meters into the sky. Hundreds of portable gas cylinders of capacities from 2 to 240 kg, and kept inside a storage room at Boral, also BLEVEd.

Power blackouts occurred after the first thundering explosion, which shattered windows. The shock wave from one of the BLEVEs uprooted a telegraphic pole that was shot away as a missile and nearly hit a woman who was standing almost 500 m away from the exploding vessel. One of a number of 30-m long cylinders blew off its mooring and rocketed through the air with a tail of flames. Upon landing it created a 2-m diameter crater in the earth, before bouncing through a wire fence into three 40-ton tanks, which were propelled into a nearby canal. The rocketing cylinder then hit and flattened an electrical substation and a panel-beating workshop before diving into the canal 300 m away from its original position.

Fortunately it all happened on a Sunday. The loss of human lives and number of injuries would have been substantially larger if the accident had occurred on a working day. In addition, a favorable wind direction contributed to limiting the spread of the fire. The chain of BLEVEs was eventually broken when Boral engineers were able to open the relief safety valves of the surviving tanks to depressurize them.

Several factory storage buildings close to the explosion sites were destroyed. The buildings that survived had doors that were thrown off their hinges, roofs that lifted and windows that shattered. The Boral plant was built in 1968 to satisfy standards that were outdated by the time of the explosion disaster in 1990.

3.6.1.4 A Supercritical Pressure BLEVE in Nihon Dempa Kogyo Crystal Inc.

Zhang et al. (2014) describe a BLEVE accident that occurred on December 7, 2009 at the Nihon Dempa Kogyo Crystal (NDK) Inc. facility in Belvidere, Illinois, United States. A 15-m high pressure vessel ruptured and as a result a number of fragments were ejected from the facility, killing a truck driver 200 m away and injuring an employee inside a building 130 m away. The US Chemical Safety and Hazard Investigation Board-concluded that stress corrosion cracking was the mechanism causing the vessel failure. After analyzing the operating conditions and the consequences, the accident was classified as a supercritical BLEVE. A consequence analysis of the accident was performed, including estimation of blast wave overpressure, fragment travel distances, and safety distances. Some important root causes, such as poor safety culture, poor management within the mother company, and poor communication between the actual facility being hit by the accident and official authorities, were also identified.

3.6.2 BLEVEs in PLG and CNG Transportation Facilities

3.6.2.1 Highway Tunnel Near Palermo, Italy, March 1996

A tank truck became involved in a car crash inside a highway tunnel. The initiating event was the skidding of a car inside the tunnel causing piling-up of cars behind it. The engine of one of the cars caught fire as a result of a collision with another car. The tank truck also entered the tunnel but had to stop about 50 m from the tunnel exit to avoid collision with the cars ahead. However, a bus just behind the tank truck arrived at high speed, skidded and crashed into the truck, causing a leakage in the upper part of the tank shell, just below the manhole. A few seconds later a "soft rumble" was heard, followed by a "hot wind" that caused serious burns to the people in the tunnel. Most of the people in the tunnel, except for five persons who had fainted as a result of the crash, managed to run out of the tunnel. Four minutes later, the tank of the tank truck BLEVEd. The resulting blast wave demolished the cars in the tunnel and killed the five persons who had been unable to escape. Dense black smoke was expelled from the tunnel exit.

3.6.2.2 Highway near Tivissa, Spain, June 2002

A tank vehicle carrying compressed natural gas (CNG) lost control on a downhill road and turned over. When it came to a halt beside a sandy slope, flames appeared between the truck cabin and the trailer due to ignition of either leaking diesel, or CNG, or perhaps both. The fire increased in size, perhaps due to involvement of CNG escaping from the PRV. There was a small explosion, followed by a strong hissing sound. Then the tank BLEVEd and the ejected CNG ignited, giving rise to a huge fireball. The accident occurred in a remote location, but it nevertheless led to one casualty and burn injuries to two persons who were about 200 m away from the blast site.

3.6.3 BLEVEs in Stationary Installations Containing Toxic but Nonflammable Chemicals

3.6.3.1 Phosgene Storage Tank, Hamburg, Germany, May 1928

A tank containing phosgene BLEVEd at a factory near the harbor area of Hamburg, Germany. About 12 tons of the deadly gas was released into the atmosphere. Eleven people were killed and 171 injured. People were affected by the gas at locations up nearly 20 km from the accident site. If the wind had blown over populated areas, the fatalities would have been much higher.

3.6.3.2 *Colorant Manufacturing Unit, Louisville, Kentucky, United States, April 2003*

A tank containing a mixture of maltodextrin and water was unintentionally overheated. On two earlier occasions, the tank had been deformed due to misapplication of vacuum, and the repairs had not been certified to meet standard requirements for structural strength. Most probably, this was the cause of the tank failure, as the internal overpressure at the time of vessel burst was less than the formal design pressure of the tank. The explosion propelled the top head of the tank to about 100 m away from the original site, and one operator lost his life. The main tank shell was ejected off its foundation and struck a 50,000 L liquid ammonia storage tank, which was knocked it sideways and ruptured, giving rise to a 12,000 kg leak. The shell then ricocheted and hit the bottom of a five-story spray drier, which toppled.

3.6.4 BLEVEs During Transportation of Toxic but Nonflammable Chemicals

3.6.4.1 *Tractor Tank Carrying Ammonia, Houston, United States, May 1976*

A tractor tank semitrailer carrying 19 tons of liquid anhydrous ammonia accidentally went through a bridge rail on an interstate highway and fell about 5 m onto a freeway, on which, at the time of the accident, the traffic was quite heavy. The tank BLEVEd and the liquid ammonia flashed off, forming a 30-m high cloud. The cloud gradually mixed with the ambient air and finally attained a width of about 300 m and a length of about 600 m. It was estimated that the evaporation of the ammonia and the subsequent mixing with air occurred within 5 min. The driver of the truck and five other people were killed. Seventy-eight people had to be taken to hospital for injury treatment, whereas about another 100 suffered less severe injuries. Five of the six casualties were due to the toxic effect of ammonia.

3.6.4.2 *Freight Train Transporting Chlorine, Montana, Mexico, 1981*

A train comprising 38 wagons, including 32 rail tank cars filled with liquid chlorine, was moving down a steep and winding valley at a 3% gradient when its brakes failed. The train derailed at over 80 km/h on a bend 350 m beyond Montana station, resulting in a pile-up that included 28 of the 32 chlorine cars. Most were badly damaged and suffered BLEVEs one after another. One tank car lost its dished end and the shell was propelled 2000 m. A second was split along its side. A third had a 0.5-m diameter hole, probably the result of an iron—chlorine fire, which could well have resulted from ignition of the cork insulation by red-hot brakes. Four other tank cars suffered damage to their valves, which were ripped off or dislodged so that they leaked. It is estimated that 100 tons of chlorine escaped in the first few minutes and 300—350 tons in all.

17 persons died, 4 in the caboose of the train and 13 from gassing. In addition about 1000 people were injured. The vegetation up the valley was bleached by the gas cloud passing up it; there was also discoloration some 50 m down the slope and up the sides for a vertical distance of about 50 m. The highest concentrations appear to have occurred in a strip 1000 m long × 40 m wide.

3.6.5 BLEVEs Involving Only Nonflammable and Nontoxic Chemicals

3.6.5.1 Liquid Nitrogen Storage Vessel in Japan, August 1992

The catastrophic failure of a storage vessel containing liquid nitrogen resulted in a BLEVE causing demolition of almost half of the factory, damaging the walls of 25 houses and destroying 39 cars, buses, and trucks, all within a 400 m radius of the original site of the storage vessel. Fragments of the vessel, including part of the 150-cm wide and 8-mm thick top head of the outer shell were ejected up to 350 m away from the original site. The estimated property loss was $5 million.

3.6.5.2 Mihama Nuclear Power Reactor, Japan, August 2004

A large pipe carrying superheated water started to leak and exploded. The resulting two-phase release of superheated water and steam scorched 11 workers. Some of these lost their lives; the others were severely injured.

3.6.6 Initiating Events Leading to BLEVEs in Case Histories

3.6.6.1 Relative Frequencies of Initiating Events

Abbasi and Abbasi (2007) were not able to trace any comprehensive, systematic record of BLEVEs that have occurred worldwide. However, from available published evidence, they deduced the relative frequencies of the initiating events leading to 88 major BLEVEs that occurred during the period 1926—2007. The results are summarized in Table 3.4. The frequency distribution given in the table refers to the initial causes leading to the very first BLEVE in an accident, irrespective of whether the accident also comprised several subsequent BLEVEs. However, more than one BLEVEs

Table 3.4 Frequencies of initiating events in 88 major BLEVEs occurring from 1926 to 2007

Fire	36%
Mechanical damage	22%
Overfilling	20%
Runaway reactions	12%
Overheating	6%
Vapor space contamination	2%
Mechanical failure	2%

From Abbasi, T., Abbasi, S.A., 2007. The boiling liquid expanding vapor explosion (BLEVE): mechanism, consequence assessment, management. Journal of Hazardous Materials, 141, 489–519.

per accident is the most common situation. Therefore, if one also considers that a preceding BLEVE in an accident was in fact the initiating event of a subsequent BLEVE in the same accident, it could well be that a preceding BLEVE is the most common trigger for BLEVEs.

3.7 PREVENTING BASIC TECHNICAL CAUSES THAT CAN LEAD TO BLEVEs

3.7.1 Systematic Application of the "Inherently Safer Plant" Concept for Eliminating/Minimizing the BLEVE Hazard

According to Shariff et al. (2014), attempts have been made at applying the "inherently safer plant" concept to design of plants to prevent and/or mitigate potential BLEVEs. One then aims at being able to evaluate and minimize consequences of potential BLEVEs at the early stage of plant design. It should then in principle be possible to develop design strategies by which the possibility of BLEVE accidents may be effectively eliminated or their potential consequences minimized to an "As Low As Reasonably Practicable" (ALARP) level. However, adequate systematic methods for adopting this approach in practice have been lacking.

In their paper, Shariff et al. propose a method to meet this need. In addition to addressing the preliminary design stage, the method also makes it possible to evaluate consequences of process modifications. They presented a model developed in an MS Excel spreadsheet, called Inherent Fire Consequence Estimation Tool, by which impacts of possible BLEVEs in terms of blast waves, thermal radiation, and missile emission could be evaluated.

3.7.2 Preventing Pressurized Liquefied Gas-containing Vessels From Becoming Exposed to Fires

3.7.2.1 Keeping PLG-containing Vessels at Safe Distances From Likely Sources of Fire

Fire engulfment is the most common basic technical cause of BLEVEs. Therefore, it is very important that PLG-containing vessels be located at sufficiently long distances from each other and from any other potential sources of fire.

3.7.2.2 Sloping of the Nearby Ground

To prevent a pool fire from occurring after an accidental spill from a fixed PLG vessel, the ground around the vessel should have a downward slope of at least 1% so to enable the spill to flow away to a safe area.

3.7.2.3 Water Barriers

Facilities for generating water mist barriers close to PLG vessels can be installed. Such barriers can capture any flammable vapor released from a PGL-vessel and disperse it without the flammable vapor getting ignited. In the case of water-soluble toxic substances like ammonia and chlorine, the water will dissolve some of the released material and thereby reduce the amount of toxic material that remains dispersed in the air.

3.7.3 Preventing Mechanical Damage of Pressurized Liquefied Gas-containing Vessels

Trucks and railroad cars carrying PLGs should be protected from accidental damage generating spills by using double-walled vessels with thermal insulation between the walls. Collisions or overturning during transportation may then damage the outer wall, without any spills occurring. It is then important to make the outer wall sufficiently strong to provide sufficient protection of the inner wall.

3.7.4 Preventing Overfilling of Vessels and Vessel Overpressure

Rigid compliance with standards for filling and weighing of vessels that may become exposed to BLEVEs, as well as for standards for relief devices, has reduced the frequency of BLEVEs due to overfilling. Relief devices can get plugged, but this can be compensated for by installing rupture disks in parallel to the relief valve.

3.7.5 Preventing Runaway Reactions

BLEVEs due to runaway reactions inside the vessel are much less common than BLEVEs due to accidental vessel damage from the outside, such as fire or missile impact. However, they do occur and precautions should be taken. Hence, all vessels and other process equipment in which runaway reactions are possible should have instrumentation for continuous monitoring of temperature and pressure. Such equipment should have facilities for counteracting excessive pressure or temperature, eg, internal cooling coils or external jackets, remotely controlled venting valves, inhibitor-injection systems, and internal deluges. There should also be alarms for control-room and field personnel when excessive pressures and/or temperature occur.

3.7.6 Preventing Destructive Exothermal Reactions in Vessels due to Reactive Impurities

Vessels containing highly reactive gases in liquefied form should be protected against contamination by foreign substances with which they can react exothermally. By installing systems for rendering inert the vapor space in pressurized vessels by nitrogen or other nonreactive gas, and/or installing explosion-suppression systems, destructive vapor space explosions can be prevented.

3.7.7 Preventing Weakening of Vessel Structures due to Fatigue, Creep, Corrosion Etc.

Proper design and testing of vessels can prevent onset of distortion and possible rupture. Periodic wall-thickness measurements, internal inspection for corrosion, acoustic emission testing to detect possible initial cracking of the container etc., should be performed to ensure the fitness of the containers.

3.7.8 Protection of Vessels From Being Exposed to Fire, Hit by Missiles or by Vessel Burial

In principle, vessels containing PLGs will be partially or fully protected from fire or missile impact if they are partially or fully buried. But such vessels are both difficult to inspect and maintain, and preventing and controlling corrosion is difficult.

3.7.9 Prevention of Excessive Superheating of Vessel Content

In distillation systems and chemical reactor nucleation devices, such as sharp-edged ceramic material or aluminum mesh, is placed inside the liquid to promote normal boiling and hence prevent superheating. Application of this technique to PLG vessels has been proposed, but well-proven established methods are not available.

3.8 PREVENTING CATASTROPHIC RUPTURING OF PRESSURE-LIQUEFIED GAS VESSELS ENGULFED IN FIRE

3.8.1 Introduction

It is almost impossible to say with certainty whether or not a vessel rupture will give rise to a BLEVE. Nor is it possible to forecast with any certainty when a vessel will give rise to a BLEVE after having ruptured. This and the further uncertainty associated with forecasting the size, range, direction, and momentum of missiles likely to become ejected from a BLEVE pose special challenges both as regards preventing a BLEVE as well as mitigating the potential damage from a BLEVE. In addition, predicting the thermal impact of fire-balls from BLEVEs present considerable challenges, as previously discussed in Section 3.8.5.

According to Abbasi and Abbasi (2007), several tragic accidents have been reported in which firefighters trying to save a liquefied petroleum gas vessel (LPG) engulfed in fire were killed by the violently expanding fireball, or by rocketing fragments generated when a vessel suddenly ruptured.

3.8.2 Thermal Insulation

The PLG containers should be thermally insulated to the extent possible. This will reduce the rate of heating of the vessel when exposed to heat load from a fire, and thereby also delay the pressure increase inside the vessel. Significant thermal insulation is obtained if the vessel wall is protected by a steel jacket containing a ceramic thermal insulation material of thickness at least 13 mm. Even steel jackets with just air inside can reduce the heating rate of the internal wall to approximately half of that of with just a single wall. Thermal insulation alone cannot prevent a BLEVE, but it may provide a delay of 4 to 5 h, which can give firefighters the time they need to remove the heat load. In fixed installations, even the vessel support system should be thermally insulated so that it does not collapse when subjected to heat. Likewise, valves, pipes, and other safety elements installed on the PLG vessel must be able to withstand the high temperatures that may occur in a crisis. The thermal insulation system should be installed in such a way that it does not interfere with the periodic inspection of the vessel surface and vessel support systems. The effectiveness of fireproofing for delaying a BLEVE also depends on whether the PRV functions correctly.

3.8.3 Cooling of the Unwetted Part of the Pressurized Liquefied Gas Vessel Wall by an Internal Liquid Spray System

Young (2004) developed a system in which a turbocharger is placed inside the PLG vessel. If the vessel becomes engulfed in a fire, the turbocharger will take liquid in the vessel from just below the liquid surface, and spray it vertically upward to cool the unwetted part of the tank. The system is illustrated in Fig. 3.7.

The turbocharger is driven by the vapor escaping through the PRV. The system is expected to prevent the unwetted part of the tank from becoming heated to the point of weakening of the metal, which would facilitate crack formation. The system can be built into most new PLG vessels, and also fitted to existing vessels, including mobile ones.

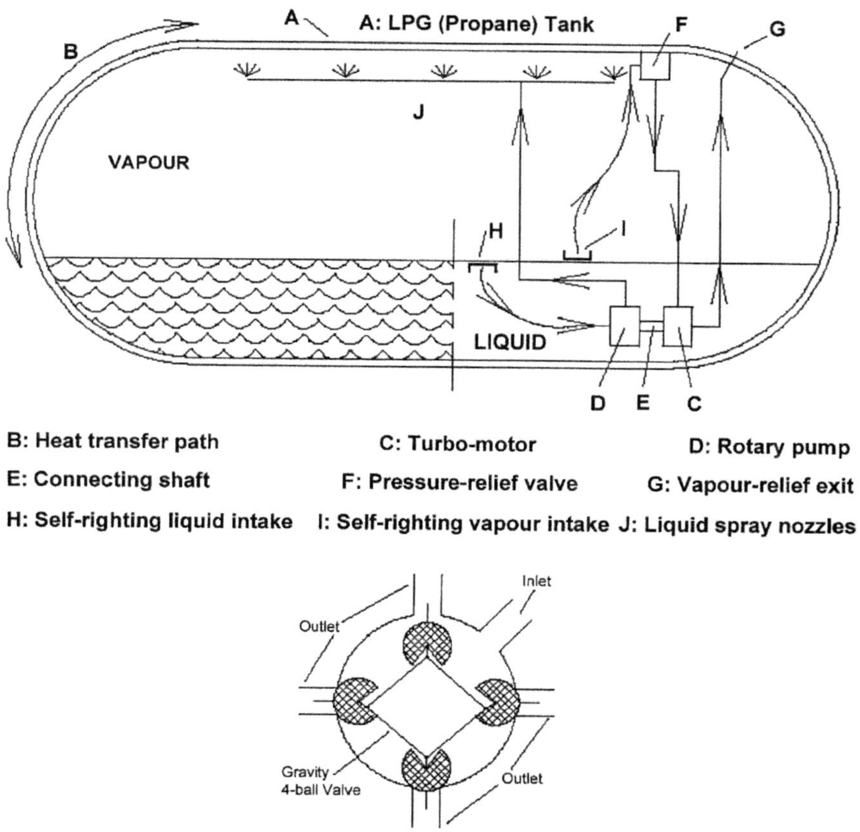

Figure 3.7 Safety system for BLEVE prevention/delay developed by Young (2004), based on cooling of the unwetted part of the PLG vessel wall by an internal liquid spray. The figure is reproduced from the paper by Abbasi and Abbasi (2007).

3.8.4 Water Deluge

The thermal load on a PGL vessel engulfed in fire can be substantially reduced by water deluge. A layer of flowing water must then be applied to the vessel as soon as possible. A water layer of sufficient thickness should then totally cover the vessel wall, not least those areas that are directly engulfed by the flame. A water flow rate of minimum 10–15 $L/m^2/min$ is recommended. If the flame is highly turbulent, flow rates even larger than 25 $L/m^2/min$ may be required. If, however, the vessel is being impinged by a jet fire, water deluge is less effective.

3.8.5 Rapid Tank Depressurization

In addition to activating the water deluge system, an effective system for rapid depressurization of the vessel, which bypasses the normal PRV, and capable of reducing the vessel pressure to half the design pressure within 15 min, should be available and activated. The released material should be eliminated in a safe manner, for example by flaring. The depressurization should not be too rapid, as it can lead to extremely low temperatures and increased fragility of the steel.

3.9 REDUCING CONSEQUENCES IF A BLEVE DOES OCCUR

3.9.1 Inherently Safer Design

As discussed by Shariff et al. (2014), implementation of the *inherently safer plant* concept can be a powerful means of reducing the BLEVE risk. It is then important that the potential consequences of a BLEVE be evaluated at the preliminary design stage, and adequate design improvements effectuated as early as possible. It may then be possible to minimize, or perhaps even eliminate, the risks associated with BLEVEs. However, the inherently safer plant concept is not always easy to implement at the preliminary design stage at which the technical details of the plant are not in place. Shariff et al. (2014) suggested a new approach to overcome this problem, using a mathematical simulation model. They were then able to assess potential BLEVE impacts in terms of overpressures, thermal radiation fluxes and missile hazards. They illustrated the feasibility of their approach by conducting a case study of a BLEVE from a propane storage vessel at the early design stage.

3.9.2 Establishment and Implementation of Adequate Safety Distances

If a vessel suffers a BLEVE within only a few minutes after the formation of the first crack, little can be done to reduce the damage it would cause. However, even if a BLEVE is delayed by several hours, challenges exist. Great care must be taken to make sure that people move to outside the safety distance, both with regard to a potential blast wave, potential missiles, and a potential major fireball. The destructive impact of a delayed BLEVE can be even more severe than from an immediate one.

Birk (1996) suggested that firefighters should not get closer to a vessel that could undergo a BLEVE than four fireball radii, but the distance should not be shorter than 90 m. He suggested that a rough estimate of the fireball radius can be obtained by the simple expression

$$R = 3 \cdot m^{0.33} \quad [3.1]$$

in which m is the mass of fuel in kg and R is the fireball radius in m.

People should be moved to a distance preferably 30 fireball radii away from vessels that could undergo a BLEVE. The distance would have to be increased downwind of the vessel. However, as the vessel size increases beyond 5 m^3 the 30 radii requirement gets too conservative and a requirement of minimum 15 radii becomes more appropriate.

If the PLG involved in a BLEVE is toxic, eg, chlorine, ammonia, methyl isocyanate or phosgene, careful attention has to be paid to its dispersion pattern both in the blast from the BLEVE and in the atmosphere farther away.

3.9.3 Preventing "Domino" Effects

The first concern toward mitigation of the damage caused by a BLEVE is to prevent the initial accident from triggering secondary, higher-order accidents. Analyses of BLEVEs that have occurred in the past reveal that most of them were not "stand-alone" accidents. Other liquid-containing vessels in the vicinity of a vessel that has BLEVEd can become exposed to heat and/or missiles from the initial event. This may then give rise to further BLEVEs. To avoid such "domino" effects, vessels that may cause BLEVEs should be kept sufficiently far away from each other to render such effects unlikely. Mechanical barriers may also be placed around vessels to limit the impact of outgoing or incoming missiles. This may require rather complex design, and careful cost/benefit considerations may be needed.

3.9.4 Fireball Suppression

If sufficient quantities of fire suppressant can be released in such a way that the suppressant becomes mixed with the flashing fuel at the moment of vessel failure, the formation of a fireball may be prevented or at least its intensity may be reduced significantly. Alternatively, the fireball, after its formation, may be wrapped in a barrier of a suitable fire suppressant. Then, the suppressant would be sucked into the fireball by the very strong prevailing air entrainment flows. As a result, the flame may become completely suppressed, or at least the fireball size would be significantly reduced. The use of water mist as a fireball suppressant is an obvious possibility but the liquid droplets may evaporate completely before the water gets sucked into the fireball. This may reduce the suppressing effect.

CHAPTER FOUR

Water-Vapor Explosions

4.1 OVERVIEW OF TOPICS COVERED

This chapter is essentially identical with the review paper by Eckhoff (2016a). The attention will mainly be paid to the extremely violent unconfined water-vapor explosions. These typically arise from accidental contact between hot melts of some metals and water. In the nuclear reactor industry, the possibility of such explosions is a concern whenever there is a possibility of molten core metal making contact with cooling water in accidental core meltdowns.

Confined/semiconfined accidental water-vapor explosions will be discussed in Section 4.3. These explosions are caused by the excess water-vapor pressure produced by heating of fully or partly confined spaces containing water/moisture. Due to the confinement, the rates of evaporation required for generating destructive and hazardous overpressures are mostly considerably lower in such cases than in the case of unconfined explosions.

4.2 UNCONFINED WATER-VAPOR EXPLOSIONS

4.2.1 Basic Nature

In the present context, an unconfined water-vapor explosion will be defined as the extremely violent generation of water vapor that can arise from accidental contact between a hot melt of some metals and liquid water. In most cases, this is a purely physical process, as pointed out by Witte et al. (1970). Years later, Bergstrøm (1995) specifically discussed water-vapor explosions resulting from contact between molten steel, ferro-alloys, and water. He reemphasized that the nature of such explosions is physical. This means that the energy driving the explosion is an extremely rapid heat transfer from the molten metal to the water. The water then vaporizes at an explosive rate whereby the cloud of water vapor generated can attain very high pressures even in fully unconfined locations. Accidental unconfined water-vapor explosions resulting in property damage and personal injuries have also occurred in paper pulping mills.

Explosion Hazards in the Process Industries
ISBN: 9780128032732
http://dx.doi.org/10.1016/B978-0-12-803273-2.00004-9

In the excellent review years after the publication by Witte et al. (1970), Meignen et al. (2014) reconfirmed that this kind of water-vapor explosion is indeed most often a purely physical process. Heat is transferred at extremely high rates from the hot, more viscous molten metal to the less viscous, cool water. The resulting explosive evaporation of the water in turn gives rise to generation of extremely high local vapor pressures before displacement of vapor into the surroundings commences. At sufficiently high local instantaneous pressures, strong shock waves are produced. This is, in a way, similar to the initiation of detonation in explosive gas mixtures (see Chapter 2), in explosive clouds of mists and dusts (see Chapters 5 and 7), and in solid explosives (see Chapter 8). Sometimes, therefore, violent unconfined water-vapor explosions are also named "thermal detonations." The very violent nature of this kind of explosion was observed experimentally by Long (1957). He conducted a series of outdoor pilot-scale experiments by pouring molten aluminum into water. He characterized the resulting water-vapor explosions as being "similar to the detonation of explosives. They appeared to occur very rapidly after the pouring of the aluminium had started. One experiment was conducted at night without artificial lights. Three observers agreed that they could not detect any flash or fire during the explosion or immediately after, except for the glow of the red,hot molten aluminium."

However, if this latter visual observation implies that there was no exothermal chemical reaction involved in the explosion, it contradicts with the findings of Epstein et al. (2000). They concluded that in very violent explosions caused by contact between hot molten aluminum and water, exothermal oxidation of the metal also contributes significantly to the violent heat transfer to the water. They based their argument on experiments showing that drops of molten aluminum were quite resistant to being fragmented by even very strong shock waves, and therefore physical heat transfer alone cannot explain the explosive development of water vapor observed in aluminum—water explosions. Exothermal chemical oxidation of the metal probably also plays a central role.

For very strong unconfined water-vapor explosions to occur purely by physical heat transfer, the time scale for transfer of heat from the molten metal to the water has to be significantly shorter than the time scale for the establishment of a shock wave moving away from the high-pressure region. Such extreme heat transfer rates cannot be explained unless it is assumed that the hot molten metal in some extremely fast way gets dispersed into microdroplets of diameters on the order of 10 μm. This crucial process has been named "fine fragmentation" and will be discussed in Section 4.2.2. The strong supersonic blast waves that can be generated under such circumstances can kill and injure people, and demolish and damage mechanical structures, eg, buildings.

For the sake of completeness, a quite different situation, in which very violent "physical" unconfined vapor explosions can occur, should be briefly mentioned. This is when liquefied natural gas (LNG) is accidentally spilled on water. In this case, the water constitutes the hot, more viscous partner (like the molten metal in the previous discussion), whereas the liquefied methane is the cold partner that evaporates at an explosive rate (like the water in the above discussion).

4.2.2 Some Early Suggestions for the Nature of the Fine Fragmentation Process in Unconfined Water-Vapor Explosions

By the time of the review by Witte et al. (1970), three hypotheses had been proposed for explaining the fine fragmentation of the hot molten metal in essentially physical water-vapor explosions.

4.2.2.1 Water Entrapment Between Initial Metal Globule and Containment Bottom

When the initial molten metal globule approaches the bottom of the water containment, some water becomes trapped between the globule and containment bottom surface. The water vaporizes violently and the explosively expanding vapor cloud blows the molten metal globule apart into the microscopic droplets required for further explosive evaporation.

4.2.2.2 Water Entrapment Inside Initial Metal Globule

According to this model, water in some way becomes trapped inside the initial molten metal globule and vaporizes immediately very violently. The rapidly expanding water vapor blows the molten metal globule apart into the microscopic droplets required for further explosive evaporation. Witte et al. did not suggest any mechanism by which the water could be trapped inside the molten metal.

4.2.2.3 Violent Nuclear Boiling of the Water

The forces generated by the transition from film boiling (see Section 4.2.4) to nucleate boiling of the superheated water close to the molten metal surface overcomes the surface tension of the molten metal, causing the molten metal to be torn apart.

4.2.3 The Weber Number *We* as a Criterion of Expected Extent of Molten-Metal Fragmentation

Witte et al. (1970) indicated the central role played by the so-called Weber number in the process of subdividing the initial large molten metal globule. In the present context, this dimensionless number is defined as:

$$We = \rho r v^2 / \sigma \quad [4.1]$$

Here, ρ is the density of the molten metal, r is the radius of the metal globule, v is the globule velocity relative to the surrounding water, and σ is the surface tension of the molten metal. Witte et al. (1970) concluded that as soon as We exceeds a critical value in the range of 5 to 10, the inertia forces will exceed the surface tension, and the lump of molten metal will be broken up into smaller entities that are more stable.

According to Lee and Frost (1987) and Pilch et al. (1981) defined five distinct mechanisms for the fragmentation of a molten-metal globule moving in water, depending on the magnitude of *We*. Illustrations of the various regimes are given in Figs. 4.1 and 4.2.

Figs. 4.1 and 4.2 are based on the assumption that the breakup of a molten-metal globule in a water flow field differs little from the breakup of a liquid globule in a gaseous flow field.

In the range of $We < 12$ (Fig. 4.1a), the breakup of the initial mother globule is due to forced vibration of the globule at its natural frequency, driven by the flow field. Experiments had confirmed that in this region of *We*, breakup results in a comparatively low number of smaller globules, and breakup times are comparatively long.

For somewhat larger *We*, in the range $12 \leq We \leq 50$ (Fig. 4.1b), the fragmentation process was described as *bag breakup*. The droplet deforms in the flow like blowing a soap bubble from a ring. A thin hollow bag is formed downstream until the bag bursts into a large number of small fragments. The remaining toroidal rim of liquid then disintegrates into a few larger fragments.

For still higher *We*, $50 \leq We \leq 100$ (Fig. 4.1c), the breakup geometry changes slightly. Apart from a thin bag anchored round a toroidal rim, a central column of liquid (stamen) is also formed along the drop axis parallel to the flow. Again, the bag disintegrates, first into small fragments with subsequent breakup of the rim and the stamen. This type of breakup is referred to as the *bag-and-stamen breakup*, or *umbrella breakup*.

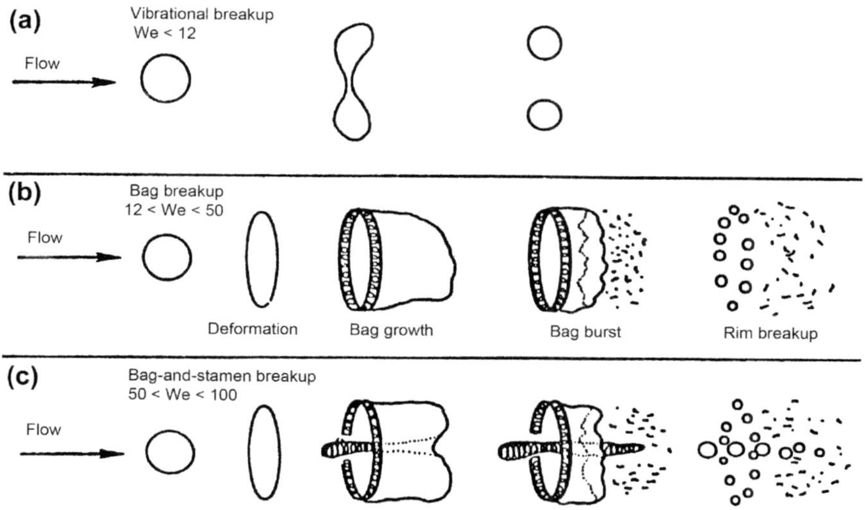

Figure 4.1 Schematic illustration of different modes of breakup of large liquid drops into smaller droplets for $We < 100$.
From Lee, J.H.S., Frost, D.L., 1987. Steam explosions – major problems and current status, paper presented at the 11th ICDERS in Poland in August 1987; after Pilch, M., Erdman, C.A., Reynolds, A.B., 1981. Acceleration induced fragmentation of liquid droplets, Report NUREG/CR-2247 prepared for the Nuclear Regulatory Commission, USA.

For even higher *We*, ie $100 \leq We \leq 350$ (Fig. 4.2a), the breakup is distinctly different from the previous configurations. Instead of the formation of a bag, a thin sheet of liquid is drawn continuously from the deformed droplet. The jet sheet then disintegrates a short distance downstream from the deformed drop. A coherent residual drop exists during the entire duration of the stripping and sheet breakup. This mode of fragmentation is referred to as *sheet stripping*.

At considerably larger *We*, ie, $We > 350$ (Fig. 4.2b), surface waves develop on the windward surface of the drop. As the amplitude grows, the wave crests are continuously eroded away by the action of the flow over the surface of the drop. This mode of breakup is referred to as *wave crest stripping*. However, longer waves also develop, and, as their amplitude grows, the droplet is divided into several large fragments increasing the surface area for the small-wavelength surface waves to grow, hence augmenting the wave crest-stripping rate. This mode is referred to as *catastrophic breakup* (Fig. 4.2c).

It should be noted that Fig. 4.1 and 4.2 merely indicate possible processes for fragmentation of the molten-metal globules at given *We* values. Actual

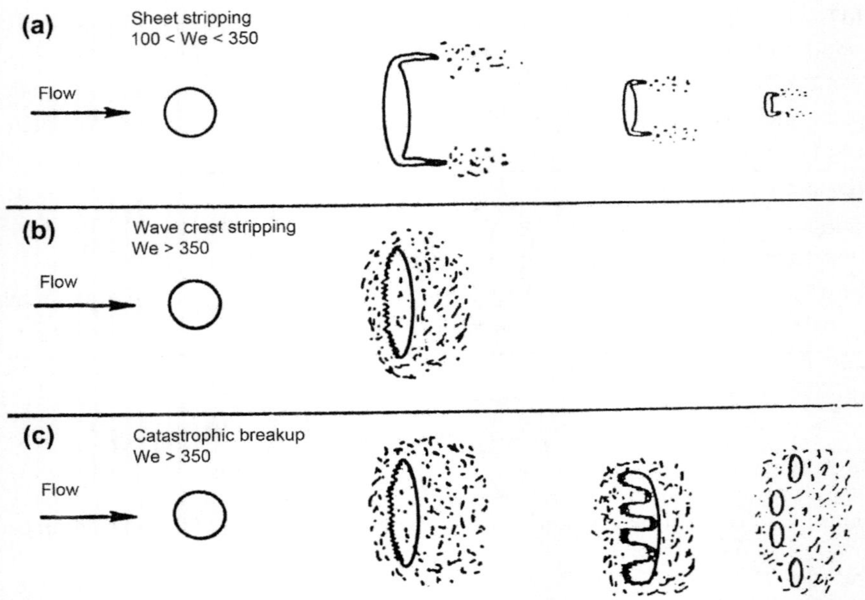

Figure 4.2 Schematic illustration of different modes of breakup of large liquid drops into smaller droplets for $100 < We < 350$ and $We > 350$.
From Lee, J.H.S., Frost, D.L., 1987. Steam explosions – major problems and current status, paper presented at the 11th ICDERS in Poland in August 1987; after Pilch, M., Erdman, C.A., Reynolds, A.B., 1981. Acceleration induced fragmentation of liquid droplets, Report NUREG/CR-2247 prepared for the Nuclear Regulatory Commission, USA.

mechanism(s) for generating the high *We* values needed for very fine fragmentation, and hence explosive water evaporation, will be discussed in the following section.

4.2.4 Initial Stage of a Four-Stage Process Leading to Violent Unconfined Water-Vapor Explosions. The Role of Film Boiling

Lee and Frost (1987) suggested that violent unconfined water-vapor explosions, caused by a sudden contact between hot molten metal and water, develop through four specific consecutive stages. These are *coarse mixing, triggering, propagation,* and finally *pressure increase* in, and *expansion* of, the water vapor.

Lee and Frost considered that the very first stage in the process is a comparatively slow formation of a dispersion of still quite coarse molten metal globules throughout the water. This occurs by vibrational breakup of the initial large globule (Fig. 4.1a). Typical resulting secondary globule sizes were assumed to be of the order of 1 cm. During the relatively slow (time scale of some seconds) process of formation of these globules throughout the water some mechanism must exist by which the globules are kept partially thermally insulated from the water. Without such a mechanism, a comparatively stable suspension of such globules would not be formed. The required thermally insulating barrier is in fact provided by the thin film of water vapor which develops when water makes contact with very hot surfaces. Because such a film has a much lower thermal conductivity than both the molten metal and liquid water, the heat transfer from the metal to the water will be comparatively slow, allowing the hot metal globules to be dispersed throughout the water on a comparatively long time scale.

The establishment of a film of water vapor on the metal globule surfaces requires that the metal temperature exceeds a critical value called the Leidenfrost temperature. For water, the Leidenfrost temperature is 273°C. Fig. 4.3 illustrates how the heat transfer between the metal and the water is reduced by the formation of the water-vapor film.

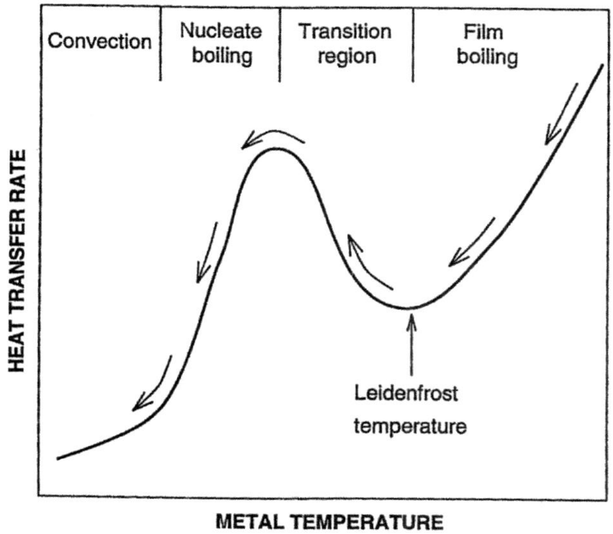

Figure 4.3 Illustration of how heat transfer between a hot metal and water is retarded at metal temperatures above the Leidenfrost temperature by formation of a water-vapor film on the metal surface.
From Bergstrøm, T.H., 1995. A review of molten metal/water reactions. In: Proceedings of the 52nd Electric Furnace Conference, November 13–16, 1994 in Nashville, USA, pp. 303–308. Published by the Iron and Steel Society, USA in 1995. ISBN:0-93289-97-5.

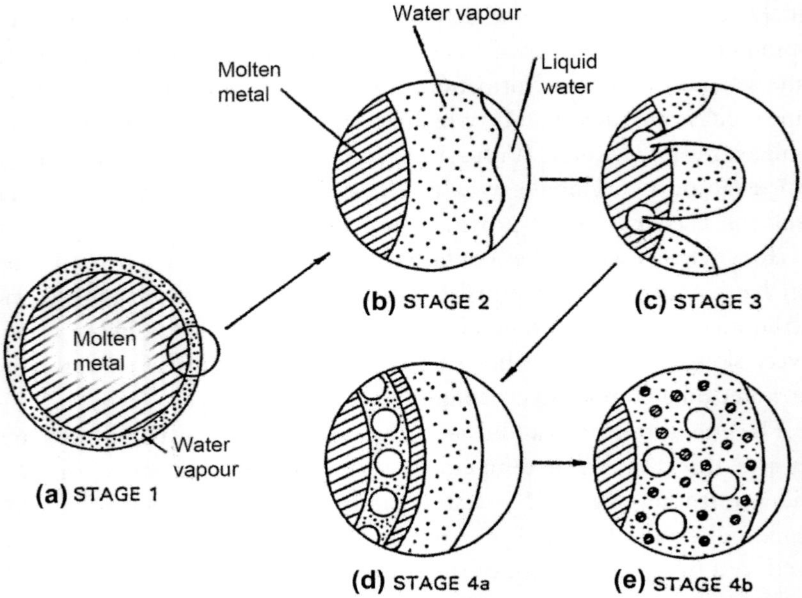

Figure 4.4 Schematic illustration of "triggering" mechanism as suggested by Kim (1985). *From Bergstrøm, T.H., 1995. A review of molten metal/water reactions. In: Proceedings of the 52nd Electric Furnace Conference, November 13–16, 1994 in Nashville, USA, pp. 303–308. Published by the Iron and Steel Society, USA in 1995. ISBN:0-93289-97-5.*

The film minimizes the flow of heat by conduction, leaving the less effective thermal radiation as the main heat transfer mechanism. The arrows along the path of the heat transfer curve in Fig. 4.4 indicate the development of the heat transfer rate with decreasing metal temperature. At the beginning of the period of contact between the metal and the water, the metal temperature will be at its maximum (to the right in Fig. 4.4), and far above the Leidenfrost temperature. Consequently, the boiling will be in the film-boiling regime. As the metal temperature decreases film boiling will continue to dominate the heat transfer, until the metal eventually attains the Leidenfrost temperature, in which the heat transfer is quite low. With a further decrease of the metal temperature, the water-vapor film will gradually disappear and the heat transfer rate will increase for some time despite the decreasing metal temperature.

4.2.5 Three Subsequent Stages of Development of Water-Vapor Explosions

Referring to even earlier reported evidence, Witte et al. (1970) emphasized that fragmentation of the initial hot molten metal globules of some millimeter diameter into tiny droplet of diameters about 10 μm is a basic prerequisite for the violent water-vapor generation experienced in genuine water-vapor explosions. This also agrees with the point of view of Lee and Frost (1987) and with the more recent the state-of-the-art review by Meignen et al. (2014). Then some *triggering* mechanism must exist that can initiate the extremely fast subdivision of the hot metal in at least one of the initial metal globules.

This mechanism must imply that the initially stable Leidenfrost vapor films surrounding the coarse molten metal globules formed in the first, coarse metal dispersion stage, in some way gets destabilized, which establishes direct contact between the molten metal and the water. This gives rise to very rapid heat transfer from the metal to the water, which, in turn, results in violent water vaporization and *fine-fragmentation* of the coarse molten metal units into the μm size metal droplets/particles required for explosive water evaporation. Abe et al. (2002) studied the fine-fragmentation process experimentally using a molten-tin—water system. They found that the size distribution of the solidified tin droplets after fragmentation was quite dependent on both the initial molten-tin temperature and the initial water temperature. For an initial molten-tin temperature of 550°C and an initial water temperature of 30°C, the size distribution of the solidified tin particles collected after the water-vapor explosion had a peak at about 500 μm. Lee and Frost suggested that the localized perturbation that initiates the collapse of the vapor film surrounding one or more of the initial metal globules could well be a local pressure increase. Another possibility suggested by Lee and Frost was hydrodynamic shear from turbulence causing destabilization of the liquid—vapor interface by the Kelvin—Helmholtz mechanism.

According to Bergstrøm (1995) and Kim (1985) suggested the detailed five-stage model of the "triggering" process illustrated in Fig. 4.4. The various stages in Fig. 4.4 are:

- Initial film boiling round the molten metal globule in the water. (Stage 1 in Fig. 4.4)
- Growth and oscillation of the water-vapor film. (Stage 2 in Fig. 4.4)
- Destabilization and collapse of the water-vapor film and formation of water jets due to Rayleigh—Taylor instability. Penetration of liquid water jets into the molten metal (Stage 3 in Fig. 4.4)
- Encapsulation of liquid water droplets inside the molten metal (Stage 4a in Fig. 4.4)
- Violent evaporation and expansion of the entrapped water, causing violent fragmentation of the molten metal (Stage 4b in Fig. 4.4).

The third crucial stage in the development of a genuine water-vapor explosion is then the *propagation* of the first triggering process throughout the dispersion of the coarse molten metal globules in the water. Meignen et al. (2014) emphasized that it is by no means obvious that the propagation phase automatically follows the triggering phase. If propagation does not occur, there will be no major violent unconfined water-vapor explosion.

Therefore, in the analysis and modeling of the conditions leading to fully developed unconfined water-vapor explosions, proper attention must be paid to the propagation phase. Lee and Frost (1987) suggested the following mechanism by which triggering may lead to propagation: The rapid local heat transfer to the water established by the "triggering" of the first coarse molten metal globule causes explosive local water-vapor formation. The shock wave and very fast small-scale hydrodynamic flow associated with the resulting vapor expansion destabilizes and causes collapse of the vapor film round neighboring molten-metal globules. In this way, the fine-fragmentation process propagates through the suspension of the coarse globules. Lee and Frost then suggested that if the process of fragmentation of the coarse molten metal units is sufficiently rapid for it to be able to follow the triggering shock wave, a coherent vaporization front will be coupled to the shock. The phenomenon is analogous with that of a detonation wave in eg, explosive premixed gas, in which the reaction front is coupled to the triggering leading shock and propagates with it. If the delay time is too long, fine fragmentation cannot follow the triggering shock wave.

The fourth and final stage suggested by Lee and Frost is the *expansion of the highly pressurized water vapor* produced in the propagation phase. Understanding this final phase is important in the evaluation of the mechanical damage to eg, buildings. Then knowledge of both the overpressure time history and the impulse of the pressure pulse caused by the water vapor are required.

4.2.6 Mathematical Modeling of Unconfined Water-Vapor Explosions

Corradini et al. (1988) presented a comprehensive review of theories and modeling of unconfined water-vapor explosions in the context of the possibility of occurrence of such explosions in light water-cooled nuclear reactors. The hazardous scenario would then be accidental reactor core meltdown and subsequent contact between the molten material and cooling water. Fletcher and Anderson (1990) conducted another comprehensive review of the state of the art at that time in modeling the propagation phase of unconfined water-vapor explosions.

The status in this field has been summarized by Meignen et al. (2014). According to these authors, the very first efforts at modeling unconfined water-vapor explosions were mainly devoted to the stage of the fine fragmentation of the initial molten-metal globules. As discussed previously, the diameters of these globules are typically a few millimeter, and very fast fine fragmentation of these droplets into very small droplets of diameters on the order of 10 μm is required for explosive water-vapor production to occur.

In the first mathematical models, it was assumed that the molten metal remained in the molten phase throughout the fine-fragmentation process. Later, however, it was realized that possible solidification of the metal during this process had also to be considered in the models. In the context of nuclear reactor meltdown hazards, it was clear that for $UO_2{-}ZrO_2$ mixtures the melt could be subject to rapid solidification during the fine-fragmentation phase, thus reducing the amount of melt participating in the water-vapor generation.

Comprehensive models will also, when appropriate, have to account for the additional effect of possible exothermal chemical oxidation of involved metals. According to Meignen et al. (2014), these issues had not been adequately considered in "academic" modeling of the fine fragmentation of the initial molten-metal globules up to the time of their review.

As pointed out by Meignen et al. (2014), modeling of the *fragmentation of the larger molten-metal globules* to microscopic droplets by hydrodynamic forces constitutes the first important part of a comprehensive model of unconfined water-vapor explosions.

The second main process to be modeled is then *the extremely fast pressure buildup in the ambient atmosphere* due to the extremely fast water vaporization caused by the fine fragmentation of the hot molten metal. Meigden et al. (2014) pointed out that the understanding of the exact mechanisms involved in this evaporation process appear comparatively poor, although significant new experimental evidence had been produced during the last few years before the publication of their paper. The new experiments had revealed that the heat transfer characteristics between the micro droplets of molten metal and the water are likely to be very close to those in film boiling (see Section 4.2.4). Apparently, classical film boiling correlations give reasonable estimates of the actual heat transfer from the hot molten-metal μm droplets to the water. Because of the extremely fine fragmentation of the hot molten metal, the contact surface between the metal and water per unit volume of water will be correspondingly very large, and the evaporation rate per unit volume of water will remain extremely high despite film boiling.

4.2.7 Case Histories of Unconfined Water-Vapor Explosions

According to Witte et al. (1970) and the Rasmussen Report (1974), water-vapor explosion accidents have occurred periodically in metal production industries. Such accidents demonstrate the magnitude of the destructive forces that can be generated in such explosions, and the circumstances giving rise to the accidents.

4.2.7.1 Metallurgical Industries

4.2.7.1.1 Mallory-Sharon Incident

In 1954, a titanium arc-melting furnace, which was water cooled, exploded at a plant in Ohio. Four people were killed and five injured. The explosion was believed to result from explosive evaporation of water entering a melting crucible.

4.2.7.1.2 Reynolds Aluminum Incident

In 1958, an aluminum—water explosion occurred in Illinois, USA. Six people were killed and 46 injured. The explosion "rocked a 25 mile area." Wet scrap metal was being loaded into a furnace when the explosion occurred.

4.2.7.1.3 Quebec Foundry Incident

This accident occurred before 1970 in a foundry building of volume approximately 0.5 million m^3. Forty-five kilograms of molten steel fell into a shallow trough containing about 0.35 m^3 of water. The resulting explosion killed one person and injured several others. The material damage to the foundry building included cracking an 8-cm-thick concrete floor, breaking 6000 panes of glass, and causing structural damage to walls and ceilings.

4.2.7.1.4 Western Foundries Incident

In 1966, according to the Rasmussen Report (1974), about 1400 kg of molten steel was being poured from an electric furnace into a tile-lined ladle. During this process a cable broke, causing the hot molten steel to be accidentally poured into a water-filled pit. The result was a violent explosion that injured three workers and tore a 55-m^2 hole in the roof of a building of about 1100 m^2 floor area. The explosion was heard about 5 km from the foundry.

4.2.7.1.5 Armco Steel Incident

In 1967, according to the Rasmussen Report (1974), an explosion occurred when molten steel fell on "damp" ground. A ladle containing several tons of molten steel had been elevated about 12 m when the ladle fell down. Six workers were killed and 24 were injured. Although the deaths and injuries were primarily attributed to burns, explosions were involved in the accident. It was concluded that sufficient moisture was present in the porous ground to trigger a series of small-scale water-vapor explosions that showered molten steel over a wide area.

4.2.7.2 Nuclear Reactor Industries

The following case histories from nuclear power plants are all taken from the Rasmussen Report (1974):

4.2.7.2.1 SPERT 1-D Reactor Accident

As part of an extensive reactor safety test program, test reactors were on purpose subjected to extreme operating conditions. An accidental water-vapor explosion resulted when a reactor was subjected to a so-called "transient" test. The reactor was equipped with pressure transducers and a pressure pulse of peak value more than 200 bars was recorded, which caused the reactor to fail. The pressure pulse occurred some 15 ms after initiation of a power excursion, which rapidly overheated the fuel rods. Both the fuel rods and the rod claddings melted, followed by dispersion of molten material into the cooling water of the reactor.

4.2.7.2.2 Canadian NRX Reactor

In 1952, at Chalk River, Ontario, Canada, during a low-power experiment, a nuclear excursion was experienced. Although the duration of the accident was less than 62 s, the damage was sufficient to result in contamination of the facility. An explosive reaction between molten uranium and water was the principal cause of damage. Whether this was merely explosive production of water vapor, or a chemical hydrogen—air explosion was also involved, was not stated in the very brief report available.

4.2.7.2.3 Borax I Reactor

In 1954, at the National Reactor Testing station in Idaho, USA, a nuclear reactor, as part of a reactor safety study, was deliberately subjected to a potentially damaging power excursion. This lasted for approximately 30 ms and produced a peak power of 19,000 Mw. The power excursion melted most of the fuel elements. The reactor tank, made of 12-mm-thick steel plate, ruptured when exposed to the very high peak pressure, probably in excess of 700 bars, resulting from the extremely fast evaporation when the water made contact with the molten metal. It was estimated that the sound of the explosion at the control station about 8 km away from the center was comparable to that from detonation of 1 kg of 40% dynamite.

4.2.7.2.4 SPERT 1-D Reactor

During the final test of a "destructive" test program with a nuclear-reactor core, damaging overpressure generation was observed. Pressure transducers recorded a pressure pulse in excess of 200 bars, which caused the destruction of the core. The pressure pulse occurred about 15 ms after initiation of the power excursion. The power excursion rapidly overheated the fuel plates, and both the plates and their metal cladding melted down. After the transient, much of the fuel that had been molten was found dispersed in the coolant.

4.2.7.2.5 SL-1 Reactor

In January 1961, a nuclear excursion occurred in the SL-1 reactor in Idaho, USA. The total energy released in the excursion was approximately 130 Mw. Of this, 50 Mw were produced in the outer fuel elements of the core. This portion of the energy was slowly transferred to the water coolant over a 2-s period, and no melting (U—Al alloy) of the outer fuel elements occurred.

In less than 30 ms, about 50—60 Mw of energy were promptly released into the cooling water from 12 heavily damaged internal fuel elements. This resulted in very rapid steam generation and excessive pressure buildup in the core. The high pressure accelerated the water above the core, producing a water hammer that hit the lid of the pressure vessel. The vessel, weighing in excess of 2 tons including its internals, had its connecting piping sheared off and was lifted approximately 3 m into the air by the momentum transferred from the water hammer. Calculations based on the mechanical deformation of the vessel indicated that about 12% of the prompt energy release, or 4.7% of the total nuclear energy release, was converted into excess water-vapor overpressure, ie, mechanical energy.

4.2.7.2.6 The 1986 Chernobyl Catastrophe

This catastrophic accident occurred 12 years after the issue of the Rasmussen report (1974). The following information is gathered from and edited on the basis of various Internet sources. On April 26, 1986, a dramatic disaster occurred at the nuclear power plant at Chernobyl to the north of the capital Kiev in Ukraine. In the disaster, one of the nuclear reactors, No. 4, was destroyed, causing the death of 30 operators and firemen within 3 months and causing several later deaths. One person was killed instantly and a second died in the hospital soon after the accident of injuries received. Within a few weeks of the disaster, 28 more people died because of excess radioactivity exposure.

On 25 April, prior to a routine shutdown, the reactor crew at reactor 4 began preparing for a test to determine how long the turbines would spin and supply power to the main circulating pumps following a loss of main electrical power supply. This test had been carried out at Chernobyl the previous year, but the power from the turbine ran down too rapidly, so new voltage regulator designs were to be tested.

A series of unfortunate operator actions, including the disabling of automatic shutdown mechanisms, preceded the attempted test early on 26 April. By the time the operator was to shut down the reactor, the reactor was in an extremely unstable condition. The interaction of very hot reactor fuel with cooling water led to fuel fragmentation, very rapid steam production and increasing water-vapor pressure. The increased pressure in turn caused partial detachment of a 1000 tons cover plate of the reactor and rupturing of the fuel channels and jamming all control rods. Violent steam production, fed by the water that had been dumped into the core following the rupture of the emergency cooling-water circuit, then spread throughout the whole reactor core. This eventually gave rise to the first major steam explosion, which also released radioactive fission products to the atmosphere.

About 2—3 s later, a second major explosion threw out hot graphite and fragments from the fuel channels. There was some dispute among experts about the nature of this second explosion, but it was probably a water-vapor explosion. It was likely to have been caused by ignition of an explosive hydrogen—air mixture stemming from hydrogen produced from zirconium—water reactions.

Apparently, the root causes of the Chernobyl disaster were poor reactor design, and some serious mistakes made by plant operators. However, in the 1991 report by the "State the Committee on the Supervision of Safety in Industry and Nuclear Power" it was emphasized that in making these mistakes the operators could not be blamed for having violated any vital operating policies and principles issued by the nuclear power station, because no such policies and principles had been presented to them.

4.2.7.3 Paper production industry

According to Witte et al. (1970), the paper industry has experienced a number of unconfined water-vapor explosions, but they have been generally less destructive in nature than those in the metals industries. Explosions typically occurred when hot molten sodium carbonate "paper smelt" was cooled in large containers of "green liquor," which is a solution in water of various inorganic compounds from the recovery boiler in the "kraft process" being a partial process in paper production.

4.3 SOME OTHER KINDS OF WATER-VAPOR EXPLOSIONS

4.3.1 Water Boiler Explosions

Accidental water boiler explosions have occurred since the invention of the steam engine. James Watt patented his engine for rotary motion in 1781. One important application was to power steam engine locomotives for trains.

In water boilers, steam is produced from a bath of liquid water that is heated to above the boiling point 100°C of water at atmospheric pressure. This is possible because the water-vapor pressure inside the confined boiler space increases correspondingly during the heating process. In normal boiler operation, the liquid water remains in the bottom part of the boiler due to gravity, while steam bubbles rise through the liquid water and are collected at the boiler top for use.

However, a steam boiler may open up to the atmosphere accidentally because of a boiler rupture or other loss of confinement. Because the water in the boiler before boiler failure is at a higher temperature (and pressure) than boiling water would have been at ambient atmospheric conditions, some of the water will flash into steam as soon as the boiler fails and the vapor pressure above the water drops to ambient. This rapid release of steam can then generate a strong blast wave that can cause great damage to both humans and property.

4.3.2 Water-Vapor Boiling Liquid Expanding Vapor Explosions

The BLEVE mechanism discussed in Chapter 3 is one way in which water vapor can evolve at explosive rates. In some cases, water boiler explosions (see Section 4.3.2) may just as well be classified as water-vapor BLEVEs. If the temperature of pressurized water in a closed vessel (boiler) exceeds the superheat limit temperature of water at atmospheric pressure, the water can, if suddenly exposed to atmospheric pressure following tank rupture, change from liquid to vapor at an extreme rate, resulting in an explosive volume increase.

Such a water-based BLEVE in fact occurred in the Mihama Nuclear Power Reactor, Japan, in August 2004. According to Abbasi and Abbasi (2007), the initiating event was the appearance of a leak in a large pipe carrying superheated water. The leak instantaneously reduced the pressure inside the pipe to atmospheric and the resulting violent nuclear boiling of the superheated water gave rise to a major water-vapor explosion. The two-phase release of superheated water and steam scorched 11 workers. Some of these lost their lives, the others were severely injured.

4.3.3 Three Other Cases of Water-Vapor Explosions in Confined/Semiconfined Systems

4.3.3.1 Explosion in Norway in a Double-Cone Mixer for SiC Powders in About 1985

The following review is based on information obtained from Lövsland (2015). The company experiencing this accident produced various particle-size grades of silicon carbide (SiC) used eg, in emery paper and grinding wheels. The essential steps in the production process were:

- crushing and fine milling
- wet classification by water in various particle-size ranges
- drying of each particle fraction in a specially constructed dryer
- homogenization of each dried fraction in a double-cone mixer as illustrated in Fig. 4.5

Due to the special construction of the dryer, it appeared that some of the wet material near the dryer bottom was not incorporated in the drying process in an adequate manner, and hence was still wet when the drying process was assumed to be completed and the content of the dryer was discharged into the double-cone mixer. The major part of the powder in the dryer that had actually been dried was quite hot.

The explosion occurred soon after the startup of the double-cone mixer. One of the cones was torn away from the rest of the mixer along the weld connecting it to the central cylindrical part and shot across the room into an opposite wall. People were around, and it was considered extremely fortunate that no persons were killed or injured.

At the outset, several alternative explosion scenarios were investigated, including a possible dust explosion. Laboratory tests confirmed, however, that even very fine SiC powder did not cause dust explosions in air.

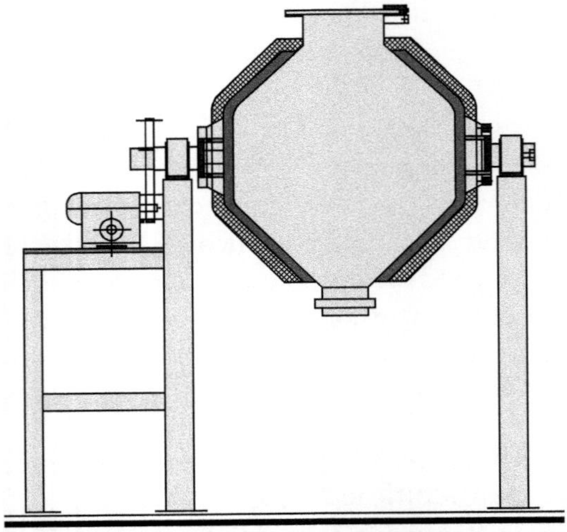

Figure 4.5 Illustration of the type of double-cone mixer that was blown open by an internal water-vapor explosion in a SiC production company in Norway in about 1985.

It was finally concluded that the explosion was in fact caused by the overpressure generated inside the sealed mixer by water vapor generated at a high rate when the hot, dry SiC powder was effectively mixed with the wet, cool SiC powder as soon as the rotation of the mixer had started.

The obvious lesson to be learned from this particular case history is that mixing of hot and dry material with cold and wet material in an enclosed mixer, can give rise to rapid and substantial pressure rise inside the mixer, and possible explosive mixer rupture. Therefore, such mixers should always be thoroughly inspected before new, hot material is admitted to them.

4.3.3.2 Water-Vapor Explosion in Norway in 1988 in an Aluminum Melting Pot

The following review is based on a short paper by Dybdahl (1989). In this explosion accident, three workers lost their lives, and one was seriously injured. The explosion occurred inside the walls and bottom of a melting pot with a capacity of 5–6 tons of molten aluminum. The explosion accident happened soon after molten aluminum at about 1000°C had been poured into the melting pot for the first time after a major renovation.

A cross-section of the melting pot is given in Fig. 4.6.

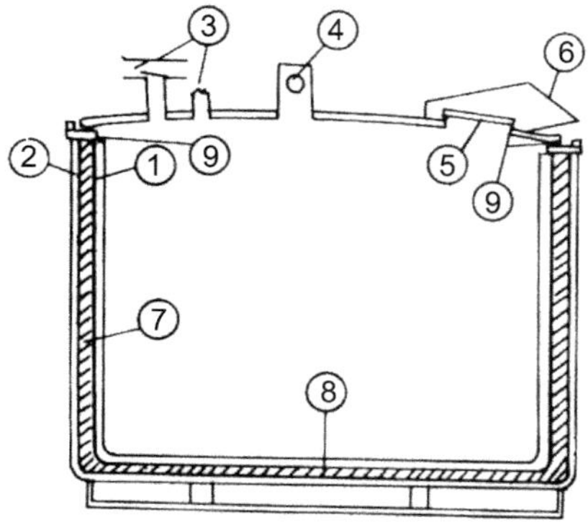

Figure 4.6 Illustration of the melting pot, of capacity 5–6 tons of aluminum, in Høyanger, Norway, which suffered a water-vapor explosion in 1988. (1) Inner mantle of cast iron. (2) Outer mantle of steel. (3) Two ejectors. (4) Hole for charging molten Al into melting pot. (5) Opening for pouring molten Al out of melting pot. (6) Chute for guiding the poured molten Al out of melting pot. (7) Precast thermal insulation type RC741. (8) Ceramic fiber plates. (9) Gaskets.
From Dybdahl, K., 1989. Vanndamp og gass sprengte digelen (Water vapour and gas burst the melting pot). Teknisk Ukeblad/Teknikk, vol. 136, no. 4, pp. 24.

The melting pot consisted of an inner mantle of cast iron, and outer steel mantle, and thermal insulation between the two. The renovation implied replacement of the worn-out inner cast iron mantle and of the thermal insulation materials in the pot walls and bottom. Unfortunately, some new materials had been used for the thermal insulation. In the walls of the melting pot, a new refractory material containing water of crystallization, and preventing proper ventilation for pressure relief, was used. The material was applied mixed with water, and after hardening some of the water remained as water of crystallization. It also appeared that the insulating material used in the bottom of the melting pot released water vapor and other gases when being heated to the high working temperatures of the melting pot.

It was concluded that this tragic explosion accident would not have occurred if the original, dry insulation materials had been used when the melting pot was renovated. The commission investigating the accident recommended that the insulation in all new or freshly renovated melting pots should be according to the original specifications for the pot. Furthermore, new and freshly renovated melting pots should be dried out at a sufficiently high temperature before being put in operation. It was also recommended that routines for melting pot inspection before and during use be improved.

4.3.3.3 *Severe Accidental Water-Vapor Explosions in a Foundry in China in 2012*

4.3.3.3.1 Venue and Consequences

The accident is described and analyzed in detail in Li and Ji (2015). The site of the explosion was an indoor molding pit of the Steel Casting Plant of the Heavy Machinery Co. Ltd., a part of the Anshan Iron and Steel Group in the Liaoning Province in northeast China. The explosion occurred on February 20, 2012. Fifteen workers lost their lives and 15 others were injured in this severe accident.

4.3.3.3.2 The Stainless-Steel Object to be Molded

The stainless-steel object to be molded was a conical ring weighing about 95 tons and supposed to constitute the foundation of a water turbine. The upper outer diameter of the ring was 6 m, the lower outer diameter 8 m, the height 2 m, and the wall thickness 0.2 m.

4.3.3.3.3 The Molding Pit

The molding pit was located inside a large factory building and had been in use for a number of years. When the pit was first made, a 4.7-m-deep hole of 13.5 × 13.5 m cross-section was first dug out of the ground. Then the hole was lined with a whole-welded steel box of 10-mm wall thickness, which was in turn lined with a 700-mm-thick concrete layer (on floor and walls). The internal dimensions of the pit were therefore a depth of about 4 m and sidewall lengths of about 12 m.

4.3.3.3.4 Penetration of Ground Water into the Pit

In the area of the molding pit, the ground-water level was about 2 m below the prevailing ground level at the location of the pit. This means that the lower part of the steel plate/concrete box in the pit, up to about 2.5 m from its external bottom, was below the ground-water level.

During normal operation, the factory building in which the pit was located was heated during wintertime. Although outdoor temperatures were typically about −25°C in January/February, indoor temperatures would be significantly above 0°C. In the case of a crack, a corroded hole, or an inadequate weld in the steel lining and a corresponding passage through the concrete wall/floor, ground water could enter the pit and moisten any sand there.

After the explosion, and after all the remaining sand had been removed from the pit, some cracks in the concrete wall were discovered. These cracks were in part below the level of the sand left in the pit before the 7-day festival break. If these cracks existed even before the explosion, they could well have been the gate admitting ground water into the pit.

4.3.3.3.5 Preparation of the New Sand Mold for Casting the Stainless-Steel Ring

To produce the sand mold for the casting of the stainless-steel ring, the workers first made a provisional base of sand at the bottom and in the corners of the pit to facilitate their entrance into the pit and the further work with producing the final sand mold. However, after having done this, the work was interrupted by a 7-day festival celebration holiday. During this break, ground water could seep into the sand in the pit through possible minor openings in the steel—concrete floor and walls of the pit. After the holiday, the workers resumed their work with completing the sand mold for the stainless-steel ring. However, because the heating of the factory hall had been minimized during the holiday break, the moistened sand inside the sand bed had become frozen and was not detected by the workers entering the pit after the holiday break.

4.3.3.3.6 Probable Sequence of Events Leading up to the Explosion

This severe accident was most probably a confined water-vapor explosion. Li and Ji (2016) suggested the following sequence of events leading up to the explosion:

The sand mold was completed and the pouring of molten stainless steel into the mold was performed just after the 7-day holiday break. However, a considerable amount of sand had been left in the molding pit from before the start of the holiday. During the holiday break, this sand became wetted by ground water seeping into the molding pit through leaks in the steel and concrete walls.

When the workers resumed their work after the break period, the sand mold was completed using freshly prepared dry molding sand. When the mold had been completed, it was filled with molten stainless steel. The mold then immediately received a substantial heat flux from the hot metal, which was conducted downward and sideways into the regions of the sand in the pit containing water. Pressurized water vapor was then generated throughout the sand bed by fast evaporation of the water trapped in the "old" sand at the bottom and in the corners of the molding pit. However, this process, and the associated buildup of water-vapor pressure throughout the sand, nevertheless required some time, and the pouring of the molten stainless steel into the sand mold could be continued without any observable problems for about 30 min.

However, the entire sand mold, with its content of molten stainless steel, was suddenly lifted 5 m into the air, just as by a piston, and shattered all around, killing 15 workers and injuring another 17.

4.4 MEASURES FOR PREVENTING AND MITIGATING WATER-VAPOR EXPLOSIONS

4.4.1 Unconfined Water-Vapor Explosions

A violent unconfined water-vapor explosion can eject, in all directions, steam, boiling-hot water, molten and solidified metal drops, and debris from previous confinement. In industry, this of course presents a considerable potential for fatalities, injuries and material damage. The information traced on specific technical and other measures for preventing various kinds of water-vapor explosions, is scarce.

In their early review, Witte et al. (1970) wrote as follows: "Interest in the vapour explosion stems from the need to prevent destructive incidents. Much research is needed to isolate the controlling parameters before effective prevention is possible."

They did, however, refer to some experiments that had shown that surface coatings, such as grease or hydrophobic paint coatings, on the bottom of water reservoirs could prevent extremely violent water-vapor explosions. This was explained in terms of the nonwetting surface preventing water from being trapped underneath the first large molten-metal globule when it approached the containment bottom on its way downward in the water.

Long (1957) investigated another preventive/mitigatory measure with a molten-aluminum—water system. In a series of outdoor pilot experiments, he noted that "the explosons[sic] were characterized by a violent bursting action, closely followed by by[sic] a loud noise and a shock wave." However, if the initial large molten-metal globule poured from the melting pot was broken up into smaller globules by passing it through a steel grid before it entered the water, no water-vapor explosion occurred. On the basis of Long's experiments it may appear, therefore, that the possibility of a water-vapor explosion can be reduced if the molten metal is deliberately fragmented into comparatively small droplets prior to contact with the water. As a technical measure to achieve this, Witte et al. (1970) suggested that "a suitable grid be installed above any sizeable water reservoir where molten metal could in some way be in the near vicinity."

As with all other kinds of explosion hazards in industry, adequate safety guidelines and adequate education and training of personnel are crucial. Adoption of the basic principles of making plants inherently safer, as summarized by eg, Srinivasan and Natarajan (2012) is also essential. In the present context, this would imply measures by which the possibility of accidental contact between hot molten metal and water is eliminated to the extent possible.

4.4.2 Confined Water-Vapor Explosions

The examples in Section 4.3.3 illustrate that the circumstances under which confined water-vapor explosions occur in practice vary considerably. It is not easy, therefore, to suggest specific, generally applicable measures for preventing such explosions. Learning "the hard way" from actual accidents still seems the main rule. However, the general recommendations in Section 4.4.3 also apply to confined water-vapor explosions.

4.4.3 General Recommendations

In order to prevent accidental water vapor explosions, availability of and living up to adequate safety guidelines are essential. To achieve this, education and training of personnel are crucial. When the potential hazard is water-vapor explosions. Whenever possible and relevant, it is also important to adopt and apply the basic principles of making plants inherently safer, as summarized by eg, Srinivasan and Natarajan (2012). In general, this would imply adoption of measures by which the possibility of accidental contact between hot molten metal and water is eliminated or to the extent possible.

4.5 CONCLUDING STATEMENT

Lee and Frost (1987) stated that water-vapor explosions were a new research area in comparison to that of classical explosion hazard areas, such as gas and dust explosions. They did, however, postulate that "in view of its importance to the metallurgical and nuclear industries, the growth of activity in this field is expected to be rapid in the future."

It may seem that their postulate has not really come to fruition. In the decades that have elapsed since 1987, only a comparatively limited number of scientific and technical papers have been published in this area. However, both metallurgical and nuclear industries are likely to remain important for a long time to come, and the water-vapor explosion hazard will therefore continue to exist.

CHAPTER FIVE

Explosions of Clouds of Combustible Liquid Droplets in Air

5.1 INTRODUCTION

Sprays and mists of combustible liquids, eg, hydrocarbons, in air at atmospheric pressure and normal temperature, with a droplet size of, say, <100 μm and a droplet mass concentration in the range 100−500 g/m^3, are explosive. This is so regardless of whether the liquid is of low or high boiling point. In the case of a low boiling-point liquid, the droplets will evaporate readily and the cloud will very soon become a mixture of combustible vapor and air. If the boiling point is high, ie, the vapor pressure at normal ambient conditions is low, the droplets will, with regard to the combustion process, behave similarly to solid particles of an organic material, which are also known to cause explosions when dispersed as clouds in air.

The phenomenon of spray/mist explosions is complex. Similar to gas and dust explosions, they are the result of two main consecutive processes, viz. generation and combustion of an explosive cloud. Ignition may be regarded as the coupling between the two main processes. Because of the partly unstable two-phase nature of a spray/mist, the cloud generation process is in itself very complex.

The terms "spray" and "mist" are used somewhat arbitrarily in the present text, although one can argue that there is a clear difference between the two. "Spray" means a cloud of liquid droplets extending from one or more nozzles (eg, accidental leaks from high-pressure process equipment), whereas "mist" denotes a cloud generated by condensation from a supersaturated fuel vapor. This implies that droplets in mists are generally smaller than typical spray droplets. With regard to accidental explosions, both "spray" and "mist" may be relevant terms depending on the actual accident situation.

Spray/mist combustion research may be motivated by two different needs. The first is to prevent and control accidental explosions, whereas the second is to improve diesel engines, gas turbines, furnaces, rockets

Explosion Hazards in the Process Industries
ISBN: 9780128032732
http://dx.doi.org/10.1016/B978-0-12-803273-2.00005-0

etc., in which liquid-fuel sprays are burnt deliberately. Published research on accidental spray/mist explosions is scarce, and it is therefore useful to consider literature related to deliberate generation, ignition, and combustion of sprays. However, the basic phenomena are the same, whether the research is conducted to prevent explosions or to improve combustion engine performance, and the reviewed work is therefore largely relevant for resolving the explosion problem. Zehr (1965) pointed out that a spray/mist and a cloud of solid particles (dust) have common features in that both consist of a finely divided dense fuel phase suspended in an oxidizer gas. However, explosive mist clouds are less stable than explosive dust clouds because collisions between droplets give rise to coalescence and transformation to fewer and larger droplets. When the droplet size gets sufficiently large, the droplet sedimentation velocity in the gas becomes significant and the droplets settle out of the cloud.

In his instructive book on combustion of liquid fuel sprays, Williams (1990) covered a wide range of topics including various properties of liquid fuels, characteristics of sprays in terms of drop size and drop velocity distributions, processes for atomization of liquid fuels, and combustion of single droplets and droplet clouds. Both experimental work and theoretical modeling are discussed. Bowen and Cameron (1999) gave an informative review of hydrocarbon aerosol explosion hazards. Förster (2000) also contains useful information on generation, ignition, and combustion of sprays/mists.

5.2 GENERATION OF CLOUDS OF LIQUID DROPLETS IN AIR

5.2.1 General

The production of mists is discussed by Green and Lane (1964) in their classical reference book on particulate clouds. According to these authors, mists are produced by condensation, accomplished by sudden cooling of high-concentration vapor clouds. A classical theory describing this kind of condensation processes is discussed by Green and Lane. Sometimes small solid particles act as condensation nuclei. Mists can also be generated by slow condensation on solid nuclei.

5.2.2 Mechanisms of Spray Generation

Valuable insight in the mechanisms of accidental spray generation can be obtained by studying the methods used for producing sprays for research purposes in the laboratory. Williams (1990) reviewed various methods for spray formation by "atomization" of liquids. When a liquid is atomized, energy is expended mainly in three ways, viz. in forming new surface, in overcoming viscous forces in changing the shape of the liquid, and in losses due to inefficient application of the energy to the liquid. Atomizing devices commonly used in generating liquid sprays may be classified in the following three main types:

- *Air-blast or aero-dynamical atomizers* in which compressed air or other gas at high velocity is used to break up liquid emerging from a nozzle, and plays a predominant part in achieving a fine degree of atomization. This kind of breakup is adopted in conventional paint spray guns, in Venturi atomizers, and in many atomizers that generate aerosols for insecticidal, disinfectant, and therapeutic purposes. It is characteristic of atomizers of this type that they give a very wide range of droplet sizes, though in some cases the range is reduced by trapping the larger droplets within the atomizer.
- *Generators depending on centrifugal action*, the liquid being fed onto the center of a rotating disc, cone, or top and centrifuged off the edge. The spray is characterized by uniformity of the main droplet size, in marked contrast to the heterogeneity of sprays produced by other methods.
- *Hydraulic or hydro-dynamical spray generator* type, in which liquid is forced through a nozzle and breaks up into droplets. Here, the disintegration depends more upon the physical properties of the liquid and the conditions of ejection from the nozzle than upon interaction between the liquid and the surrounding gas. Probably the most successful hydraulic atomizer, and indeed the only one which has application for fine atomization, is the swirl chamber atomizer used in agricultural spraying equipment, oil-fired furnaces, internal combustion engines, and gas turbines. The swirl is produced by leading the liquid tangentially into the chamber and forcing it to spray out through a central orifice of small diameter.
- *Special atomizers*. These include the electrostatic atomizer which breaks up liquid by the action of electrostatic forces and the acoustic atomizer which utilizes high-intensity sonic or ultrasonic vibrations.

Lefebvre (1989) summarized the challenges in spray generation as follows:

> *The subject of atomization and sprays is one that has attracted the attention of many research workers and has been the focus of numerous theoretical and experimental studies. However, our knowledge of the atomization process is far*

from complete. The physics is not well understood, the available data and correlations for drop size distributions are sometimes of questionable validity, and there is little agreement between the various investigators as to the exact relationships between liquid properties, nozzle dimensions, and mean drop size. These comments are especially true for pressure-swirl atomizers. Many factors contribute to this unsatisfactory situation. They include the great complexity of the atomization process, differences in the design, size, and operating conditions of the nozzles tested, and the difficulties involved in accurate drop size measurement. These difficulties include the very large number of drops in a spray, the high and varying velocity of the drops, the wide range of drop sizes encountered in most practical sprays, and the changes in drop size with time due to evaporation and coalescence.

5.2.3 Coalescence of Drops in Sprays/Mists

A comprehensive review of the state of the art several decades ago was given by Green and Lane (1964). Much of this information is still valid. The same applies to the comprehensive review by Zebel (1966), who considered three main categories of coagulation mechanisms. The first is thermal coagulation, driven by Brownian movement of the droplets, and the second is coagulation influenced by interdroplet forces (van der Waals and electrical/magnetic forces). The third category is coagulation under the influence of external force fields (electrical and magnetic, gravitational and centrifugal, sonic, and laminar and turbulent flow fields).

Förster (1990) argued that accidental generation of large turbulent explosive clouds of sprays/mists of organic liquids of high boiling points is not very likely. The reason is that the mean droplet—droplet distance in the explosive range is on the order of only 10 droplet diameters, which in a turbulent cloud makes fast coalescence of the small droplets to larger ones highly probable. The larger droplets will then "rain out" and the fuel concentration in the cloud will fall below the explosive range. Förster confirmed experimentally that, with a high-boiling-point liquid (octanol), it was indeed very difficult to generate an explosive spray in a 1 m^3 explosion chamber, unless a highly sophisticated spray nozzle system was adopted.

5.3 COMBUSTION OF CLOUDS OF LIQUID DROPLETS IN AIR (SPRAY/MIST)

5.3.1 General

Zehr (1965), discussing the explosion hazard, emphasized that a mist and a dust cloud have common features in the sense that both consist of a finely divided dense phase suspended in a gas. However, as said in Section 5.2.3, mist clouds of high mass densities are less stable than high-density dust clouds because collisions between small drops inevitably give rise to coalescence and transformation of the cloud into one of fewer and larger drops. When the drop size gets sufficiently large, the drop sedimentation velocity in the gas becomes significant and the drops settle out of the cloud.

As pointed out by Zehr, the ignition and combustion of droplets of organic liquids, and of particles of solid organic materials, are similar. The thermal energy from an ignition source, or from nearby burning droplets/particles, first causes the liquid of the droplet/particle to evaporate partly or fully (via pyrolysis for liquids of high boiling points and organic solids). Further heating causes the vapor to ignite and burn, either as individual diffusion flames around each droplet/particle, or as a more or less homogeneous premixed vapor/air flame. Faeth (1987) discussed the theoretical description of mixing, transport, and combustion in sprays and mists. In particular, advances concerning the description of drop—turbulence interactions were reviewed.

In his instructive book, Williams (1990) covered a wide range of topics relevant to spray and mist explosions. This includes various properties of liquid fuels, characteristics of sprays in terms of drop size and drop velocity distributions, processes for atomization of liquid fuels, and combustion of single droplets and droplet clouds. Both experimental work and theoretical modeling are discussed.

Understanding the details of the ignition and combustion of individual liquid droplets is a premise for understanding the combustion of collectives of such droplets, ie, spray and mist combustion. It is, however, outside the scope of the present text to discuss these aspects any further.

5.3.2 Laminar or Close-to-Laminar Flame Propagation in Sprays/Mists

In their classical investigation, Burgoyne and Cohen (1954) developed a method whereby sprays and mists of controlled uniform drop size could be prepared from pure liquids. Using tetralin as the fuel, it was possible, within limits, to study the effect of the drop size on the combustion properties of a suspension of liquid drops in air. Through a study of limits of flammability, nitrogen dilution limits, and burning velocities, it was shown that the mechanism of flame propagation was completely transformed over the drop-size range from 7 to 40 μm. Thus, below about 7 μm the suspension behaved like a premixed gas, whereas above 40 μm the drops burnt individually, in their own air envelope, one burning drop igniting adjacent ones, thus spreading combustion. At intermediate sizes, behavior was transitional. A practical consequence of this transformation was that the lower concentration limit for flame propagation decreased and the rate of burning increased with increasing drop size.

In a later study, Burgoyne (1963) discussed the concept of "concentration" of the fuel in this context. The following conclusions were drawn:

- For suspensions formed by uncontrolled condensation of saturated vapor (mists), the drop size is small and the lower limit of flammability, measured as mass of liquid per unit volume of air, is essentially the same as that of the vapor. For saturated hydrocarbons, this is on the order of 0.045 g/m^3.
- For suspensions formed mechanically by liquid breakup, the drop size is considerably larger, and in specifying lower-limit concentrations account must be taken of the distinction between "static" and "kinetic" concentrations, arising from the sedimentation of the drops.

Burgoyne (1963) also pointed out that in very coarse sprays, with drop diameters between 0.6 and 1.4 mm, normal flame propagation becomes impossible although burning drops may carry flame downward. However, the presence of smaller droplets in such suspensions may, depending on their concentration, make upward flame propagation possible. Furthermore, the entry of a fast-moving explosion flame (eg, from a pipe run) into a chamber containing a droplet suspension that is normally noninflammable due to large drop size, flame propagation may nevertheless occur, due to the shattering effect on the drops of the pressure waves preceding the explosion flame.

Published work on exact burning velocities of explosive sprays/mists is scarce. This may in part be due to the rather complex experiments required, in which fuel concentration and droplet size have to be controlled independently. In addition, with low-boiling-point liquids, droplet evaporation during droplet cloud generation introduces further complications. Fig. 5.1 gives some results from two independent studies of close-to-laminar flame propagation of tetralin droplets in air (tetralin is a cyclic organic compound of molecular weight 132, boiling point 207°C, and flash point 71°C).

As Fig. 5.1 shows, the two independent investigations gave somewhat diverging results for droplet diameter between 10 and 20 μm. However, both studies conclude that there are two different combustion regimes of

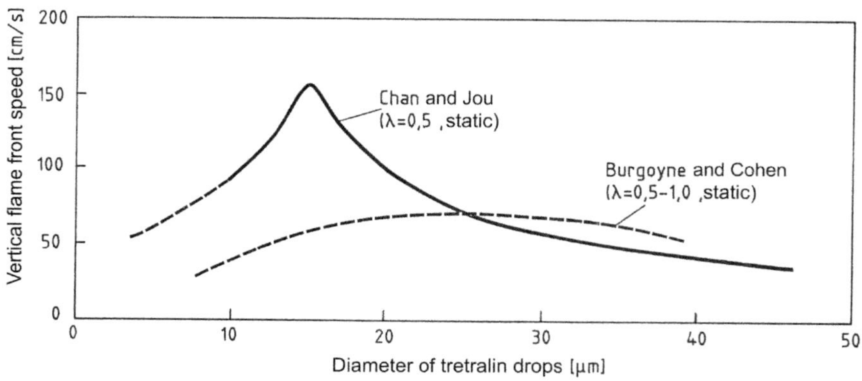

Figure 5.1 Experimental flame speeds (vertical upward flame propagation) in suspensions of mono-sized tetralin droplets in air, as a function of droplet size.
From Burgoyne, J.H., Cohen, L., 1954. The effect of drop size on flame propagation in liquid aerosols. Proceedings of the Royal Society of London. Series A 225, 375–392 and Chan, K.-K., Jou, C.-S., 1988. An experimental and theoretical investigation of the transition phenomenon in fuel spray deflagration. Part 1: the experiment. Fuel 67, 1223–1227.

the tetralin droplets, depending on droplet size. For sizes <10 μm, the droplets evaporate completely before combustion gets under way, and the flame propagation process resembles that of premixed gaseous fuel/air. For sizes >50 μm, however, each droplet burns individually, and flame propagation is more discrete, from droplet to droplet. The distinct peak in vertical burning velocity that is exhibited by the data of Chan and Jou (1988) occurs in the region of drop sizes between these two regimes.

Bowen and Cameron (1999) reviewed several mathematical models for predicting laminar burning velocities in clouds of liquid fuels. They all agree that for droplet diameters >30 μm burning velocities cannot exceed that of the equivalent homogeneous vapor/air mixture. However, for the transition range of 7–15 μm droplet diameters, the various model predictions are inconsistent. Some models do predict a significant increase of the burning velocity in the transition region, but Bowen and Cameron called for further experimental data against which the models can be tested.

Apart from the comparatively high peak value observed by Chan and Jou (1988) in the transition region, Fig. 5.1 does suggest that close-to-laminar burning velocities of sprays and mists of organic liquids in air are on the same order as laminar burning velocities of saturated gaseous hydrocarbon–air mixtures.

5.3.3 Turbulent Flame Propagation in Sprays/Mists

Fig. 5.2 shows some results from flame propagation experiments in turbulent clouds of kerosene droplets in air. The Reynolds numbers were on the order of Re $\approx 10^5$. Kerosene is a petroleum distillate with boiling point range 180–300°C, and flash point range 38–66°C. By increasing the flow velocity in the channel (Re is proportional to the flow velocity) by a factor of 3, the turbulent burning velocity (net velocity in relation to unburnt cloud ahead of flame front) with 90 μm droplets increased by about the same factor. In addition, when the droplet size is reduced to 50 μm and further to 30 μm, the turbulent burning velocity increased systematically with flow velocity, but the relative increase gets smaller with decreasing droplet size.

Fig. 5.3 shows the variation of the turbulent burning velocity with the kerosene/air ratio, for three different droplet sizes. For the smallest droplets the maximum burning velocity occurred at about stoichiometric composition ($\lambda = 1.0$), whereas the peak seems to be shifted systematically toward higher kerosene/air ratios ($\lambda > 1.0$) as the droplet size increases.

Fig. 5.4 shows that with the prevailing flow conditions and range of λ values studied, the turbulent burning velocities for 30 μm droplets were nearly identical for the three different fuels, kerosene, toluene and decalin.

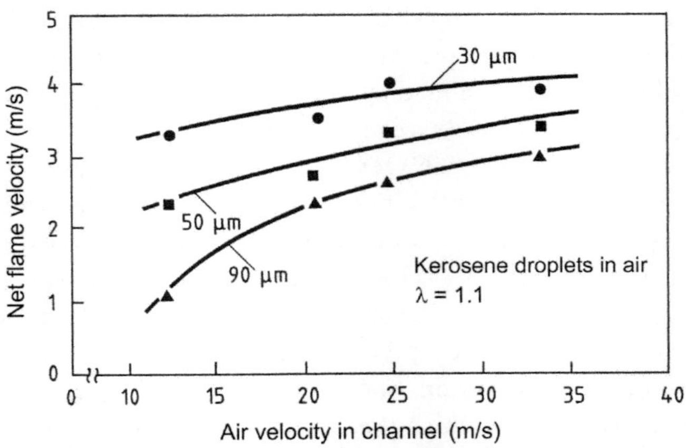

Figure 5.2 Influence of flow-generated turbulence on the burning velocity of suspensions of kerosene droplets in air, for various mean droplet sizes. (Kerosene is a petroleum distillate with boiling point range 180–300°C, and flash point range 38–36°C.) Slightly overstoichiometric fuel concentration (equivalence ratio $\lambda = 1.1$). Horizontal flow of droplet suspension in a channel of rectangular cross-section (width 30 cm, height 10 cm).
From Richards, G.A., Lefebvre, A.H., 1989. Turbulent flame speeds of hydrocarbon fuel droplets in air. Combustion and Flame 78, 299–307.

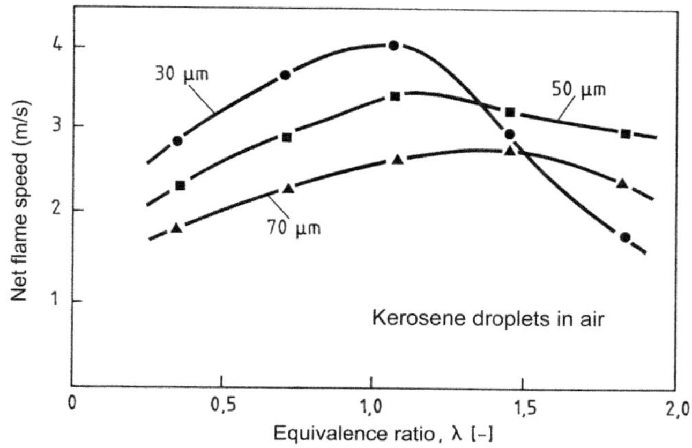

Figure 5.3 Influence of fuel/air ratio and droplet size on burning velocity of suspensions of kerosene droplets in air, flowing at 25 m/s in a horizontal channel of rectangular cross-section (width 30 cm, height 10 cm).
From Richards, G.A., Lefebvre, A.H., 1989. Turbulent flame speeds of hydrocarbon fuel droplets in air. Combustion and Flame 78, 299–307.

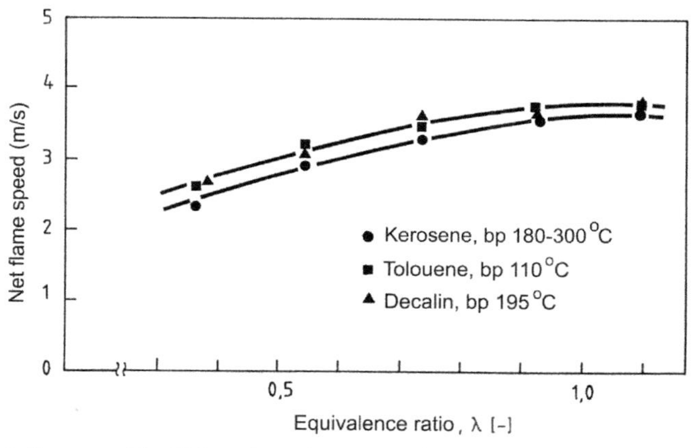

Figure 5.4 Influence of fuel/air ratio on burning velocity of suspensions of 30 µm diameter droplets in air of three different liquid fuels, flowing at 20 m/s in a horizontal channel of rectangular cross-section (width 30 cm, height 10 cm).
From Richards, G.A., Lefebvre, A.H., 1989. Turbulent flame speeds of hydrocarbon fuel droplets in air. Combustion and Flame 78, 299–307.

However, other data obtained by Richards and Lefebvre (1989) indicate that for larger-droplet sizes the burning velocity tends to decrease with increasing boiling point of the liquid. The complete relationships between burning velocity on the one hand, and degree of turbulence, droplet size, fuel/air ratio etc. on the other, are very complex.

Hansen and Wilkins (2004) conducted a series of interesting spray explosion experiments in a straight horizontal vented laboratory channel of length 2.5 × 0.3 m cross-section. They found that:

- Sprays of hexane droplets (high volatility) in air exploded at least as violently as premixed propane/air.
- The violence of spray explosions remained high also for overstoichiometric droplet concentrations.
- The mounting of a baffle in the channel raised the maximum explosion pressure by factor of eight due to intense flow-induced turbulence (cfr. Section 2.1.6 of Chapter 2).
- Under conditions (intense turbulence) causing droplet breakup, even sprays of high-flashpoint oils exploded violently.
- Worst-case spray concentrations increased with increasing droplet size.

5.3.4 Maximum Constant-Volume Adiabatic Explosion Pressures in Sprays/Mists

Experimental data presented by Förster (2000) indicate that the maximum pressures generated in constant-volume adiabatic combustion of clouds of droplets of some organic liquids in air is on the order of 5 bar (g).

5.3.5 Detonations in Sprays/Mists

Sprays and mists of combustible liquids in air can detonate, depending on fuel chemistry, drop size and liquid mass concentration. This is not surprising in view of the fact that both homogeneous mixtures of combustible gases and air and clouds of some organic dusts in air can propagate detonation waves. Bull et al. (1979) performed experiments to determine the marginal conditions for spherical detonation in unconfined fuel/air sprays. The sprays were generated in a 5 m^3 plastic foil tent by sonic air-blast atomizers producing measured droplet sizes in the range 5—30 μm. Detonation of the spray cloud was initiated by detonating plastic high-explosive charges positioned inside the cloud. Particular attention was paid to the study the influence of fuel volatility on the detonability of the spray. N-hexane and n-dodecane were used to represent extremes of very high and very low vapor pressure fuels. Fuels having intermediate volatilities were synthesized by mixing the two fuels in appropriate proportions.

It was found that n-hexane sprays detonated within fuel—air mass concentration limits similar to those for gaseous detonations, although the measured detonation front velocities, as for planar two-phase detonations, were somewhat lower than the calculated Chapman—Jouquet (CJ) values, even with droplet diameters smaller than 50 μm. However, sprays of n-dodecane and n-decane did not detonate, even when using 0.5 kg initiator charges. It was concluded that this was because with these fuels the minimum effective equivalence ratio necessary for detonation propagation was not being attained behind the shock wave within the limiting induction period available. It was further concluded that propagation of self-sustained detonations in n-alkane sprays in the spherical mode, with droplet sizes larger than 10 μm, may require the presence of a certain quantity of fuel vapor prior to ignition.

Smeets (1985) performed experimental spray detonation studies in a 50-cm-diameter vertical shock tube. The sprays were stoichiometric clouds of droplets in air of methanol, ethanol, propanol, hexanol, decanol, or decane. The droplets had a defined narrow-size distribution with a mean diameter of 350 μm. It was found that even in sprays of low vapor pressure fuels, like n-decane and hexanol, self-sustained detonations can exist. However, a minimum concentration of fuel vapor is required, in addition to the droplets, as also indicated by Bull et al. (1979), discussed previously. By measuring several physical quantities within the reaction zone, using laser Doppler techniques, the processes of liquid fuel fragmentation, evaporation, turbulent mixing, and combustion could be observed in detail. From the data recorded it was concluded that for sprays of 350-μm-diameter droplets, the controlling mechanism for the reaction is turbulent mixing of the vapor originating from the droplets, with air.

Tang et al. (1986) provided an explanation of the difference between the theoretically predicted and the experimentally observed detonation velocities in clouds of droplets of low-vapor-pressure liquids in air or oxygen. In an experimental investigation, mono-sized n-decane droplets of diameter 400 μm were allowed to settle vertically in air or oxygen at ambient pressure and temperature in a vertical shock tube. A shock wave was emitted vertically downward into the droplet suspension, initiating a self-sustained detonation wave in the cloud. The propagation of this wave was monitored using pressure switches and pressure transducers.

Measured detonation velocities were then compared with theoretical predictions. The 0th order analytical prediction was just a CJ calculation for premixed gas. The first-order prediction considered the wall momentum and heat-transfer losses, whereas the second-order refinement also considered the effect of incomplete combustion of the droplets. When all these refinements were taken into account, quite satisfactory agreement between experiment and theory was obtained for lean mixtures, but there were quite substantial differences for very rich mixtures. In the latter case, the experimental velocities reached a peak somewhat on the rich side and then decreased very little for richer mixtures. The conclusion was that for rich mixtures, the amount of energy released in determining the CJ plane is that corresponding to the maximum-velocity condition, ie, that released by a near-stoichiometric mixture.

Förster (2000) also gives references to published work on detonation of sprays and mists. Experiments have confirmed that clouds of combustible liquid droplets in air can detonate in the same way as premixed clouds of gaseous fuels and air. Typical measured detonation velocities are in the range 1300–1600 m/s, ie, somewhat lower than typical values for saturated hydrocarbon gas–air mixtures.

5.4 IGNITION OF CLOUDS OF LIQUID DROPLETS IN AIR (SPRAY/MIST)

5.4.1 Ignition by Hot Surfaces

The problem is of the same nature as discussed previously with hot-surface ignition of gases and vapors. However, it has not been possible to trace any standardized test methods for measuring minimum ignition temperatures of sprays/mists. Fig. 5.5 gives some results from hot-surface ignition of suspensions of JET-A fuel droplets flowing through a vertical furnace. JET-A is a petroleum distillate similar to kerosene (see previous discussion). Note that the vertical axis in Fig. 5.5 covers only 120°C. This figure also gives corresponding data for premixed propane–air, and it is seen that the JET-A cloud ignites at significantly lower furnace-wall temperatures than the propane air under the prevailing conditions.

A comparison of the T_{min} values for propane–air in Fig. 5.5 with those in Fig. 2.17 gives further support to the view that T_{min} for gases and vapors determined by standard flask tests are highly conservative for many practical situations.

Fig. 5.6 shows that T_{min} for a cloud of combustible liquid droplets increased systematically with the velocity of the cloud across the hot surface. As expected, the trend was the same for propane–air, although less pronounced.

5.4.2 Ignition by Electric/Electrostatic Sparks/Discharges

Explosive sprays/mist can be ignited by electric spark and arcs in the same way as premixed gaseous fuel–air mixtures can. In principle, therefore,

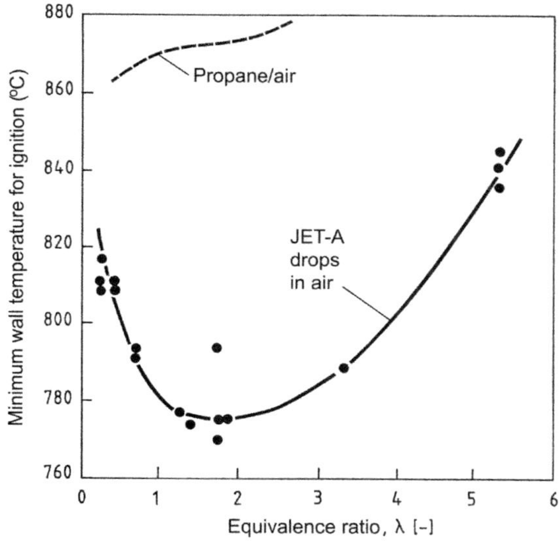

Figure 5.5 Influence of fuel/air ratio on the minimum temperature of a hot wall in a vertical furnace required for igniting a suspension in air of JET-A fuel droplets flowing at 1 m/s. Droplet diameters <50 μm. Comparison with ignition of propane—air in same apparatus and at same mean flow velocity.

From Graves, C.B. et al., 1986. Ignition of a fuel spray by a hot surface. AIAA Journal 24, 1994—2001.

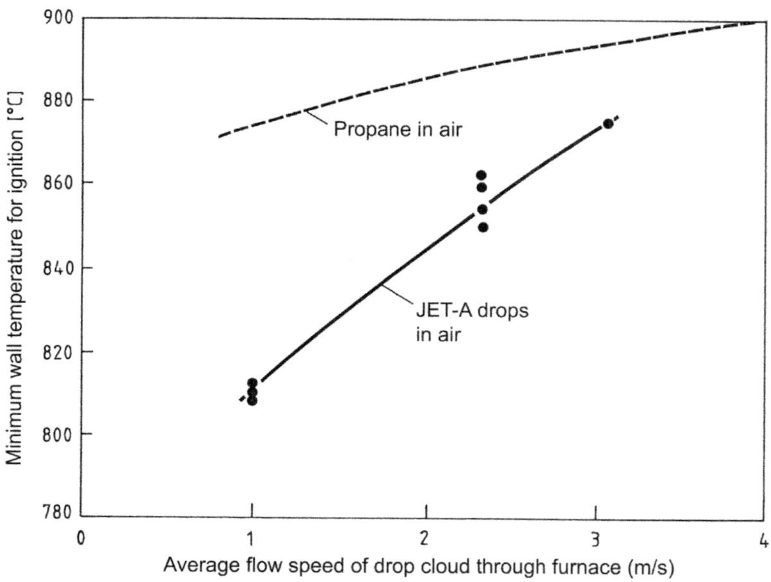

Figure 5.6 Influence of average flow speed on the minimum temperature of a hot wall in a vertical furnace required for igniting a suspension in air of JET-A fuel droplets. Droplet diameters <50 μm. Stoichiometric fuel/air ratio. Comparison with ignition of propane—air in same apparatus.

From Graves, C.B. et al., 1986. Ignition of a fuel spray by a hot surface. AIAA Journal 24, 1994—2001.

the concepts of minimum ignition energy (MIE) and quenching distance are just as valid for sprays/mists as for premixed gases. However, in the case of spray/mists the experimental determination is considerably more difficult.

Fig. 5.7 gives a set of data showing the influence of spark-gap length on minimum ignition energy of a suspension of kerosene droplets in air (see data for kerosene in caption of Fig. 5.2). The data in Fig. 5.7 indicate that the MIE in this particular case, with the droplet suspension flowing at a high velocity past the spark gap, was about 20 mJ and the quenching distance 3–4 mm.

Ballal (1976) performed detailed measurements of the electric-spark ignition characteristics of kerosene sprays injected into a flowing air stream. Measurements of spark discharge parameters (breakdown voltage, optimum spark duration, energy released in the gap) and ignition characteristics (optimum spark gap, minimum ignition energy, lean ignition limit) were performed. Three different spray generators (atomizers) were used at air velocities up to 38 m/s, producing fuel droplet diameters in the range 20–110 μm. The initial pressure was varied between 0.2 and 0.9 bar (abs.), the initial temperatures between 228 and 350K, and the equivalence ratios (nondimensional mass of fuel/volume of air ratios) between 0.3 and 0.9. The results showed that optimum spark duration for ignition

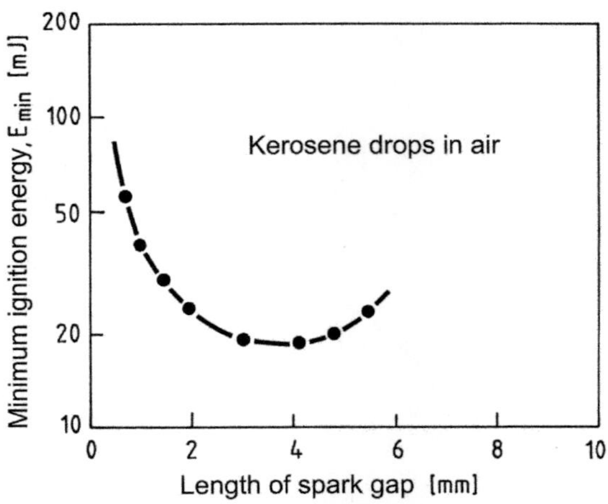

Figure 5.7 Influence of spark-gap length on minimum ignition energy of a suspension of kerosene droplets in air at atmospheric pressure and normal temperature. Flow velocity of suspension past spark gap 3 m/s, and equivalence ratio $\lambda = 0.65$.
From Rao, K.V.L., Lefebvre, A.H., 1976. Evaporation characteristics of kerosene sprays injected into a flowing air stream. Combustion and Flame 26, 303–309.

increased with decreasing initial pressure and spray cloud velocity, and with increasing droplet size. MIE decreased markedly with decreasing fuel-droplet size.

Fig. 5.8, from the subsequent work by Ballal and Lefebvre (1978), shows the variation of MIE for clouds of heavy-oil droplets in air, with the equivalence ratio λ in the range of $\lambda < 1$, and the mean droplet diameter in the range 40–150 μm. It seems that the decrease of MIE with droplet diameter seizes at about 10 μm. As Fig. 5.8 shows, MIE for 40 μm droplets of heavy oil was only 2 mJ.

Fig. 5.9 gives experimentally determined MIEs for clouds in air of 100 μm droplets of a range of different liquid fuels, as a function of the equivalence ratio λ. Spalding's "mass transfer number" B, as defined in the figure caption, is used as parameter. B is an amplified function of vapor pressure P_A of the liquid, and Fig. 5.9 shows quite clearly that MIE decreases systematically with B, or P_A.

Later Ballal and Lefebvre (1979) extended their studies to measurement of MIE over wide ranges of initial pressures, velocities, turbulence intensities, equivalence ratios, mean droplet sizes, and fuel volatilities. Model predictions showed satisfactory agreement with the experimental data.

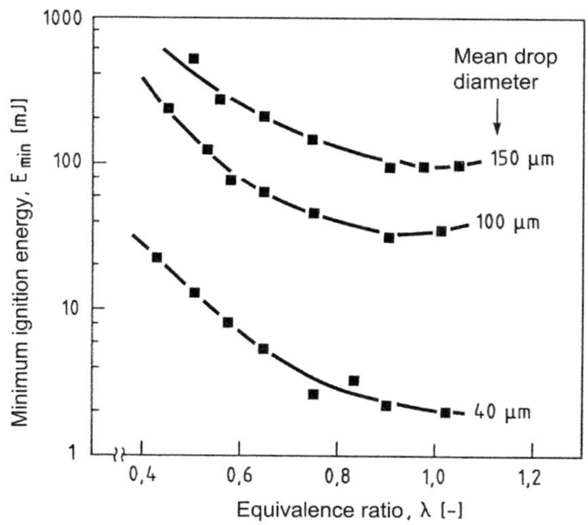

Figure 5.8 Influence of the equivalence ratio λ (fuel/air ratio) and mean droplet diameter on the minimum ignition energy of close-to-quiescent suspensions of heavy-oil droplets in air at atmospheric pressure and normal temperature.
From Ballal, D.R., Lefebvre, A.H., 1978. Ignition and flame quenching of quiescent fuel mists. Proceedings of the Royal Society London A364, 277–294.

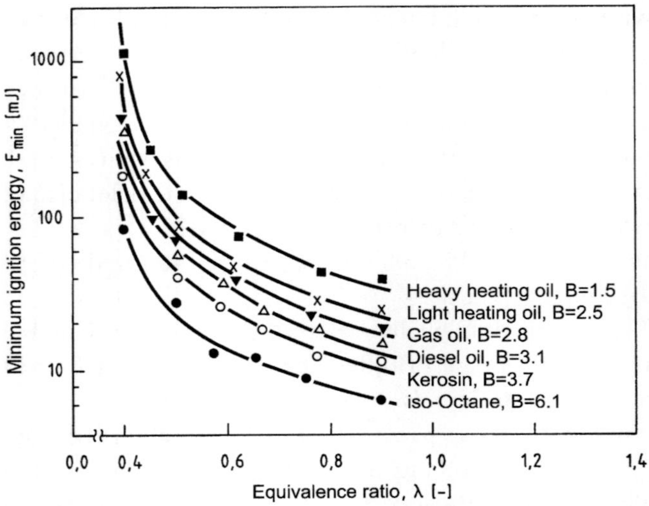

Figure 5.9 Influence of the equivalence ratio λ (fuel/air ratio) on minimum ignition energy of close-to-quiescent suspensions of 100 μm droplets of various combustible liquids in air at atmospheric pressure and normal temperature. B is Spalding's dimensionless "mass transfer number" defined as $B = P_A/(P_{tot} - P_A)$, in which P_A is the vapor pressure of the combustible liquid, and P_{tot} the total pressure.
From Ballal, D.R., Lefebvre, A.H., 1978. Ignition and flame quenching of quiescent fuel mists. Proceedings of the Royal Society London A364, 277–294.

5.4.3 Other Ignition Sources

Explosive clouds of liquid droplets in air can also be ignited by fast (adiabatic) compression (diesel engines). Baev et al. (1984) determined induction times for ignition of liquid-fuel sprays at high temperatures and pressures. The apparatus was a closed cylindrical chamber of 150-mm-diameter and length, filled with air at the desired temperature and pressure. The end plates were fitted with thick quartz windows. The fuel was injected into the chamber during 3–4 ms via a special nozzle system. The induction time for ignition was defined as the time interval from onset of fuel injection to onset of the sharp rise in pressure due to combustion in the chamber.

It may also be expected that explosive clouds of liquid-fuel droplets in air can be ignited by certain types of metal sparks and thermite reactions from single impacts. However, it has not been possible to trace any experimental data illustrating this.

5.4.4 Lack of Standard Test Methods for Ignitability and Explosivity of Sprays/Mists

It has not been possible to trace any standardized methods for ignitability and explosivity of clouds of combustible-liquid droplets in air.

5.5 CASE HISTORIES OF MIST/SPRAY EXPLOSIONS

The main source of this chapter is the review by Eckhoff (1991). Open reports of case histories of accidental mist/spray explosions in the process industries are scarce. Therefore, some spray/mist explosions that have occurred in other contexts have been included in the present section.

5.5.1 Spray/Mist May Have Been Involved in Some Reported "Vapor Cloud" Explosions

Eichhorn (1973) refers, without giving any details or references, to a report, probably from the 1950s, of a mist explosion in the chemical industry, which had prompted an experimental and literature study of mist combustion. The experimental work suggested that mists of flammable liquids in air can propagate flames, and that vapor pressure of the liquid is a poor index for judging mist flammability. Spray/mists of chlorinated liquid compounds, however, can be self-quenching.

Quite often the clouds that gave rise to reported vapor cloud explosions were generated by rapid release of combustible liquids from pressurized process equipment. However, such a process may, in addition to producing genuine vapor, also generate sprays of fine liquid droplets that may have played a significant part in the explosion development. The Flixborough disaster in the United Kingdom in 1974 may be one example (see Section 2.3.4 of Chapter 2). Kletz, as quoted by Anonym (1975), pointed out that an investigation conducted by ICI suggested that of all the cyclohexane that exploded only one-eighth appeared as true vapor, whereas most of the rest formed a liquid spray. He also emphasized that the spray should be regarded as being just as dangerous (explosive) as the genuine premixed vapor/air.

5.5.2 Oil Mist Explosion in a Compressor Test Loop

This explosion, discussed by Schmitt (1973), occurred in 1959 in a compressor test facility at Ingersoll-Rand Co.'s plant in Phillipsburg, New Jersey, United States. The test of a centrifugal compressor was being conducted in accordance with standard procedure, which involved the use of a closed loop for containment and recirculation of the gas used during the test. The test was nearly completed when, after approximately 6 h of test operation, an explosion occurred. The inlet side of the loop pipe ruptured with great violence, claiming the lives of six men, severely injuring six others, and imposing minor injuries on 24 additional men.

The explosion was described as a heavy flash fire accompanied by shock waves. Small localized fires and smoke followed. The floor plates of the platform immediately over the test loop were ripped loose. Heavy brick walls in the immediate vicinity were blown out. The roof of the building immediately over the blast was damaged and moved approximately 30 cm. The lightweight side walls of the building were torn off. Many windows were broken, some of these being located as far as 150 m away from the explosion center. The test loop, the seal oil system and most of the equipment on the lower level were destroyed or damaged beyond repair. The coupling spacer between the gear and the compressor was shattered, and the coupling end of the gear pinion was twisted off and thrown outside the building.

The centrifugal compressor involved was the high-pressure unit of a system employed for compressing natural gas in oil field repressuring service. Floating-ring-type oil seals were used to contain the gas within the test loop in which the shaft passed through the casing. This design combined the seal with the radial bearings. The sealing medium was oil supplied at a pressure of 3–3.5 bar above the gas pressure in the loop. A small quantity of oil was permitted to pass through the seal for lubrication.

The investigating committee concluded that regardless of the details of the design of this type seal, extremely small leakages of oil into the gas stream are unavoidable. As a result of the high air velocities, which tend to atomize and mix the oil droplets with the gas flow in such a manner that they cannot be effectively drained, the oil spray in the air tends to concentrate. The definite evidence of a combustion-type explosion confirmed that there must have been enough hydrocarbon oil present within the loop system to produce an explosive oil–air mixture throughout the loop system. Oil had been found to be present when blowing out low points in the piping between test runs. The destructive pressures were produced by heat liberated by the combustion of fine oil droplets mixed with the air being circulated by the compressor. The pressure produced by an explosion is approximately proportional to the initial absolute pressure of the explosive mixture. For mixtures of common hydrocarbon compounds and air, the ratio of maximum explosion pressure to initial pressure is typically about 8.

5.5.3 Spray/Mist Explosions in Crank Cases in Large Diesel Engines

Crank case explosions, their origin and nature, and means of prevention and mitigation, were discussed in detail by Minkhorst (1957) in a comprehensive literature survey. Engines fitted with cranks are normally equipped with a lubrication system by which oil is supplied to the bearings by pressurization. To protect the environment from oil spray and to prevent loss of oil, the crank system is usually fully enclosed by a case. Explosive spray clouds may be generated within the case by oil being flung into the air from rotating parts. In addition, mists of finer droplets may be generated by evaporation of oil in the hotter parts of the crank case and subsequent circulation to and condensation in the colder parts. If ignition occurs, the ignition source would normally be a hot surface, generated eg, by a faulty bearing or a cylinder liner in which the clearance between piston and liner is too narrow. In the early stages, before the mist concentration has reached explosive levels, the hot surface may act as a mist generator by enhancing evaporation of oil that is subsequently condensed to mist in the colder parts of the system. If the temperature of the hot surface is very high, the oil may be subject to chemical cracking, and highly explosive vapors may also be formed.

A severe crank case explosion, discussed by Minkhorst (1957), occurred in September 1947 in all the four engines of the ship "Reina del Pacifico." The first warning of a potential hazard was given by some maintenance people who had noticed that a cylinder liner of one of the engines was overheated. Somewhat later a strong explosion occurred in the same engine, which subsequently propagated into the other three engines of the ship. Four distinct explosions were heard at very short intervals. The engines blew up, and 28 men were killed and 23 injured. Minkhorst also discussed a number of experimental investigations related to initiation and propagation of crank case explosions, and to various possible means of preventing and mitigating such explosions.

According to Nordahl (1991), crank case explosions in ship engines still occur. In fact, the frequency of such explosions seems to have increased rather than decreased since the time when Minkhorst conducted his investigation. The chain of events leading to the explosions seems to be the same as in the past, the ignition source being unintentionally heated surfaces. Large 12-cylinder ship engines of up to 60,000 Hp have crank cases of 18–20 m length and 150 m^3 volume. It seems that crank case explosions occur more frequently with 4-stroke engines than with 2-stroke ones.

5.5.4 Major Oil Spray/Mist Explosion in a Transformer Room of a Hydroelectric Power Station

This accident, in which three men were killed and several others injured, was described by Johnsen and Holte (1973). Further details were given by Schjelderup (1990). The accident occurred in 1973 in the hydroelectric power station at Tonstad, Sirdal, in Norway. The entire station is located underground. Fig. 5.10 gives a top view of the entire underground installation, and side view of one section.

Fig. 5.11 gives a cross-section of the oil-filled cable junction box, and Fig. 5.12 a cross-section of the main underground hall with the control rooms and a transformer with junction box on top in a separate room.

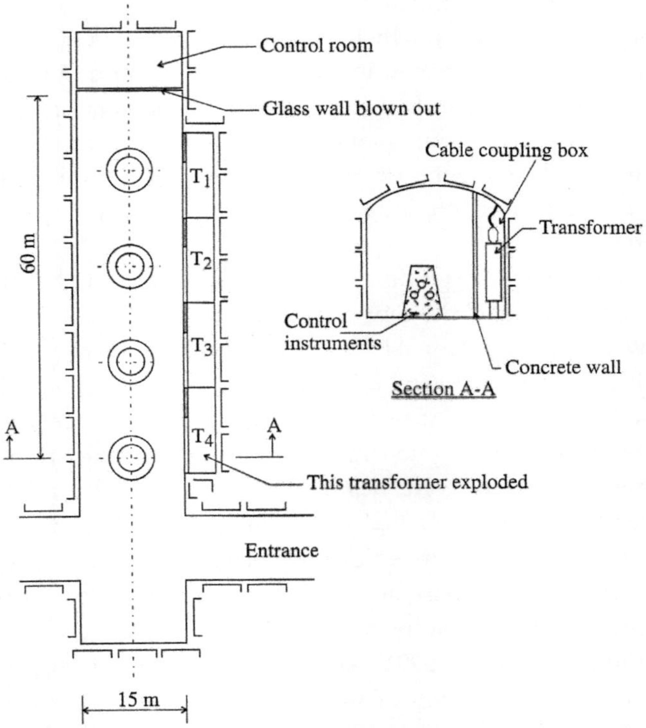

Figure 5.10 Top view, with side view of one section, of the underground hydroelectric power plant at Tonstad, Sirdal, Norway, with vertical section through the main hall, the control instruments above the dynamos, and the transformer in a separate room on the side of the main hall.
From Schjelderup, S., 1990. Private Communication to R.K. Eckhoff From S. Schjelderup, SiraKvina Kraftselskap, Norway.

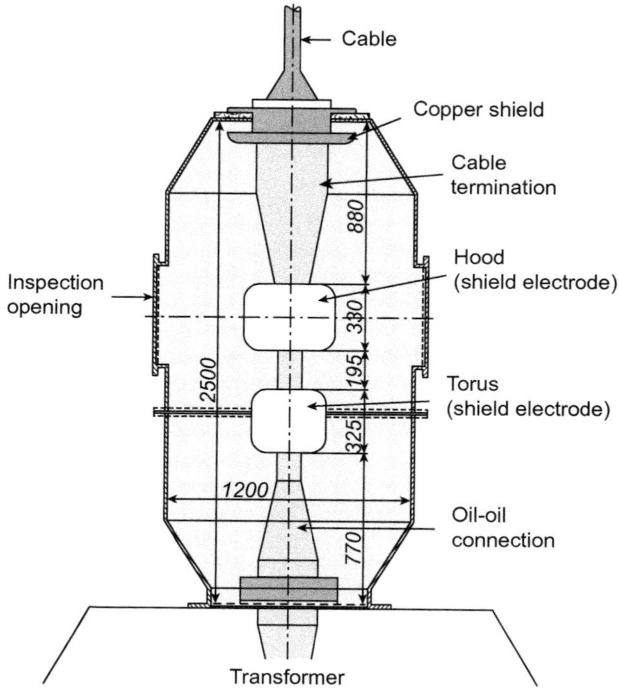

Figure 5.11 Cross-section of the oil-filled cable junction box that exploded in the underground hydroelectric power plant at Tonstad, Sirdal, Norway.

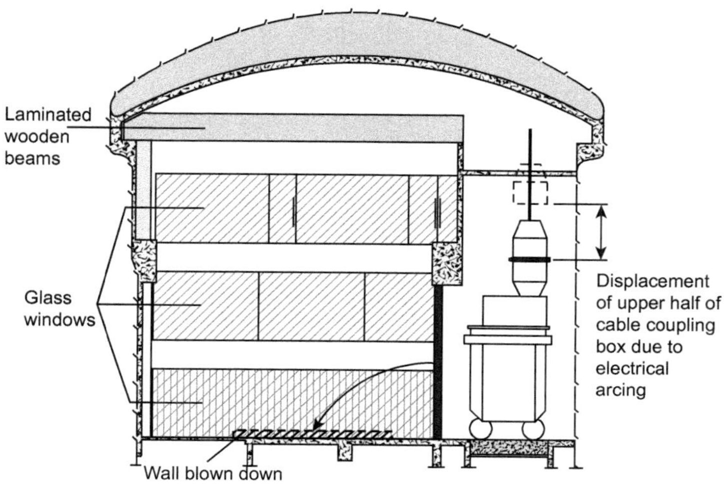

Figure 5.12 Cross-section of the main hall of the underground hydroelectric power plant at Tonstad, Sirdal, Norway, with the separate transformer concrete cell to the right.

The origin of the event was a flash-over inside a 2 m^3 oil-filled cable junction box of one of the transformers. The probable reason for the flash-over was the presence of significant dissolved quantities of a sulfur-containing cement in the oil, which, in combination with moisture, reduced the flash-over resistance of the oil. The substantial thermal power of the flash-over arc gave rise to a sudden and substantial rise of the internal pressure in the junction box. This caused the box to rupture and the oil to be expelled as a fine spray throughout the unvented concrete cell of volume 800 m^3, in which the high-voltage transformers were located. The quantity of oil in the 2 m^3 junction box was more than sufficient to supply the entire 800 m^3 volume of the transformer cell with an oil mist cloud of the most explosive concentration. The mist cloud exploded immediately (see Figs. 5.13 and 5.14), ignited by the arc in the junction box, and the wall of the transformer cell was blown out. A strong blast wave swept through the large engine hall and blew in the windows of the central control room. Dense smoke penetrated the entire system including the road tunnels to the open, rendering the rescue operation very difficult.

Figure 5.13 Photo of debris after the oil spray/mist explosion in the main hall of the underground hydroelectric power station at Tonstad, Sirdal, Norway, illustrated in Fig. 5.10. The photo also shows the conical cabinets of the control instruments on top of the turbine/generator assemblies.

Figure 5.14 Photo of debris and damage in the control room after the oil spray/mist explosion in the underground hydroelectric power station at Tonstad, Sirdal, Norway.

5.6 MEANS OF PREVENTING AND MITIGATING MIST/SPRAY EXPLOSIONS IN THE PROCESS INDUSTRIES

5.6.1 Preventing and Limiting Size of Explosive Clouds

Although the physics, not least the fluid dynamics, of mists and sprays differs appreciably from the physics of homogeneous gases, it is customary to apply the principles developed for gases even to mists/sprays. These are described in Section 2.4.2 of Chapter 2. This may be a reasonable approach in the case of liquids of low boiling points, in the case of which the droplets will evaporate quickly once released into the open air. However, in the case of liquids of higher boiling points, the approach of adopting the gas principles may not always be justified. On the other hand, published specific guidance for mists/sprays is scarce.

5.6.2 Preventing and Controlling Ignition Sources

The same overall comment as given in Section 5.6.1 applies. Guidance for explosive gases is given in Sections 2.4.3 and 2.4.4 of Chapter 2.

5.6.3 Mitigatory Measures

The same overall comment as in Section 5.6.1 applies. Guidance for explosive gases is given in Section 2.4.5 of Chapter 2.

CHAPTER SIX

Gas and Dust Explosions Caused by Smoldering Combustion in Powder Layers and Deposits

6.1 BURNING VELOCITIES/RATES IN POWDER LAYERS AND DEPOSITS

6.1.1 Dust Layers

Propagation of combustion through layers and deposits of most dusts will occur very slowly compared to the propagation of a flame through an explosive gas/vapor cloud. This is because the ratio of combustible material/oxygen in a dust layer/deposit is about two orders of magnitude larger than the stoichiometric ratio. Very limited supply of oxygen to support the combustion process, and the heat sink effect of all the excess material that does not take part in the combustion, will limit both the rate or propagation of the combustion process, and the temperature in the reaction zone. Therefore, for most dusts, the combustion zone will normally be a smolder or glow rather than an open flame, and typical velocities of the combustion wave will be several orders of magnitude lower than laminar burning velocities in premixed gases and dust clouds.

Fig. 6.1 shows some data reported by Palmer (1973) from experiments with glowing/smoldering combustion in horizontal layers of beech dust.

The layers, of various thicknesses, were ignited at one end and the velocity of the propagation smolder/glow front along the layer was measured as a function of layer thickness, particle size, and the velocity of a horizontal airflow along the surface of the layer. Fig. 6.1 exhibits the following trends:

- The observed smolder/glow velocities were in the range of 0.05–0.5 m/h, as opposed to the 0.5 m/s for laminar burning velocities of premixed hydrocarbon gas/air, ie, a difference by a factor on the order of 10^4.
- The smolder/glow velocity along the layer increases markedly if an airflow of velocity of some m/s is passed across the layer in the same direction as the propagation of the smolder/glow.

Explosion Hazards in the Process Industries
ISBN: 9780128032732
http://dx.doi.org/10.1016/B978-0-12-803273-2.00006-2

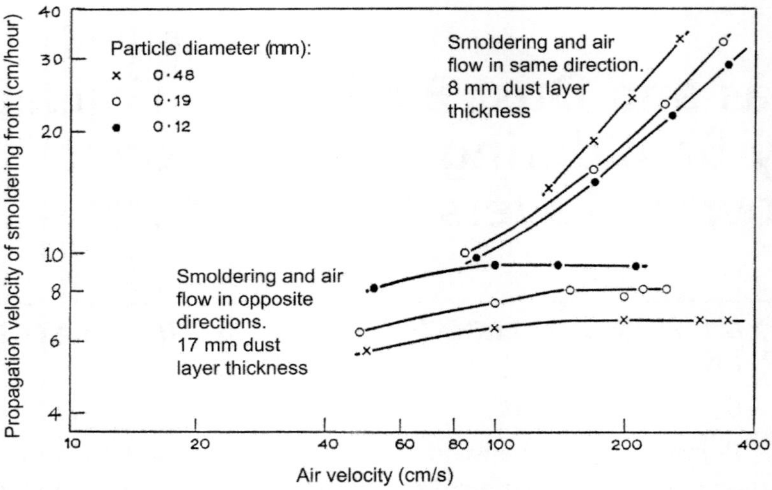

Figure 6.1 Influence of the air velocity along a layer of beech sawdust on the rate of smoldering-front propagation along the layer.
From Palmer, K.N., 1973. Dust Explosions and Fires. Chapman and Hall, London, ISBN:412 09430 4.

- If the direction of the airflow is opposite to that of the propagation of the smolder/glow, there is no significant influence of the airflow in the velocity range 0.4–4 m/s investigated.
- The influence of particle size on the effect of the airflow is somewhat complex. With the airflow in the same direction as the smolder glow propagation, the latter increases with increasing particle size, whereas with the air-flow in the opposite direction, the highest smolder/glow velocities were found for the smallest particles. However, in the range investigated the influence of particle size was only modest.
- With high air velocities across the layers, in the direction of the propagation of the glow/smolder, the temperatures in the reaction zone inside the layers were quite high, approaching 1000°C.
- However, in the absence of an airflow across the layer, there was no visible sign of smolder or glow at the surface of the layer, and the temperature of the smoldering/glowing reaction zone inside the layer was in the range 550–620°C.
- There was little influence of the material of the surface on which the dust layer was resting (asbestos, metal wood), on the velocity of propagation of the smolder/glow in the layer.

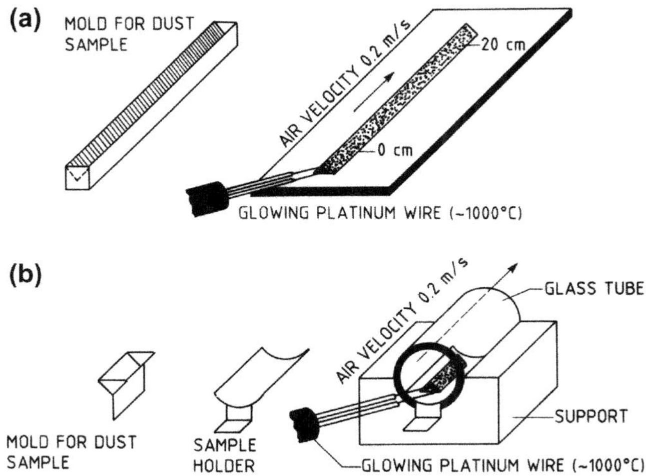

Figure 6.2 Semiquantitative laboratory method for testing of the ability of a horizontal ridge of a given powder/dust to propagate combustion. (a) shows the set-up for tests at normal ambient temperature. (b) shows the set-up for tests at elevated ambient temperatures.
From Eckhoff, R.K., 2003. Dust Explosions in the Process Industries, third ed. Gulf Professional Publishing/Elsevier (first ed. published in 1991, second ed. in 1997), ISBN:0-7506-7602-7.

Fig. 6.2 illustrates a standardized method for characterizing the ability of layers of powders/dusts to propagate smolder, glow, and in some cases even an open flame.

The foundation of this method was laid by Lütolf (1971). A German standard was produced by Verein deutscher Ingenieure (1988). A nearly identical European standard was being finalized in 2015. For tests at ambient temperature, a ridge of the dust of triangular cross-section is placed on a ceramic plate, and one of the ends is ignited using a white-glowing platinum wire as shown in Fig. 6.2a. For tests at elevated temperatures, the sample holder shown in Fig. 6.2b is used. The sample placed in a glass tube heated to the desired temperature. A small airflow of about 0.2 m/s through the glass tube must be ensured.

The dusts are classified according to their ability to propagate combustion when deposited as a layer. The test sample is a 2-cm wide and 20-cm long ridge-shaped dust layer resting on a ceramic plate. The dust ridge is ignited at one end using either a gas flame or a glowing platinum wire at 1000°C. The definitions of the various combustibility classes are approximately as follows:

Class 1: No self-sustained combustion

Class 2: Local combustion of short duration

Class 3: Local sustained combustion, but no propagation

Class 4: Slow smoldering combustion propagating along whole sample length

Class 5: Relatively slow open flame or "fireworks reaction" propagating along whole sample length

Class 6: Fast/explosive combustion along whole sample length

6.1.2 Deep Dust Deposits

Fig. 6.3 gives the results from investigation of smoldering/glow propagation in deep deposits of beech dusts, also reported by Palmer (1973).

In this case, the thickness of the dust deposits was much larger than that of the layers discussed previously, ranging from 40 mm to nearly 1000 mm, and the main measurement taken was the time from initiation of the

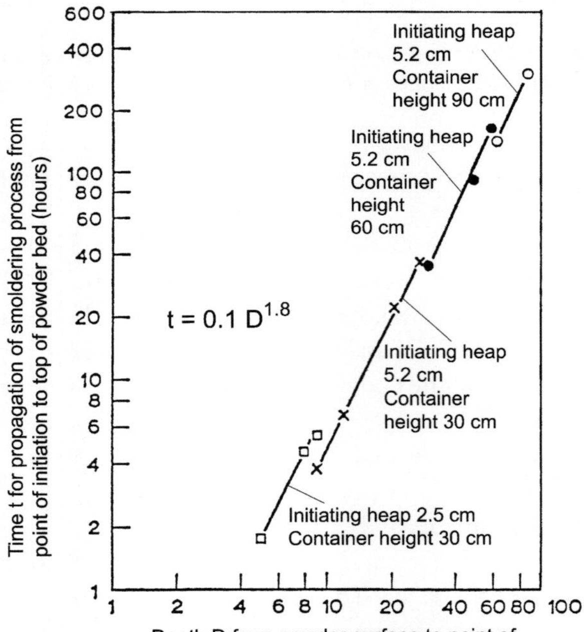

Figure 6.3 Influence of the depth, from the top of a deposit of wood sawdust to the point of initiation of smoldering/glow inside the deposit, on the time from initiation to arrival of the smoldering/glow front at the deposit top.
From Palmer, K.N., 1973. Dust Explosions and Fires. Chapman and Hall, London, ISBN:412 09430 4.

smolder/glow at the bottom of the deposit, till the smolder/glow front reached the deposit top. The following was found:

- The average velocity of propagation of the smolder/glow front, from initiation at the deposit bottom, to arrival at the deposit top, decreased systematically with increasing deposit depth D, according to the approximate relation:

$$t = 0.1D^{1.8} \tag{6.1}$$

  in which t is in hours and D in cm.
- With D equal to 85 cm, the propagation of the smolder/glow front from the bottom to the top of the deposit took about 300 h, ie, 12–13 days. The average rate of propagation was about 2.8 mm/h, which is about 18 times less than the lowest layer propagation velocities of about 5 cm/h given in Fig. 6.1.
- However, as soon as the smolder/glow has reached the top of the deposit, it will start to propagate sideways a higher velocity typical of horizontal propagation in thin layers.

Smolder/glow processes can propagate inside dust deposits of considerably larger dimensions than those of the experiment illustrated in Fig. 6.3. The combustion propagation velocities will then be even lower than in that experiment. The times needed for the smolder/glow front to reach the borders of a very large powder/dust deposit can be up to several years.

The main reason why smolder/glow processes inside large powder/dust deposits are so slow is lack of supply of oxygen to the exothermal chemical reaction. The oxygen has to be transported to the reaction zone by diffusion through the narrow channels between the particles in the deposit. The reason why the exothermal process is nevertheless maintained is the inherent thermal insulation capability of the powder/dust deposit. The very limited amount of heat generated per unit time in the chemical reaction is used effectively for heating the next layer of dust, adjacent to the combustion zone, without being lost into the farther surroundings. The basic principle is the same as that of the general theory of ignition discussed in Section 2.2.2 in Chapter 2 and illustrated in Fig. 2.13.

The high thermal insulation power of powders/dusts not only promotes development and sustainment of smolder/glow in powder/dust deposits. It also presents substantial challenges in the efforts to effectively extinguish smoldering/glow processes in large powder/dust deposits, eg, in silos. Most often, the main problem is not just to extinguish, ie, to stop the exothermal reaction, but to cool down the hot material to normal temperatures. This is confirmed by several of the case histories discussed in Section 6.4.

6.2 INITIATION OF COMBUSTION IN POWDER/DUST LAYERS AND DEPOSITS

6.2.1 Minimum Ignition Temperature of Dust Layers

The apparatus shown in Fig. 6.4 consists of an electric hot plate, a temperature control unit, three thermocouples and a two-channel recorder.

The hot plate is kept at a given temperature, which is read by one of the thermocouples and displayed on one of the recorder channels. The second thermocouple is used for regulating the plate temperature. When a test is performed, a metal ring of internal diameter 100 mm and height either 5 or 15 mm is first placed on the surface of the plate. The powder/dust sample to be tested is then placed on the part of the hot plate that is inside the metal ring and carefully leveled off to the height of the ring. The third thermocouple is placed in the sample through holes in the metal ring. The sample temperature is displayed on the second recorder channel. Typical outcomes of a test are illustrated in Fig. 6.5.

Originally, it was considered that the maximum temperature in the dust sample had to exceed the hot-plate temperature by more than 20°C for the test to be recorded as ignition. However, it was later realized that a considerably higher temperature rise, on the order of at least 100°C would be more

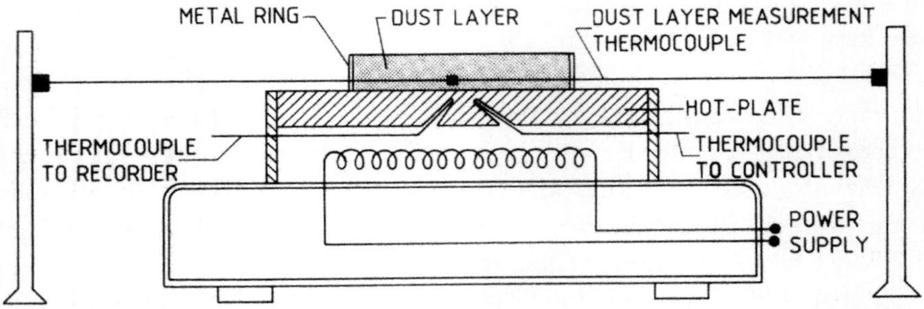

Figure 6.4 Internationally standardized method for determining the minimum hot-plate temperature required for initiating smoldering or flaming combustion of layers of combustible powders/dusts.
From Eckhoff, R.K., 2003. Dust Explosions in the Process Industries, third ed. Gulf Professional Publishing/Elsevier (first ed. published in 1991, second ed. in 1997), ISBN: 0-7506-7602-7.

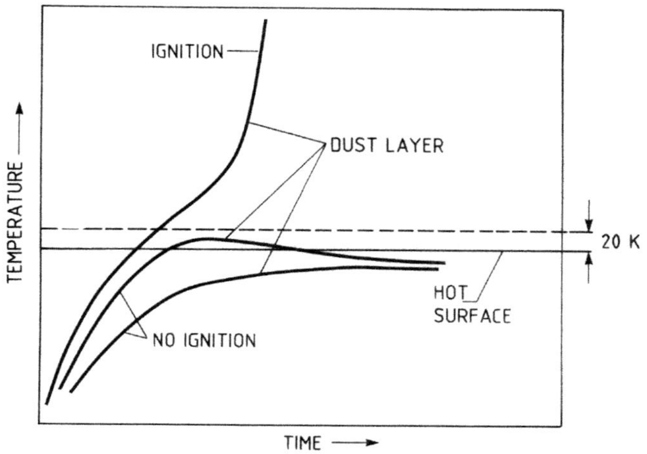

Figure 6.5 Illustration of the three possible outcomes of the test illustrated in Fig. 6.4. *From Eckhoff, R.K., 2003. Dust Explosions in the Process Industries, third ed. Gulf Professional Publishing/Elsevier (first ed. published in 1991, second ed. in 1997), ISBN: 0-7506-7602-7.*

appropriate. To determine the minimum ignition temperature for the layer thickness tested, repeated tests are conducted at different hot-plate temperatures until the minimum temperature for ignition has been identified. It is important to note that the minimum hot-plate ignition temperature decreases systematically with increasing dust-layer thickness. If the values for two different layer thicknesses have been determined, simplified theory enables estimation of the values for other thicknesses.

6.2.2 Minimum Ignition Temperature of Dust Deposits

In this case, one considers a powder/dust deposit surrounded by air, and the question asked is what minimum temperature of the air surrounding the deposit is required for development of self-ignition inside the deposit. The type of apparatus used for determining this kind of minimum ignition temperature is illustrated in Fig. 6.6.

The powder/dust sample is placed in a metal gauze basket that is placed inside a heating chamber through which a given flow of preheated air is circulated. The temperature of the air surrounding the basket and the temperature inside the powder sample are measured and the difference recorded. If the temperature in the powder sample increases beyond that of the surrounding air, but not higher than 400°C, the phenomenon is

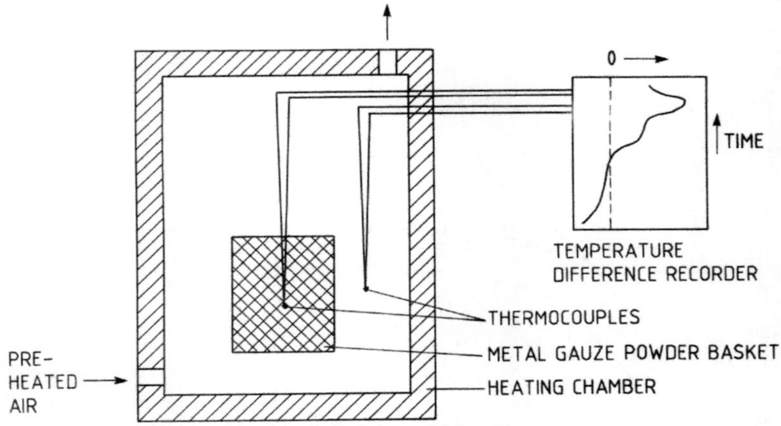

Figure 6.6 Illustration of method for determining the minimum temperature of the surrounding air for initiation of smoldering combustion in a dust sample contained in a metal gauze basket.
From Eckhoff, R.K., 2003. Dust Explosions in the Process Industries, third ed. Gulf Professional Publishing/Elsevier (first ed. published in 1991, second ed. in 1997), ISBN: 0-7506-7602-7.

named self-heating. Temperature rises beyond 400°C are named self-ignition. As indicated above, the aim of the test is to determine the minimum temperature of the air surrounding the sample, which gives rise to self-ignition. However, this temperature is not a constant for a given powder, but decreases systematically with increasing volume of the powder sample tested. Fig. 6.7 illustrates the linear relationship most often obtained between the minimum ignition temperature and the volume of the powder sample when plotting the experimental data in a double-logarithmic sheet. This enables extrapolation to larger volumes than those actually tested.

Instruments that are more sophisticated for detailed studies of the rate of reaction as a function of temperature, under adiabatic conditions, are also available, including *accelerating rate calorimeters* (*ARCs*), which are essentially highly computerized adiabatic calorimeters. During an ARC experiment, the sample is maintained in a near-to-perfect adiabatic condition, whereas time, temperature, and pressure data are automatically collected and stored. The data can then be processed by computers. In addition to ARC, *differential scanning calorimetry* (*DSC*) is also in use. In DSC, the rate at which heat must be transferred to or from the test sample, to make the sample attain the same temperature as an inert reference sample, is measured.

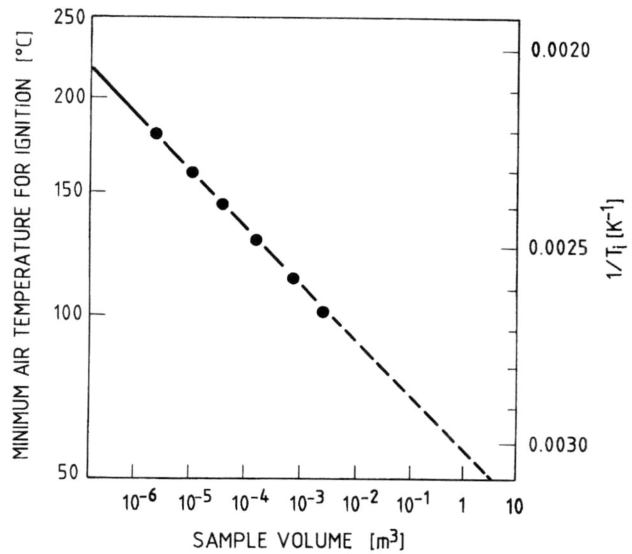

Figure 6.7 Example of results from the experiment illustrated in Fig. 6.6, for a given powder/dust, using a range of powder sample volumes.
From Eckhoff, R.K., 2003. Dust Explosions in the Process Industries, third ed. Gulf Professional Publishing/Elsevier (first ed. published in 1991, second ed. in 1997), ISBN: 0-7506-7602-7.

6.3 EXPLOSION INITIATION PROCESSES BY SMOLDERING/GLOWING MATERIAL

Smolder/glow processes in powder/dust deposits are in themselves a significant problem in some process industries. However, in some cases these processes are only the first step in an escalating chain of events that can lead to both gas and dust explosions. Figs. 6.8 and 6.9 illustrate how smoldering powder/dust layers/deposits can give rise to dust explosions.

Fig. 6.10 illustrates a more complex scenario, whereby the smolder/glow process in a silo generates carbon monoxide (CO), which diffuses upward through the powder/dust bed and mixes with the air in the empty space above the bed to form an explosive mixture. This mixture is then, in turn, ignited by the glow front as soon as it reaches the top of the powder/dust bed. Further escalation to a secondary dust explosion can occur if dust layers have accumulated in area above the silo top. Case histories confirming this chain of events are given in Section 6.4.

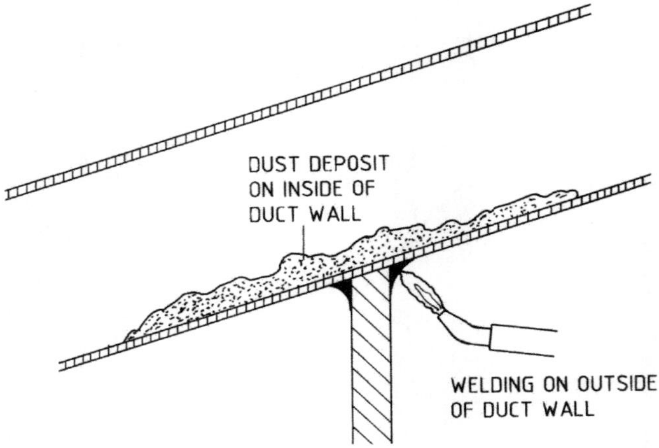

Figure 6.8 Initiation of combustion in a powder/dust layer inside a steel duct, by a welding operation on the outside of the duct. With an explosive dust cloud flowing through the duct, the glowing/flaming dust layer can initiate a dust explosion there. *From Eckhoff, R.K., 2003. Dust Explosions in the Process Industries, third ed. Gulf Professional Publishing/Elsevier (first ed. published in 1991, second ed. in 1997), ISBN:0-7506-7602-7.*

6.4 CASE HISTORIES OF ACCIDENTAL SMOLDERING COMBUSTION IN SILOS, RESULTING GAS EXPLOSIONS, AND ADOPTED EXTINCTION PROCEDURES

6.4.1 Fire and Subsequent Explosion in a Silo Plant in Stavanger, Norway (1985)

This accident, described by Braaten (1985), was not primarily a dust explosion, but an explosion of combustible gases released from a solid organic material during smoldering combustion in a silo cell. The cause of events was in accordance with Fig. 6.10. The explosion occurred in a modern reinforced-concrete silo complex used for storage of various feedstuffs. Pellets of Canadian rapeseed flour had been stored in one of the silos for some time when it was discovered that the material in the bottom part of the silo had become packed to a solid mass and could not be discharged through the silo exit. Sometime later, one week before the explosion, flames were observed in the silo. The fire brigade was called and covered the pellets in the silo with foam from above. Various unsuccessful attempts were then made at discharging the pellets mass at the silo bottom. During this phase,

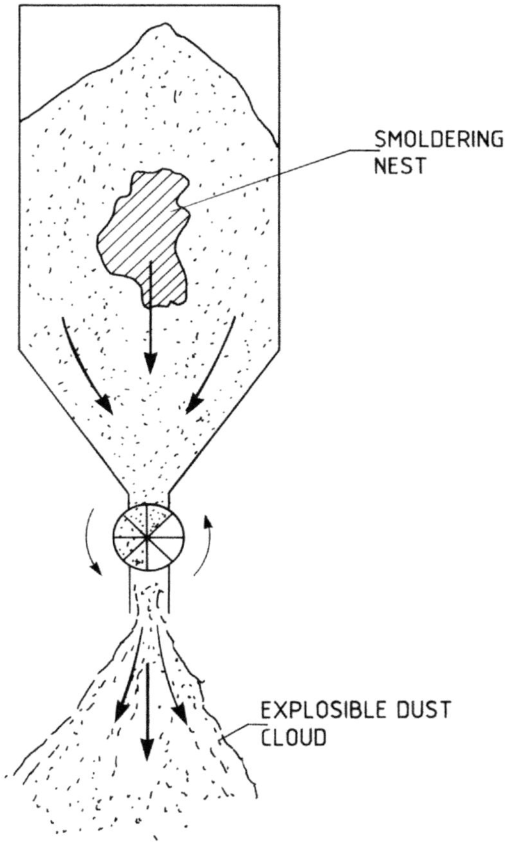

Figure 6.9 Initiation of dust explosion in cloud generated by discharge of powder/dust from silo/hopper, by smoldering nest in powder/dust.
From Eckhoff, R.K., 2003. Dust Explosions in the Process Industries, third ed. Gulf Professional Publishing/Elsevier (first ed. published in 1991, second ed. in 1997), ISBN: 0-7506-7602-7.

there was considerable development of smoke, which mixed with the air not only in the silo cell in question, but also in the silo loft above the cells. It is probable that the smoke contained combustible gases, eg, CO, and that the strong explosion that occurred just after the top of the pellets had been covered with foam once more, was mainly a gas explosion. However, any dust deposits in the loft may also have become involved. The entire roof of the building was blown up, and debris was thrown into the surrounding area. Because the explosion occurred in the middle of the night (0300), and just after the fire brigade had left, nobody was killed or hurt.

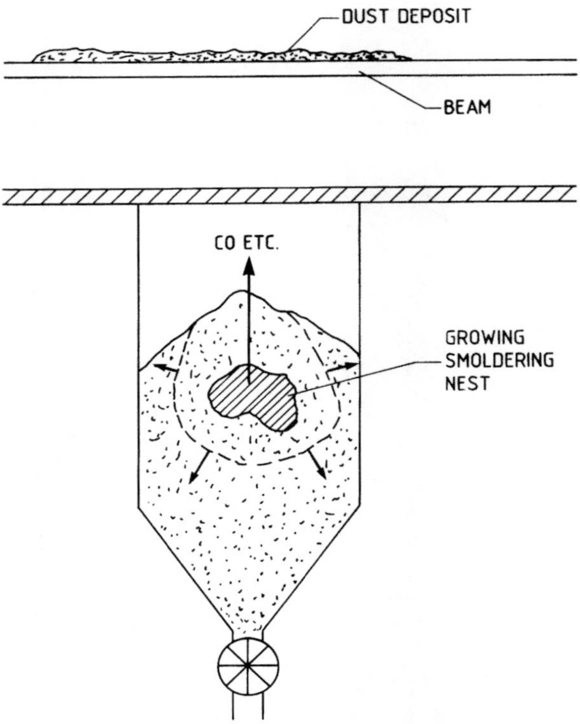

Figure 6.10 Initiation of explosion in a gas/air mixture developed by a smoldering nest inside powder/dust in a silo, and accumulated above the powder/dust in the silo, when the smoldering front reaches the top of the powder/dust.
From Eckhoff, R.K., 2003. Dust Explosions in the Process Industries, third ed. Gulf Professional Publishing/Elsevier (first ed. published in 1991, second ed. in 1997), ISBN: 0-7506-7602-7.

6.4.2 Multiple Silo Fires and Resulting Gas Explosions in a Large Storage Facility in Tomylovo in the Knibyshev Region in Russia (1987–89)

This extensive series of explosions was of the same nature as the explosion discussed in Section 6.4.1. A report of the event was given by Borisov and Gelfand (1989). The large storage facility for grain and feedstuffs, shown in Fig. 6.11, consisted of four sections of 60 silo cells each, ie, 240 silo cells altogether. Each cell was of 3 × 3 m square cross-section and 30-m height.

The first explosion occurred in December 1987 in a silo cell containing moist sunflower seed, which was not supposed to be stored in such silos due to the risk of self-heating. However, this had nevertheless been done, and

Figure 6.11 The concrete silo complex in Tomylovo, Knibyshev region, Russia, which was damaged by a series of smoldering gas explosions in the silo cells.
Courtesy of Borisov, A., Gelfand, B., 1989. Private Communication to R.K. Eckhoff from A. Borisov and B. Gelfand. Russian Academy of Science, Moscow.

the resulting self-heating developed into extensive smoldering decomposition during which methane and carbon monoxide was produced and mixed with the air in the empty top part of the silo, above the powder bed surface. It is reasonable to believe that the primary explosion was in this mixture of explosive gas and air, and that the ignition source was the smoldering combustion when it penetrated to the powder bed top surface, as illustrated in Fig. 6.10. However, in addition, dust deposits on the internal silo walls and roof may well have been entrained by the initial blast and become involved in the explosion.

This was only the first of a large series of 20 to 30 subsequent explosions that took place in the same facility, in one silo cell after the other, during 1988 and 1989. There are two main reasons for this continued explosion activity in the silo complex. The most important is the heat transfer from a silo cell in which smoldering combustion is taking place, to the neighboring cells. Such heat transfer was facilitated by the large contact surface area between the cells provided due to the square cross-section of the cells. The second main reason for the repeated explosions was that sunflower seed was not the only material in the facility that was not supposed to be stored there. Some of the silo cells contained buckwheat and wheat grain of higher moisture contents than the maximum permissible limits for storage in such facilities.

During the period of repeated explosions, a series of attempts were made at breaking the unfortunate chain of events. Cells were opened at the top for inspection. However, this admitted fresh air to the smoldering mass and enhanced the combustion process. Attempts were made at quenching and cooling the powder mass by liquid nitrogen, but this was only partly successful.

It was finally decided to demolish the entire facility using explosives.

6.4.3 Fire and Explosion in Pelletized Wheat Bran in a Silo Cell at Nord Mills in Malmö, Sweden, in 1989

A cross-section of the silo cell is shown in Fig. 6.12.

The course of events, as recorded by Templin (1990), was as follows:

Saturday 28 January, 0700: The night shift stopped the production for the weekend according to schedule, and all activity in the grain silo plant terminated.

Saturday 28 January, 1000: According to Nord Mills' safety procedures, the safety guard team made its inspection round through the entire plant. Nothing special was observed. No persons were encountered.

Saturday 28 January, 2300: A bang, muffled by the noise of strong winds, was heard in the neighborhood, but no action was taken.

Sunday 29 January, 0930: During its scheduled inspection round, the safety guard team discovered fragments of shattered windowpanes spread over the entire yard. Inspection of the roof of the silo building revealed that the roof of an intermediate star cell had blown up, and that dense smoke was emerging from the open cell top. (A photo of the scene is given in Eckhoff (2003).) The height of the cell that exploded was about 36 m and its cross-sectional area about 20 m^2. Most of the silo was empty, the pelletized wheat bran occupying only the first 7 m above the cell bottom.

Sunday 29 January, rest of day: Fire brigade and other personnel were called, and the entire plant area was cordoned off. About 2000 kg of gaseous carbon dioxide was pumped into the burning silo from above

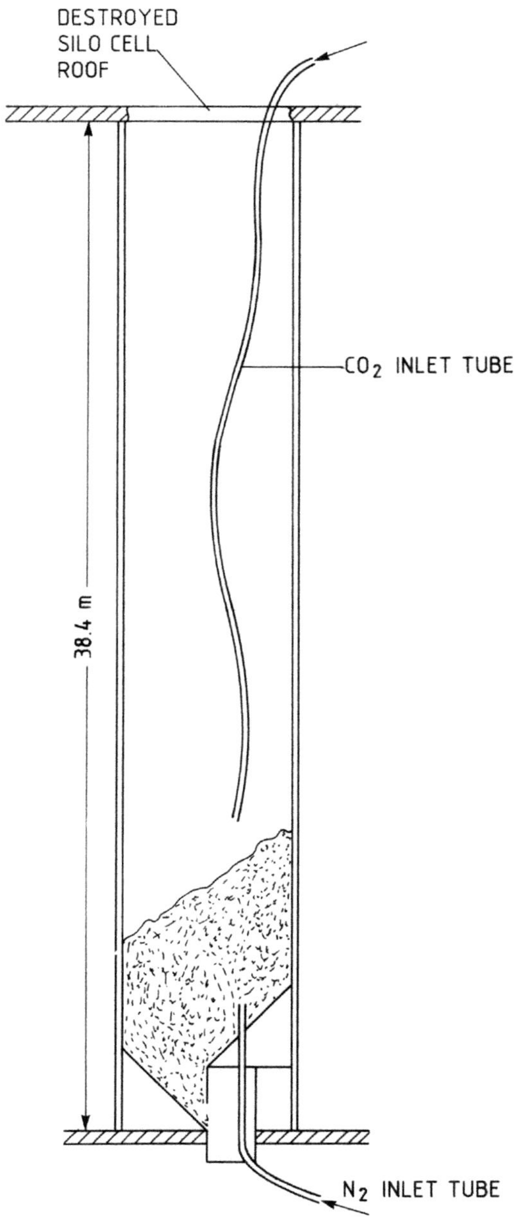

Figure 6.12 Cross-section of silo in Malmö, Sweden, in which a destructive explosion occurred in gas (CO) developed from smoldering combustion of wheat bran pellets in 1989. The inlets for subsequent supply/injection of carbon dioxide and nitrogen for extinction and cooling are also indicated.
Courtesy of Templin, G., 1990. Private Communication to R.K. Eckhoff from G. Templin, Nord Mills, Malmö, Sweden.

through a long vertical pipe extending right down to the surface of the smoldering pellets (see Fig. 6.12).

Monday 30 January, early morning: The discharge valve at the cell bottom was removed, and discharge of the pellets mass, using a mobile suction unit, was started. This gave rise to increased smoke production, and at 0330 more carbon dioxide was loaded into the silo cell from above.

Tuesday 31 January: The discharge operation was interrupted. Carbon dioxide was emerging through the bottom silo exit, and more was loaded into the silo at the top.

Wednesday 1 February: More carbon dioxide was loaded into the silo at the top. From 03.00 to 12.50 the smoke development was enhanced by vibrations due to operation of another silo cell. The smoke temperature just above the pellets was 96°C, and just above the silo top 45°C.

Thursday 2 February—Wednesday 8 February: Smoke development and temperature rise was suppressed temporarily by loading several tonnes of carbon dioxide into the silo from the top, but there was only slow permanent progress. Temperature rise was observed in the material stored in the four larger adjacent silo cells.

Thursday 9 February—Saturday 11 February: Holes were drilled through the silo bottom and at intervals a total of several tonnes of nitrogen were pumped into the pellets from below, whereas carbon dioxide was charged from above.

Monday 13 February—Wednesday 15 February: Some 6000 kg of carbon dioxide and 3000 kg of N_2 was injected into the burning pellets. Temperatures in the burning and adjoining cells and contents of oxygen, CO, and CO_2 in the gas above the pellets, were monitored regularly.

Monday 20 February: The smoldering combustion in the wheat bran pellets had finally been brought to an end.

This case history illustrates that fighting smoldering combustion in large silo complexes is not only a matter of quenching, or terminating the oxidation reaction. It is also, indeed, a matter of cooling massive bulks of poor heat conductors to a temperature level at which the combustion process will not start again once air is readmitted to the system.

6.4.4 Extinction of Smoldering Fire in a Fishmeal Silo in Norway in 1992 Using Water

During the struggle with trying to fight the smoldering fires in the silo complex in Tomylovo, Russia (see Section 6.4.2) it was agreed that the use of water was not feasible. The arguments were that limited quantities of water would probably enhance the self-heating process rather than quench it, and use of extensive quantities would increase the load on the silo walls (hydrostatic pressure) and cause collapse of the entire structure.

In view of this, it is interesting to consider the use of water for extinguishing a smoldering fire in a fishmeal storage silo in Norway a few years later. In this case, described by Kroken (1997), the silo was cylindrical, of height 20 m and diameter 4.5 m. The bottom hopper was wedge-shaped and fitted with a horizontal screw conveyor for discharging of the meal. The sides of the hopper were at 60 degrees against the horizontal plane.

The entire sequence of events was as follows:

- The fresh fishmeal from the production plant was filled into a mixing silo for being recycled and conditioned, before being transferred to a storage silo.
- To prevent self-heating in the meal during the conditioning period, recycling by discharging the meal from the silo bottom via the screw conveyor, and returning it to the silo at the top, was started.
- After some days, it was discovered that the meal discharge operated by funnel flow rather than by mass flow, ie, most of the material was not taking part in the recycling process.
- It was realized that digging out the material in the "dead" zones in the silo would be necessary, but due to lack of time, this work was not started until 4 weeks later. This made it possible for self-heating to develop in the stored fishmeal.
- When trying to discharge the meal from the silo bottom after the 4-week break, bridging of the meal across the entire silo cross-section was discovered.
- Digging-out was started from the silo top. When material had been removed down to the level of 2 m above the transition between the cylindrical silo wall and the hopper, the digging had to be stopped. The three main reasons were smoke development, that the meal was too hot to stand on and handle, and that the meal had become very hard (a steel spear could only be forced c.10 cm into meal by hand).
- A nitrogen atmosphere was established in silo above the meal surface.
- The fire brigade was called. However, they did not know how to deal with the matter, and left.
- The nitrogen atmosphere in the silo was kept for 3 weeks, but the smoldering fire inside the meal did not extinguish.
- It was then decided to try to fill water into the silo. First, information was collected about the strength of the bottom/hopper part of the silo. It was concluded that the bottom part could just about take the hydrostatic pressure that might result from soaking the meal with water.

- Water was filled into the silo. "Cracking" sounds were heard from the hopper section.
- After 10 days, the silo bottom was opened, and the water was drained out.
- Workers were admitted into the silo again from the top to resume the digging-out work. The meal was then just like concrete and had to be removed by using pick-axes and hammers and chisels.
- After some digging-out, regions with smoldering meal were once more encountered, and the digging had to be interrupted again.
- The silo was once more sealed, and the remaining fishmeal soaked with water.
- After 10 more days, the water was drained out, and the digging-out resumed. This time it was possible to empty the silo completely.
- Outward buckling of the hopper walls was observed.

The total time of the silo-emptying process, from the onset of the efforts to extinguish the smoldering fire to completion, was about 2 months. This story illustrates that it remains to develop adequate procedures for extinction of fires in silos.

A recent book edited by Krause (2009) contains a chapter reviewing methods for fighting silo fires. The book also contains some additional case histories.

6.5 MEASURES FOR PREVENTING EXCESSIVE SELF-HEATING IN SILOS

6.5.1 Temperature Measurement to Detect Self-Heating/Self-Ignition

Temperature sensors positioned inside the bulk of material stored in silos have been used for detecting the onset of self-heating for many decades. However, in larger silos, it is impossible in practice to cover the entire silo volume by temperature sensors, and self-heating and even smoldering fires may develop without being detected by the array of sensors available.

6.5.2 Early Detection of Tracer Gases From Self-Heating

More recently, increasing focus has been on the use of very sensitive sensors for detecting the onset of generation of decomposition gases from self-ignition processes. Zockoll (1996) described the development of a new system for early detection of self-heating/self-ignition in deposits of organic powders, with special reference to milk powder in spray dryers,

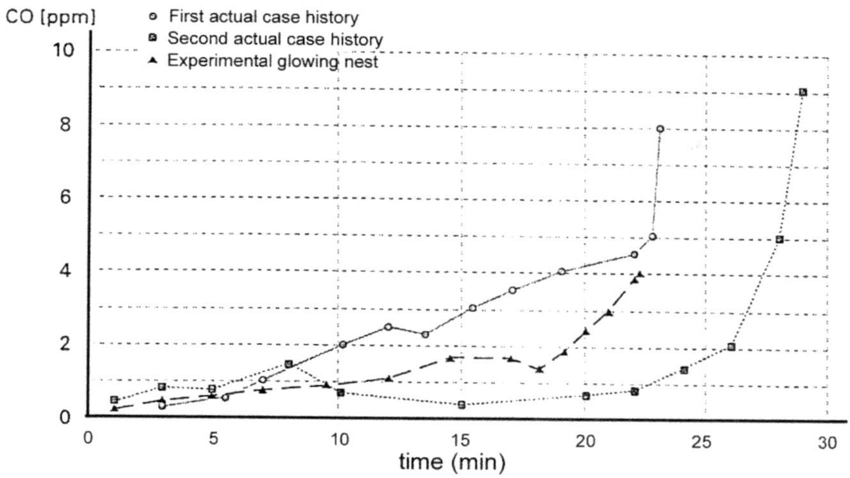

Figure 6.13 CO concentration development with time in two actual industrial cases of accidental self-ignition, and in one controlled self-ignition experiment.
From Zockoll, C., 1996. Früherkennung von Bränden durch CO-Detektion. VDI-Berichte 1272, 411–427, VDI-Verlag GmbH, Düsseldorf, Germany.

based on detection (infrared light) of low concentrations of CO in the 1–10 ppm range. Fig. 6.13 gives some results from Zockoll's work, showing how the measured CO concentrations increase with time. As soon as unacceptably high concentrations are detected, alarms are automatically activated, and measures to handle the problem are taken. In the case of early self-heating, recirculation of the material in the silo can be an appropriate measure (see Section 6.5.3).

Kohl and Kelleter (1996) described a multisensor system for detection of gases developed by slow smoldering fires in lignite. H_2 and CO were monitored continually, using electrochemical sensors. It was found that the ratio of the measured concentrations of the two gases could provide information about the development of the smoldering fire. When implementing systems for early detection of combustion gases, it is important that the supplier of the detection system and the user work closely together to ensure optimal system design for any given application.

6.5.3 Recirculation of Material in Silo by Discharging at Bottom and Refilling at Top

This method requires adequate silo design (mass flow as opposed to core flow) to ensure that the entire material becomes involved, not only a central core. As illustrated by the case history described in Section 6.4.4, recirculation can fail in smoothing out temperature differences in silos if only part of the stored material is participating in the recirculation process (see Fig. 7.69 in Chapter 7.).

6.5.4 Mathematical Modeling/Numerical Simulation of Self-Heating/Self-Ignition Processes

It is anticipated that in the future the likelihood of onset of self-heating in any given industrial situation can be assessed by numerical simulation. Carson (1996) suggested such a method for assessing the tendency of combustible dusts to self-heat and start to burn spontaneously in a given practical situation, comprising various laboratory-scale tests combined with mathematical modeling. The test methods included differential thermal analysis (DTA), thermogravimetric analysis (TGA), isothermal oven tests, and adiabatic calorimetry. Krause and Hensel (1996) presented a numerical method by which nonsteady temperature fields in dust deposits can be computed. This enables numerical analysis of a number of practical cases of self-heating/self-ignition that cannot be analyzed using the classical thermal explosion theory of Frank—Kamenetzki. Krause and Schmidt (2001) studied experimentally critical thermal conditions that may lead to initiation of smoldering processes, or to further development of such processes, once initiated. This type of evidence is important for further development of mathematical simulation models.

Krause et al. (2006) produced a quite comprehensive numerical model for simulating fires in bulk materials and powder deposits. However, this model was based solely on heat conduction and molecular diffusion. At that time the authors were unable to include convection of the gas in the pores between the particles. Therefore the "chimney effect" by which air is sucked into the powder bed from the outside could not be taken into account. However, more recently Krause et al. (2009) presented an improved model that accounts for this effect.

CHAPTER SEVEN

Dust Explosions

7.1 WHAT IS A DUST EXPLOSION?

7.1.1 Introduction

The phenomenon named *dust explosions* is in fact quite simple and easy to envisage in terms of daily life experience. Any solid material that can burn in air will do so with a violence and speed that increases with increasing degree of subdivision of the material. Fig. 7.1a illustrates how a piece of wood, once ignited, burns slowly, releasing its heat over a long period of time. When cut in small pieces, as illustrated in Fig. 7.1b, the combustion rate increases because the total contact surface area between wood and air has been increased. In addition, the ignition of the wood has become easier. If the subdivision is continued right down to the level of small particles of sizes on the order of 0.1 mm or less, and the particles are suspended in a sufficiently large volume of air to give each particle enough space for its unrestricted burning, the combustion rate will be very fast, and the energy required for ignition very small. This is illustrated in Fig. 7.1c. Such a burning dust cloud is a dust explosion.

In general, the dust cloud will be easier to ignite and burn more violently the smaller the dust particles are, down to some limiting particle size that depends on the type of dust material. If such an explosive combustion of a dust cloud takes place inside process equipment or workrooms, the pressure in the fully or partly enclosed explosion space may rise rapidly and the process equipment or building may burst, and life, limb, and property can be lost.

A more comprehensive account of the various aspects of dust explosions is given by Eckhoff (2003), which is also the main source of the summary given in the present chapter.

7.1.2 Materials That Can Give Dust Explosions

As with gas/vapor and spray/mist explosions, dust explosions also generally arise from rapid release of heat due to the chemical reaction:

$$\text{fuel} + \text{oxygen} \rightarrow \text{oxides} + \text{heat} \qquad [i]$$

Explosion Hazards in the Process Industries
ISBN: 9780128032732
http://dx.doi.org/10.1016/B978-0-12-803273-2.00007-4

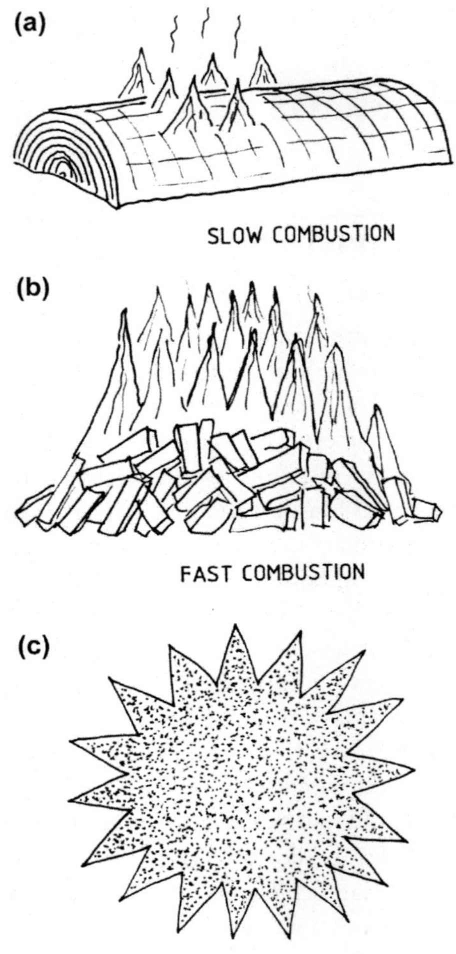

Figure 7.1 Illustration of the increase of the combustion rate of a given mass of combustible solid increases with increasing subdivision.
From Eckhoff, R.K., 2003. Dust Explosions in the Process Industries, third ed. Gulf Professional Publishing/Elsevier. (first ed. published in 1991, second ed. in 1997). ISBN:0-7506-7602-7.

In some special cases, metal dusts can also react exothermally with nitrogen or carbon dioxide, but most often oxidation by oxygen is the heat-generating process in a dust explosion. This means that only materials that are not already stable oxides can give rise to dust explosions. This excludes substances such as silicates, sulfates, nitrates, carbonates, and phosphates, and therefore dust clouds of Portland cement, sand, limestone, etc. cannot give dust explosions.

The materials that can give dust explosions include:

- Natural organic materials (grain, wood, linen, sugar, etc.).
- Synthetic organic materials (plastics, organic pigments, pesticides, pharmaceuticals, etc.).
- Coal and peat.
- Metals (aluminum, magnesium, titanium, zinc, iron, etc.).

The heat of combustion of the material is an important parameter because it determines the amount of heat that can be liberated in the explosion. However, when comparing the various materials in terms of their potential hazard, it is useful to relate the heat of combustion to the amount of oxygen consumed. This is because the gas in a given volume of dust cloud contains a limited amount of oxygen, which limits how much heat can be released in an explosion per unit volume of dust cloud. In Table 7.1 the heats of combustion of various substances, per mole of oxygen consumed, are given. Ca and Mg top the list, with Al closely behind. Si is also high up on the list with a heat of combustion per mole of oxygen about twice the value of typical natural and synthetic organic substances and coals.

Table 7.1 is in accordance with the experience that the temperatures of flames of dusts of metals like Al and Si are very high compared with those of flames of organic dusts and coal dusts.

Table 7.1 Heats of combustion (oxidation) of various substances per mole O_2 consumed

Substance	Oxidation product(s)	O_2 (kJ/mole)
Ca	CaO	1270
Mg	MgO	1240
Al	Al_2O_3	1100
Si	SiO_2	830
Cr	Cr_2O_3	750
Zn	ZnO	700
Fe	Fe_2O_3	530
Cu	CuO	300
Sucrose	CO_2 and H_2O	470
Starch	CO_2 and H_2O	470
Polyethylene	CO_2 and H_2O	390
Carbon	CO_2	400
Coal	CO_2 and H_2O	400
Sulfur	SO_2	300

From Eckhoff, R.K., 2003. Dust Explosions in the Process Industries, third ed. Gulf Professional Publishing/Elsevier. (first ed. published in 1991, second ed. in 1997). ISBN:0-7506-7602-7.

7.1.3 Factors Influencing the Ignitability and Explosivity of Dust Clouds

7.1.3.1 Overview

A fairly comprehensive list of factors influencing the ignitability and explosivity of dust clouds may look as follows:

- Chemical composition of the dust, including its moisture content
- Chemical composition, and initial pressure and temperature of the gas phase
- Distributions of particle sizes and shapes in the dust, determining the specific surface area of the dust in the fully dispersed state
- Degree of dispersion, or agglomeration, of dust particles, determining the effective specific surface area available to the combustion process in the dust cloud in the actual industrial situation
- Distribution of dust concentration in the actual cloud
- Distribution of initial turbulence in the actual cloud
- Possibility of generation of explosion-induced turbulence in the still-unburnt part of the cloud (location of ignition source important parameter)
- Possibility of flame-front distortion by other mechanisms than turbulence
- Possibility of significant radiative heat transfer (highly dependent on flame temperature, which in turn depends on particle chemistry)

Bullet points 1, 2, 3, and 9 can be regarded as basic parameters of the explosive dust cloud. Factors 4–8 are, however, influenced by the actual industrial dust cloud generation process and explosion development. These in turn depend on the nature of the industrial process (flow rates, etc.) and the geometry of the system in which the dust cloud burns. The location of the ignition point is another parameter that can play an important role in deciding the course of the explosion.

In view of the wide spectrum of dust cloud concentrations, degrees of dust dispersion, turbulence, and locations of potential ignition sources in industry, a correspondingly wide spectrum of possible dust cloud ignition sensitivities and combustion rates must be expected. This complex reality of the process industry is also shared by laboratory experimentation and represents a constant challenge in the design of adequate experiments and interpretation of experimental results.

VDI (1988) presented an overview of experimental methods used at that time for assessing various combustibility and ignitability properties of explosible dust clouds.

7.1.3.2 "Hybrid Mixtures"

Bullet point No. 2 in Section 7.1.3.1 also embraces the special case in which the air contains a fraction of a combustible gas/vapor. Originally, as in the comprehensive text by Bartknecht (1993), focus was primarily on mixtures in which the combustible gas/vapor fraction was smaller than the lower explosibility limit (LEL) of the gas/vapour itself. Such mixtures of air, an explosive concentration of the dust, and a < LEL fraction of a combustible gas/vapor, were named "hybrid" mixtures. It is not clear who was the first researcher to study the ignitability and explosibility of this kind of mixture. However, 130 years ago Engler (1885) conducted experiments in a wooden explosion box of volume of about 30 L, and essentially open at the bottom. The box was filled with a mixture of air methane of the desired methane concentration, and a cloud of fine charcoal dust, which was unable to give dust explosions in pure air, was introduced at the container top by a vibratory feeder. Engler observed that methane concentrations as low as 2.5 vol% made clouds of the charcoal dust explosible, whereas the methane/air alone without the dust, did not burn unless the gas content was raised to 5.5 to 6 vol%. One generation later, Engler (1907) described a simple laboratory-scale experiment by which this "hybrid" effect could be demonstrated. The original sketch of the apparatus and a detailed description of the experimental procedure are given in Eckhoff (2003).

Adding small percentages of combustible gas to the air most often also reduces the minimum explosible dust concentration (MEC). Cardillo and Anthony (1978) studied this effect experimentally using propane mixed with the air. He found that clouds of iron dust responded to the propane addition in the same systematic way as organic dusts. For no propane in the air, MEC for the iron dust was 200 g/m^3, whereas for 1 vol% propane it was 100 g/m^3.

The influence of small fractions of methane in the air on the minimum electric spark energy (MIE) for igniting clouds of coal dusts was investigated systematically by Franke (1978). He measured substantial reductions in the MIE, by factors on the order of 100, when the methane content in the air was increased from 0 to 3 vol%. Pellmont (1979) found the same substantial effect on MIE of mixing propane with the air as Franke.

One of the first systematic investigations of the "hybrid" effect on dust explosion violence was probably conducted by Nagy and Portman (1961). When a coal dust was dispersed in various methane/air mixtures, they found a marked increase of the maximum rate of pressure rise $(dP/dt)_{max}$ in the explosions with increasing methane content in the air.

In one experiment $(dP/dt)_{max}$ was less than 100 bar/s without methane, and nearly 500 bar/s with 2% methane in the air.

Addai et al. (2015) published results from an experimental investigation of the minimum ignition temperature (MIT) and lower flammability limit/minimum explosible concentration (LEL/MEC) of hybrid mixtures of a combustible dust and a combustible gas/vapor in air. They adopted a more general, expanded hybrid-mixture concept comprising all mixtures/clouds in air of combustible dust particles and combustible gas/vapor, irrespective of fuel concentrations. The experiments were performed in a modified Godbert–Greenwald (GG) furnace. The study comprised five dusts, three gases, and three solvent vapors. For the dusts alone the experimental protocol was in accordance with the European Standard (EN) 50281-2-1, whereas for the gases, solvent vapors, and mixtures of these with the dusts, the protocol had to be modified slightly. It was found that the MITs of the gases and solvent vapors decreased significantly when a combustible dust, of concentration below the LEL of the dust alone, was added. A significant decrease of MIT was also found if the gas/vapor cloud was mixed with a combustible dust that could not at all be ignited alone in the GG furnace. Example: MIT of toluene vapor/air decreased from 540 to 455°C when a small fraction of *Lycopodium* particles was mixed into the vapor/air cloud. It was also found that a hybrid cloud of a combustible dust at a concentration < MEC and a combustible gas at a concentration < LEL could be ignited in the GG furnace. Example: The MEC of clouds of a given dust in air decreased from about 300 to about 40 g/m^3 when propane of a concentration below LEL for propane/air alone was added to the cloud.

7.2 COMBUSTION OF DUST CLOUDS IN AIR

7.2.1 Explosive Concentration Ranges of Dust Clouds in Air

The explosive combustion of dust clouds, as illustrated in Fig. 7.1c cannot take place unless the dust concentration, ie, the mass of dust per unit volume of dust clouds, is within certain limits. This is analogous with the combustion of homogeneous mixtures of gaseous fuels and air, for which the upper and lower flammability limits are well established. Fig. 7.2 shows the explosive range for a typical natural organic material, such as maize starch, in air at normal temperature and atmospheric pressure. This range is quite narrow, extending over less than two orders of

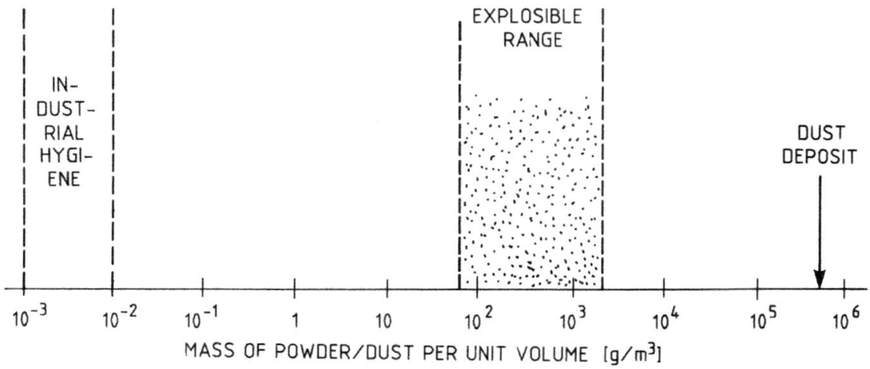

Figure 7.2 Range of explosive dust concentrations in air at normal temperature and atmospheric pressure for a typical natural organic dust (maize starch), compared with typical range of maximum-permissible dust concentrations in the context of industrial hygiene, and a typical density of deposits of natural organic dusts.

magnitude, from 100 g/m^3 on the lean side to 2–3 kg/m^3 on the rich one. The explosivity limits differ somewhat for the various dust materials. For example, zinc powder has a minimum explosive concentration in air of about 500 g/m^3.

Explosive dust clouds have a high optical density, even at the LEL. This is illustrated, as shown in Fig. 7.2, by the fact that the range of maximum permissible dust concentrations that is specified in the context of industrial hygiene in working atmospheres, are three to four orders of magnitude lower than minimum explosive dust concentrations. This means that the unpleasant dust-concentration levels that can sometimes occur in the general working atmosphere of a factory, and which is the concern of industrial hygiene authorities, are far below the concentration levels that can propagate dust flames. Therefore, the MEC corresponds to dust clouds of high optical densities, which are unlikely to occur regularly in workrooms of factories. Fig. 7.3 illustrates the high optical densities of explosive dust clouds, based on a rule of thumb by Intelmann, and quoted by Zehr (1965): If a glowing 25-W light bulb is observed through 2 m of coal-dust cloud, the bulb cannot be seen through dust concentrations exceeding 40 g/m^3, ie, about half of the MEC.

It then follows that the dust clouds in which dust explosions are primarily initiated are practically always found inside process equipment, such as mills, mixers, screens, dryers, cyclones, filters, bucket elevators, hoppers, silos,

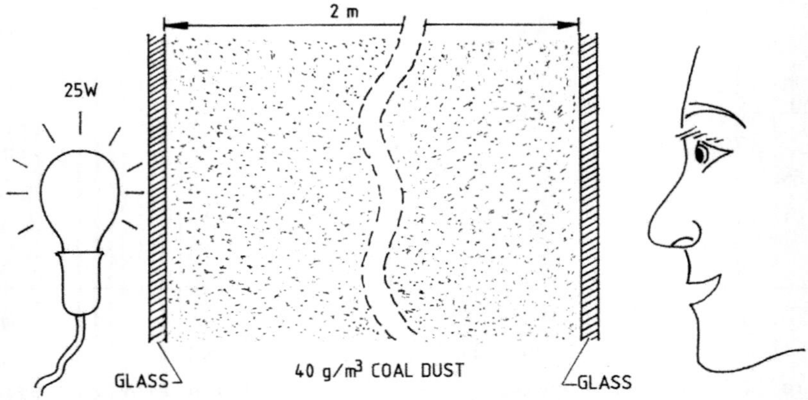

Figure 7.3 A cloud of 40 g/m^3 of coal dust in air is so dense that a glowing 25-W light bulb can hardly be seen through a dust cloud of 2-m thickness.
From Eckhoff, R.K., 2003. Dust Explosions in the Process Industries, third ed. Gulf Professional Publishing/Elsevier. (first ed. published in 1991, second ed. in 1997). ISBN:0-7506-7602-7.

aspiration ducts, and pipes for pneumatic transport of powders. Such explosions, initiated by some ignition source, are called *primary* explosions.

This reveals an important difference between primary dust and gas explosions. In the case of dusts, the primary explosive clouds practically always occur inside process equipment, ie, in areas where electrical apparatus is scarce or absent. In the case of gases, however, the main hazard is primary gas explosions outside process equipment in which gas from accidental leaks gets mixed with air, and explosive atmospheres are generated, and in which a wide range of potential ignition sources may be found, including various electrical apparatuses.

7.2.2 Flame Propagation Processes in Dust Clouds

7.2.2.1 Basic Differences Between Premixed Gas Flames and Flames in Dust Clouds

Leuschke (1965) pointed out some characteristic differences between a laminar premixed gas flame and a laminar dust flame. The reaction zone in the dust cloud is generally thicker than in premixed gas, typically in the range 5–100 mm. When discussing dust flames, it is useful to distinguish between two flame types. The first, the Nusselt type, is controlled by diffusion of oxygen to the surface of individual particles, where a heterogeneous chemical reaction takes place. In the second type, the volatile flame type, the rate of gasification, pyrolysis, or devolatilization is the controlling process, and the chemical reaction takes place mainly in a homogeneous gas phase. In Nusselt-type flames, the greater thickness of the combustion zone as compared with that of premixed-gas flames, results from the slower rate of molecular diffusion, compared to diffusion in homogeneous premixed gases. With the volatile flame type, the greater flame thickness is due to a preheating zone, in which combustible gases and vapors (volatiles) are driven out of the particles ahead of the flame. Mixtures of such volatiles and air are just as explosive as explosive mixtures of gaseous hydrocarbons and air. The combustion of the remaining solid-char particles occurs subsequently at a slower rate in the tail of the flame, and therefore the volatile flame in clouds of coals and organic dusts is also, in fact coupled to a Nusselt-type flame.

In the case of metals, low-melting-point materials may oxidize in the vapor phase, but due to the oxide film around each particle, this does not result in a homogeneous metal vapor/air flame. Because of the large heat of combustion per mole oxygen of, for example aluminum and magnesium dust, compared with organic dusts, the temperature of the burning particles is very high, and thermal radiation plays a central role in the transfer of heat in the combustion wave. Radiative heat transfer is also supposed to play a role in coal-dust flames. However, because the thermal radiation is proportional to the fourth power of the temperature, the role of thermal radiation in coal-dust flames is probably less important than in aluminum- and magnesium-dust flames. Leuschke (1965) conducted an illustrative series of experiments demonstrating the importance of radiative heat transfer in metal-dust flame propagation, using the experimental set-up illustrated in Fig. 7.4.

Two transient dust clouds were generated simultaneously on the two sides of a double-glass window, one being ignited immediately by a gas flame. It was then observed whether the radiation from the burning cloud was able to ignite the other cloud. Only flames of Zr, Ti, Al, and Mg were able to produce sufficient radiation to ignite the second cloud. The specific role of thermal radiation in the propagation of dust flames and buildup of the explosion pressure varies with the dust type, and more research is needed to quantify the radiative contribution under various circumstances.

Another difference between flame propagation in a premixed gas and dust cloud has been elucidated by Goral et al. (1988). They studied upward propagation of flames in a lean methane/air mixture to which had been added inert particles (sand). It was found that the upward flame velocity

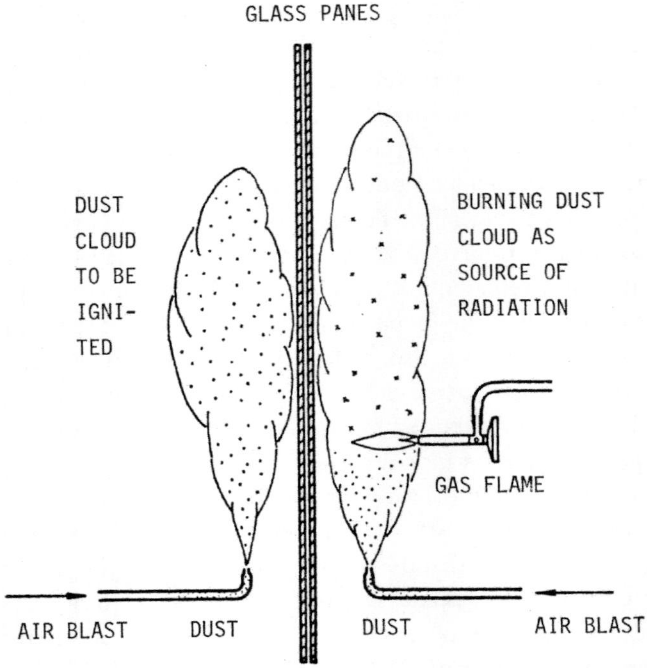

Figure 7.4 Experiment demonstrating the ignition of a cloud of metal dust in air by radiation from a burning cloud of the same dust, through a double-glass window. *From Leuschke, G., 1965. Beitrage zur Erforschung des Mechanismus der Flammenausbreitung in Staubwolken. Staub 25, 180–186.*

increased with increasing sand grain size. The effect was mainly attributed to the enhanced combustion due to the microturbulence generated in the wake of the falling particles, although thermal radiation effects were also assumed to play a role.

7.2.2.2 Influence of Dust Chemistry

There are two aspects to consider, namely the thermodynamics of the explosion, and the kinetics. The thermodynamics is concerned with the amount of heat that is liberated during combustion, the kinetics with the rate at which the heat is liberated.

The maximum rate with which the explosion pressure rises in closed-bomb experiments is a frequently used relative measure of the violence to be expected from explosions of a given dust.

Fig. 7.5 shows how the maximum rates of pressure rise of starch (potato and maize starch) are systematically higher than for protein (two fish

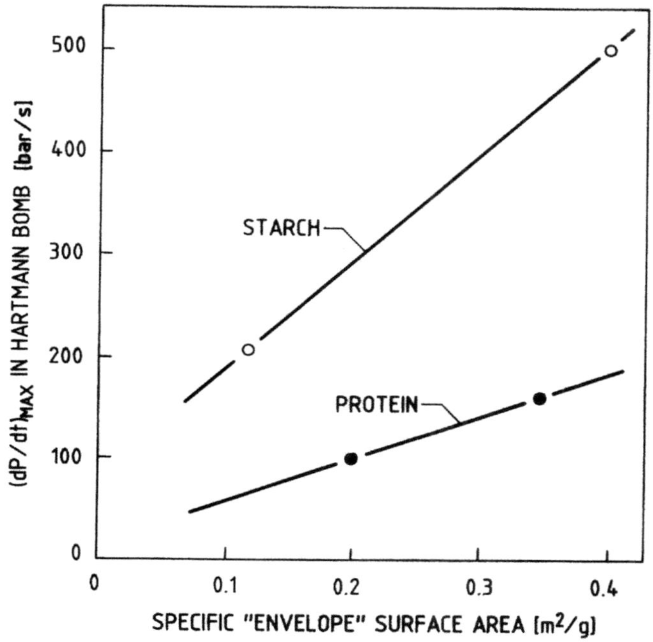

Figure 7.5 Influence of chemistry (starch or protein) and specific surface area of natural organic materials on maximum rate of pressure rise in closed 1.2 L Hartmann bomb.

From Eckhoff, R.K., 1977/1978. Pressure development during explosions in clouds of dusts from grain, feedstuffs and other natural organic materials. Fire Research 1, 71–85.

powders with fat removed) for the same specific surface area. The nitrogen compounds in the protein are probably in some way slowing down the combustion process.

Another example of the influence of dust chemistry on the explosion kinetics is shown in Fig. 7.6.

The heats of combustion of PVC and polyethylene are not very different. As Fig. 7.6 shows, closed-bomb experiments also give about the same maximum pressure for very small particle sizes. However, the chlorine in the PVC causes a quite dramatic drop in the rate of heat release as the median particle size increases beyond about 20 μm. Due to the very slow combustion, P_{max} for PVC also drops much faster as the particle size increases than for polyethylene. The retarding influence of chlorine on the combustion process most probably is of the same nature as that of the halogens in the Halons, which were extensively used for explosion and

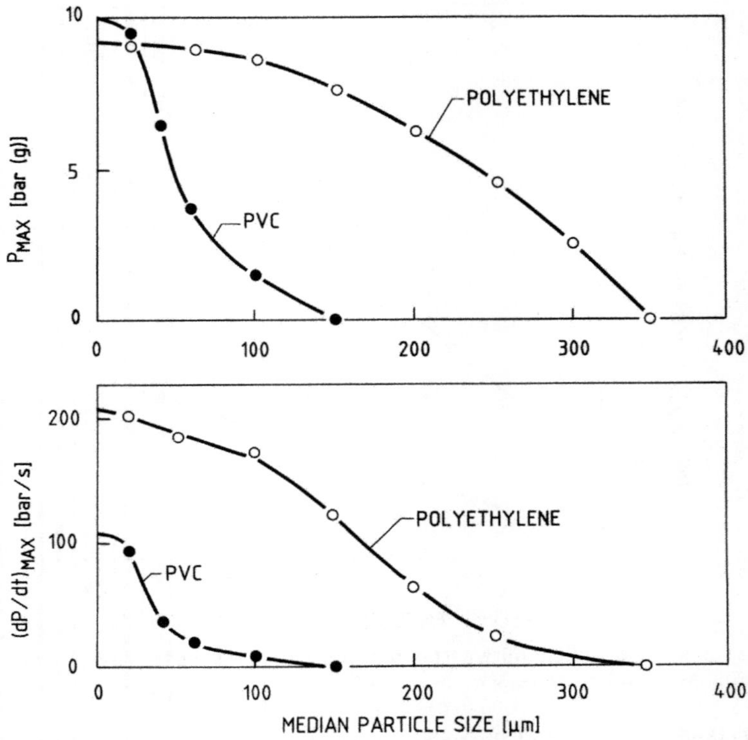

Figure 7.6 Influence of chlorine in molecule of dust material on maximum explosion of pressure rise in the 1 m^3 standard ISO (International Organization for Standardization) vessel, for various particle sizes.
From Bartknecht, W., 1978. Explosionen — Ablauf und Schutzmassnahmen. Springer-Verlag. ISBN:0-387-08675-7.

fire suppression before the negative influence of such materials on the global environment was fully realized.

7.2.2.3 Influence of Moisture

Moisture in the dust reduces both ignition sensitivity and explosion violence of dust clouds markedly. If safety measures against electric spark ignition are based on MIE data for a given dust moisture content, it is essential that this moisture content is not subsided in practice. The influence of dust moisture on the MIT of dust clouds is less marked than on MIE. The specific role of moisture in reducing both ignition sensitivity and explosion violence of clouds of organic dusts is complex. First, evaporation and heating of water represents an inert heat sink. Second, the water vapor mixes with the pyrolysis gases in the preheating zone of the combustion wave, and makes the gas mixture less reactive. A third factor is that moisture increases the interparticle cohesion of the dust and prevents dispersion into primary particles.

7.2.2.4 Influence of Particle Size

Figs. 7.5 and 7.6, in addition to illustrating the influence of dust chemistry on the dust cloud combustion rate, also exhibit a clear dependence of the explosion violence on the particle size/specific surface area for the four dust types shown. This is a general trend for most dusts. However, this trend does not continue indefinitely as the particles get smaller. In the case of coal and organic materials, pyrolysis/devolatilization always precedes combustion, which primarily occurs in the homogeneous gas phase. The limiting particle size, below which the combustion rate of the dust cloud does not increase any more, depends on the ratios between the time constants of the three consecutive processes: devolatilization, gas phase mixing, and gas phase combustion. Particle size primarily influences the devolatilization rate. Therefore, if the gas phase combustion is the slowest of the three steps, increasing the devolatilization rate by decreasing the particle size does not increase the overall combustion rate. For coals, it was found that the limiting particle diameter is on the order of 50 μm. However, for materials yielding gaseous pyrolysis products that are more reactive than volatiles from coal, eg, due to unsaturated gaseous compounds, one would expect the limiting particle size to be smaller than for coal. For natural organic compounds such as starch and protein, the limiting particle diameter is probably not much smaller than about 10 μm, whereas for reactive dusts such as some organic dyes, it may well be considerably smaller.

For metals, in particular those at the top of Table 7.1, the limiting particle size, below which there is no longer any increase in ignition sensitivity and explosion violence, is considerably smaller than for most organic materials. This is because these metals do not devolatilize or pyrolyze, but melt, evaporate, and burn as discrete entities. In such cases the aggravating influence of particle size may continue even below 1 μm.

In Section 7.3.6.2.5, the pronounced decrease of MIE with decreasing particle size is discussed.

7.2.3 Close-to-Laminar Flame Propagation in Dust Clouds

Laminar 20-mm diameter burner flames of clouds in air of *Lycopodium* and polyvinyl alcohol were studied by Kaesche-Krischer and Zehr (1958) and

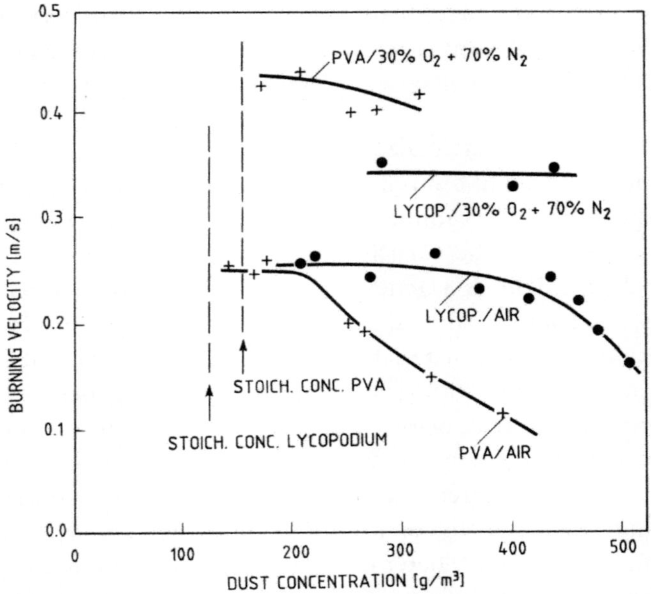

Figure 7.7 Burning velocities of flames of *Lycopodium* and polyvinylalcohol dust (<60-μm particle diameter) flames as functions of dust concentration. The two vertical dotted lines indicate the stoichiometric concentrations of the two dusts in air only.
From Kaesche-Krischer, B., Zehr, J., 1958. Untersuchungen an Staub/Luft-Flammen. Zeitschrift für Physikalische Chemie Neue Folge 14, 5/6 and Kaesche-Krischer, B., 1959. Untersuchungen an vorgemischten, laminarer Staub/Luft-Flammen. Staub 19, 200–203.

Kaesche-Krischer (1959). The burning velocity, defined as the ratio of airflow and flame-cone area, was determined photographically from the height of the flame cone.

Some results are given in Fig. 7.7. *Lycopodium*/air flames of dust concentrations lower than 180 g/m^3 and higher than 500 g/m^3 were difficult to stabilize (stoichiometric concentration $\approx$125 g/m^3). The appearance of a stabilized *Lycopodium*/air flame was very similar to that of a rich hydrocarbon/air flame, ie, a blue flame front followed by a more or less luminous soot tail. Approximate thermocouple measurements of flame temperatures gave about 1800K for a 180 g/m^3 flame, and 1100K for a 500 g/m^3 flame. Fig. 7.7 shows the measured burning velocities as a function of the dust concentration. In the range 180–300 g/m^3, the burning velocity of *Lycopodium* flames has a maximum value of about 0.25 m/s. The corresponding concentration range for the polyvinyl alcohol (PVA) dust was 140–220 g/m^3

Fig. 7.7 also shows that an increase of the oxygen content in the gas phase from 21 to 30 vol% of air, gave a significant increase of the measured burning velocities for both dusts, in accordance with expectations. The photographs provided by Kaesche-Krischer and Zehr (1958) indicate typical thicknesses of quasilaminar *Lycopodium* flames in air of a few mm.

Kaesche-Krischer and Zehr implied that the differences in the concentration ranges giving the highest burning velocities for the two dusts were due to a higher volatile content in the PVA than in *Lycopodium*, assuming that the flame essentially propagates through a homogenous mixture of volatiles and air.

In a comprehensive survey of a number of investigations on the propagation of laminar pulverized coal dust/air flames, Smoot and Horton (1977) discussed factors influencing experimentally determined burning velocities, flame temperatures, and flame thicknesses. Most experiments have been performed on stabilized-dust flames in burners of various kinds. Due to heat losses by radiation from the hot dust particles and conduction, typical stabilized burner flames will have temperatures that are lower than the adiabatic flame temperature. In principle heat losses can be avoided by using burners of very large diameters, or equipped with walls having temperature and emissivity profiles matching those of the flame. Smoot and Horton found large differences in burning velocities observed by various investigators which could not be explained in terms of variations in dust properties or dust concentration. They considered incomplete dispersion of fine cohesive dusts, as illustrated in Fig. 7.8, as the main source of error.

The experimental results of Smoot and Horton in Fig. 7.9 illustrate how improved dispersion of a fine coal dust increased the burning velocity by at least 50%.

Nonhomogeneity is an inherent feature of any real dust cloud, and this also influences the pattern of flame propagation through such clouds. This not least complicates the experimental determination of the minimum explosive dust concentration. Informative experiments have been conducted in quite large industrial-scale equipment. The work of Palmer and Tonkin (1971) is a good example. Their apparatus is illustrated in Fig. 7.10.

In the experiments illustrated by Fig. 7.10, the dust was introduced at the top of the tube by a screw feeder and dropped into a vibrating 20-cm diameter and 15-cm high dispersing cylinder hanging immediately underneath the screw exit. After having passed through the perforated bottom

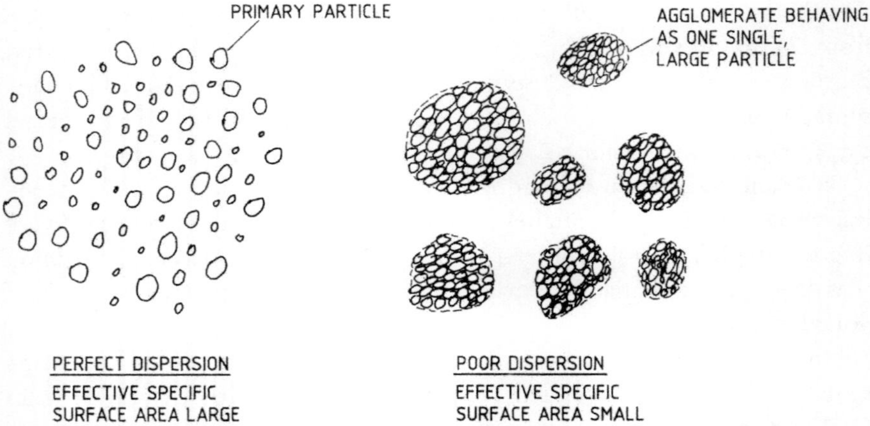

Figure 7.8 Illustration of perfectly dispersed dust cloud and cloud consisting of agglomerates of much larger effective particle sizes than those of the primary particles.
From Eckhoff, R.K., 2003. Dust Explosions in the Process Industries, third ed. Gulf Professional Publishing/Elsevier. (first ed. published in 1991, second ed. in 1997). ISBN:0-7506-7602-7.

of the cylinder, the dispersed dust settled freely under gravity through the entire length of the tube until finally being collected in a bin at the bottom end. Dust concentration and flame propagation could not be measured in the same test, but had to be determined in separate tests at nominally

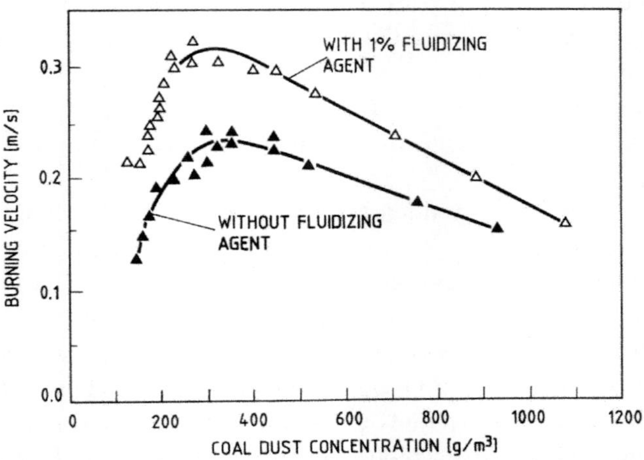

Figure 7.9 Effect of very fine SiO_2-fluidizing agent (Acrosil) on the burning velocity of an air suspension of 10 μm, 28% volatile content Sewell coal dust.
From Smoot, L.D., Horton, M.D., 1977. Propagation of laminar coal-air flames. Progress in Energy and Combustion Science 3, 235–258.

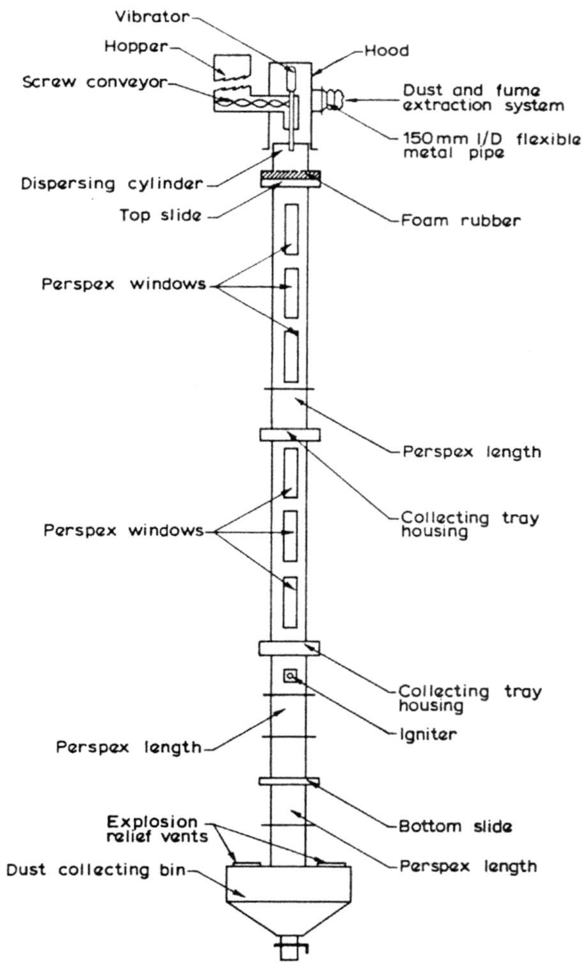

Figure 7.10 Vertical large-scale explosion tube facility for flame propagation studies. Internal tube diameter 0.25 m. Total tube length 5.2 m. Ignition point 1.5 m above bottom. *From Palmer, K.N., Tonkin, P.S., 1971. Coal dust explosions in a large-scale vertical tube apparatus. Combustion and Flame 17, 159–170.*

identical dust cloud generation conditions, ie, rotating speed of the feeding screw conveyor and vibration mode of the dust disperser. The dust concentration was measured gravimetrically. A manually operated sliding tray was inserted into the tube like a gate valve about 3.5 m below the tube top. By simultaneously closing the tube at the top by a conventional sliding gate valve, the volume of dust cloud between the top valve and the tray was trapped. By dividing the amount of dust finally settled out on the tray by the volume between the tray and the top valve, the average dust concentration in this section of the tube could be obtained. Immediately before performing an explosion test, the dust feed was stopped and the bottom end of the tube closed by a gate valve located just below the ignition zone. The ignition source was a propane flame generated by injecting a small pocket of a propane/air mixture into the bottom region of the explosion tube and igniting by means of an electric spark located at the tube axis.

By means of this apparatus the average length of upward vertical flame propagation away from the ignition source could be determined as a function of the average dust concentration in the measurement section.

7.2.4 Maximum Pressures Generated From Constant-Volume Adiabatic Combustion of Dust Clouds in Air

In Table 7.1, calcium, magnesium, and aluminum top the list with 1100–1300 kJ/mol oxygen consumed. The lowest values are 300 kJ/mol oxygen for copper and sulfur. It would be expected that this difference will to some extent be reflected in the maximum pressure of explosions, when performed adiabatically at constant volume. Zehr (1957) made some calculations of the maximum pressures to be expected under such conditions. In Fig. 7.11 his results have been plotted against data from experiments in either 1 m^3 or 20 L closed bombs.

For aluminum and magnesium, Zehr only indicated that the theoretical values would be larger than 10 and 13.5 bar (g), respectively. This is because at high temperatures the two metals will also react exothermally with the nitrogen in the air. Theoretical maximum pressures based on the assumption that reaction with oxygen is the only contributor to heat production therefore gives lower values than the experimental ones. In view of the uncertainties involved in the calculations, Fig. 7.11 suggests a fair correlation between theoretical and experimental data, with the theoretical results being somewhat higher than the experimental ones for the organic dusts. This would be expected because of the idealized assumptions of stoichiometry and complete oxidation of all fuel, on which the calculations were based.

The upper part of Fig. 7.12 shows experimental data confirming that the maximum adiabatic constant-volume explosion pressure (absolute [abs]) for dust clouds is in fact proportional with the initial pressure (abs) of the unburnt cloud, as for explosive mixtures of combustible gases. For the maximum rate of pressure rise, the lower part of Fig. 7.12 indicates a similar simple linear correlation. This is, however, more difficult to explain.

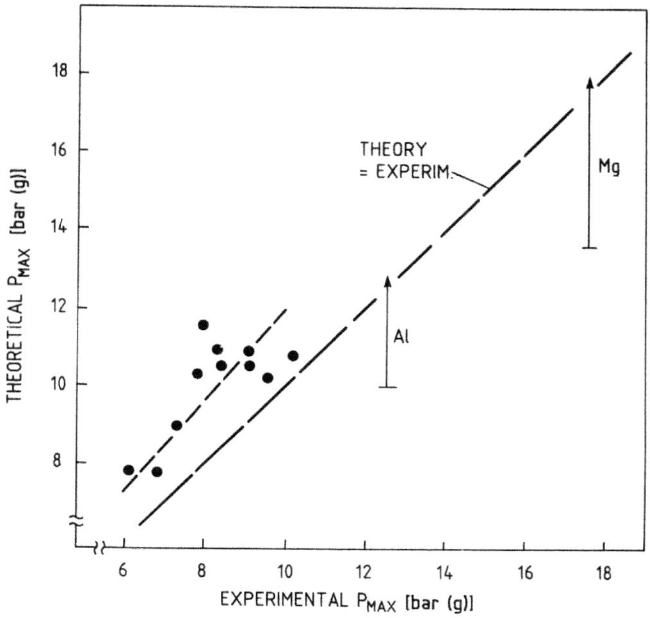

Figure 7.11 Correlation between P_{max} from experiments in 1 m^3 or 20 L closed vessels, and theoretical adiabatic constant-volume P_{max} values calculated by Zehr (1957).

7.2.5 Turbulent Flame Propagation in Dust Clouds

7.2.5.1 Basic Features of Closed-Bomb Studies of Turbulent Dust Flames

Much of the published experimental work on turbulent dust explosions has been conducted in closed vessels of the type illustrated in Fig. 7.13.

The explosion vessel of volume V_1 and initial pressure P_1 is equipped with a dust-dispersion system, a pressure sensor, and an ignition source. In most equipment the dust-dispersion system consists of a compressed-air reservoir of volume $V_2 << V_1$, at an initial pressure $P_2 >> P_1$. In some apparatuses, the dust is initially placed on the high-pressure side of the dispersion air valve, as indicated in Fig. 7.13, whereas in other apparatuses it is placed downstream of the valve. Normally, the mass of dispersion air is not negligible compared with the initial mass of air in the main vessel. This causes a significant rise of the pressure in the main vessel once the dispersion air has been discharged into the main vessel. In some investigations, this is compensated for by partial evacuation of the main vessel prior to dispersion so that the final pressure after dispersion completion, just prior

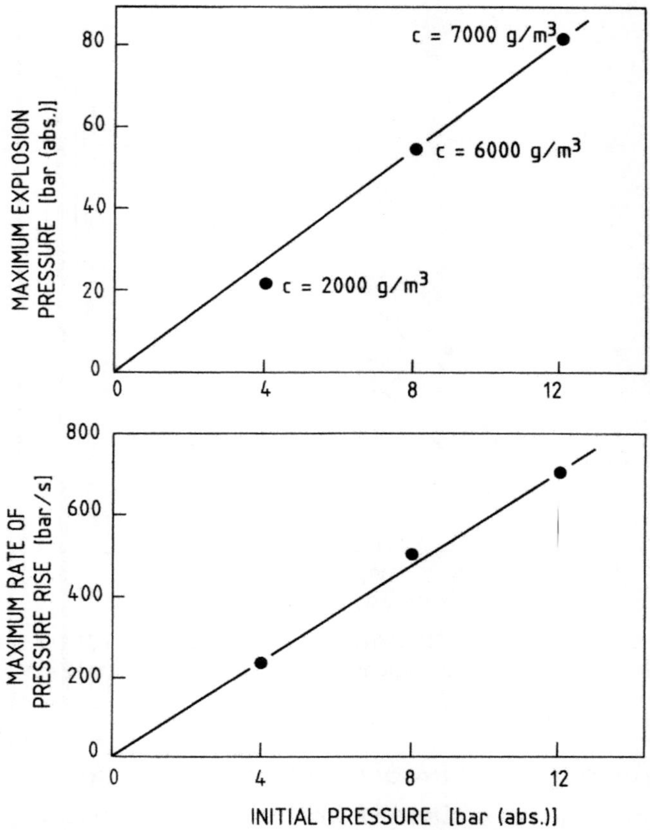

Figure 7.12 Influence of initial pressure on maximum pressure and maximum rate of pressure rise in explosions of clouds of subbituminous coal dust in air in a 15-L closed bomb. The median particle size by mass of the dust equals 100 μm.
From Pedersen, G.H., Wilkins, B.A., 1988. Explosibility of Coal Dusts and Coal Dust/Limestone Mixtures at Elevated Initial Pressures. Report No. 88/02101. Christian Michelsen Institute (now GexCon AS), Bergen, Norway.

to ignition, is atmospheric. This is important if absolute data are required, because the maximum explosion pressure for a given dust at a given concentration is approximately proportional to the initial absolute air pressure. Both the absolute sizes of V_1 and V_2 and the ratio between them vary substantially from apparatus to apparatus. The smallest V_1 used are on the order of 1 L, whereas the largest that has been traced is 250 m^3. The design of the dust-dispersion system varies considerably from apparatus to apparatus. A number of different nozzle types have been developed with the aim to break up agglomerates and ensure homogeneous distribution of the dust in the

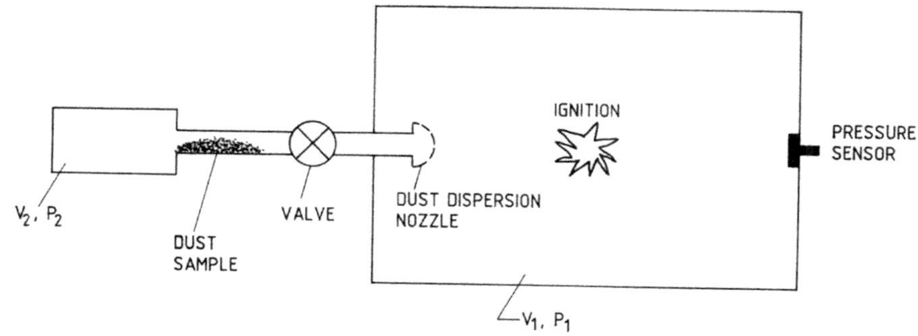

Figure 7.13 Schematic illustration of the type of apparatus commonly used in closed-vessel turbulent dust explosion experiments.
From Eckhoff, R.K., 2003. Dust Explosions in the Process Industries, third ed. Gulf Professional Publishing/Elsevier. (first ed. published in 1991, second ed. in 1997). ISBN:0-7506-7602-7.

main vessel. The ignition source has also been a factor of considerable variation. In some of the earlier investigations, continuous sources like electric arcs or trains of electric sparks, and glowing resistance wire coils were used, but more recently it has become common to use short-duration sources initiated at a given time interval after opening of the dust-dispersion valve. These sources vary from electric sparks via exploding wires to various forms of electrically triggered chemical ignitors.

An important inherent feature of all apparatus of the type illustrated in Fig. 7.13 is that the dispersion of the dust inevitably induces turbulence in the main vessel. The level of turbulence will be at maximum during the first period of the dispersion air being blown into the main vessel. After the main, strong flow of dispersion air into the main vessel has terminated, the turbulence decays with a time constant that decreases with increasing V_1. Hence, both the strength of the dispersion air blast and the delay between opening of the dust-dispersion value and ignition have strong influences on the state of turbulence in the dust cloud at the moment of ignition, and consequently also on the violence of the explosion. The situation is illustrated in Fig. 7.14.

The common primary result from a closed-bomb dust explosion experiment is a trace of the explosion pressure as a function of time, as illustrated in Fig. 7.15. Two parameters are normally of prime interest, viz. the maximum explosion pressure, ie, the peak value of the recorded pressure—time trace, and the maximum rate of pressure rise, ie, the steepest gradient of the rising part of the pressure—time trace.

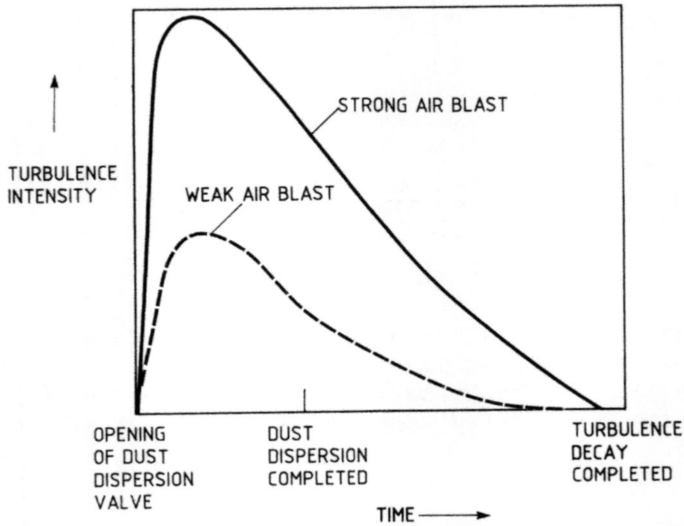

Figure 7.14 Illustration of generation and decay of turbulence during and after dispersion of dust in an apparatus of the type illustrated in Fig. 7.13. (Note: a common way of quantifying turbulence intensity is by the root mean square (rms) of the turbulent velocity fluctuations as measured by eg, laser doppler anemometry).

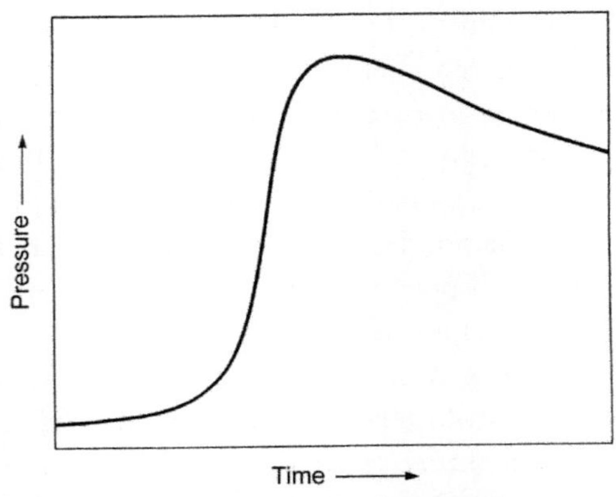

Figure 7.15 Illustration of typical trace of explosion pressure versus time during a dust explosion in a closed vessel.

7.2.5.2 Closed-Bomb Experiments

Fig. 7.16 gives an example of a set of results from a complete series of closed-bomb tests of one given dust. Tests are carried out over a wide range of nominal dust concentrations (mass of dust dispersed divided by vessel volume). Furthermore, due to inevitable scatter, several tests must be

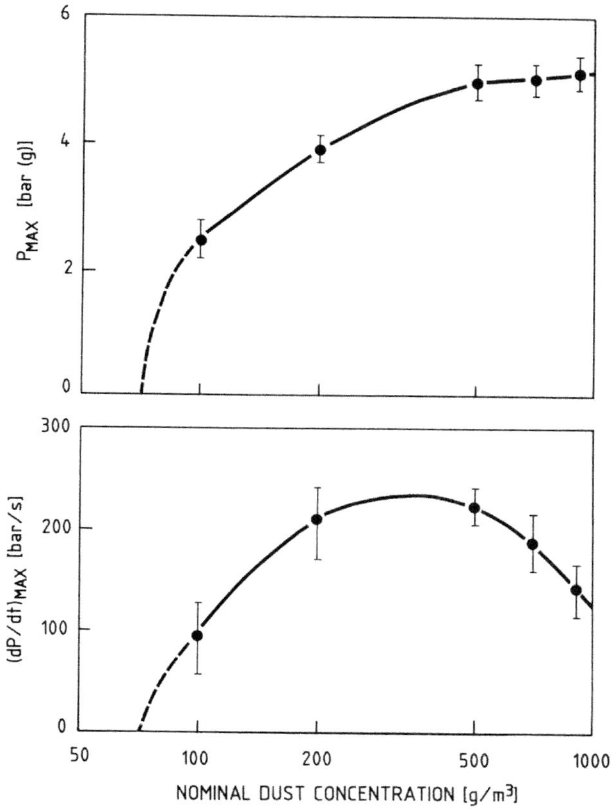

Figure 7.16 Influence of nominal dust concentration on maximum explosion pressure and maximum rate of pressure rise in dust explosions in air in the 1.2-L Hartmann bomb. Dust: maize starch containing 11% moisture. Bars indicate ±1 standard deviation. *From Eckhoff, R.K., Fuhre, K., Pedersen, G.H., 1985. Vented Maize Starch Explosions in a 236 m^3 Experimental Silo. Report No. 843307-2, (December). Christian Michelsen Institute (now GexCon AS), Bergen, Norway.*

performed at each nominal dust concentration. Normally, the dust concentrations that give the highest values of the maximum pressure and maximum rate of pressure rise are not the same. The test results are the highest mean values obtained for each of the two parameters.

The data in Fig. 7.17 illustrate the influence of the ignition delay on the explosion development in a cloud of *Lycopodium* in air in a 1.2-L Hartmann bomb.

As can be seen there is little difference between the maximum explosion pressure obtained with a delay of 40 and 200 ms, whereas the maximum rate of pressure rise is drastically reduced, from 430 to 50 bar/s, ie, by a factor of

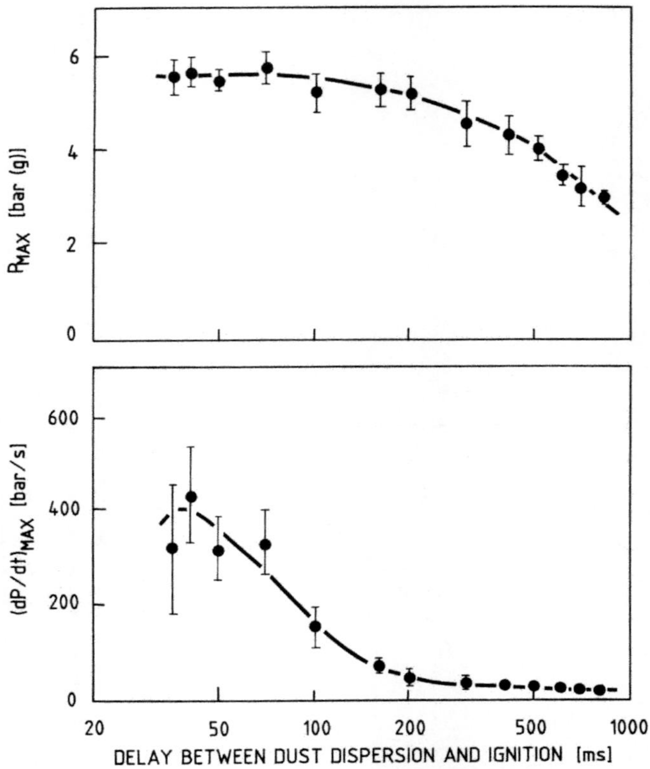

Figure 7.17 Influence of initial turbulence on explosion rate of a dust cloud. Experiments with 420 g/m^3 of *Lycopodium* in air in the 1.2-L Hartmann bomb. Five experiments per delay. Bars indicate ±1 standard deviation.
From Eckhoff, R.K., 1977/1978. Pressure development during explosions in clouds of dusts from grain, feedstuffs and other natural organic materials. Fire Research 1, 71–85.

almost 10. There is little doubt that this is due to the reduced initial turbulence in the dust cloud at the large ignition delays. With increasing ignition delay beyond 200 ms, the maximum explosion pressure is also reduced, most probably because the dust has started to settle out of suspension before the ignition source is activated.

As would be expected, the same kind of influence of ignition delay as shown in Fig. 7.17 is found in all experiments of the type illustrated in Figs. 7.13 and 7.14. One of the first researchers to observe this effect was Bartknecht (1971) when performing dust explosion experiments in a 1-m^3 explosion vessel equal to that illustrated in Fig. 7.18. As the ignition delay was increased from the lowest value of about 0.3 to about 1 s,

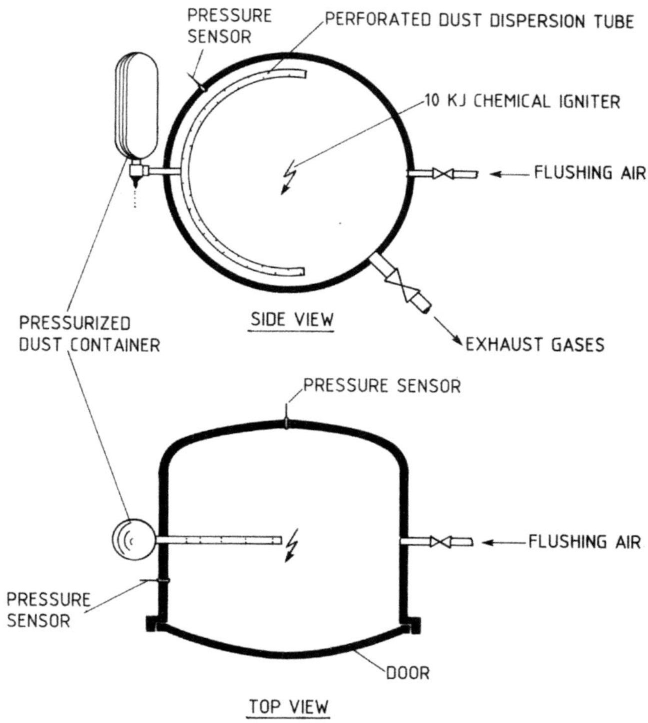

Figure 7.18 One cubic meter closed standard vessel, at first specified by the ISO (1985), for determination of maximum explosion pressures and maximum rates of pressure rise of dust clouds in air.

Bartknecht found a marked decrease of $(dP/dt)_{max}$, whereas P_{max} was comparatively independent of the ignition delay, for both dusts. If the ignition delay was increased further, however, there was a marked decrease even for P_{max}.

For several decades the 1.2-L Hartmann bomb, originally developed by I. Hartmann at the US Bureau of Mines in the 1930s, was used as a standard apparatus in many countries for assessing maximum explosion pressures and maximum rates of pressure rise of various dusts. However, because of the small bomb size, the P_{max} values obtained are normally significantly lower than those attained in larger volumes for the same dusts. The ISO, and later also the European Union, standard apparatus for measuring $(dP/dt)_{max}$ and P_{max} for various dusts is illustrated in Fig. 7.18. The apparatus and test method is essentially identical with the apparatus and procedures developed by Bartknecht (1971).

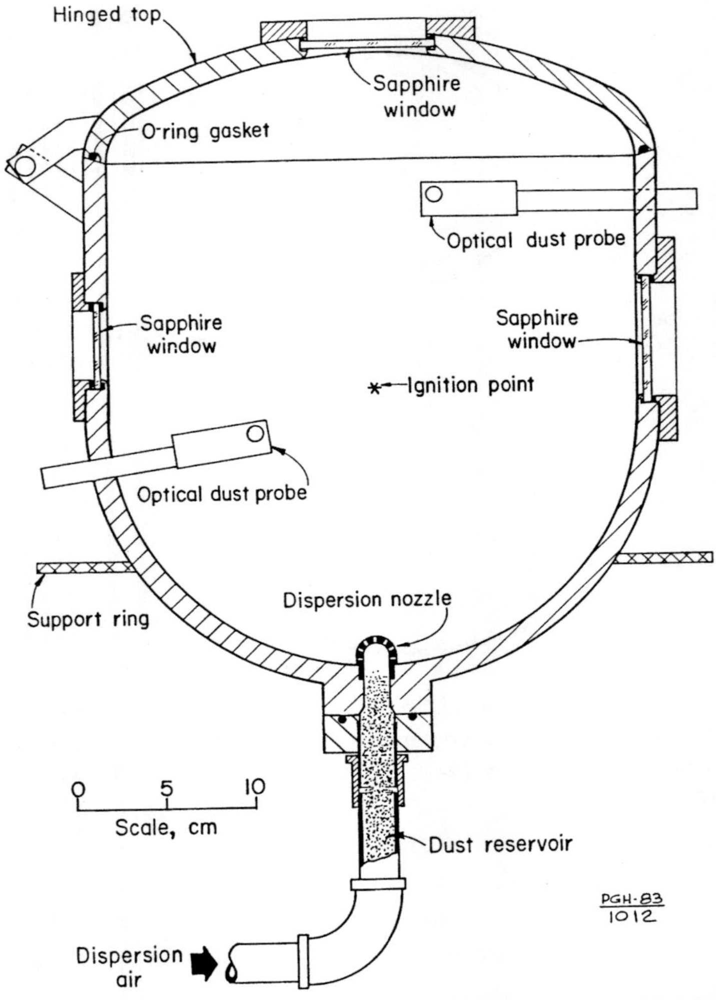

Figure 7.19 Cross section of US Bureau of Mines 20-L explosion vessel for determination of maximum explosion pressure, maximum rate of rise of explosion pressure, minimum explosive dust concentration, and other parameters of explosive dust clouds. *From Cashdollar, K.L., Hertzberg, M., 1985. 20 litre explosibility test chamber for dusts and gases. Review of Scientific Instruments 56, 596–602.*

Based on the extensive studies by Bartknecht and Siwek, a smaller vessel of 20-L volume is now widely used for determining so-called K_{St} values that are assumed to equal to those that would have been obtained by means of the proper 1-m^3 ISO standard. K_{St} is defined as the product of the maximum rate of pressure rise determined in the 1-m^3 ISO standard vessel and the cube root of the vessel volume. If other, smaller vessels are used, the dust-dispersion system, the ignition source strength and the ignition delay must be tuned in such a way that the products of the measured maximum rates of pressure rise (bar/s) and the cube roots of the vessel volumes (m) equal the K_{St} values that would have been measured for the same dusts in the 1-m^3 ISO standard test.

Fig. 7.19 shows the 20-L closed vessel constructed by US Bureau of Mines. This vessel design gives easy access to the interior of the vessel for cleaning, etc. The original dust-dispersion system shown in the figure is not compatible with the ISO standard. However, as shown by Skjold (2003), compatibility can be obtained for this chamber by using an alternative dust-dispersion system developed in Switzerland for the 20-L Siwek bomb.

7.2.5.3 Acceleration of Dust Explosions in Elongated Systems

Coal mines essentially consist of long galleries of large length-to-diameter ratio (L/D). Since the onset of systematic research on the propagation of coal dust explosions in mines, large-scale experimental galleries have been a main tool of investigation. The tests in experimental coal mines in the United Kingdom at about 1890, was probably the first of this kind.

Fischer (1957) reported results from coal-dust explosion experiments in a 260-m long experimental coal-mine gallery of equivalent-circle cross-sectional diameter of 3.2 m, ie, L/D of about 80. The main purpose of these experiments was to investigate whether deposits of stone dust on shelves in the upper part of the gallery cross section would prevent the

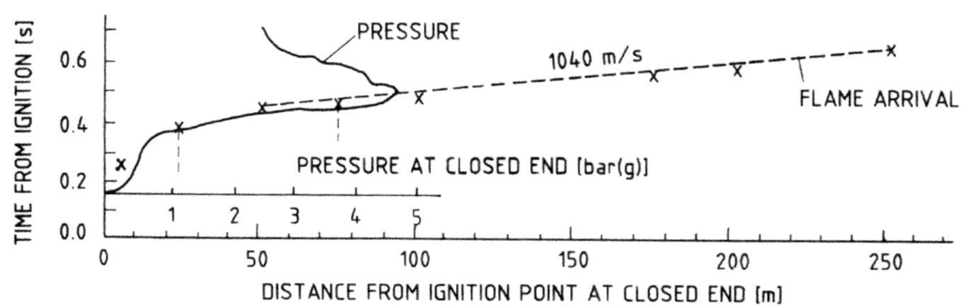

Figure 7.20 Time of arrival of bituminous coal dust/air flame as a function of distance from ignition point at closed end of gallery of length 260 m and diameter 3.2 m. Pressure at closed end as a function of time. Nominal average dust concentration 500 g/m^3. *From Fischer, K., 1957. Ablauf von Kohlenstaubexplosionen. Zeitschrift für Elektrochemie 61, 685–692.*

propagation of coal-dust explosions in the gallery. However, it appeared that under certain circumstances this stone dust had little effect, and very violent flame acceleration phenomena were observed, as shown in Fig. 7.20.

The coal-dust explosion was initiated by an explosion of 40 m^3 methane/air at the upstream closed end of the gallery. The gas was ignited by black powder ensuring both efficient ignition and violent combustion of the gas. The blast from the gas explosion in turn swept up the coal-dust layer of 4 kg per m length of gallery on the floor and initiated the self-sustained dust explosion down the entire length of the gallery. The most striking feature of Fig. 7.20 is the very constant flame speed of 1040 m/s measured from about 50 m from the closed end right to the open tube end 200 m further down. Fischer associated this with "some kind of detonation" (see next subsection). The pressure versus time ratio was recorded only at the upstream closed end of the gallery, because the explosion was so violent that all the measurement stations further down the gallery were destroyed. As can be seen, the peak pressure at the closed end was about 5 bar (g). It would be anticipated that the pressures further down the gallery were considerably higher.

Jost and Wagner (in Freytag, 1965) have illustrated the various characteristic phenomena occurring during acceleration of premixed gas flames in long one-end-open tubes. There are good reasons for assuming that their overall picture, as reproduced in Fig. 7.21, also applies to dust clouds.

The only major difference is that a dust cloud needs to be generated by raising dust deposits into suspension. This means that stage 1 and possibly also stage 2 in Fig. 7.21, the ignition and laminar propagation of the initial flame, respectively, may not be relevant for dust flames. As already discussed, Fischer (1957) used a turbulent gas flame for initiating the dust entrainment and explosion. However, once the dust explosion gets under way, the blast wave generated by it will entrain dust further downstream as already discussed. Therefore all stages of Fig. 7.21, from stage 3 and downward, will apply even to dust clouds. The essential reason for the flame acceleration is turbulence generated in the unburnt cloud ahead of the flame due to wall friction when the cloud is pushed toward the open tube end by the expansion of the part of the cloud that has burnt. The cloud combustion rate increases as soon as the flame front reaches the turbulent region of the unburnt cloud. This in turn increases the expansion rate of the combustion products and therefore also the flow rate of the unburnt cloud further ahead. The result is an even higher turbulence level and further increase of the combustion rate.

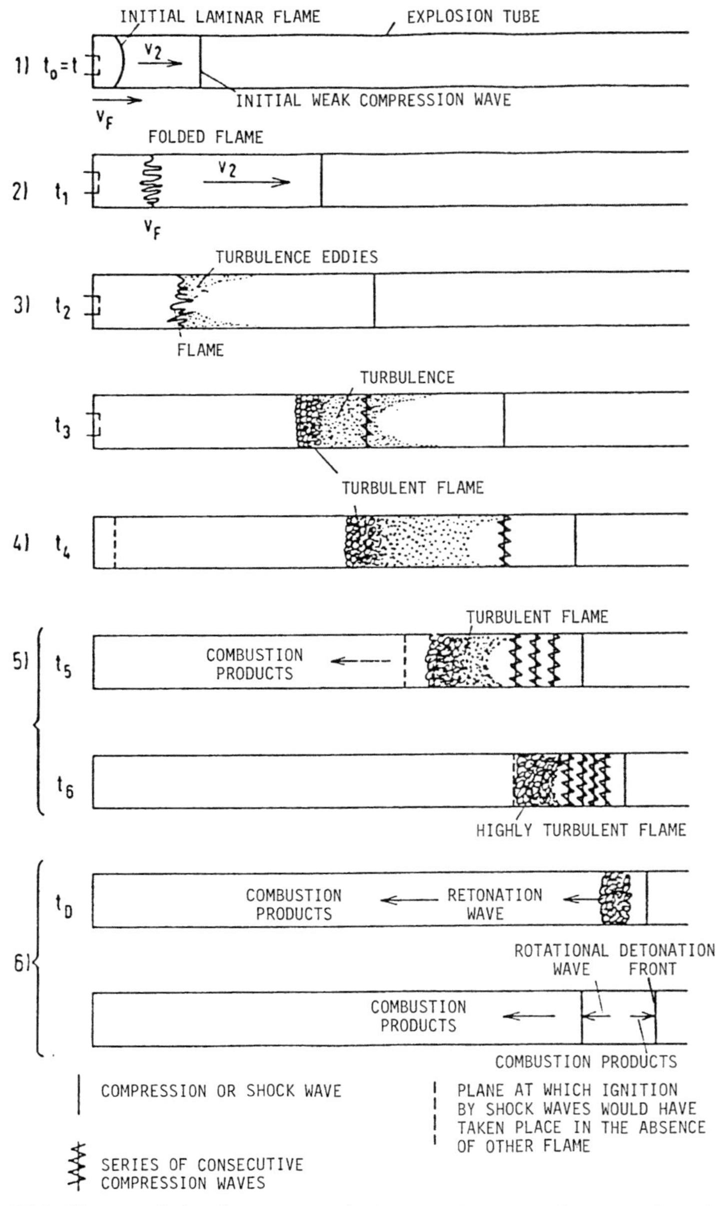

Figure 7.21 Characteristic phenomena during acceleration of gas or dust flames in one-end-open long tubes, from laminar combustion via turbulent combustion to detonation. V_F, flame speed; V_2, velocity of the unburnt gas or dust cloud ahead of the flame. *From Jost and Wagner, in Freytag, H.H., 1965. Handbuch der Raumexplosionen. Verlag Chemie, Weinheim, Germany.*

During all these stages, compression waves will be emitted and propagate toward the open tube end. Because the cloud ahead of the flame is heated due to adiabatic compression, each wave will propagate at slightly higher velocity than the previous one. Ultimately, therefore, they will all catch up with the initial wave and form a strong leading shock front. The turbulent flame front will also, due to the positive feedback mechanism of combustion rate, flow rate, increased turbulence, and turbulence-enhanced increased combustion rate, eventually catch up with the leading shock wave. If the leading shock is sufficiently strong, a switch can occur in the mechanism of flame propagation. Instead of heat being transferred by turbulent diffusion behind the leading shock wave, the dust cloud may become ignited in the highly compressed state inside the leading shock. If the induction time of ignition is sufficiently short, the chemical reaction zone and the propagating shock wave then become closely coupled and propagate through the cloud at constant velocity. This is detonation. However, as already mentioned, flame propagation at a constant high speed will not necessarily have to be a classical detonation, but can also be a high-speed turbulent deflagration supported by wall friction-induced turbulence.

7.2.6 Detonation of Dust Clouds

7.2.6.1 Qualitative Description of Detonation

As for premixed gases (see Section 2.1.7 of Chapter 2), detonation in an explosive dust cloud is a singular, extreme mode of flame propagation through the cloud. The transfer of heat from the burning to the unburnt cloud is then no longer by diffusion, which is characteristic of the deflagration mode of explosion propagation discussed so far. Instead, heat is transferred by extreme, very fast compression of the unburnt cloud captured in a shock wave driven through the cloud at supersonic speed by the explosion itself. The detailed mechanism of ignition and combustion inside the shocked detonation front is still a subject of research.

Detonation in a dust cloud can only be initiated by a sufficiently strong shock wave. In principle such a wave can be generated inside the explosive cloud in two different ways. The first is by, eg, detonating an explosive charge or discharging a very energetic electric spark inside the cloud. The second is by gradual buildup of a strong shock front by turbulent acceleration of the explosion itself as discussed in Section 7.2.5.3 and illustrated in Fig. 7.21 (Deflagration-to-Detonation Transition, DDT.)

7.2.6.2 Experiments in Narrow Ducts and Large-Scale Galleries

Various workers have found clear evidence of proper dust detonations even in ducts of quite small diameters. For example, Kauffman et al. (1982, 1984) demonstrated that a self-sustained steady detonation wave propagated through clouds of oats and wheat grain dust in air, inside a vertical laboratory-scale duct of 6.35 × 6.35-cm square cross section and 6-m length. The dust was charged into the duct at the top at a mass rate giving the desired dust concentration during gravity settling down the tube. The dust explosion was initiated by a local hydrogen/oxygen explosion at the bottom tube end. At an oat-dust concentration of 250–270 g/m^3 (slightly below stoichiometric) the measured detonation front velocity was 1540 m/s, which is somewhat lower than the theoretical Chapman–Jouguet velocity at stoichiometric concentration of about 1800 m/s. The highest-measured peak pressure was about 24 bar, quite close to the theoretical C-J pressure at stoichiometric concentration of 22.4 bar.

Kauffman et al. (1984) also investigated the upper and lower dust concentration limit for detonations of oat dust in air in their laboratory-scale vertical tube. They found that detonations could only be initiated, even with very vigorous ignition sources, within the narrow concentration range of approximately 200–450 g/m^3.

Gardner et al. (1986) provided further conclusive evidence of genuine detonations being possible in clouds of high-volatility coal dusts in air, and that such detonations can be brought about by turbulent flame acceleration (DDT), as illustrated in Fig. 7.21. The experimental arrangement consisted of a 20 m^3 ignition chamber connected to a 42-m long straight cylindrical test duct of diameter 0.6 m, which was essentially open at the downstream end. To start the experiments, air was blown through the system at a volumetric rate giving an average velocity of 20–30 m/s in the duct. Dust was then fed into the air stream just upstream of the 20 m^3 chamber to give the desired dust concentration, ranging from 30 to 850 g/m^3, in the explosion chamber as well as in the 42-m long duct. The dust cloud was ignited in the 20 m^3 chamber by a flame jet or a chemical ignitor. An extreme experimental transient peak pressure value of 81 bar (g) was measured in these experiments. The reason for high transient pressure peaks is that at the onset of detonation there is always a regime in which the detonation wave is overdriven before settling down to the C-J conditions. During this transient period, the detonation pressure can exceed the C-J value considerably.

7.2.6.3 *Unconfined Dust Cloud Detonations*

Tulis (1984) provided direct experimental evidence of a self-sustained detonation wave being able to travel through an unconfined dust cloud in air. In these experiments a very fine aluminum-flake powder of specific surface area 3—4 m^2/g was used. The unconfined cloud was about 10-m long and about 1-m high. When this cloud was initiated at one end with a 2.3 kg high-explosive charge, stable, self-sustained detonation was achieved. The average velocity of the detonation wave was 1750 m/s, and the peak pressures in excess of 28 bar. The corresponding calculated C-J values were 1850 m/s and 26 bar. This close agreement between experiment and theory seems to confirm that the phenomenon observed was in fact a proper, unconfined dust cloud detonation.

7.2.7 Primary and Secondary Dust Explosions

One important objective of dust explosion control is to limit dust explosions in process equipment to the process unit in which they were initiated. This first explosion inside a process unit is called the *primary* explosion. A central concern in all dust explosion protection efforts is to avoid *secondary* explosions outside process equipment due to entrainment and ignition of dust layers accumulated there by the blast wave and flame from the primary explosion, as illustrated in Fig. 7.22.

As shown in Fig. 7.2, there is a gap of about two orders of magnitude between the maximum explosive dust concentration and the bulk density of dust layers and heaps. The consequence of this with regard to the volume of dust cloud that can be produced by a given volume of dust layer, is illustrated in Fig. 7.23.

The simple relationship between the dust concentration c and the bulk density of the dust layer ρ_{bulk}, the layer thickness h, and the height H of the dust cloud produced from the layer, is:

$$c = \rho_{bulk}(h/H) \qquad [7.1]$$

If a dust layer of thickness h on the internal wall of a cylindrical duct of diameter D is dispersed homogeneously over the whole tube cross section, one has:

$$c = \rho_{bulk}(4h/D) \qquad [7.2]$$

In the case of a tube diameter of 0.2 m, typical of many dust-extraction ducts in industry, a layer thickness of only 0.1 mm is sufficient for generating a dust concentration of 1000 g/m^3 with a dust of bulk density 500 kg/m^3. In general, dispersible dust layers in process plants represent a potential hazard of extensive secondary dust explosions, which must be reduced to the extent possible.

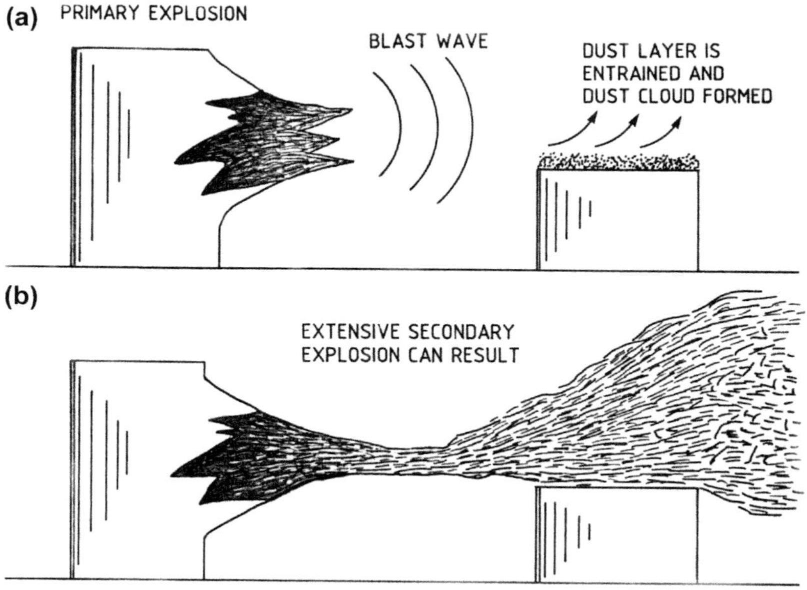

Figure 7.22 Illustration of how the blast wave from a primary explosion entrains and disperses a dust layer, which is subsequently ignited by the primary dust flame. (a) Pressure wave and flame jet are ejected from exploding vessel. (b) The flame jet ignites secondary dust clouds from dispersion of dust layers in workrooms.

From Eckhoff, R.K., 2003. Dust Explosions in the Process Industries, third ed. Gulf Professional Publishing/Elsevier. (first ed. published in 1991, second ed. in 1997). ISBN:0-7506-7602-7.

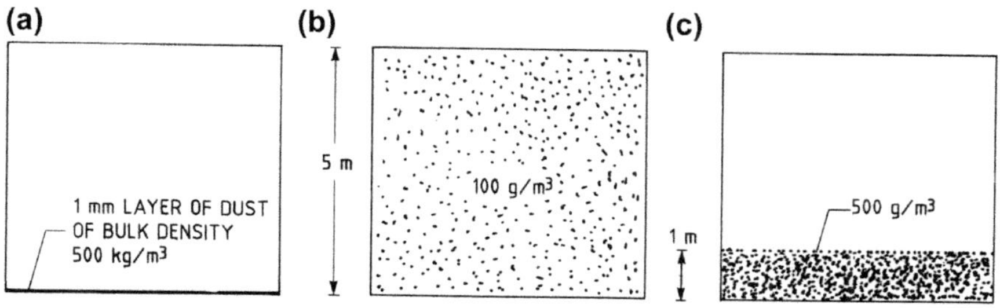

Figure 7.23 Illustration of the potential hazard of even thin dust layers. A 1 mm layer of a dust (a) of bulk density 500 kg/m^3 will generate a cloud of average concentration 100 g/m^3 if dispersed in a room of 5-m height. (b) Partial dispersion up to only 1 m gives 500 g/m^3.

From Eckhoff, R.K., 2003. Dust Explosions in the Process Industries, third ed. Gulf Professional Publishing/Elsevier. (first ed. published in 1991, second ed. in 1997). ISBN:0-7506-7602-7.

7.3 IGNITION OF DUST CLOUDS IN AIR

7.3.1 Introduction

A combustible dust cloud will not start to burn unless it becomes ignited by a source of heat of sufficient strength. The most common ignition sources are:

- Smoldering or burning dust
- Open flames (welding, cutting, matches, etc.)
- Hot surfaces (hot bearings, dryers, heaters, etc.)
- Burning metal particles or hot spots from mechanical impacts
- Electrical discharges and arcs

In addition, there are some more sophisticated potential ignition sources including:

- Laser light
- Adiabatic compression and shock waves
- Ultrasonic waves

There is considerable variation in the ignition sensitivity of clouds in air of various types of dusts, due to variations in chemical composition including moisture content and particle size and shape. To quantify the ignition sensitivity of dust clouds and dust deposits, when exposed to various kinds of ignition sources, a range of laboratory-scale test methods have been developed.

7.3.2 Smoldering or Burning "Nests"

As discussed in Chapter 6, experience has shown that combustible dusts, when deposited in heaps or layers, may under certain circumstances develop internal smoldering combustion nests of high temperatures. This is due to the porous structure of dust deposits, which gives oxygen access to the particle surface throughout the deposit, and also makes the heat conductivity of the deposit low. The initial oxidation inside the deposit may sometimes be due to the deposited dust or powder having initially a higher temperature than intended or permissible. However, some natural vegetable materials of high fat and or moisture content may develop spontaneous combustion even from normal ambient temperatures, due to biochemical activity. If such a reaction zone, often called a "smouldering[sic] nest", makes contact with an explosive dust cloud, a dust explosion may be initiated.

In other cases, the dust deposit is a layer resting on a heated surface that supplies the heat needed to trigger self-heating in the layer. Such surfaces can be overheated bearings, heaters in workrooms, light bulbs, walls in dryers, etc. Dust deposits on heated surfaces may prevent normal cooling by forming an insulating layer. This may give rise to an undesirable temperature rise in the surface, which will, in turn, further increase the probability of ignition of the dust. In general, the minimum temperature of a hot surface that is required for the dust layer to self-ignite decreases with increasing thickness of the dust layer. A dust explosion may result if such a layer makes contact with an explosive dust cloud.

Pinkwasser (1985, 1986) studied the possibility of dust explosions being initiated by smoldering lumps ("nests") of powdered material that was conveyed through a process system. The object of the first investigation (1985) was to disclose the conditions under which smoldering material that had entered a pneumatic conveying line would be extinguished, ie, cooled to a temperature range in which the risk of ignition in the downstream equipment was no longer present. In the case of >1 kg/m^3 pneumatic transport of screenings, low-grade flour and C3 patent flour, it was impossible to transmit a 10 g smoldering nest through the conveying line any significant distance. After only a few m, the temperature of the smoldering lump had dropped to a safe level. In the case of lower dust concentrations, between 0.1 and 0.9 kg/m^3, ie, within the most explosive range, the smoldering nest could be conveyed for an appreciable distance, but no ignition was ever observed in the conveying line.

In the second investigation by Pinkwasser (1986), smoldering nests of 700°C were allowed to fall freely through a 1-m tall column containing dust clouds in air of 100–1000 g/m^3 of wheat flour or wheat starch. Ignition was never observed during free fall. However, in some tests with nests of at least 25-mm diameter and weight at least 15 g, ignition occurred immediately after the nest had come to rest at the bottom of the test column. This indicates that a smoldering nest falling freely through a dust cloud in a silo without disintegrating during the fall, has a higher probability of igniting the dust cloud at the bottom of the silo than during the fall.

This result agrees with the findings of Alfert et al. (1989), who studied the ignition of dust clouds in a 22-m tall silo of diameter 3.7 m by falling smoldering nests. It was found that nests of low mechanical strength disintegrated during the fall and generated a large fireball that ignited the dust cloud. With mechanically stable nests, ignition of the dust cloud occurred either some time after the nest had come to rest at the silo bottom, or when the nest became disintegrated when impacting with the silo bottom. However, as soon as the nest had come to rest at the silo bottom, it could also become covered with dust before ignition of the dust cloud got under way.

7.3.3 Open Flames

The flames of welding and cutting burners are sufficiently powerful to initiate explosions in any dust cloud that is at all able to propagate a self-sustained flame. The cutting burner flame is particularly hazardous because it supplies excess oxygen to the working zone. All codes and regulations for preventing dust explosions contain strict requirements to the safety precautions that have to be taken when performing hot work in areas containing combustible dust. Smoking should be prohibited in areas where combustible dusts exist. A burning wooden match develops about 100 J/s of thermal energy. This is more than sufficient for initiating explosions in most combustible dust clouds.

A pragmatic test method for assessing whether a dust cloud in air can be ignited by an open flame is illustrated in Fig. 7.24. The test is based on the consideration that a welding-torch flame is among the strongest ignition

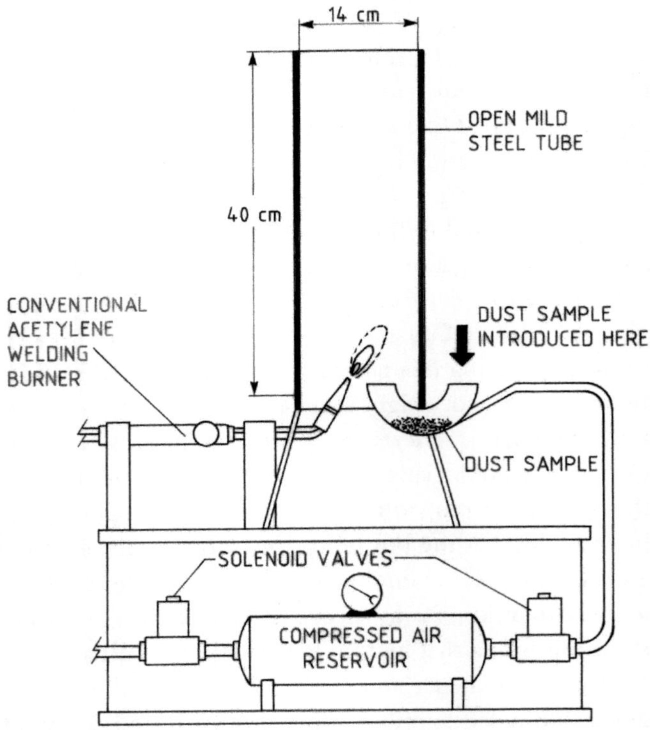

Figure 7.24 Welding torch ignition test apparatus used in Norway (Chr. Michelsen Institute/GexCon AS) for assessing whether or not a dust cloud can be ignited by an open flame.
From Eckhoff, R.K., 2003. Dust Explosions in the Process Industries, third ed. Gulf Professional Publishing/Elsevier. (first ed. published in 1991, second ed. in 1997). ISBN:0-7506-7602-7.

sources encountered in industrial practice. A vertical tube of length 40 cm and diameter 14 cm, open at both ends, was fitted with a U-shaped dust-dispersion tube and an acetylene/oxygen welding burner. A sample of the powder to be tested was placed at the bottom of a U-shaped dispersion tube, and a controlled blast from a compressed air reservoir dispersed the dust into a cloud in the tube in which it was immediately exposed to the hot flame from the welding torch. The amount of powder and the dispersion air pressure was varied to produce optimal conditions for ignition. When a dust flame occurred, its maximum height, color, and apparent violence were assessed by the observer.

7.3.4 Hot Surfaces

Besides being able to ignite dust layers, hot surfaces can also initiate dust explosions directly by contact between the dust cloud and the hot surface. However, the minimum hot-surface temperatures needed for this are generally considerably higher (typically 400–500°C for organic dusts) than for initiating self-ignition in dust layers.

In the United States, the ignition temperature of dust clouds in contact with a hot surface was traditionally determined in the GG furnace, which is illustrated in Fig. 7.25.

In this apparatus the internal surface of a vertical cylindrical ceramic tube, open at the lower end, is kept at a known, constant temperature, and a sample of the powder is dispersed as a dust cloud into the tube from above by means of a blast of air. The automatically controlled temperature of the internal wall of the tube is changed in steps and the experiment repeated until the minimum temperature for ignition has been identified. During the 1970s, the International Electrotechnical Commission (IEC) investigated the performance of the GG furnace through a comprehensive round-robin test series involving several central test laboratories in Europe and the United States. The influences of a number of details of the apparatus itself and of the experimental procedure were studied. The resulting, improved GG furnace was then adopted by IEC as a standard method for determining minimum ignition temperatures of dust clouds.

7.3.5 Heat From Mechanical Impacts

The literature on dust explosions is sometimes confusing when discussing ignition of dust clouds by heat from mechanical impacts. This is reflected in the use of terms such as "friction" or "friction sparks" when categorizing ignition sources. To clarify the situation, it seems useful to distinguish between friction and impact.

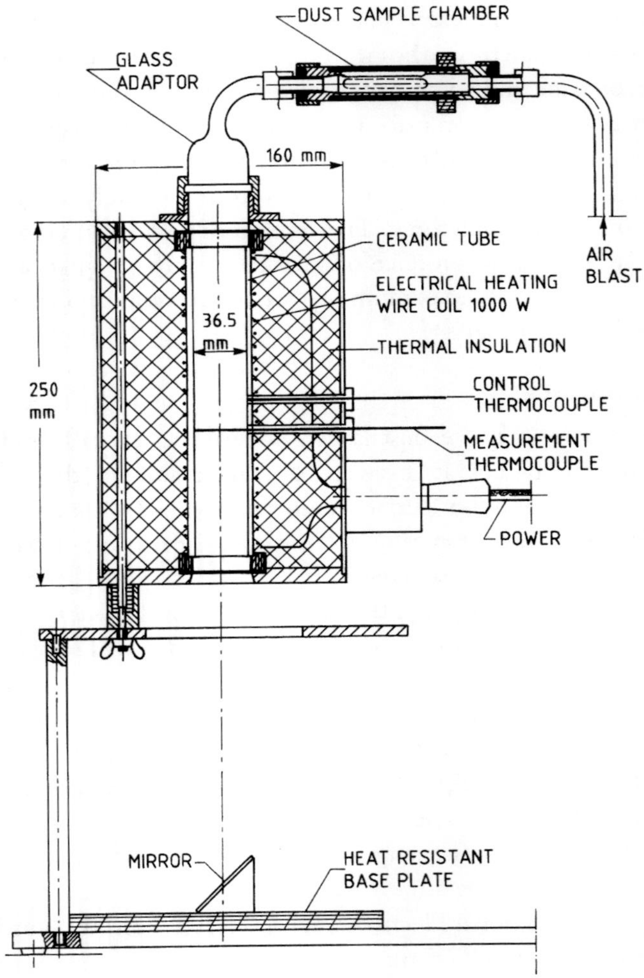

Figure 7.25 Godbert-Greenwald (GG) furnace for determination of the minimum ignition temperature of dust clouds.
From Eckhoff, R.K., 2003. Dust Explosions in the Process Industries, third ed. Gulf Professional Publishing/Elsevier. (first ed. published in 1991, second ed. in 1997). ISBN:0-7506-7602-7.

Friction is a process of fairly long duration whereby objects are rubbed against each other and heat is gradually accumulated. This produces hot surfaces, and in some cases inflammation, for example when an elevator or conveyor belt is slipping.

Impact is a short-duration interaction between two solid bodies under conditions of large transient mechanical forces. Small fragments of solid material may be torn off, and if made of metal, they may start burning in air due to the initial heat absorbed in the impact process. In addition, local "hot spots" may be generated at the points of impact. In some cases the impact may occur repeatedly at one specific point, for example when some misplaced stationary object inside a bucket elevator is repeatedly hit by the buckets. This may gradually generate a hot spot of sufficient size and temperature to ignite the dust cloud directly.

A practical mechanical impact situation may be as follows: A steel bolt accidentally enters the top of a large concrete silo during filling of the silo with maize starch. The bolt falls down into the nearly empty silo and hits the concrete wall near the silo bottom at a velocity of 25–30 m/s and metal particle sparks are generated. A dense, explosive cloud of maize starch occupies the region where the impact occurs. Will the cloud ignite? A test method for investigating this problem is shown in Fig. 7.26.

Experiments using this apparatus indicated that ignition of clouds of natural organic materials by metal sparks from single impacts, where steel is the metal component, seems much less likely than believed by many in the past. However, if the metal is titanium or zirconium, ignition can occur quite readily. The thermite reaction

$$2Al + Fe_2O_3 \rightarrow Al_2O_3 + 2Fe + \text{heat} \qquad [7.3]$$

is often mentioned as a potential ignition source from impacts involving aluminum and rust. However, if a lump of normal soft aluminum collides with a rusty steel surface, a thermite reaction will not necessarily take place. In fact, due to the softness of the aluminum, the result is often just a thin smear of aluminum on top of the rust. However, if this sandwich of aluminum and rust is given a blow by a third hard object, eg, a steel hammer, a thermite flash capable of igniting dust clouds can be produced. The same applies to a rusty surface that has been painted with aluminum paint, if the pigment content of the paint is comparatively high.

7.3.6 Electric Sparks and Arcs; Electrostatic Discharges

7.3.6.1 Introduction

It has been known for more than 100 years that electric sparks and arcs can initiate dust explosions. The minimum spark energy required for ignition varies with the type of dust, the effective particle size distribution in the dust cloud, the dust concentration and turbulence, and the spatial and temporal distribution of the energy in the electric discharge or arc. For many

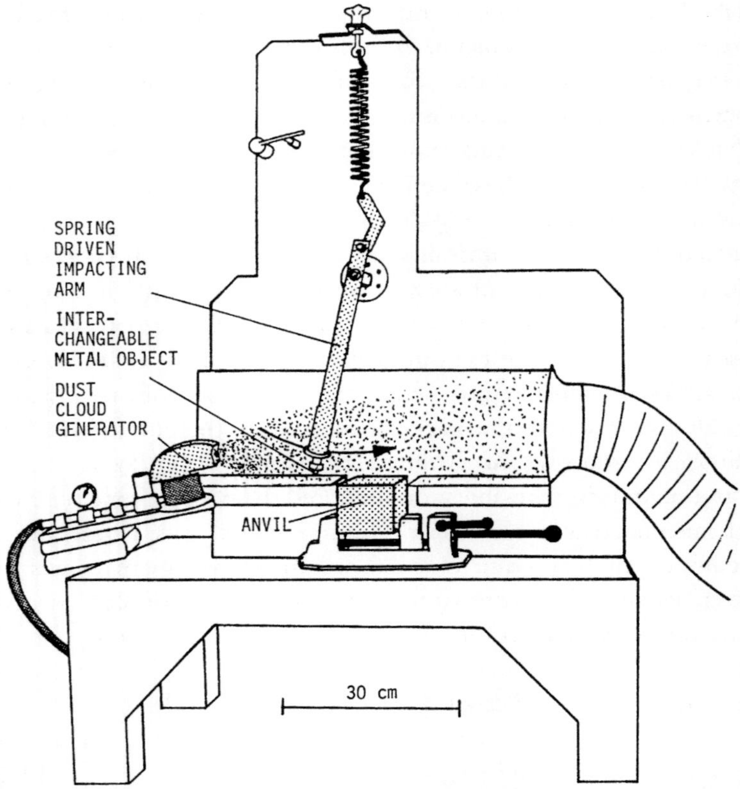

Figure 7.26 Apparatus for determining the sensitivity of dust clouds to ignition by single accidental mechanical impacts.
From Pedersen, G.H., Eckhoff, R.K., 1987. Initiation of grain dust explosions by heat generated during single impact between solid bodies. Fire Safety Journal 12, 153–164.

decades it was thought that the electric spark energies needed for igniting dust clouds in air were generally much higher, by one or two orders of magnitude, than the minimum ignition energies for gases and vapors in air. However, it is now generally accepted that clouds in air of many dusts can be ignited by spark energies in the range 1–10 mJ, and with some dusts even in the range 0.1 to 1 mJ, which is the range typical of minimum ignition energies of many gases and vapors in air.

7.3.6.2 Inductive and Capacitive Spark Discharges

7.3.6.2.1 General

It is useful to distinguish between *inductive sparks/arcs* generated when live electric circuits are broken, either accidentally or intentionally (eg, in switches), and *capacitive sparks* caused by release of accumulated electrostatic charge on electrically conducting objects, across a spark gap.

In the former case, if the points of rupture are separated at high speed, transient inductive sparks are formed across the gap as illustrated in Fig. 7.27.

If the current in the circuit prior to rupture is i and the circuit inductance L, the theoretical spark energy, neglecting external circuit losses, will be $\frac{1}{2}Li^2$. As an example, a current of 10 A and L equal to 10^{-5} H gives a theoretical spark energy of 0.5 mJ. This is too low for igniting most dust clouds in air. However, larger currents and/or inductances can easily give incendiary sparks. Sometimes rupture only results in a small gap of permanent distance. This may result in a hazardous stationary arc if the circuit stays live.

Capacitive spark discharges occur when charge that has accumulated on an electrically conducting, unearthed, object is discharged to earth across a small air gap. In process industries producing and/or handling powders, electrostatic charging is generally triboelectric, which implies transfer of electrons between objects of different electron affinity when being separated after having first been in contact. In a process plant this occurs during handling and transport of powder and dust whenever the powder/dust and the process equipment make contact and are subsequently separated.

An electrostatic spark discharge results if the discharge occurs between two electrically conducting electrodes. The spark-gap distance must then be sufficiently short to allow breakdown and spark channel formation at the actual voltage difference between the charged object and earth.

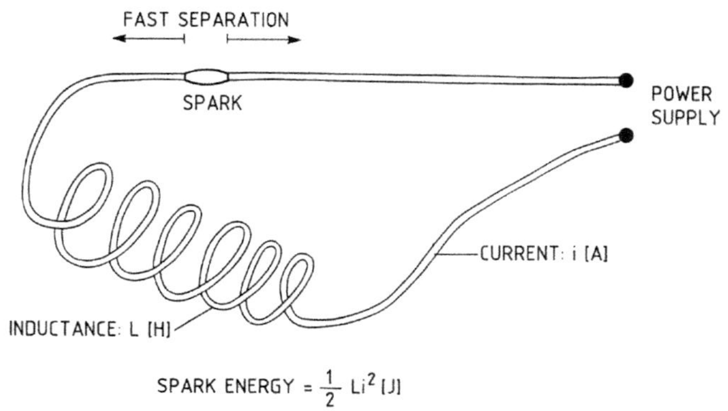

Figure 7.27 An inductive spark or "break flash" is generated when a live electric circuit is suddenly broken and the points of rupture are separated at high speed.
From Eckhoff, R.K., 2003. Dust Explosions in the Process Industries, third ed. Gulf Professional Publishing/Elsevier. (first ed. published in 1991, second ed. in 1997). ISBN:0-7506-7602-7.

On the other hand, for the spark to become incendiary, the gap distance must be sufficiently long to permit the required voltage difference to build up before breakdown of the gap occurs. The theoretical spark energy, neglecting external circuit losses, equals $\frac{1}{2}CU^2$, where C is the capacitance of the unearthed, charged process item with respect to earth, and U is the voltage difference. Fig. 7.28 illustrates a practical situation that could lead to a dust explosion initiated by an electrostatic spark discharge.

Glor (1988) estimated typical approximate capacitances to earth for various objects encountered in the process industry. The values given in Table 2.11 in Section 2.2 of Chapter 2 (ignition of gases) are equally valid in the context of dust explosions. This kind of capacitance value is used for estimating the maximum theoretical spark energies $\frac{1}{2}CU^2$ than can result in industry when discharging a nonearthed metal object of capacitance C at an initial voltage U, to earth.

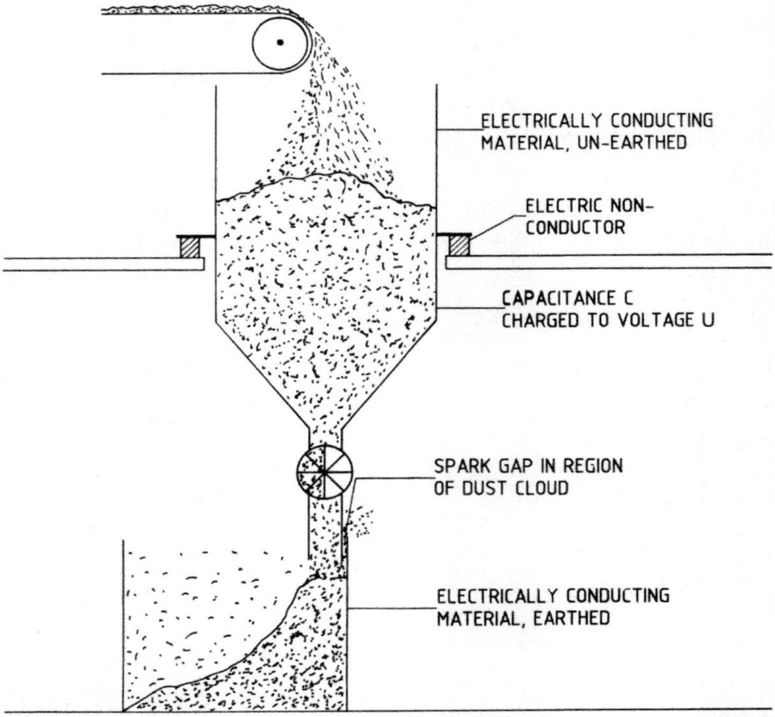

Figure 7.28 Illustration of a practical situation that could lead to a dust explosion initiated by an electrostatic spark discharge.
From Eckhoff, R.K., 2003. Dust Explosions in the Process Industries, third ed. Gulf Professional Publishing/Elsevier. (first ed. published in 1991, second ed. in 1997). ISBN:0-7506-7602-7.

7.3.6.2.2 The Minimum Ignition Energy (MIE) of a Dust Cloud

The MIE of a given dust in air is a commonly used measure of the ease with which the cloud is ignited by electric sparks and electrostatic discharges. A typical apparatus for laboratory determination of MIE of dust clouds is illustrated in Fig. 7.29. An appropriate quantity of the dust is placed in the dispersion cup at the bottom of the 1.2-L plastic cylinder and dispersed by a blast of air deflected by a conical "hat" as indicated. A spark of the desired energy $\frac{1}{2}CU^2$ is discharged across the electrodes synchronously with the appearance of the transient dust cloud in the spark-gap region. It is recorded visually whether or not ignition takes place. MIE is the lowest spark energy that gives at least one ignition in a given number of trials, typically 10 or 20, at the same spark energy.

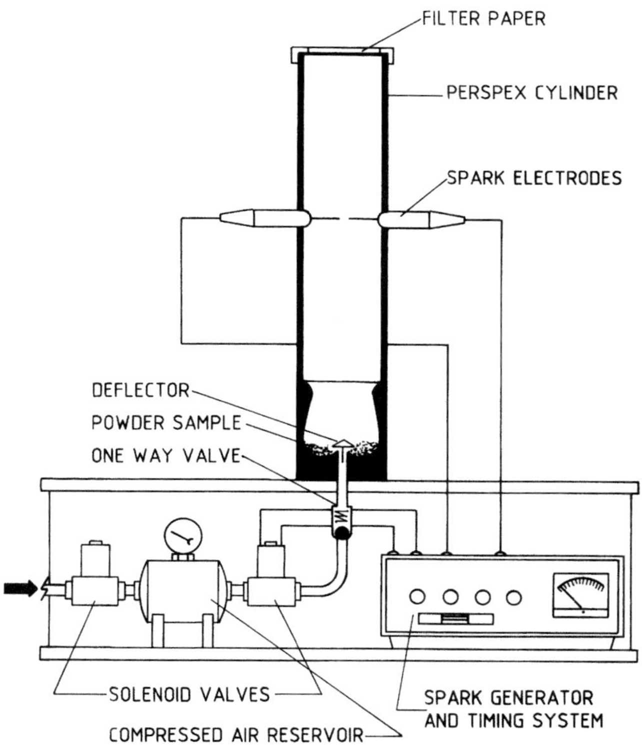

Figure 7.29 Illustration of an apparatus type commonly used for experimental determination of minimum electric spark ignition energies (MIE) of explosive dust clouds in air.

From Eckhoff, R.K., 2003. Dust Explosions in the Process Industries, third ed. Gulf Professional Publishing/Elsevier. (first ed. published in 1991, second ed. in 1997). ISBN:0-7506-7602-7.

Fig. 7.30 illustrates various electrical circuits that have been used to generate the electric spark discharges in MIE tests.

In the original circuit of US Bureau of Mines, a substantial part of the capacitor energy was lost in the transformer. In the case of the direct

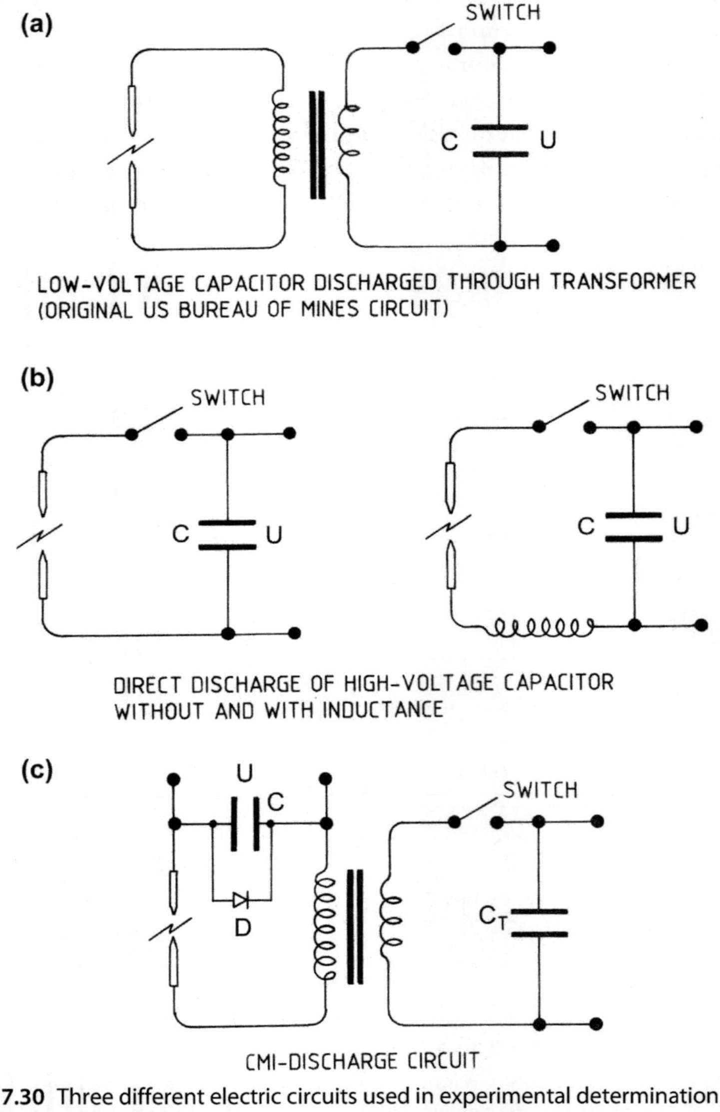

Figure 7.30 Three different electric circuits used in experimental determination of minimum electric spark ignition energies (MIE) of explosive dust clouds in air.
From Eckhoff, R.K., 2003. Dust Explosions in the Process Industries, third ed. Gulf Professional Publishing/Elsevier. (first ed. published in 1991, second ed. in 1997). ISBN:0-7506-7602-7.

high-voltage discharge circuit, the switch element needed for synchronization of dust dispersion and spark discharge requires special considerations. In the Chr. Michelsen Institute (CMI) circuit, a trigger transformer feeding 2–3 mJ into the main discharge is used for triggering the discharge. Other synchronization principles are also in use, including a fast-moving earthed electrode.

As discussed by Eckhoff (2003, 2016b), there can be a strong influence of the spark discharge time on MIE for dust clouds, in particular for spark energies exceeding 1 J. In standard testing for MIE optimal discharge, time for ignition is achieved by introducing a 1–2 mH inductance in the discharge circuit, and this feature is compulsory in the current international (IEC) and European Standard methods for MIE determination for dust clouds. This inductance is to be removed when the purpose of the test is to assess the sensitivity to ignition by accidental electrostatic spark discharges in industrial plants. A major limitation of the standard test apparatuses commonly used is that they cannot generate sparks of energies significantly lower than 2–3 mJ. Efforts have been made, therefore, to develop generators for producing synchronized sparks of energies down to the order of 0.1 mJ. One successful solution was presented by Olsen et al. (2015).

7.3.6.2.3 Range of MIEs for Dust Clouds

Minimum electric spark energies (MIE) for ignition of dust clouds vary, as already mentioned, with dust type, particle size, and other factors. For a long time right up to the 1970s, it was thought that the absolute lower limit of MIEs of dust clouds in air was on the order of 10 mJ, ie, about two orders of magnitude larger than the typical MIE range most gases and vapors in air. However, it is now generally accepted that clouds of many dusts in air have MIEs in air well below almost all the spark energy values in Table 2.11 in Chapter 2. In fact, the single-screw value of 10 μJ in the table may be the only exception. Bartknecht (1993) showed that MIEs of clouds of various dusts in air span over at least 7–8 decades, from perhaps as low as 10 μJ in the lower end to beyond 1 kJ in the upper.

7.3.6.2.4 Conservative Ignition Curves for Dusts Clouds Based on MIE

For capacitive circuits, conservative ignition curves are obtained directly from the equation

$$\mathrm{MIE} = \tfrac{1}{2}CU^2 \qquad [7.4]$$

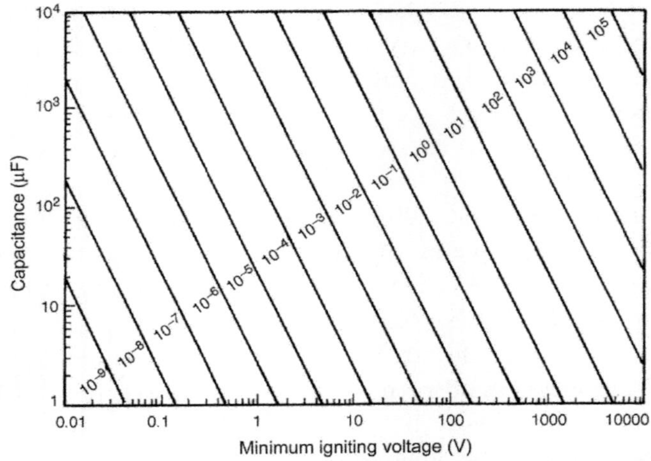

Figure 7.31 Theoretical conservative capacitive ignition curves for dust clouds based on the equation $½CU^2 = MIE$. The numbers attached to the straight lines are the respective MIE values in J.
From Eckhoff, R.K., 2003. Dust Explosions in the Process Industries, third ed. Gulf Professional Publishing/Elsevier. (first ed. published in 1991, second ed. in 1997). ISBN:0-7506-7602-7.

Charts showing this graphically, for a range of MIE values, are given in Fig. 7.31. In the case of inductive circuits with $L > 1$ mH, conservative ignition curves are obtained directly from the equation

$$MIE = ½Li^2 \quad [7.5]$$

For $L < 1$ mH the ignition current will be essentially independent of L, and the circuit will be essentially resistive. Charts showing conservative ignition curves for inductive circuits, for a range of MIE values, are given in Fig. 7.32. For resistive ignition circuits, Eckhoff (2002) has suggested that an equation of the form $I = A \cdot MIE^B/U^2$, in which A and B are empirical constants, may provide a first-order conservative estimate. Graphs representing this equation for preliminary values of A and B are given by Eckhoff (2003). More research is needed to establish more reliable resistive-circuit ignition data in the range of higher MIEs.

7.3.6.2.5 Parameters Influencing the Minimum-Electric-Spark-Ignition Energy of Dust Clouds

Turbulence in the dust cloud increases the effective MIE and therefore provides a safety factor. This is illustrated by the work of Yong Fan Yu (1985),

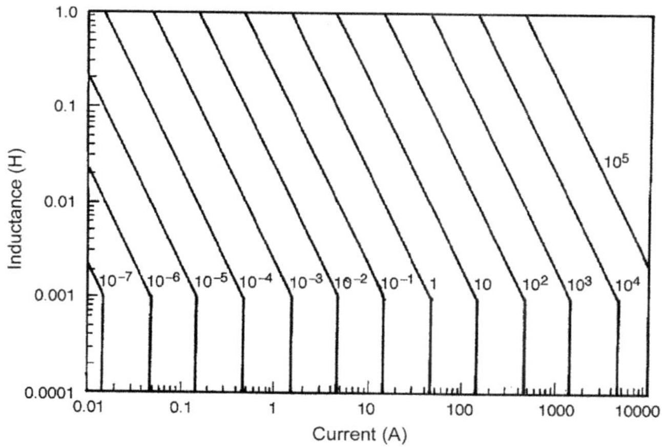

Figure 7.32 Theoretical conservative inductive ignition curves for dust clouds based on the equation $\frac{1}{2}Li^2 = MIE$, valid for $L > 1$ mH. The numbers attached to the straight lines are the respective MIE values in J.
From Eckhoff, R.K., 2003. Dust Explosions in the Process Industries, third ed. Gulf Professional Publishing/Elsevier. (first ed. published in 1991, second ed. in 1997). ISBN:0-7506-7602-7.

who was unable to ignite turbulent explosive clouds of wheat-grain dust in a container at the exit of a pneumatic transport pipe, even with highly incendiary "soft" electric sparks of energies in excess of 1 J. In the type of test illustrated in Fig. 7.29, the turbulence of the dust cloud at the moment of ignition can be varied by varying the delay between dust cloud generation and ignition. The longer the delay is the lower is the turbulence. Fig. 7.33 gives some results from this kind of experiment.

Fig. 7.34 shows how particle size influences the MIE for three different dusts. The vertical scale is logarithmic, and it is seen that the effect is very strong. Kalkert and Schecker (1979) developed a theory indicating that MIE is proportional to the cube of the particle diameter, as also illustrated in Fig. 7.34 by their theoretical prediction of the relationship for polyethylene.

Moisture in the dust reduces both ignition sensitivity and explosion violence of dust clouds. Fig. 7.35 illustrates the influence of dust moisture on MIE. The vertical axis is logarithmic, and it is seen that the effect is quite significant. If safety measures against electric spark ignition are based on MIE data for a given dust moisture content, it is essential that this content is not subsided in practice.

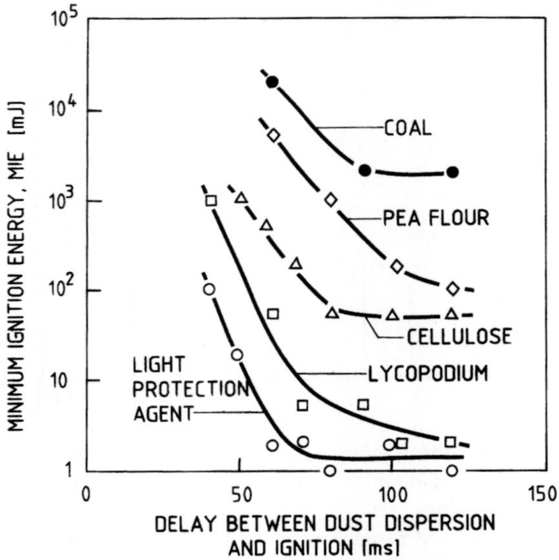

Figure 7.33 Illustration of the influence of initial turbulence of explosive dust clouds on the minimum electric spark energies required for ignition.
Data from Glarner, T., 1984. Mindestzündenergie — Einfluss der Temperatur. VDI-Berichte 494, 109—118. VDI-Verlag GmbH, Düsseldorf.

7.3.6.3 Other Electrostatic Discharges Than Sparks

7.3.6.3.1 Overview

Glor (1988) and Lüttgens and Glor (1989) distinguished between five different types of electrostatic discharges in addition to sparks, namely:

- Corona discharges
- Brush discharges
- Propagating brush discharges
- Discharges along the surface of powder/dust in bulk
- Lightning-like discharges

7.3.6.3.2 Corona Discharges

This discharge type is illustrated in Fig. 2.33 in Section 2.2 of Chapter 2. It occurs between a single earthed-metal electrode of very small radius of curvature, such as a needle, a sharp corner or a sharp edge, and a charged nonconducting object. Corona discharges take place at much lower field strengths than brush discharges and the discharged energies will be much

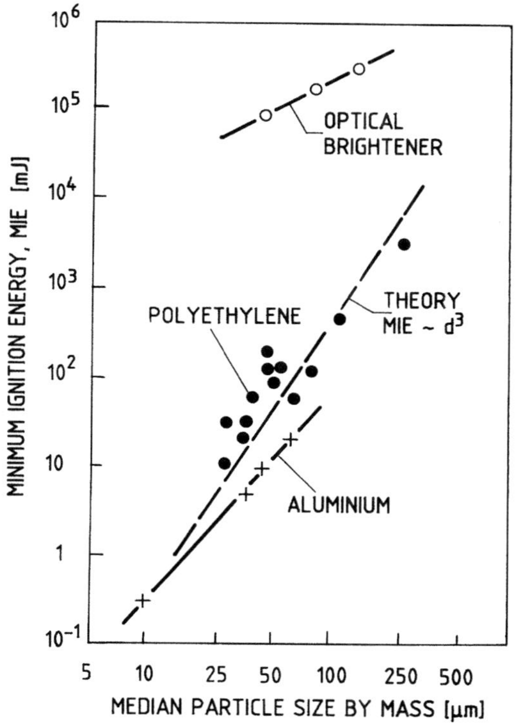

Figure 7.34 Minimum experimental electric spark ignition energies (MIE) of clouds in air of an optical brightener, polyethylene, and aluminum, as functions of median particle size *(From Bartknecht, W., 1987. Staubexplosionen – Ablauf und Schutzmassnahmen. Springer-Verlag. ISBN:0-387-16243-7.)* and theoretical line for polyethylene *(From Kalkert, N., Schecker, H.-G., 1979. Theoretische Überlegungen zum Einfluss der Teilchengrösse auf die Mindestzündenergie von Staüben. Chemie Ingenieur Technik 51, 1248–1249.)*

lower. It appears that the possibility of igniting dust clouds in air by corona discharges can be ruled out.

7.3.6.3.3 Brush Discharges

A brush discharge is illustrated in Fig. 2.34 in Section 2.2 of Chapter 2. It occurs between a single curved, earthed-metal electrode (radius of curvature 5–50 mm) and a charged nonconducting surface (plastic, rubber, deposited dust). It is well known that brush discharges can ignite explosive gas mixtures. However, according to Glor (1988), no ignition of a dust cloud in air by a brush discharge had by that time been demonstrated, not even in sophisticated laboratory tests using very ignition-sensitive dusts. The

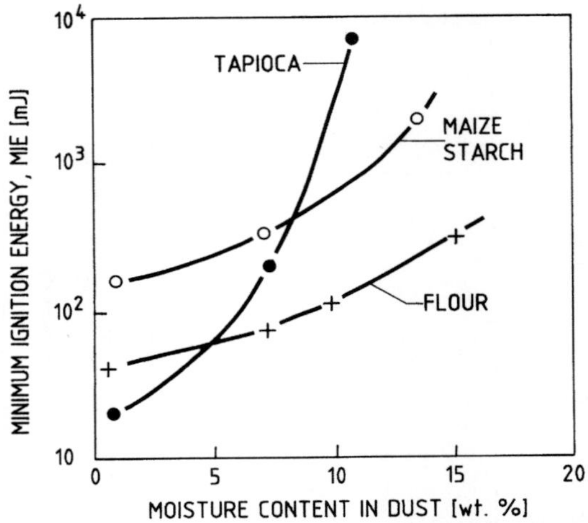

Figure 7.35 Influence of dust moisture content on minimum electric spark ignition energy (MIE) for three dusts.
From van Laar, G.F.M., Zeeuwen, J.P., 1985. On the minimum ignition energy of dust/air mixtures. Archivum Combustionis 5, 145–159.

conclusion that brush discharges cannot ignite dust clouds in air is also strongly supported by the experiments by Larsen et al. (2001).

It must be emphasized, however, that this conclusion does not apply if the powder/dust contains significant quantities of combustible solvents. This is because the solvent vapor can increase the ignition sensitivity of the dust cloud substantially, even when the vapor concentration in the air is well below the LEL of the vapor alone. Such clouds of a combustible dust suspended in a < LEL homogeneous mixture of a combustible gas/vapor and air have traditionally been named "hybrid" mixtures.

7.3.6.3.4 Propagating Brush Discharges

This type of discharge, which is illustrated in Figs. 2.35 and 2.36 in Section 2.2 of Chapter 2, can, however, initiate dust explosions. Propagating brush discharges will normally have substantially higher energies than ordinary brush discharges. They occur if a double layer of charges of opposite polarity is generated across a thin sheet (<8-mm thickness) of a nonconducting material (Glor, 1988). The reason for the high discharge energy is that the opposite charges allow the nonconductor surfaces to accumulate much higher charge densities before becoming saturated than if the sheet had been charged on only one of the faces.

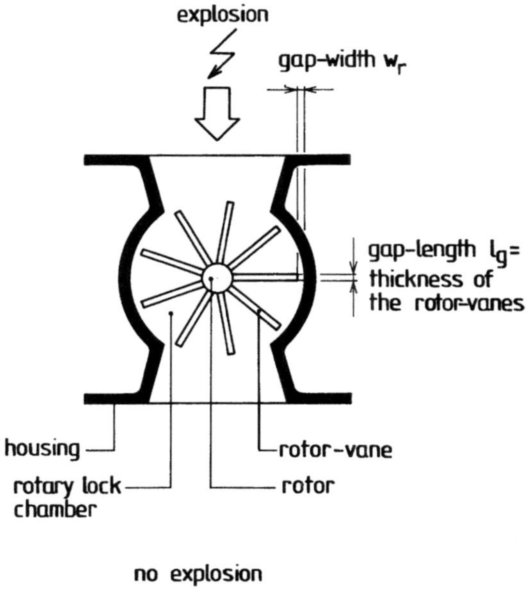

Figure 7.36 Illustration of a practical flame jet transmission situation for burning dust clouds: a rotary lock with a dust explosion occurring on one side.
From Schuber, G., 1988. Zünddurchschlagverhalten von Staub/Luft Gemischen und Hybriden-Gemischen. Publication series Humanisierung des Arbeitslebens, vol. 72. VDI-Verlag, Düsseldorf.

Glor pointed out that in principle close contact of one of the faces of the sheet with earth is not necessary for obtaining a charged double layer. By adding electrons to one side of a plastic sheet and removing electrons from the other, the same result can be obtained. However, in practice, earth on one side is the most common configuration. In the context of powders and dusts, Fig. 7.36 may illustrate pneumatic transport of a powder/dust in a steel pipe with an internal electrically insulating plastic coating. Due to the rubbing of the powder against the plastic, charge is accumulated on the internal face of the plastic coating. The high mobility of the electrons in the steel causes buildup of a corresponding charge of opposite polarity on the outer face of coating in contact with the steel. If a short passage between the two oppositely charged faces of the plastic coating is provided, either via a perforation of the coating or at the pipe exit, a propagating brush discharge can result.

Lüttgens and Glor (1989) gave an example of a dust explosion that was initiated by a propagating brush discharge. Acrylic powder was transported pneumatically in a 50-mm diameter plastic pipe outdoors, and rainwater and snow provided the earthed electrically conducting shield on the outer surface of the pipe.

Glor (1988) identified five typical situations which may lead to propagating brush discharges during transport and handling of powders:

- High-velocity pneumatic transport of powder through an electrically insulating pipe, or a conductive pipe with an insulating internal coating
- Use of inspection windows of glass or Plexiglass in pneumatic transport pipes
- Continuous impact of powder particles onto an insulating surface (eg, a coated dust deflector plate in the cyclone of a dust separator)
- Fast movement of conveyor or transmission belts made of an insulating material, or of a conductive material coated with an insulating layer of high dielectric strength
- Filling of large containers or silos made of insulating material (eg, flexible intermediate bulk containers) or of metallic containers or silos coated internally with an insulating layer of high dielectric strength

7.3.6.3.5 Electrostatic Discharge Along the Surface of Powder/Dust in Bulk

This type of discharge may occur if strongly nonconducting powders are blown or poured into a large container or silo. When the charged particles settle in a heap in the container, very high space-charge densities may be generated and luminous discharges may propagate along the surface of the powder heap from its base to its top. However, under realistic industrial conditions only very large particles, of 1−10 mm diameter, are likely to generate spark discharges due to this process. It further seems that very high specific electrical resistivity of the powder is also a requirement ($>10^{10}$ Ω m) which probably limits this type of discharge to coarse plastic powders and granulates. The maximum equivalent spark energy for this type of discharge has been estimated to the order of 10 mJ, but little is known about the exact nature and combustibility of these discharges. Because of the large particle size required to generate the charge, these particles are unlikely to give dust explosions, and therefore a possible explosion hazard must be associated with the simultaneous presence of an explosive cloud of an additional, fine-dust fraction. Glor (1988) pointed out that the probability of occurrence of this discharge type increases with increasing charge-to-mass ratio in the powder, and increasing mass filling rate.

7.3.6.3.6 Lightning Type Discharge

Lightning discharges may in principle occur within an electrically insulating container with no conductive connection from the interior to the earth. There is, however, no evidence of lightning discharges having occurred in dust clouds generated in industrial operations.

7.3.7 Jets of Hot Combustion Products

The basic process is the same as for explosive gas mixtures, as described in Section 2.2.7 of Chapter 2. The maximum experimental safe gap (MESG) may be defined as the largest width of a slot that will just prevent transmission of a flame in a gas or dust cloud inside an enclosure to a similar gas or dust cloud on the outside. This definition is somewhat vague and raises several questions. It neither defines the length of the slot, nor the explosion pressure inside or the volume of the enclosure. Therefore, MESG is not a fixed constant for a given explosive cloud, but depends on the actual circumstances. MESG for dust clouds is of some importance in practice, although not in the context of electrical equipment enclosure design, and therefore needs to be assessed.

Schuber (1988, 1989) investigated the influence of various parameters on MESG for dust clouds. The practical situation addressed was possible transmission of a dust flame from one side of a rotary lock to the other, as illustrated in Fig. 7.36.

The research apparatus used by Schuber is illustrated in Fig. 7.37. Explosive dust clouds of desired concentrations were generated simultaneously from compressed dust reservoirs in both the large vessel (1 m^3) and the smaller vessel (40 L) mounted inside the large one. The cloud in the smaller vessel was subsequently ignited, and it was observed whether the cloud in the large vessel was ignited by the flame jet being transmitted through the annular gap in the wall of the smaller vessel.

Schuber found that MESG increased with increasing initial turbulence in the dust clouds. This is in harmony with the increase of the minimum electric spark energy for ignition of both gases and dust clouds with increasing turbulence. To ensure conservative results, Schuber conducted his experiments with comparatively low initial turbulence in the dust clouds. He correlated his experimental MESG values with the product of minimum electric spark ignition energy and the dimensionless MIT, as shown in Fig. 7.38. Because (TI + 273)/273 is in the range between two and three for most of the dusts tested, the double–logarithmic correlation in Fig. 7.38 is essentially between MESG and MIE.

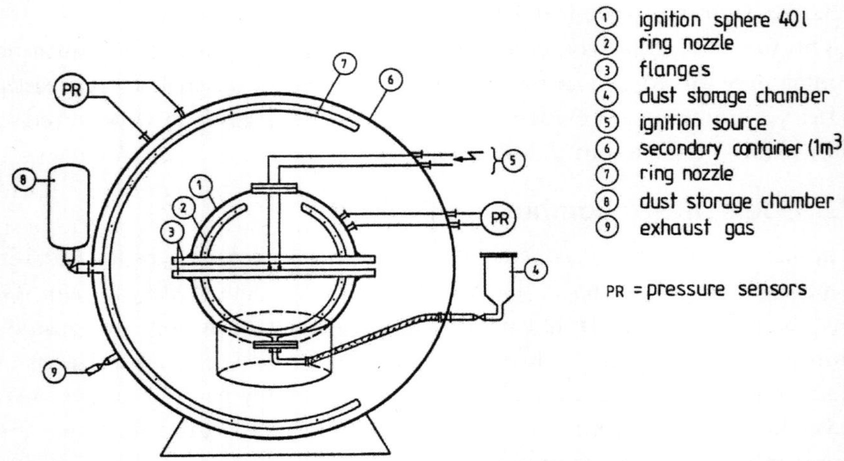

Figure 7.37 Illustration of the experimental setup comprising a split and flanged 40-L explosion vessel mounted inside a 1 m^3 explosion vessel. The annular gap for possible flame transmission is between the two flanges on the two halves of the 40-L sphere.
From Schuber, G., 1989. Ignition breakthrough behaviour of dust/air and hybrid mixtures through narrow gaps. Proceedings of the 6th International Symposium on Loss Prevention and Safety Promotion in the Process Industries, Oslo, Norway.

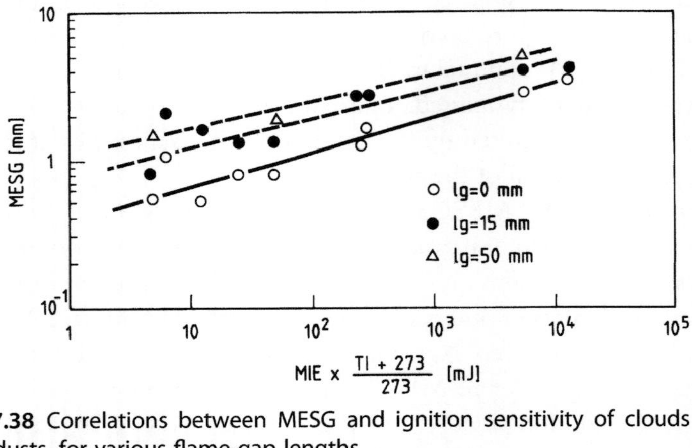

Figure 7.38 Correlations between MESG and ignition sensitivity of clouds in air of various dusts, for various flame-gap lengths.
From Schuber, G., 1989. Ignition breakthrough behaviour of dust/air and hybrid mixtures through narrow gaps. Proceedings of the 6th International Symposium on Loss Prevention and Safety Promotion in the Process Industries, Oslo, Norway.

As already pointed out the original motivation for Schuber's work was the uncertainty related to the ability of rotary locks to prevent transmission of dust explosions. Therefore, he also carried out industrial-scale experiments with a setup as illustrated in Fig. 7.36, with a rotary lock mounted between two vessels in which explosive dust clouds were generated simultaneously. The dust cloud on the one side was then ignited and it was observed if transmission of flame occurred to the extent that the dust cloud on the other side was also ignited. On the basis of numerous experiments, Schuber (1989) constructed a nomograph for predicting maximum permissible gaps between the edges of the rotary lock blades and the housing for preventing flame transmission (see Eckhoff, 2003).

7.3.8 Shock Waves

Shock waves, ie, very steep and strong pressure wave fronts traveling at speeds exceeding the speed of sound, can be generated by industrial accidents, eg, by pressure vessel failure. An informative analysis of shock wave ignition of dust clouds was given by Wolanski (1990). His illustration of the particle interaction with the convective flow behind the incident shock wave is shown in Fig. 7.39.

At the first moment of interaction, the stationary particle is subjected to the supersonic flow. A bow shock is formed near the particle; and because the temperature of the gas between the particle and the bow shock is close to the stagnation temperature, the particle is heated rapidly. At the same time, the drag force causes the particle to accelerate and the temperature

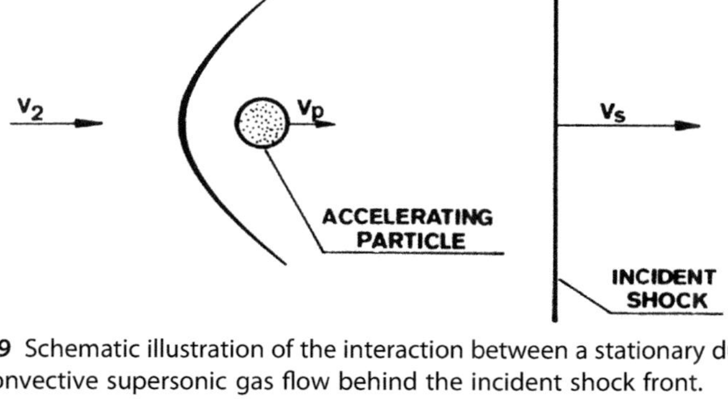

Figure 7.39 Schematic illustration of the interaction between a stationary dust particle and the convective supersonic gas flow behind the incident shock front.
From Wolanski, P., 1990. Deflagration and Detonation Combustion of Dust Mixtures. Paper published by the American Institute of Aeronautics and Astronautics, pp. 3–31.

to decrease. The rate of particle acceleration depends on particle size and other properties. Because very small particles accelerate rapidly to supersonic speeds, the stagnation temperature at the particle front exists only for an extremely short time. Larger particles accelerate more slowly, thus remaining exposed to the higher stagnation temperature between the bow shock and the particle for a longer time. However, the larger particles require more heat for ignition. Wolanski suggests, therefore, that depending on the dust type, there is an optimal particle-diameter range, which provides the most favorable conditions for ignition. In fact, in incident shock waves some dust particles can be more easily ignited than can even a hydrogen/oxygen mixture. According to Wolanski, experimental data do confirm that the ignition delay of dust particles behind the incident shock depends on particle diameter; and the optimum diameter for ignition varies with the nature of the particle. For organic particles the optimal diameter is on the order of some 10 μm. Particles of diameters of only a few μm, or larger than 100 μm are usually more difficult to ignite in an incident shock than particles of intermediate sizes.

7.3.9 Light Radiation

Ignition experiments using laser light can provide basic information not only about dust cloud ignition processes, but also about flame propagation processes in dust clouds. Wel et al. (1994) used a simple, modified Semenov theory for autoignition (no temperature gradients inside initially heated volume), for using the experimental laser-light-pulse ignition data to predicted minimum ignition temperatures and energies of dust clouds. The predicted values were in approximate agreement with minimum ignition temperatures and energies measured directly. Zevenbergen (2002) reviewed various aspects of ignition of dust clouds by laser light.

Proust (1996), using a Neodymium-yttrium-aluminum garnet continuous wave (Nd-YAG CW) laser, investigated experimentally the extent to which a beam of laser light is able to ignite an explosive dust cloud. He was unable to find a generally valid correlation between the ease with which a dust cloud could be ignited directly by a laser beam, and the minimum electrical spark ignition energy of the same dust cloud. Proust also conducted experiments in which ignition occurred indirectly via a small solid target inside the dust cloud that had first been heated by the laser beam. He found that the probability of a given laser beam igniting a given dust cloud increased markedly with this setup compared with direct laser beam ignition. Proust (2002) extended these experiments by using both continuous and pulsed radiation, and both targets of layers of noncombustible material, eg, iron oxide, and layers of the same material as in the dust cloud to be ignited. He was able to quantify minimum laser beam powers impinging on the target required for igniting dust clouds by the heat absorbed by a solid target. The parameters included the laser beam diameter, the duration of the irradiation, the target material (combustible/noncombustible), and the type of dust (starch, *Lycopodium*, lignite, sulfur, Acrylonitrile-Butadiene-Styrene [ABS], and aluminum). An inherent complicating element in this kind of experiment is that the laser light has to travel through some length of dust cloud before impinging on the target. The part of the beam power absorbed by the dust cloud is a function of both the length of the light path through the dust cloud, the dust concentration, and the material and geometry of the dust particles. Fig. 7.40 shows a set of Proust's results for pulsed laser radiation.

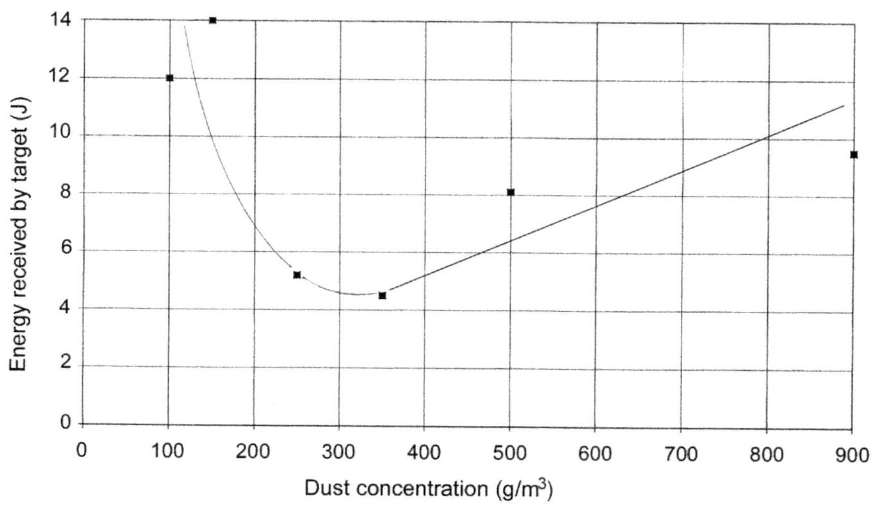

Figure 7.40 Deduced relationship for the minimum energy of a very short laser pulse, hitting a solid inert iron oxide target inside an explosive cloud of starch particles in air, required for igniting the cloud, as a function of the starch concentration in the cloud. *From Proust, C., 2002. Laser Ignition of Dust Clouds. Paper Presented at 4th International Symposium on Hazards, Prevention and Mitigation of Industrial Explosions, Bourges, France, October 21–25, 2002. In: Sochet, I. (Ed.), http://www.google.com/url?q=http://jp4.journaldephysique.org/&sa=U&ved=0ahUKEwi3x5aAtKPKAhUPkI4KHdWdC4sQFggUMAA&usg=AFQjCNEn_4nAZ0R3MDj6mZt9sJSr8Oz3ZQ, vol. 12, Pr7, pp. 79–88.*

The MIEs in Fig. 7.40 were deduced from experimental graphs of the minimum-radiated laser energy received by the target required for igniting the dust cloud, as a function of the time from onset of radiation to ignition. By extrapolating the almost linear relationships found to zero time, an estimate of the MIE with very short laser pulses hitting an inert iron oxide target could be deduced. The MIEs found in this way, ranging from 4 to 14 J are substantially higher than the minimum electric spark ignition energy of 10 mJ found for the starch used in a standard MIE test.

7.4 CASE HISTORIES OF DUST EXPLOSIONS

7.4.1 Motivation for Selection

All the case histories given in the following are taken from the book by Eckhoff (2003). His close cooperation with Norwegian industry in investigating accidental dust explosions for 25 years gave this author access to detailed information on several such explosions that had occurred in Norway through the years. It was natural, therefore, to include some of this information even in the present book.

Some well-known dust explosions in other countries, which have been described extensively elsewhere in the open literature, have not been included. This, for example, applies to the catastrophic wheat flour explosion in the Roland Mill in Bremen, Germany, which has been discussed in detail by the Fire and Police Authorities of Bremen (1979). It also applies to many of the major dust explosions in the United States after 1975, which have been discussed in detail by Kauffman (1982, 1989) and Kauffman and Hubbard (1984).

7.4.2 Historical Perspective: Wheat Flour Explosion in Turin in 1785

The dust explosion hazard has probably been recognized in Europe for several centuries, but the wheat flour explosion in Turin in 1785 seems to have been the first accident of this kind that was investigated extensively. When the Academy of Science of Turin heard about Morozzo's investigations, they asked him to prepare a written account of his findings. However, only very rarely are details of Count Morozzo's fascinating account from 1795 mentioned in modern literature. It is considered appropriate, therefore, to start this sequence of dust explosion case histories by quoting some selected sections of the full original account (see Eckhoff, 2003).

Fig. 7.41 is an attempt at reconstructing a possible scene of the accidental explosion in the Turin bakery based on the original written account below.

On the 14th of December, 1785, about six o'clock in the evening, there took place in the house of Mr. Giacomelli, baker in this city, an explosion which threw down the windows and window-frames of his shop, which looked into the street; the noise was as loud as that qf a large cracker, and was heard at a considerable distance. At the moment of the explosion, a very bright flame, which lasted only a few seconds, was seen in the shop; and it was immediately observed, that the inflammation proceeded from the flour warehouse, which was situated over the back shop, and where a boy was employed in stirring some flour by the light of a lamp. The boy had his face and arms scorched by the explosion; his hair was burnt, and it was more than a fortnight before his burns were healed. He was not

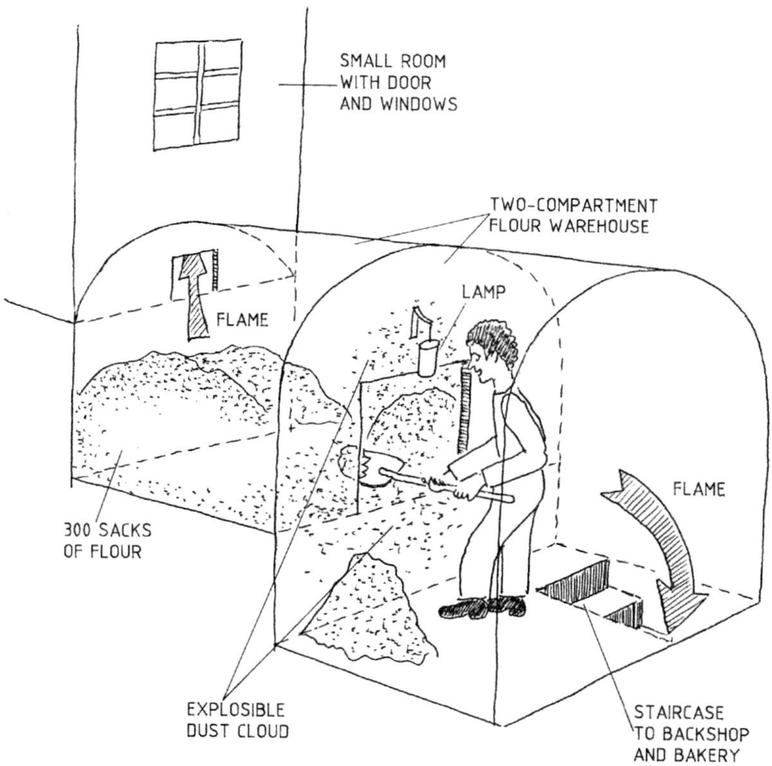

Figure 7.41 Reconstruction of possible scene of wheat flour explosion in Mr. Giacomelli's bakery in Turin, Italy, on December 14, 1785, as described by Count Morozzo (1795).
From Eckhoff, R.K., 2003. Dust Explosions in the Process Industries, third ed. Gulf Professional Publishing/Elsevier. (first ed. published in 1991, second ed. in 1997). ISBN:0-7506-7602-7.

the only victim of this event; another boy, who happened to be upon a scaffold, in a little room on the other side of the warehouse, seeing the flame, which had made its passage that way, and thinking the house was on fire, jumped down from the scaffold, and broke his leg.

The flour warehouse, which is situated above the back shop, is six feet high, six feet wide, and about eight feet long. It is divided into two parts, by a wall; and arched ceiling extends over both, but the pavement of one part is raised about two feet higher than that of the other. In the middle of the wall is an opening of communication, two feet and a half wide, and three feet high; through it the flour is conveyed from the upper chamber into the lower one.

The boy, who was employed in the lower chamber, in collecting flour to supply the bolter below, dug about the sides of the opening, in order to make the flour fall from the upper chamber into that in which he was; and, as he was digging, rather deeply, a sudden fall of a great quantity took place, followed by a thick cloud, which immediately caught fire, from the lamp hanging to the wall, and caused the violent explosion here treated of.

The flame pushed itself in two directions; it penetrated, by a little opening, from the upper chamber of the warehouse, into a very small room above it, where, the door and window frames being well closed and very strong, it produced no explosion; here the poor boy, already mentioned, broke his leg. The greatest inflammation, on the contrary, took place in the smaller chamber, and, taking the direction of a small staircase, which leads into the back shop, caused a violent explosion, which threw down the frames of the windows which looked into the street. The baker himself, who happened then to be in his shop, saw the room all on fire some moments before he felt the shock of the explosion.

The baker told me that he had never had flour so dry as in that year (1785), during which the weather had been remarkably dry, there having been no rain in Piedmont for the space of five or six months: indeed, he attributed the accident which had happened in his warehouse to the extraordinary dryness of the corn.

The phenomenon, however, striking at the time it happened, was not entirely new to the baker, who told me that he had, when he was a boy, witnessed a similar inflammation; it took place in a flour-warehouse, where they were pouring flour through a long wooden trough, into a bolter, while there was a light on one side; but, in this case, the inflammation was not followed by an explosion.

The wheat flour explosion in Mr. Giacomelli's bakery in Turin was a comparatively minor one, but there is still much to learn from Count Morozzo's analysis. The considerations related to the low moisture content of the flour due to dry weather are still highly relevant. The same applies to the observation of a primary explosion causing a secondary explosion by entrainment of dust deposits.

7.4.3 Three Grain Dust Explosions in Norway Between 1970 and 1988

7.4.3.1 Wheat Grain Dust, Stavanger Port Silo, June 1970

The explosion occurred in Norway's largest and newly built import grain silo in Stavanger on a hot and dry summer day. Fortunately, no persons were killed, but some workers suffered first-degree burns. Although the extent of flame propagation was considerable, the material damage was moderate due to the comparatively strong reinforced-concrete structure of the buildings and the venting through existing openings.

The entire event lasted for a period of about 25–30 s, during which a sequence of six or seven distinct explosions were heard. In the middle of this sequence there was an interval of 10–12 s. The flame propagated a total distance of about 1500 m, through a number of bucket-elevators, horizontal conveyors, ducting, filters and rooms in the building. Dust explosions occurred in six of the large, cylindrical storage silos of 2000 m^3 volume each, in one slightly smaller silo, in seven of the slimmer, intermediate silos of capacities 400 or 1000 m^3, in one 150 m^3 silo, and in seven small loading-out silos of capacities 50 m^3 each. The six largest silos had no venting, whereas the explosions in the single, slightly smaller silo, and in all the intermediate and loading-out silos, were vented two through 0.4-m manholes, which had their covers flung open. It is interesting to note that only one silo was damaged in this incident, namely one of the six unvented, large storage silos, which had its roof blown up, as shown in Fig. 7.42. It is thus clear that the maximum explosion pressures in all the other 21 silos, vented and unvented, were lower than about 0.2 bar (g), which would be required to blow up the actual type of silo roof.

Almost all the windows, except those in the office department, were blown out, as was a large provisional light wall at the top of the head house. The legs of all of the five bucket elevators of 0.65 × 0.44 m cross section were torn open from bottom to top. The dust extraction ducts were also in part torn open.

The ignition source and its location were never fully identified. However, two hypotheses were put forward. The first was self-ignition of dust deposits in the boot of a bucket elevator in which the explosion was supposed to have started. The self-ignition process was initiated by a bucket that had been heated by repeated impacts until it finally loosened and fell into the dust deposit in the elevator boot. The second hypothesis was that the chain of events leading to ignition started with welding on the outside

Figure 7.42 Damaged silo roof after the wheat grain dust explosion in Stavanger in June 1970.
From Eckhoff, R.K., 2003. Dust Explosions in the Process Industries, third ed. Gulf Professional Publishing/Elsevier. (first ed. published in 1991, second ed. in 1997). ISBN:0-7506-7602-7.

of the grain-feeding duct leading to one of the elevator boots. Due to efficient heat transfer through the duct wall, self-heating could have been initiated in a possible dust deposit on the inside of the duct wall. Lumps of the smoldering deposit could then have loosened and subsequently become conveyed into the elevator boot, initiating an explosion in the dust cloud there.

7.4.3.2 Wheat Grain Dust in New Part of Stavanger Port Silo, October 1988

This event was described by Olsen (1989). Because of effective explosion mitigation by venting and automatic explosion suppression, the extent of and damage caused by the explosion were minor. There were neither fatalities nor injuries. The incident deserves attention, however, because the chain of events leading to explosion initiation was identified, and because the incident illustrates that properly designed measures for explosion mitigation are effective.

The explosion occurred in a bucket elevator head immediately after termination of transfer of Norwegian wheat grain between two silo cells. At the moment of explosion the transport system was free of grain. In this new part of Stavanger Port Silo, the bucket elevator legs are cylindrical and mounted outdoors, along the wall of the head house. A number of vents are located along the length of the legs. In the explosion incident the vent covers on the elevator leg involved were blown out, which undoubtedly contributed to reducing the extent of the explosion. In addition, automatic explosion suppression equipment had been installed and was operating both at the bottom and top of the elevator. The explosion caused no significant material damage, either by pressure or by heat. Fig. 7.43 illustrates the head of the bucket elevator in which the explosion occurred.

Because of a slight offset, the steel cover plate for the felt dust seal for the pulley shaft touched the shaft and became heated by friction during operation of the elevator. The hot steel plate in turn ignited the felt seal, from

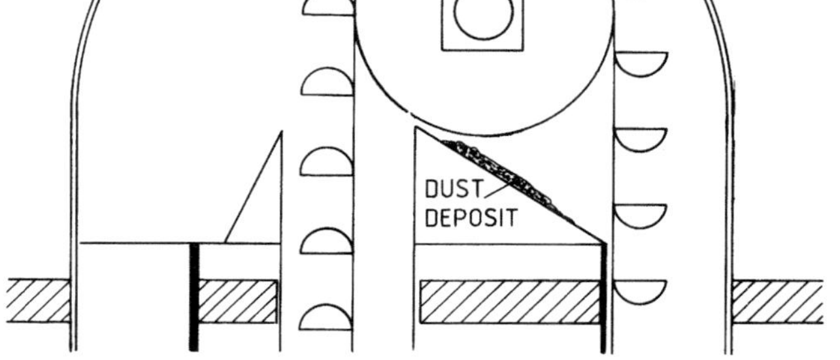

Figure 7.43 Schematic illustration of head of a bucket elevator in the new part of Stavanger Port Silo, where the minor 1988 wheat grain dust explosion was initiated. *From Olsen, O., 1989. Private Communication to R.K. Eckhoff. From Olsen, O., Stavanger Port Silo, Norway and Eckhoff, R.K., 2003. Dust Explosions in the Process Industries, third ed. Gulf Professional Publishing/Elsevier. (first ed. published in 1991, second ed. in 1997). ISBN: 0-7506-7602-7.*

which one or more glowing fragments dropped into the wheat grain dust deposit on the inclined surface below, initiating smoldering combustion in the deposit. Just after the elevator had stopped, there was presumably still enough dust in the air to be ignitable by the smoldering dust, and to be able to propagate a flame. Alternatively, some of the smoldering dust may have slid down the inclined surface and become dispersed and transformed into an exploding dust cloud. Just after the explosion, some smoldering dust was still left on the inclined plate below the elevator pulley.

7.4.3.3 Barley/Oats Dust Explosion in Head House of Silo Plant at Kambo, June 1976

This explosion, described by Storli (1976), caused considerable material damage, but due to fortunate circumstances there were neither fatalities nor significant injuries. The dust involved was from Norwegian barley or oats.

The explosion probably started in a bucket elevator and was initiated by burning/glowing material from an overheated hammer mill. The primary explosion escalated into a secondary explosion throughout the head house, whereby most of the head-house front wall was blown outward, as shown in Fig. 7.44.

Two of the bucket elevators had bulged out along the entire length and the dust extraction ducting was torn apart, which gave rise to the secondary explosion. Because the floors were supported by the wall, and the connections between wall and floors were weak, the entire wall sheet was pushed out at a quite low explosion pressure, leaving the floors unsupported at the front.

After the explosion, the entire head house was reconstructed. The floors were supported by a rigid framework, and should an explosion occur again, the lightweight wall elements can serve as vent covers, without weakening the support of the floors.

7.4.4 Catastrophic Linen Dust Explosion in Harbin, China in 1987

7.4.4.1 General Outline

In the middle of the night, at 2.39 am, on March 15, 1987, the spinning section of the large linen textile plant in Harbin, P.R. China, was afflicted with a catastrophic dust explosion. The losses were substantial. Out of the 327 women and men working night shift in the spinning section when the explosion occurred, 58 workers lost their lives and 177 were injured. Thirteen thousand square meter of factory area were demolished.

Figure 7.44 Damaged silo head house after a grain dust explosion at Kambo, Norway, in June 1976.
From Eckhoff, R.K., 2003. Dust Explosions in the Process Industries, third ed. Gulf Professional Publishing/Elsevier. (first ed. published in 1991, second ed. in 1997). ISBN:0-7506-7602-7.

The explosion accident has been discussed in detail by Xu Bowen (1988) and Zhu Hailin (1988). Xu Bowen et al. (1988) reconstructed a possible course of the explosion development on the basis of a seismic recording of the explosion by the State Station of Seismology, located only 17 km from the linen textile plant.

7.4.4.2 Explosion Initiation and Development, Scenario 1

Fig. 7.45 illustrates the 13,000 m^2 spinning section throughout which the explosion swept, and the possible locations and sequence of the nine successive explosions that comprised the event, according to Xu Bowen (1988) and Xu Bowen et al. (1988). These workers based their reconstruction of the explosion on three independent elements of evidence. First, they

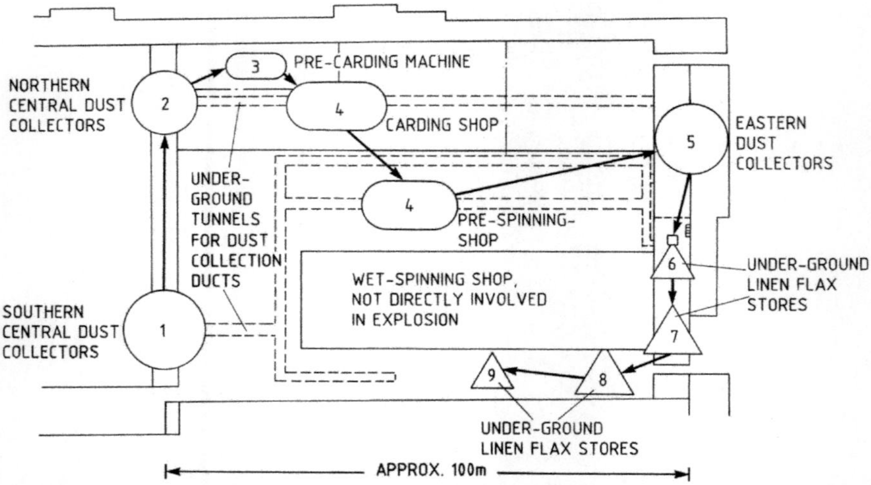

Figure 7.45 Schematic illustration of the 13,000 m^2 spinning section of the Harbin Linen Textile Plant, P.R. China, that was afflicted with a catastrophic dust explosion on March 15, 1987. *Numbered circles, ovals, and triangles* indicate location and sequence of a postulated series of nine successive explosions.
From Eckhoff, R.K., 2002. Minimum ignition energy (MIE) — a basic ignition sensitivity parameter in design of intrinsically safe electrical apparatus for explosive dust clouds. Journal of Loss Prevention in the Process Industries 15, 305–310.

identified the location of the various explosion sites throughout the damaged plant. Second, they ranked the relative strengths of the local explosions by studying the extent and nature of the damage. Third, they arranged the various local explosions in time by means of the relative strengths of the nine successive explosions, identified by decoding the seismic recording of the event.

Fig. 7.46a shows a direct tracing of the amplitude-modulated seismic signal actually recorded 17 km from the explosion site. Fig. 7.46b shows the sequence of nine energy pulses impacting on the earth at the location of Harbin Linen Textile Plant, deduced from the signal in Fig. 7.46a. Finally, Fig. 7.46c shows the theoretical prediction of the seismic signal to be expected from the sequence of explosions in Fig. 7.46b. The agreement between the (a) and (c) signals is striking, which supports the validity of the postulated energy impact pulse train (b).

According to Xu Bowen et al. (1988), the explosion was initiated in one of the nine units in the central dust-collector system. All nine units were connected by ducting. The ignition sources were not identified, but an electrostatic spark was considered as one possibility, a local fire or glow

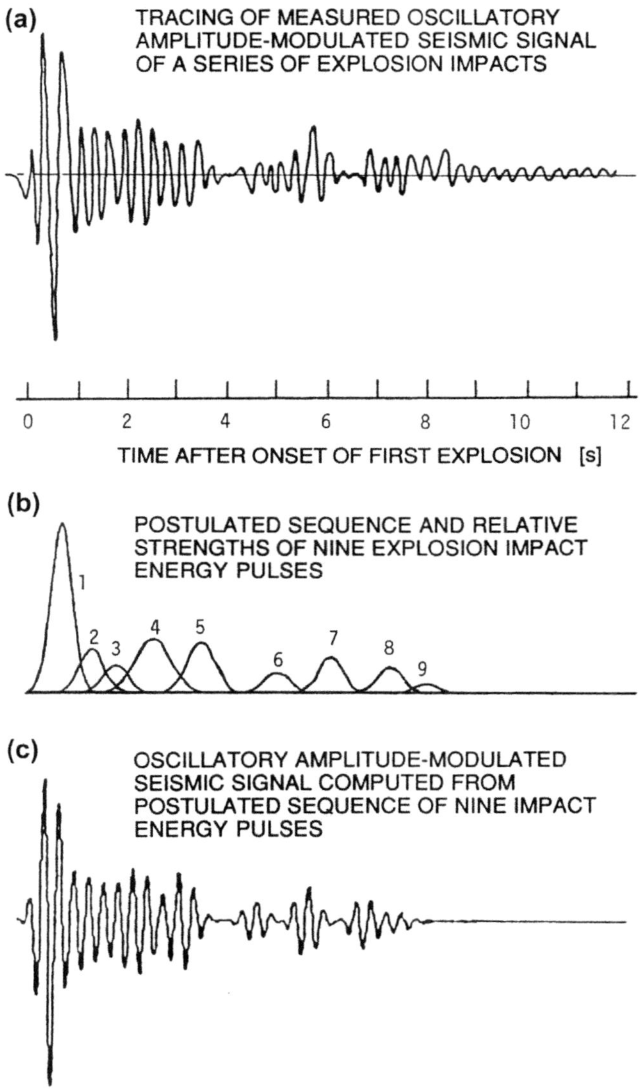

Figure 7.46 Sequence of nine impact energy pulses from nine successive dust explosions in the Harbin Linen Textile Plant, Harbin, P.R. China, March 15, 1987, postulated on the basis of a seismic record of the event.
From Eckhoff, R.K., 2003. Dust Explosions in the Process Industries, third ed. Gulf Professional Publishing/Elsevier. (first ed. published in 1991, second ed. in 1997). ISBN:0-7506-7602-7.

as another. The initial flame was transmitted immediately to the next dust-collecting unit and both units (1) indicated in Fig. 7.45 exploded almost simultaneously, giving rise to the first major impact pulse in Fig. 7.46b. The explosion then propagated through the other seven dust-collecting units in the central collecting plant (2) indicated in Fig. 7.45, and into the precarding area. Here the blast wave preceding the flame had generated an explosive dust cloud in the room, which was ignited by the flame jet from the dust collectors (3). The room explosion propagated further to the carding and prespinning shops (4), and right up to the eastern dust collectors, where another distinct explosion (5) occurred. The final four explosion pulses were generated as the explosion propagated further into the underground linen flax stores, where it finally terminated after having travelled a total distance of about 300 m. The chain of nine explosions lasted for about 8 s.

7.4.4.3 Explosion Initiation and Development, Scenario 2

This alternative scenario originates from the investigation of Zhu Hailin (1988), who found evidence of an initial smoldering dust fire caused by a live 40 W electrical portable light lamp lying in a flax dust layer of 6–8 cm thickness in a ventilation room. He also found evidence of flame propagation through the underground tunnels for the dust-collection ducting. On the basis of his analysis, Zhu suggested that the explosion was initiated in the eastern dust collectors (5) indicated in Fig. 7.45, from which it transmitted to the nine units of the central dust-collecting plant (1) and (2) via the ducting in the underground tunnels. Severe room explosions were initiated when the ducting in the tunnel ruptured, and the resulting blast dispersed large quantities of dust in the workrooms into explosive clouds that were subsequently ignited. From the eastern dust collectors the explosion also propagated into the underground flax stores.

7.4.4.4 Final Remark

The investigation of the Harbin disaster exposed the great difficulties in identifying the exact course of events of major explosions creating massive damage. In addition to causing pain and grief, loss of life also means loss of eyewitnesses. Besides, the immediate need for firefighting and rescue operations changes the scene before the investigators can make their observations. In addition, the explosion itself often erases evidence, eg, of the ignition source. However, the Harbin disaster unequivocally demonstrated the possible dramatic consequences of inadequate housekeeping in industrial plants where fine explosible dust is generated.

7.4.5 Major Silicon Dust Explosion in Bremanger, Norway in 1972

In this serious explosion accident, five workers lost their lives and four were severely injured. The explosion occurred in the milling section of the silicon production plant, rupturing or buckling most of the process equipment and blowing out practically all the wall panels of the building. Fig. 7.47 shows a detailed view of the extensive damage. Eye-witnesses reported that the dust flame was very bright, almost white. This is in accordance with the fact that the temperature of silicon dust flames, as of flames of aluminum and magnesium dust, is very high due to the large amounts of heat released in the combustion process per mole of oxygen consumed (see Table 7.1). Because of the high flame temperature, the thermal radiation from the flame was

Figure 7.47 View of the extensive demolition of the silicon grinding plant caused by the silicon dust explosion at Bremanger, Norway, October 1972.
From Eckhoff, R.K., 2003. Dust Explosions in the Process Industries, third ed. Gulf Professional Publishing/Elsevier. (first ed. published in 1991, second ed. in 1997). ISBN:0-7506-7602-7.

intense, which was a main reason for the very severe burns that nine of the workers suffered.

The investigation after the accident disclosed a small hole in a steel pipe for conveying silicon powder from one of the mechanical sieves to a silo below. An oxygen/acetylene cutting torch with both valves open was found lying on the floor about 1 m from the pipe with the hole. According to Kjerpeseth (1990), there was strong evidence of the small hole having been made by means of the cutting torch just at the time when the explosion occurred. The interior of the pipe that was perforated had probably not been cleaned prior to the perforation. At the moment of the explosion, part of the plant was closed down due to various repair work. However, the dust-extraction system was operating. In view of the high temperature and excessive thermal power of the cutting torch, and not least the fact that it supplied pure oxygen to the working zone, a layer of fine dust on the internal pipe wall may well have become dispersed and ignited as soon as the torch had burnt its way through the pipe wall. The blast from the resulting primary silicon dust explosion then raised dust deposits in other parts of the plant into suspension and allowed the explosion to propagate further until it eventually engulfed the entire silicon grinding building. The grinding plant was not rebuilt after the explosion.

7.4.6 Major Aluminum Dust Explosion at Gullhaug, Norway in 1973

The main source of information concerning the original investigation of the accident is Berg (1989). The explosion occurred during working hours, just before lunch, while 10 workers were in the same building. Five of these lost their lives, two were seriously injured, two suffered minor injuries, whereas only one escaped unhurt. A substantial part of the plant was totally demolished, as illustrated by Fig. 7.48.

The premix preparation plant building was completely destroyed. Debris was found up to 75 m from the explosion site. The explosion was followed by a violent fire in the powders left in the ruins of the plant and in adjacent storehouse for raw materials. The explosion occurred when charging the 5.2 m^3 batch mixer, illustrated in Fig. 7.49.

It appeared that about 200 kg of very fine aluminum flake, sulfur, and some other ingredients had been charged at the moment of the explosion. The total normal charge of the formulation in question was 1200 kg. The upper part of the closed vertical mixing vessel was cylindrical, and the lower part conical. The feed chute was at the bottom. The internal mixing device

Figure 7.48 Scene of total demolition after aluminum dust explosion in the premix plant of a slurry explosives factory at Gullaug, Norway, in August 1973.
From Eckhoff, R.K., 2003. Dust Explosions in the Process Industries, third ed. Gulf Professional Publishing/Elsevier. (first ed. published in 1991, second ed. in 1997). ISBN:0-7506-7602-7.

consisted of a vertical rubber-lined screw surrounded by a rubber-lined earthed-steel tube. The powders to be mixed were transported upward by the screw, and when emerging from the top outlet of the tube, they dropped to the surface of the powder heap in the lower part of the vessel. There they became mixed with other powder elements and eventually re-transported to the top.

The construction materials of the mixer had been selected so as to eliminate the formation of metal-particle sparks. This was probably why both the screw and the internal wall of the surrounding earthed-steel tube had been lined with rubber.

During operation the 5.2 m^3 vessel was flushed with nitrogen, the concentration of oxygen in the vessel being controlled by a direct-reading oxygen analyzer at the vessel top. According to the foreman's statement, the oxygen content at the moment of explosion was within the specified limit. After the explosion, the heavy central screw part of the mixer, with the mixer end cap, was retrieved about 12 m away from the location of the mixer prior to the explosion. More detailed investigation of the part of the screw that was shielded by the steel tube, revealed that the screw wings had been deformed bidirectionally as if an explosion in the central part had expanded violently both upward and downward. This evidence was considered a strong indication of the explosion having in fact been initiated inside the steel tube surrounding the screw. The blast and flame from this primary

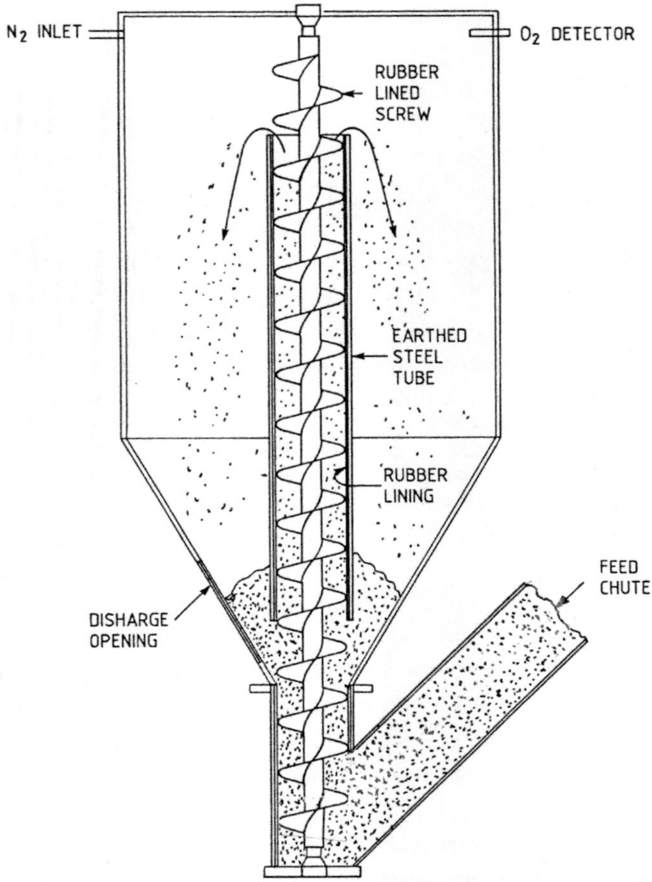

Figure 7.49 Cross section of the mixer used for production of dry premix for slurry explosives at Gullaug, Norway, in 1973.
From Eckhoff, R.K., 2003. Dust Explosions in the Process Industries, third ed. Gulf Professional Publishing/Elsevier. (first ed. published in 1991, second ed. in 1997). ISBN:0-7506-7602-7.

explosion, in turn, generated and ignited the dust cloud in the main space inside the mixer. Finally the main bulk of the powder in the mixer was thrown into suspension and brought to ignition when the mixer ruptured, giving rise to a major explosion in the workrooms.

Subsequent investigations revealed that clouds in air of the fine aluminum flake powder involved were both extremely sensitive to ignition and exploded extremely violently. The minimum electric spark ignition energy (MIE) was probably significantly lower than 1 mJ. The maximum rate of pressure rise in the Hartmann bomb was 2600 bar/s. Both these values are extreme. The thickness of the aluminum flakes was about 0.1 μm, which corresponds to a specific surface area of about 7.5 m^2/g.

The investigation further disclosed that the design of the nitrogen inerting system of the mixer was inadequate. First, the nitrogen flow was insufficient to enable reduction of the average oxygen concentration to the specified maximum level of 10 vol% within the time allocated. Secondly, even if the flow had been adequate, both the nitrogen inlet and the oxygen concentration probe were located in the upper part of the vessel (see Fig. 7.49), which rendered the measured oxygen concentration unreliable as an indicator of the general oxygen level in the mixer. It is highly probable that the oxygen concentration in the lower part of the mixer, and in particular in the space inside the tube surrounding the screw, was considerably higher than the measured value. This explains why a dust explosion could occur in spite of low measured oxygen concentration.

The final central concern of the investigators was identification of the probable ignition source. In the reports from 1973 it was concluded that the primary explosion in the tube surrounding the screw was probably initiated by an electrostatic discharge. However, this conclusion was not qualified in any detail. In more recent years the knowledge about various kinds of electrostatic discharges has increased considerably. It now seems highly probable that the ignition source in the 1973 Gullaug explosion was a propagating brush discharge, brought about by the high charge density that could accumulate on the internal rubber lining of the steel tube surrounding the screw, because of the earthed electrically conducting backing provided by the steel tube (see Section 2.2.6.2 of Chapter 2 and Section 7.3.6.3).

7.4.7 Major Polyethylene Dust Explosion, Kinston, North Carolina, United States, in 2003

7.4.7.1 Introduction

On January 29, 2003, a dust explosion occurred at the West Pharmaceutical Services, Inc. plant in Kinston, North Carolina, United States. Six workers lost their lives and 38 were injured, including two firefighters. Because of the number of deaths and injuries, the US Chemical Safety and Hazard Investigation Board (CSB) launched an investigation to determine the root and contributing causes of the explosion and to make recommendations to prevent similar occurrences. The present account is based on the comprehensive CSB (2003) report.

7.4.7.2 Company and Process

Being founded in 1923, West is one of the world's largest manufacturers of closures and components for sealing drug vials and prefilled syringes. The headquarters are in Lionville, Pennsylvania, United States. The company has approximately 4000 employees working in eight facilities in North America and 10 in Europe and Asia.

The West Kinston plant had manufactured rubber drug delivery components for syringe plungers, septums, and vial seals since 1975. The rubber compounding process in use at the time of the dust explosion was started up in 1987, following a major expansion and automation project. This process was basically similar to other rubber manufacturing processes, such as tire production. Production operations included rubber compounding, moulding, and extrusion. Raw materials were prepared in another area of the plant. The production was semi-continuous, producing sequential batches and operating 24 h per day, 5 or 6 days per week. At the time of the explosion, 264 employees and 35 full time contract workers were employed at the Kinston plant.

7.4.7.3 Location and Layout of the Kinston Plant

The Kinston plant was located in a light industrial business park adjacent to the regional airport. Two private residences and the local Humane Society shelter were each located about 300 m from the facility. The plant area illustrated in Fig. 7.50 was approximately 13,500 m^2, and primarily single story. However; some of the rubber compounding equipment was located in a 18-m high, two story area, which is hatched in Fig. 7.50. The plant housed two main operations, viz. rubber compounding and product finishing. In the finishing process the compounded rubber was molded and pressed into stoppers and plungers.

7.4.7.4 Rubber Compounding Process

7.4.7.4.1 Overview

An outline of the plant, showing the compounding area in relation to the other facilities, is given in Fig. 7.50. The automated rubber compounding system consisted of two separate production lines, each with a mixer, a roller, and a "batchoff" machine. The compounding was done in batches, and the purpose of the batchoff machine was to cool, coat, and fold the strips of rubber from a compounded batch. The batchoff machines used were of a common design utilized in various industries.

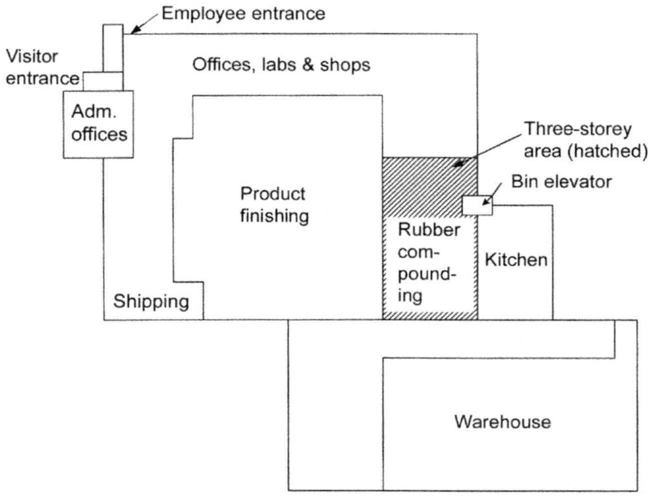

Figure 7.50 Plan of the West Pharmaceutical Services Inc., production plant in Kingston, North Carolina, United States, prior to the dust explosion January 29, 2003. *From CSB, 2003. Dust Explosion, West Pharmaceutical Services, Inc. Report No. 2003-07-I-NC. Chemical Safety and Hazard Investigation Board (CSB), Washington, DC, USA.*

7.4.7.4.2 Raw Material Preparation

The raw material preparation for the rubber compounding process was taking place in the "kitchen." This was a process area to the side of the rubber compounding area and separated from it by a concrete masonry firewall. Solid materials were weighed and loaded into bins. A roller conveyer transported the bins to an elevator, where they were lifted to the second floor of the compounding structure.

7.4.7.4.3 Mixing/Kneading

A conveyor carried the bins further to the mixers, where the ingredients were compounded. Ingredients were generally loaded into the mixers through an open hatch on the side. However, bulk powders used in large portions, such as calcined clay, were pneumatically transferred to weigh hoppers and automatically dropped into the mixers. Each mixer had two opposing rotors that meshed, pulled, and sheared the components to create a uniform mix. The frictional heat generated by the mixer facilitated this process. Mineral oil was used as a plasticizer for the rubber blends and was piped directly to the mixers. After the batch was loaded, the operator closed the feed door and engaged the mixer with automatic controls.

The kneading action of the mixer caused frictional heating of the rubber. Although chilled cooling water flowed through the kneading rotors, the varying speed of the rotors and the duration of the mixing phase largely controlled the temperature of the rubber. The process temperature was held by automatic controls to below the onset temperature for vulcanization (the rubber was vulcanized later during the forming process, when the finished products were shaped by molding).

When the batch had reached a predefined condition of treatment in the mixer, a door on the bottom of the mixer was opened automatically to allow the rubber to drop through a chute into a bucket located in the milling area on the ground floor.

7.4.7.4.4 Rolling, Trimming, Cooling, and Drying

Once compounded in the mixers, the rubber was dropped through chutes to the ground floor, where rollers smoothed it into sheets of roughly uniform thickness. The sheets were then cut into strips, which entered the batchoff machine, where they were cooled, coated, and folded. Then the strips were trimmed, ie, they were dipped into a tank containing a slurry of very fine polyethylene powder in water. The polyethylene powder acted as an antitack agent and had an average particle size of 12 μm, ie, it was very fine. The water slurry also cooled the product to prevent premature vulcanization.

7.4.7.4.5 Air Drying of Trimmed Strips

After leaving the dip tank, the rubber passed in front of a series of air fans. The fans drew air from the room and blew it across the rubber strips to enhance drying. At the exit of the batchoff, the rubber was folded. Finally, the dried strips were stacked for shipment, or for molding in the finishing area of the plant. After drying almost all of the polyethylene powder in the slurry coating adhered to the strips, but a small amount became airborne.

7.4.7.5 Housekeeping Standards

7.4.7.5.1 General

The plant management knew that the compounding process could create dusty conditions. Therefore, local exhaust ventilation (LEV) ducts had been installed at the compound mixers and in certain areas of the kitchen, primarily to limit employee exposure to airborne nuisance dusts. The LEV ducts transported the captured dust to collectors located outdoors. Efforts were also made to prevent dust accumulation in work areas by a having a continuous house-keeping program. A cleaning staff worked around the clock vacuuming and wiping up dust to minimize visible accumulation on exposed surfaces. Because the plant manufactured products for pharmaceutical use, keeping the facility fee of dust was given high priority.

7.4.7.5.2 Batchoff Machines

The batchoff machines were sources of fugitive emissions of combustible dust. Twelve fans blew air across the rubber strip to cool and dry it as it passed through each machine. Some portion of the antitack agent was carried by air currents from the machine into the room, where it tended to settle on surfaces. The cleaning crew continuously wiped and vacuumed the dust from surfaces so that the area was generally free of visible accumulation. However, there was no organized cleaning program for surfaces of beams, conduits, and other features above the ceiling, where dust was known to accumulate due to the design of the dust-extraction system of the machines. Partition walls partially enclosed the batchoff machines to separate them from other areas. Regular housekeeping was conducted around the machines, and dehumidifiers/filters associated with the enclosures removed some dust from the air. Witness statements and photographs submitted by the Kinston plant indicated that visible accumulation of dust in the milling area, even around the batchoff machines, was minimal. Management focused on the extent and effectiveness of housekeeping in working areas, and the effort was a matter of facility pride.

7.4.7.5.3 Hidden Areas Not Covered by the Housekeeping Program

However, although the cleaning crew continuously cleaned the areas around the equipment, several employees told CSB investigators that there was a layer of dust on top of the suspended ceiling, above the room where the rolling mills and batchoff machines were located. Accumulation was reported to be widespread but heaviest in the areas directly above these machines. Accounts on the thickness of dust layers varied. Some employees claimed that dust accumulations of 6 mm were common, but other witnesses described heavier accumulations, such as 13 mm or more. One individual who had performed a maintenance job above the ceiling in the months prior to the incident recalled seeing as much as 50 mm of powder in some areas. Another person, who had been above the ceiling 2 weeks before the explosion, estimated an accumulation of up to 13 mm across 90% of the ceiling area. However, the investigation by the company management concluded that the overall thickness of dust accumulations ranged from 3 to 6 mm.

The area above the ceiling also contained pneumatic conveying lines for the calcined clay and other high-volume noncombustible powders used in the mixers. Because these lines were reported by employees to have leaked on at least one occasion, it is possible that some of the dust accumulation above the ceiling was noncombustible.

7.4.7.6 Outline of the Accidental Explosion

Interviews conducted by CSB investigators indicate that the operations on the day of the explosion were as usual. The explosion occurred abruptly at 1328 on January 29. No one of all the people interviewed recalled to have noticed any sights, sounds, or smells that would have indicated a problem prior to the explosion. Employees throughout the plant heard the explosion, which some of them described as sounding like a "rolling thunder." After seeing the exterior side panels being blown off the second story of the compounding building witnesses outside saw a huge fireball and a rising smoke cloud above the building.

Inside the facility, employees had made different observations. Those most distant from the compounding area saw lights flickering off, and ceiling tiles and debris being blown about. Some workers saw a bright flash and felt either a pressure wave or a vacuum effect that knocked them off their feet. The entire facility was affected to some extent, though explosion damage was most severe in the rubber compounding and milling areas. Fig. 7.51 shows the plant after the explosion. The elevated part of the structure in the background is what was left of the two-story rubber compounding section (see Fig. 7.50). As can be seen, the damage was extensive.

A Kinston police officer on patrol about 1 km south of the plant noticed smoke above the tree lines that surrounded the facility. He immediately contacted his dispatch to inquire if a controlled burn was taking place at the airport. Seconds later, he observed the smoke rising into a cloud about 100 m into the air, and he reported the explosion, which was heard as far as 40 km away. He immediately proceeded to the scene and began helping victims out of the facility to safety.

7.4.7.7 Fires Following the Explosion

Fires began to develop throughout the facility. The sprinkler system designed to mitigate incipient fires within the plant was rendered inoperable from the outset of the incident because the explosion broke feeder lines to

Figure 7.51 Photo of the West Pharmaceutical Services Inc., production plant in Kingston, North Carolina, United States, after the dust explosion January 29, 2003. *From CSB, 2003. Dust Explosion, West Pharmaceutical Services, Inc. Report No. 2003-07-I-NC. Chemical Safety and Hazard Investigation Board (CSB), Washington, DC, USA.*

the system. Emergency responders reported hearing water freely flowing into the structure. The largest and most persistent fire, lasting for two days, developed in the warehouse (Fig. 7.50). Rubber and other raw materials were stored in a warehouse, and the thermal effect from the explosion most probably reached this area and initiated the fire. Eventually, the entire warehouse was fully engulfed in flames involving the large volume of stored baled and strip rubber. Heat from the fire caused most of the steel framing to yield and collapse. Some of the rubber continued to smolder and flare up for about a week.

Mineral oil was stored in two 28 m^3 plastic tanks located between the kitchen and the warehouse (Fig. 7.50). These tanks failed, spilled their contents, and burnt to the ground. Two additional but smaller plastic tanks containing mineral oil, located near the warehouse, also failed and contributed fuel for the fire. The concrete masonry retention walls around the tanks failed and did not prevent the burning oil from spreading.

7.4.7.8 Fatalities and Injuries

The six people who were killed were working on the ground level of the plant. Three were near one of the mills and its batchoff machine. The fourth, who died several weeks after the incident, was working at another batchoff machine. The force of the blast pushed the fifth victim east into the kitchen, and falling objects on the finishing side of the plant fatally injured the sixth person. The majority of the fatal injuries were either by thermal burns or by ejected objects or collapsing walls. When police officers entered the facility, one of the victims was trapped under a fallen girder near the end of a batch-off machine. However, because of the advancing fire attempts at rescuing this person failed, and he died at the scene.

Immediately after the explosion, many employees were dazed or buried under debris. Responders and other employees equipped with flashlights assisted them out of the plant to triage areas. A few workers clung to the exposed frame of the building's second story and were later rescued by fire-fighters.

One student was injured when windows were broken at a school about 1 km away. Businesses located in the same industrial park as the Kinston plant were damaged, and windborne burning debris ignited fires in wooded areas as far as 3 km away. One home located nearby was damaged slightly, and at least two families were evacuated as a precautionary measure.

7.4.7.9 Facility Damage and Relocation of Production

The explosion and ensuing fire heavily damaged the compounding section of the Kinston facility. Fig. 7.51 shows the extent of damage. All exterior sheathing on the compounding structure was destroyed and masonry block walls were knocked down. The warehouse collapsed, and the remaining building structure was rendered mostly unusable.

Fourteen months after the explosion, the company relocated their production to an industrial facility several km south of the destroyed plant. Some equipment that was not used in the compounding process was salvaged from the original plant and is in use at the new location, and much of the workforce was rehired. When the CSB report was written, however, the destroyed facility and the compounding machinery was not in use.

The relocated production plant is not compounding rubber. Instead, rubber strips are being produced by contract manufacturers or at other West facilities and are shipped to Kinston for molding.

7.4.7.10 Analysis of Explosion

7.4.7.10.1 Fuel for Explosion

Because no other material capable of producing such a large explosion was present or used at the plant, CSB concluded that accumulated fine polyethylene dust above the suspended ceiling tiles was the main fuel in the explosion. Other possibilities were investigated, but were found not credible. However, CSB were unable to conclusively determine what specific mechanism dispersed the polyethylene dust to create the primary explosive cloud.

7.4.7.10.2 Center of Explosion

CSB further concluded that the primary, strong explosion occurred in the compounding section of the plant, as illustrated in Fig. 7.52. Fig. 7.53 shows a photograph of the extensive damage in what was assumed by CSB to be the center of the explosion.

The location of the highest pressure was determined to be close to Roller 1 on the first floor of the three-storey compounding section, as indicated in Fig. 7.52. Force vectors derived from observed damage to the building and surrounding equipment indicate that the largest pressure developed in the compounding section and emanated spherically outward. CSB investigators also recovered ceiling tiles debris from the suspended roof above the ground

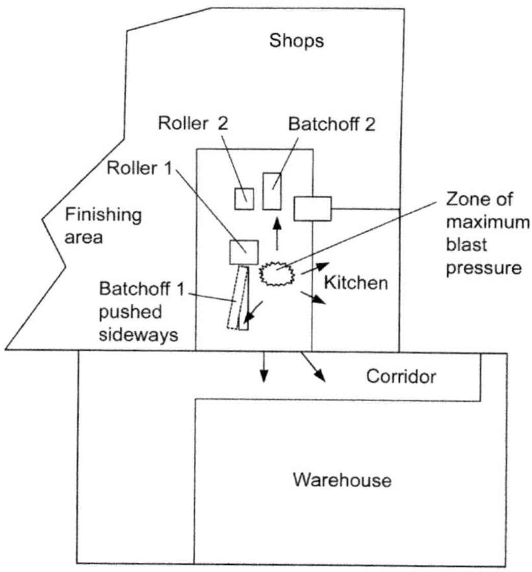

Figure 7.52 Plan of part of the West Pharmaceutical Services Inc., production plant in Kingston, North Carolina, United States, showing the center of the dust explosion January 29, 2003.
From CSB, 2003. Dust Explosion, West Pharmaceutical Services, Inc. Report No. 2003-07-I-NC. Chemical Safety and Hazard Investigation Board (CSB), Washington, DC, USA.

Figure 7.53 Photograph of the probable center of the dust explosion in the West Pharmaceutical Services Inc., production plant in Kingston, North Carolina, United States, January 29, 2003.
From CSB, 2003. Dust Explosion, West Pharmaceutical Services, Inc. Report No. 2003-07-I-NC. Chemical Safety and Hazard Investigation Board (CSB), Washington, DC, USA.

floor of the compounding section. Nearly all of the tiles appeared to be burnt and splattered on top, but not on the bottom, which had faced the ground floor room below. Furthermore, some of the fluorescent light fixture pans recovered from the Roller 1 area had been flattened from above, as if they had been forcefully driven downward to the concrete floor. These two pieces of evidence give further support to the theory that the explosion occurred within the confined space above the suspended ceiling. On the basis of independent quantitative computational fluid dynamics (CFD) simulations it was concluded that the explosion occurred on the first floor of the compounding section, and that the explosion pressure from this area was the source of the extensive blast damage throughout the facility.

7.4.7.11 Initiating Events Giving Rise to the Explosive Dust Cloud and Its Ignition

CSB concluded that the accumulation of combustible dust, mainly fine polyethylene dust, above the suspended ceiling, was the most important safety issue in the West incident. Because it is virtually impossible to eliminate all ignition sources from an industrial setting, preventing the accumulation of fuel provides the best protection against fires and explosions. The extent of damage to the Kinston facility made it extremely difficult to definitively determine the initial events that dispersed the polyethylene dust and ignited it. CSB was unable to determine whether any of the following scenarios may have been the actual initiating event:

- Overheating of a batch of rubber and subsequent ignition of the vapors produced by thermal decomposition
- Ignition of the dust layer by an overheated electrical ballast or light fixture
- Ignition of the dust layer by an electrical spark from an unidentified electrical fault
- Unsettling of dust in a cooling air duct for an electric motor and subsequent ignition of the dust by the motor

Instead, the investigators focused on the most pertinent hazard, ie, the possibility of accumulation of combustible dust despite a systematic housekeeping program, and considered the initiating event as a matter of secondary importance.

7.4.7.12 Previous Incident in the Plant

CSB were informed that in an earlier maintenance operation involving welding, polyethylene powder in proximity to the batchoff machine had ignited, but the fire self-extinguished. This incident demonstrated that the powder was ignitable. However, there was no documented investigation of this incident.

7.4.7.13 Root Causes

CSB concluded that the following root causes were responsible for the explosion and subsequent fire in the Kingston plant of West Pharmaceutical Services Inc. in 2003:

- The company did not perform an adequate engineering assessment of the use of powdered zinc stearate and polyethylene as antitack agents in the rubber batchoff process.
- The company's engineering management systems did not ensure that relevant industrial fire safety standards were consulted.
- The company's management systems for reviewing Material Safety Data Sheets (MSDSs) did not identify combustible dust hazards.
- The hazard communication program at the Kinston plant did not identify combustible dust hazards or make the employees aware of such.

7.4.7.14 Some Recommendations Given by CSB to West Pharmaceuticals

- Revise policies and procedures for new material safety reviews. In particular: Use the most recent versions of MSDSs and other technical hazard information.
- Fully identify the hazardous characteristics of new materials, including relevant physical and chemical properties, to ensure that those characteristics are incorporated into safety practices, as appropriate. Include an engineering element that identifies and addresses the potential safety implications of new materials on manufacturing processes.
- Develop and implement policies and procedures for safety reviews of engineering projects. In particular: Address the hazards of individual materials and equipment and their effect on entire processes and facilities, and consider hazards during the conceptual design phase, as well as during engineering and construction phases.
- Cover all phases of the project, including engineering and construction performed by outside firms.
- Identify and consider applicable codes and standards in the design.
- Identify other production plants within the company that use combustible dusts. Ensure that they incorporate applicable recognized safety precautions. In particular: Ensure that penetrations of partitions, floors, walls, and ceilings are sealed dust tight, and ensure that spaces inaccessible to housekeeping are sealed to prevent dust accumulation.
- Improve hazard communication programs so that the hazards of combustible dust are clearly identified and communicated to the employees. In particular: Ensure that the latest, updated versions of codes of practice are in use and that employees receive training on all updated information.

7.4.8 Catastrophic Sugar Dust Explosion in Georgia, United States in 2008

The following account is based on the comprehensive report by CSB (2009).

7.4.8.1 Overview of the Explosion/Fire Catastrophe

On February 7, 2008, at about 7.15 pm, the Imperial Sugar manufacturing plant in Port Wentworth, Georgia, United States, suffered a major dust explosion catastrophe. Fourteen workers lost their lives; 36 workers suffered serious burns and injuries, some of which resulted in permanent, life-altering conditions. Approximately 85 other workers who were at the facility at the time of the catastrophe were uninjured.

Fig. 7.54 shows an overview of the massive destruction following the series of sugar-dust explosions and subsequent fires.

The first in a series of several consecutive explosions occurred in the enclosed steel conveyor belt under three large concrete silos for storing and conditioning granulated sugar, illustrated in Figs. 7.55 and 7.56. Seconds later, several massive secondary dust explosions propagated throughout the entire granulated- and powdered-sugar packing buildings, bulk sugar-loading buildings, and parts of the raw sugar refinery. Violent fireballs erupted from the penthouse on top of the silos, bucket elevator structures, and surrounding buildings. The silo penthouse housed the belt conveyors for charging granulated sugar to the three silos at the top.

On the same day, shortly before the explosion, the new chief executive officer (CEO) of the Imperial Sugar Company was touring the facility together with three employees. He had joined Imperial Sugar Company as a chief operating officer in February, 2007, and had been promoted to the CEO position on January 29, 2008, only nine days before the explosion catastrophe. Walking through the refinery toward the south packing building, the group of four were startled by what sounded and felt like a heavy roll of packing material dropped from a forklift somewhere in the packing building. A loud explosion 3—5 s later knocked them backward, and debris

Figure 7.54 Photograph of the material damage resulting from a series of sugar-dust explosions and subsequent fires at the Imperial Sugar manufacturing plant in Port Wentworth, Georgia, United States, on February 7, 2008.
From CSB, September 2009. Sugar Dust Explosion and Fire. Report No. 2008-05-I-GA. Chemical Safety and Hazard Investigation Board (CSB), Washington, DC, USA.

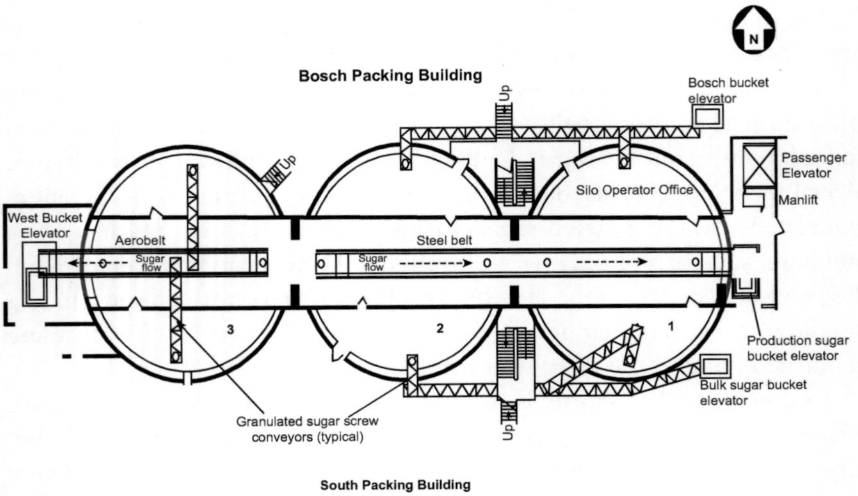

Figure 7.55 Illustration of the three concrete silos of diameters 12 m, heights 32 m, and capacities 2300 tonnes in the Port Wentworth sugar refinery plant for conditioning and storing granulated sugar.
From CSB, September 2009. Sugar Dust Explosion and Fire. Report No. 2008-05-I-GA. Chemical Safety and Hazard Investigation Board (CSB), Washington, DC, USA.

was thrown through the large packing building doorway. The security guard at the gatehouse, and other workers in nearby buildings, also heard a loud explosion. Outside, massive flames and debris erupted above the packing buildings and silos.

Eight centimeter thick concrete floors in the south packing building were lifted and buckled from the explosive force of several successive explosions fueled by sugar dust, as they progressed through the packing buildings and into adjacent buildings. The wooden roof on the palletizer room was shattered and blown into the bulk-sugar rail loading area. The explosions and subsequent fires destroyed the sugar packing buildings, palletizer room, and silos, and severely damaged the bulk train car loading area and parts of the sugar refining process areas.

Workers in the packing buildings received little or no warning before walls, equipment, and furniture were thrown about. Superheated air burned exposed flesh. Workers attempting to escape struggled to find their way out of the smoke-filled, darkened work areas. Debris littered the passageways. Some exits were blocked by collapsed brick walls and other debris. The fire sprinkler system failed because the explosions ruptured the water pipes.

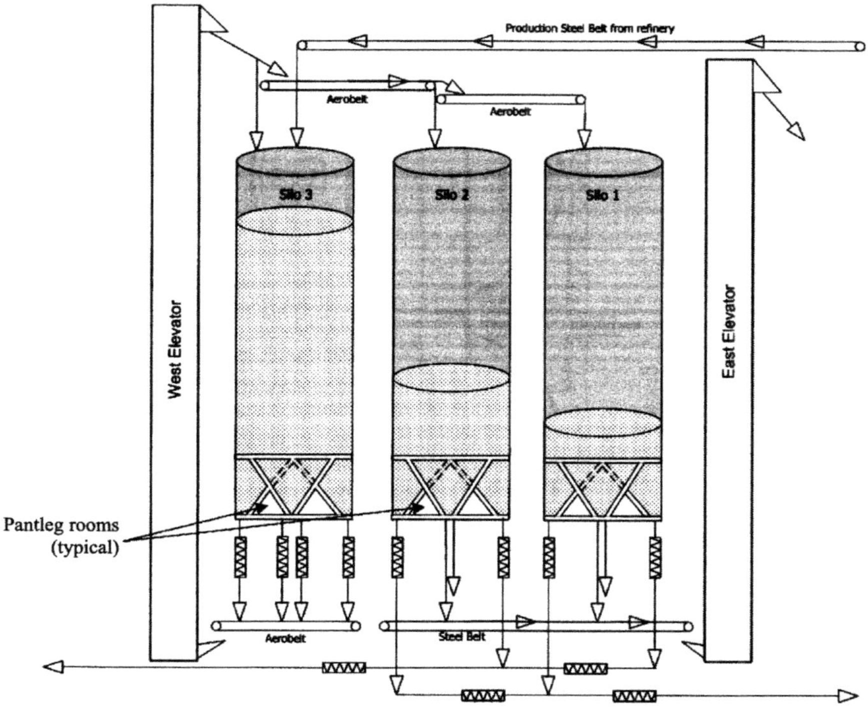

Figure 7.56 Cross section of the three concrete silos and associated conveying equipment for storage of granulated sugar at the Imperial Sugar manufacturing plant in Port Wentworth, Georgia.
From CSB, September 2009. Sugar Dust Explosion and Fire. Report No. 2008-05-I-GA. Chemical Safety and Hazard Investigation Board (CSB), Washington, DC, USA.

Intense fireballs advanced through the entire north and south packing and palletizer buildings as sugar dust, shaken loose from overhead surfaces, ignited. Heaving and buckling floors opened large crevasses. The piles of granulated and powdered sugar that had accumulated around equipment rained down and intensified the fires burning below. Fireballs advanced through enclosed screw conveyors and ignited fires in the refinery and bulk sugar building hundreds of feet from the packing buildings where the incident had begun. Violent fireballs erupted from the facility for more than 15 min as spilled sugar and accumulated sugar dust continued fueling the fires. The major fires in the buildings were extinguished the next day. However, the silo fires continued to smolder for seven more days before they were finally extinguished.

7.4.8.2 *Description of the Sugar Refinery Plant*

7.4.8.2.1 Brief History of Production, Housekeeping, and Explosion/Fire Incidents

The construction of the plant began in the early 1900. Production of granulated sugar from "raw" sugar started in 1917. Over the years, the facility was upgraded by adding refining and packaging capacity, raw sugar and product warehouses, and upgrading the steam/electric power plant. When the explosion and fire catastrophe happened in 2008, the Port Wentworth plant had been in operation for more than 80 years without experiencing any devastating dust explosion.

However, housekeeping had obviously been a problem. Photographs dating from the 1970s through 2007, as well as internal correspondence dating to the early 1960s, confirmed that significant sugar dust and spilled sugar in the packing buildings and silo penthouse had been long-standing problems. As late as in 2007, quality audits and worker injury reports documented spilled sugar at "knee deep" levels in some areas.

Workers did recall local explosions in the milling room on the fourth floor in the 1960s, but they did not cause any widespread damage. More recently small fires sometimes had occurred in the buildings, but they were quickly extinguished without escalating to major incidents. Less than two weeks before the 2008 catastrophe a small explosion in a dry dust collector on the roof of the packing building damaged the dust collector, but was safely vented through the deflagration vent panels. The dust collector had not yet been reinstalled at the time of the 2008 catastrophe.

During the investigation in 2008 operators reported that the buckets inside bucket elevators sometimes broke loose and fell to the bottom of the elevator. They could recall one incident, some 10 years earlier, in which a falling bucket was thought to be the ignition source of a local fire involving a bucket elevator in the raw sugar warehouse.

7.4.8.2.2 The Sugar Refinery Plant at Time of the 2008 Catastrophe

Refined sugar was stored in three concrete silos (see Figs. 7.55 and 7.56) and then transferred to the bulk sugar truck and train loading area, to the packing buildings, and to the powdered sugar production plant, which was located in the south packing building. Packaged products were palletized, and transferred to a warehouse for distribution to customers.

Dozens of screw conveyors, bucket elevators, and horizontal conveyor belts transported granulated sugar throughout he packing buildings. Granulated sugar was scooped from a pit, or directly flowed into the buckets at the base of the bucket elevator and poured out of the bucket into a discharge chute at the top of the elevator housing. The refined sugar, either granulated or powdered, entered the top feed opening at one end. The rotating helical screw conveyor pushed the sugar to the opposite end of the conveyor where it exited through a bottom discharge chute. Although the bucket elevator screw conveyors were enclosed, they were not adequately sealed to prevent release of sugar dust into the work areas. Because the large open work areas were not equipped with airborne dust-removal equipment, sugar dust accumulated on overhead conduits, piping, ceiling beams, lights, and equipment. The closed granulated-sugar screw conveyors located throughout the facility were not equipped with dust-removal equipment and were not designed to safely vent overpressure outside the building if combustible dust inside the enclosure ignited.

7.4.8.2.3 Granulated Sugar Storage Silos

Three concrete silos, each of diameter 12 m, height 32 m, and capacity 2300 tonnes, conditioned and stored the granulated sugar produced in the refinery. The silos are illustrated in Figs. 7.55 and 7.56 (new). Each of the silos rested on an elevated circular concrete foundation above the packing building floor. Two belt conveyors were located under the silo floors inside a 40-m long tunnel of height 2 m and width 3.7 m. Granulated sugar from the refinery entered silo No. 3 and then was transferred into silos 1 and 2. The sugar was transferred from silos 1 and 2 to the bulk sugar building, powdered sugar mills, specialty sugar production equipment, and granulated-sugar packing machines. Sugar exited silo 3 into an Aerobelt conveyor in the silo tunnel, which discharged the sugar into the west bucket elevator pit. An Aerobelt conveyor is a special dust-tight conveyor belt assembly with a fiber belt traveling on an air cushion. The conveyor is equipped with dust-collector attachment ports for removing airborne dust inside the enclosure, which in the present case were connected to wet dust collectors. The bucket elevator lifted the sugar to the penthouse above the silo tops and onto another series of conveyor belts that loaded silos 1 and 2. Granulated sugar was discharged from the two silos through two 0.5-m diameter penetrations in the silo floors onto a steel conveyor belt in the silo tunnel. Separate 0.5-m diameter floor openings fed sugar to the screw conveyors under the north and south quadrants of each silo. Various conveyors and bucket elevators distributed the sugar to the powdered sugar mills, the packaging equipment, and the bulk sugar building. Granulated sugar was loaded into semitrailers and railcars in the bulk sugar building.

7.4.8.2.4 Conveyor Belt for Silos 1 and 2

A 0.8-m wide, 24-m long steel conveyor belt in the silo tunnel transported granulated sugar from silos 1 and 2 into the pit of the packaging production bucket elevator at the east side of silo 1. Similar steel belt conveyors were located in the penthouse above the silos.

For many years, granulated sugar on these conveyors was exposed to possible contamination from debris that could fall onto the sugar. In 2007, a stainless-steel frame with top and side panels to fully enclose each belt assembly had been installed. This was done to protect the granulated sugar from falling debris and reduce the possibility of intentional contamination. The top and side panels of the steel frame could be removed to provide access for cleaning and maintenance. Although sugar dust was generated as the sugar flowed onto the belt and when flow blockages caused sugar to spill off the belt, these new enclosures were not equipped with a dust-removal system or dust-explosion vents.

During more than 80 years of operation without this new enclosure, the sugar dust released inside the much larger open volume of the silo tunnel most probably never accumulated to concentrations above the minimum explosible concentration (MEC). However, after the steel belt was enclosed, sugar dust was contained and remained suspended inside the unventilated enclosure. Because the volume of the enclosure was only 10% of the volume of the tunnel, airborne sugar dust inside the enclosure could easily accumulate to concentrations above the MEC. Furthermore, multiple potential dust cloud ignition sources were identified inside the enclosure.

7.4.8.2.5 Bosch Packing Building

The four-story Bosch packing building located on the north side of the silos was a steel-frame structure on a concrete floor, covered with corrugated steel panels. Large doorways at the southeast corner on each floor gave forklift and personnel access to the south packing building. A steel-framed stairway enclosed in a concrete masonry block wall was located between silos 1 and 2 to provide access to each floor and the roof.

Granulated sugar was transported by screw conveyors and a bucket elevator to the screens inside the "hummer room" located on the roof of the building. The hummer room contained granulated-sugar sieves, which removed lumps and separated the sugar into standard grain sizes. Screw conveyors transported the screened sugar to feed hoppers on the third floor above each packaging machine. The second floor contained granulated-sugar packaging machines and plastic packaging supplies. Conveyors transported the packaged sugar westward to the palletizer room.

7.4.8.2.6 South Packing Building

The south packing building was a four-story, steel-frame structure with 8-cm thick poured concrete floors and brick exterior walls. The three large concrete silos formed the northern wall. A stairway enclosed by steel-frame/brick walls was located between silos 1 and 2 to provide access to the various floors. The east end of the packing building opened into the refinery areas. Heavy, semitransparent plastic strip curtains for containing airborne dust were installed on the large passageways.

The second and third floors housed granulated, powdered, and packaging equipment for soft (brown) sugar. Screw conveyors and bucket elevators transported the granulated sugar to hoppers above the packaging equipment throughout the packing building. Belt and roller conveyors transported the packaged sugar products to the palletizer room, some passing through large openings in the floors and west wall. Local explosions and fires could easily spread to adjacent areas through openings in the floor and walls.

The equipment for production of powdered sugar from granulated sugar was located on the fourth floor of the building. A small amount of about 3% of cornstarch was added to the granulated sugar to prevent the sugar from clumping. Then the mixture was fed into the grinding mills, where the sugar/cornstarch mixture was pulverized to the specified grain size. Then the powdered sugar was discharged into hoppers above the packaging machines. The grinding mills were connected to the dry dust collector system to remove airborne sugar and cornstarch dust.

7.4.8.2.7 Sugar Dust Spillage and Control

Dust-extraction systems were attached to the packaging machines, the sugar screening equipment, and the mills for powdered-sugar production, to remove sugar dust. The dust-extraction ducts were connected to dry dust collectors. However, it was found that the maintenance of the dust-collection system had been poor. In addition, some equipment was significantly undersized or incorrectly installed. Some dust ducts were found to be partially, and in some locations, completely filled with sugar dust.

The plant had hired an outside consultant to evaluate airflows, pressure drops, and other operating parameters on both the dry and the wet dust-collection systems. The report identified numerous design and maintenance deficiencies. Because the report was delivered only a few days before the 2008 catastrophe, there had been no opportunity to review or act on it.

In addition, during the investigation following the 2008 catastrophe, it was found that sugar conveying and processing equipment were not adequately sealed to prevent spillage of significant quantities of sugar onto the adjacent floors. Less than 2 months before the catastrophe, an internal inspection by company supervisors and quality assurance personnel learnt that many tonnes of spilled sugar had to be removed from the floors at intervals and returned to the refinery for reprocessing. Packaging operators and other employees also reported significant amounts of sugar dust escaping the packaging equipment into the working areas.

The systems for conveying maize starch to be mixed with the granulated sugar, the grinding mills, and the powdered-sugar packaging machines all generated significant quantities of sugar dust and maize starch in the work area. Workers reported that airborne sugar dust and spilled sugar in the powdered-sugar processing and packaging work areas were a constant problem and that significant accumulations were often seen on equipment and on the floor. One worker told that he used a squeegee to clear a path on the floor through spilled powdered sugar to get to equipment he operated during his shift.

7.4.8.3 Possible Steps of Development of the 2008 Catastrophic Explosion

The CSB (2009) concluded that the 2008 dust explosion catastrophe consisted of a combination of one primary and multiple secondary dust explosions. They further concluded that the secondary explosions would have been highly unlikely, had the company performed routine maintenance on the sugar conveying and packaging equipment to minimize dust releases and sugar spillage, and promptly removed accumulated dust and spilled sugar. When sugar dust and sugar escaped the equipment in the packing buildings, adequate housekeeping should have been effectuated to remove accumulations from elevated horizontal surfaces and floors before they reached hazardous levels.

The investigations revealed that the primary dust explosion occurred in the tunnel under the three large storage silos (see Figs. 7.55 and 7.56). This explosion then sent overpressure waves into the three silo pant-leg rooms, and further out into the first floor of the Bosch building. Each pant-leg room had up to four 0.5-m diameter holes through the floor into the tunnel below. Explosion pressure waves could then propagate between silos 1 and 2 and exit into the south stairwell, blowing the brick walls out into the packing building. Sugar dust accumulations on elevated surfaces and floors around the packaging equipment were then dispersed into explosive dust clouds that propagated the explosion(s) further through the plant.

Big fireballs extended above the entire facility, fueled by sugar dust dislodged from overhead equipment by the pressure waves ahead of the propagating dust flame. The pressure waves also violently heaved concrete floors upward throughout the south packing building.

7.4.8.4 Consequences of Enclosing the Steel Belt Conveyors

Workers had experienced that conveying granulated sugar on the steel belt conveyor in the tunnel under silos 1 and 2 generated some sugar dust. They had also reported that big sugar lumps sometimes got trapped in a silo outlet pipe and blocked the movement of the sugar on the belt. Then sugar would spill off the belt and generate clouds of airborne dust. Before the steel belt conveyor was enclosed, the dust was released directly into the much larger volume of the tunnel. The airflow through the tunnel would then tend to keep the dust concentration in the air low. Hence, before the belt conveyor got enclosed, the concentration of airborne sugar dust in the tunnel would probably not accumulate to explosive concentrations.

However, in 2007, the granulated-sugar belt conveyors in the penthouse and the tunnel under the silos were enclosed to address sugar contamination concerns. However, the hazards associated with generating and accumulating combustible dust inside the new enclosure were not adequately addressed. A dust-removal system to ensure that sugar dust did not reach the MEC inside the enclosure was not installed. The volume of the new steel belt enclosures was 10 times smaller than the silo tunnel volume, and airborne sugar dust inside the unventilated enclosures could easily accumulate to above the minimum explosive concentration.

Furthermore, the enclosures were not equipped with explosion vents to relieve possible explosion overpressures safely out of the building. This could have been accomplished for the enclosure on the belt conveyors in the penthouse, but accommodation of an explosion venting system for the enclosed belt conveyor in the silo tunnel would have been difficult, because the conveyor was located on the ground floor in the center of the building far away from any exterior wall or roof.

7.4.8.5 Possible Ignition Sources

Combustible dust can be ignited by electrical sparks or static discharge, hot surfaces, open flames, or friction-induced hot surfaces or sparks. Controlling possible ignition sources wherever combustible dust is present helps minimize the chance of a fire or explosion.

7.4.8.5.1 Open Flames and Hot Surfaces

Imperial Sugar had policies and procedures to control open flames. Workers were permitted to smoke only in designated areas, which were away from the production or packaging equipment. The hot-work-permit procedure required the workers to control combustible materials and post a fire watch whenever they performed welding or other hot work.

Photographs taken in 2006 showed large electric motors covered with sugar, a condition that could cause the motors to overheat and possibly cause ignition of airborne sugar dust. It was assumed that rather poor housekeeping procedures probably contributed to, or directly resulted in, some of the small sugar fires that had sometimes occurred. Workers told that, occasionally, small incipient-stage fires occurred in the packing buildings when sugar or packaging materials was ignited by a hot electrical device or a hot bearing (incipient-stage fire is a fire that is at its initial stage and can be controlled or extinguished by portable fire extinguishers). However, none of these earlier incidents ever resulted in a large fire or involved airborne dust. The CSB ruled out that an open flame could have been the initial ignition source of the 2008 catastrophe.

7.4.8.5.2 Potential Ignition Sources Inside the Steel Belt Enclosure

The CSB concluded that the initiating event was most probably ignition of an explosive concentration of sugar dust within the steel belt enclosure under silos 1 and 2 (see Figs. 7.55 and 7.56). Several possible ignition sources were identified during examination of the damaged steel belt and discussions with operations and maintenance personnel. However, the extent of the destruction from the explosions and long-burning afterfires prevented identification of a likely ignition source for the primary explosion.

7.4.8.5.2.1 Electric Spark Ignition The four limit switches on the steel belt conveyor were located inside the newly installed enclosure, and therefore they were exposed to sugar-dust clouds. Examination of two of the four switches that survived the fires showed that they were rated explosion proof. Therefore, it was concluded that these switches most probably were not the ignition source of the primary dust explosion.

7.4.8.5.2.2 Metal Particle Sparks A rather unlikely ignition source would be metal sparks from sliding contact between a steel belt and a jammed support wheel or a steel enclosure frame. Experiments by Dahn (1986) showed that sparks from sliding contact between objects of carbon steel are unlikely to ignite combustible dusts unless the contact speed exceeds 9.2 m/s. In the present case, speed of the steel conveyor belt was estimated to be less than 1.5 m/s which is probably significantly lower than the minimum speed required for igniting the dust at hand.

7.4.8.5.2.3 A Hot Surface the Most Likely Ignition Source For clouds of sugar dust in air the minimum ignition temperatures range from 360 to 420°C depending on the test apparatus used and other factors. Explosive concentrations of airborne sugar dust was probably always present when sugar was being transported inside the enclosed steel belt conveyors, especially when blockages caused the granulated sugar to spill off the steel belt.

According to operators, bearings on the steel belt roller supports sometimes got very hot due to malfunction. Such a hot bearing inside an enclosed steel belt conveyor might be able to ignite airborne sugar dust, and was considered the most likely ignition source of the primary explosion leading to the 2008 sugar-dust explosion disaster. In addition, sugar layers on sufficiently hot surfaces will decompose and release combustible gases and vapors, which could increase the likelihood of accidental ignition.

7.4.8.6 Summary of Basic Causes of the 2008 Explosion/Fire Catastrophe

- Sugar- and cornstarch-conveying equipment was not designed or maintained to minimize the release of sugar dust into the work area
- Inadequate housekeeping practices resulted in significant accumulations of combustible sugar dust on the floors and elevated surfaces throughout the packing buildings
- Airborne combustible sugar dust accumulated above the minimum explosible concentration inside the newly enclosed steel belt assembly under silos 1 and 2
- An overheated bearing in the steel belt conveyor most likely ignited a primary dust explosion
- The primary dust explosion inside the enclosed steel conveyor belt under silos 1 and 2 led to massive secondary dust explosions and fires throughout the packing buildings
- The 14 fatalities were most likely the result of the secondary explosions and fires

- Imperial Sugar emergency evacuation plans were inadequate. Emergency evacuation drills were not conducted, and prompt worker notification to evacuate in the event of an emergency was inadequate
- The secondary dust explosions, rapid spreading of the fires throughout the facility, and resulting fatalities would likely not have occurred if Imperial Sugar had enforced routine housekeeping policies and procedures to remove sugar dust from overhead and elevated work surfaces and remove the large accumulations of spilled sugar throughout the packing buildings
- The Port Wentworth facility risk assessment performed by Zurich Services Corporation in May 2007 and the report submitted to Imperial Sugar management did not adequately address combustible-dust hazards.

7.4.9 Catastrophic Aluminum-Alloy Dust Explosion in China in 2014

7.4.9.1 Overview

The following account is based on the extensive paper by Li et al. (2016a). On August 2, 2014 a catastrophic dust explosion occurred in a large industrial plant for polishing various metal parts, in Kunshan, China. When the explosion occurred the plant was polishing aluminum-alloy wheel hubs for the car industry. 75 people lost their lives immediately and another 185 were injured. Subsequently, 71 of the seriously injured also died, which increased the total loss of lives to 146. The direct economic loss of was 351 million yuan. This is probably one of the most serious dust-explosion catastrophes in human history apart from a number of very major coal-dust explosion disasters in coal mines.

7.4.9.2 The Plant That was Struck by the Explosion

7.4.9.2.1 The Building

The explosion occurred in a two-story reinforced concrete-frame structure process building of length 44 m (from north to south) and width 24 m (from east to west). The two floors comprised a basement with concrete floor and a first floor above it. The height between floor and ceiling was 4.5 m in both floors. The total floor area (both floors) was 2112 m^2. The roof of the building (and hence also of the first floor) was a light steel beam structure covered by steel panels. The external and internal walls of the building were of brick. The two floors were connected by open staircases at each end (north and south) of the building. On both sides of the eastern wall there was a 4 m $\times$ 4 m steel-panel sliding door leading to the outside.

7.4.9.2.2 Process Equipment Layout

The 32 polishing production lines (16 lines on the basement and 16 lines on the basement) were arranged in parallel in the south—north direction, as shown in Fig. 7.57 Along each line there were 12 working stations, as also indicated in Fig. 7.57.

On the day before the accident, a total of 29 production lines were in operation, 13 on the first floor and 16 on the first floor; 348 workers were on duty. Polishing operations were conducted manually as shown in Fig. 7.58. Electric grinding guns were the main tools. According to the finish smoothness required, different grades of grinding heads and/or emery papers were used.

7.4.9.2.3 Dust-Collection System and Processing Equipment

Eight sets of similar dust-collection systems were installed in March 2006 outside the main factory building to serve the polishing processes on the two floors. Therefore, every two production lines shared a set of dust collection systems. Every two pieces of dust-extraction ducts were merged by a T-joint into one main 450 mm Φ main duct leading to the bag filter. Hence, each dust extraction system collected the dust from 48 work locations.

According to the design of the bag filters in the dust collectors, the bags were supposed to be cleaned by mechanical shaking at intervals. However,

Figure 7.57 Production lines with work stations.
From Li, G., Yang, H., Yuan, C., Ge, S., Eckhoff, R.K., 2016. A catastrophic aluminium-alloy dust explosion in China in 2014. Journal of Loss Prevention in the Process Industries 39, 121—130.

Figure 7.58 Workers on duty at the work stations.
From Li, G., Yang, H., Yuan, C., Ge, S., Eckhoff, R.K., 2016. A catastrophic aluminium-alloy dust explosion in China in 2014. Journal of Loss Prevention in the Process Industries 39, 121–130.

after the explosion accident survivors told that the automatic filter-bag shaking system had been out of operation since a long time before the accident, due to breakdown of the electric driving motor. Instead workers had to clean the bags manually every morning before commencing work by vibrating the carriage of the bags. This process was called "shaking ash."

The airflow for each of the eight main dust-extraction lines was produced by a suction fan mounted on the clean side of the bag filter unit, and all eight airflows were joined in a main discharge duct to the atmosphere.

No special requirements addressing a possible dust-explosion hazard had been enforced on selection and installation of all the electrical equipment used in the plant. Neither the dust collectors, the dust-extraction ducting, nor all electrical sockets and power distribution cabinets had been adequately earthed.

7.4.9.3 Explosion Development

The series of strong explosions occurred in the morning when normal hub polishing activity had been going on for half an hour. A survivor stated that he was polishing his second hub at the moment of the first explosion. A video camera located outside another factory building about 500 m away from the building that was stuck by the explosion revealed a sequence of several explosions lasting for about 5–7 s, including a distinct series of eight successive explosions. These eight explosions could be identified on the video recording as violent "mushroom-shaped" dust/smoke clouds being expelled abruptly from each of the eight dust collectors. All the windows in the first floor of the building that exploded were shattered and blown to the outside, and the window frames were completely deformed. Two-thirds of the southern wall of the building collapsed. Window frames were blown out of the eastern wall, and two air conditioners located at this wall became partly detached from the wall. The blasts also lifted some of the light-color steel roof of the second floor, and broke all the windows, surprisingly, the glass scattered on the ground inside. Two-thirds of the southern wall of the building collapsed. Some of the damage is shown in Fig. 7.59.

Almost all process equipment in the workshop was destroyed. Fig. 7.60 shows the total damage of the process lines on the second floor.

The dust-extraction ducts inside the building were intact, but the duct connected to the #1 filter suffered serious damage and got detached from the T-joint. A duct branch within the workshop was broken into three parts with the middle part of it having been torn apart. The dust filters (#1, #3, #5, and #7) collecting dust from the first floor were more severely damaged than those (#2, #4, #6, and #8) collecting dust from the second floor. In particular, dust filter #1 was blown entirely apart.

The dust-extraction ducts inside the building were intact, but the duct connected to the #1 filter suffered serious damage and became detached from the T-joint. As Fig. 7.61 illustrates, the system of dust-extraction ducts in the production halls was torn apart by the explosion.

Figure 7.59 Collapsed and destroyed eastern wall and air conditioner.
From Li, G., Yang, H., Yuan, C., Ge, S., Eckhoff, R.K., 2016. A catastrophic aluminium-alloy dust explosion in China in 2014. Journal of Loss Prevention in the Process Industries 39, 121–130.

Figure 7.60 Destroyed process lines on the second floor.
From Li, G., Yang, H., Yuan, C., Ge, S., Eckhoff, R.K., 2015. A catastrophic aluminium-alloy dust explosion in China in 2014. Journal of Loss Prevention in the Process Industries (in press).

Figure 7.61 Detailed view of damage of dust-extraction ducting.
From Li, G., Yang, H., Yuan, C., Ge, S., Eckhoff, R.K., 2015. A catastrophic aluminium-alloy dust explosion in China in 2014. Journal of Loss Prevention in the Process Industries (in press).

7.4.9.4 Probable Path of Explosion Development

According to the evidence found on site, the explosion started in filter #1 connected to the first floor in the factory building. The blast wave, followed by burning dust cloud, then propagated upstream inside the dust-extraction duct and entered the first-floor working lines. From there it exited into the main process hall via all the dust-extraction hoods above all the working tables, and eventually ignited the dispersed deposited dust on the floor near both the #1 and #2 production lines, causing major escalating secondary explosions. Then these explosions and associated dust flames propagated the entire length from north to south along the first floor and were sucked into the dust-extraction ducts leading to the #3, #5, and #7 dust filters, triggering three further major filter explosions. The explosion also propagated up to the second floor via a staircase and swept through that floor as well.

7.4.9.5 Probable Ignition Source of the Initial Primary Explosion

Two significant details were noted. First, the dust-collecting barrel of this filter had its bottom completely blown out. Second, another smaller hole was found in the barrel wall that was probably not caused by the explosion but rather by corrosion over a long period prior to the explosion. The filters and the barrels underneath were located outdoors. It had been raining heavily for 2 days before the explosion accident, and it still rained lightly at the time of the accident. Water may have entered the barrel through the corroded hole and moistened the aluminum-alloy dust in the barrel. This water probably was a decisive factor in the process leading to self-ignition of the contaminated aluminum-alloy dust in the barrel according to the reaction equation:

$$2Al + 6H_2O = 2Al(OH)_3 + 3H_2 + \text{heat} \qquad [7.6]$$

Laboratory experiments conducted after the explosion accident, which support this hypothesis, are described by Li et al. (2016). An important detail must be emphasized. The aluminum-alloy dust in the barrel was contaminated by some organic material from the wheel hub grinding/polishing process (polishing waxes, etc.). This made the dust "fluffy" with a much lower bulk density than in deposits of pure dry aluminum-alloy dust. Therefore, the deposit had also a much lower thermal conductivity than would be expected for pure aluminum-alloy dust of the same particle-size distribution.

Another significant detail was observed during these laboratory experiments: After the self-ignition process had developed fully inside the wet contaminated metal dust deposit, open flames lasting for about 10 s were observed on top of the deposit surface. It seems reasonable to assume that these flames were due to burning of the hydrogen released according to Eq. [7.6] possibly mixed with pyrolysis gases from the organic matter in the deposit. Such flames may have played a central role in igniting the cloud of aluminum-alloy dust falling down from the filters in the dust collector during the shaking of the filters.

7.4.9.6 Recommended Actions for Preventing and Mitigation Similar Explosion Disasters

General ignorance of the potential risk of dust explosions, in industries producing fine metal dusts as a small-quantity waste product only, was identified as an important root cause of this catastrophic accident. Even in such plants excessive deposits of waste dust can build up over a long time on floors, shelves, beams, and process equipment. Therefore, good regular housekeeping to remove layers of accumulated waste dust is a most effective and practical way of preventing and mitigating serious secondary metal dust explosions in such plants. In addition, ignition source prevention and explosion isolation between dust-collecting systems and workrooms are important means for minimizing the risks of such explosions.

7.5 MEANS OF PREVENTING AND MITIGATING DUST EXPLOSIONS IN THE PROCESS INDUSTRIES

7.5.1 Overview

Table 7.2 gives an overview of the various means that are presently known and in use.

They can be divided in two main groups, namely means for preventing explosions and means for their mitigation. The preventive means can again be split in the two categories, prevention of ignition sources and prevention of explosive/combustible clouds. One central issue is whether only preventing ignition sources can give sufficient safety, or whether it is also necessary in general to employ additional means of prevention and/or mitigation. In the following sections, the various means of prevention and mitigation listed in Table 7.2 will be discussed separately.

7.5.2 Preventing Explosive Dust Clouds

7.5.2.1 Inerting by Adding Inert Gas to the Air

7.5.2.1.1 Outline of Basic Principle

European Committee for Standardization (CEN) (2004a) issued guidelines for design of systems for preventing gas/vapor and dust explosions by inerting. For a given type of dust suspended in air, and a given type of added inert gas, there is a "limiting oxygen concentration" (LOC) in the air/gas mixture below which the dust cloud is unable to propagate a self-sustained flame.

Table 7.2 Overview of means for preventing and mitigating dust explosions in the process industries

Prevention		
Preventing explosive dust clouds	**Preventing ignition sources**	**Mitigation**
Inerting by N_2, CO_2, and rare gases	Smoldering combustion in dust, dust flames	Reduce expl. cloud size
Intrinsic inerting	Other types of open flames (eg, hot work)	Partial inerting Isolation (sectioning)
Inerting by adding inert dust	Hot surfaces	Venting
Dust concentration outside explosive range	Electric sparks and arcs, electrostatic discharges	Pressure-resistant construction
	Heat from mechanical impact (metal sparks and hot spots)	Automatic suppression Good housekeeping (dust removal/cleaning)

As the oxygen content in the gas is gradually reduced from that of air, ignitability and explosivity of the dust cloud are also gradually reduced, until ultimately flame propagation becomes impossible at the LOC. By keeping the oxygen content below this limit throughout the process system, the possibility of dust explosions is excluded. Four types of inert gases are in common use for this purpose:

- Carbon dioxide
- Water vapor
- Nitrogen
- Rare gases

In addition, as described in subsequent Section 7.5.2.1.2, inert gas mixtures produced in the plant itself, eg, combustion gases, can in some cases be used for inerting. In the past, halogenated hydrocarbons (halons) were also used, but due to the environmental problems caused by these substances, this is no longer permitted. The choice of inert gas depends on several considerations, such as availability and cost, possible contaminating effects on products, and effectiveness. In the case of dusts of light metals, such as aluminum and magnesium, exothermic reactions with CO_2 and also in some situations with N_2 are known, and the use of rare gases may have to be considered.

7.5.2.1.2 Intrinsic Inerting

In some cases, the special protective method "intrinsic inerting" can be a good solution. This implies that the required quantity of inert gas is produced in the plant itself, eg, by controlled combustion in a hot-gas generator and recirculation of the gas. Such hot gases mainly consist of nitrogen, carbon dioxide, and water vapor. The residual concentration of oxygen is kept at a sufficiently low level to ensure inert conditions. However, normally such combustion gases are not clean enough for use in connection with food and feed materials, pharmaceuticals, etc.

7.5.2.1.3 Design of Gas Inerting Systems

The design of gas inerting systems depends on whether the process is continuous or of the batch type, the strength of the process equipment, and type and source of inert gas. Two main principles are used for establishing the desired atmosphere in the process:

- Pressure variation method
- Flushing method

The *pressure variation method* operates either *above* or *below* atmospheric pressure. In the former case, the process equipment, initially filled with air at atmospheric pressure, is pressurized to a given overpressure by inert gas. When good mixing of air and inert gas has been obtained, the process equipment is vented to the atmosphere and the cycle repeated until the oxygen content has been reduced to the desired level. In the second case, the process equipment is first evacuated to a given underpressure, and then filled up with inert gas to atmospheric pressure. This cycle is repeated the number of times required for reducing the oxygen content to the desired level.

The *flushing method* is used if the process equipment has not been designed for the significant pressure increase or pressure reduction required by the pressure variation method. There are two extreme cases of the flushing method, viz. the *replacement method* (plug flow) and the *through-mixing method* (stirred tank). To maintain plug flow, the flow velocity of inert gas into the system must be low (<1 m/s) and the geometry must be favorable for avoiding mixing. In practice, this is very difficult to achieve, and the stirred tank method, using high gas velocities and turbulent mixing, is normally employed. It is essential that the instantaneous through-mixing is complete over the entire volume, otherwise pockets of unacceptably high oxygen concentrations may form.

7.5.2.2 Dust Concentration Outside Explosive Range

In principle, one could avoid dust explosions by running the process in such a way that explosive dust concentrations are avoided. In practice, however, this is difficult in most cases, because the dust concentration inside process equipment most often varies in unpredictable and uncontrollable ways. On the other hand, maintaining the powder/dust in the settled state by avoiding entrainment or fluidization in the air is one way of ensuring that the dust concentration is either zero or well above the upper explosive concentration. Good process design can significantly reduce the regions in which explosive dust concentrations occur, as well as the frequencies of their occurrence. One example is the use of mass-flow silos instead of the traditional funnel-flow type.

Nevertheless, it may be possible to actively keep the dust concentration below the lower explosive limit in some cases. One such situation is in dust-extraction ducts, another is in cabinets for electrostatic powder coating, and the third is spray dryers. One essential requirement for controlling dust concentration is that the concentration can be adequately measured. In situ methods based on light attenuation or backscattering of light have been found most suitable. The use of dust control in dust-extraction systems is most likely to be successful if a small dust fraction is to be removed from a coarse main product, eg, grain dust from grain, or plastic dust from pellets. By monitoring dust concentrations and controlling airflows, the desired level of dust concentration can be maintained. However, if the air velocities are too low to prevent dust deposition on the internal walls of the ducting over time, dust explosions may nevertheless be able to propagate through the ducts.

Dust entrainment and formation of explosive dust clouds by the air blast preceding a propagating primary dust explosion may also occur in mixers, conveyors, etc. in which sufficient quantities of fine dust are present as layers/deposits. This means that explosion control by controlling the concentration of suspended dust is only feasible for preventing the primary explosion initiation, and not for preventing secondary explosions.

7.5.2.3 Adding Inert Dust

This principle is used in coal mines, by providing sufficient quantities of stone dust either as a layer on the mine gallery floor, or on shelves, etc. The blast that will always precede the flame in a dust explosion will then entrain the stone dust and coal dust simultaneously and form a mixture that is noncombustible in air, and the flame, when arriving, will become quenched. In industries other than mining, adding inert dust is seldom applicable due to product contamination and other problems.

7.5.3 Preventing Ignition Sources

7.5.3.1 Introduction

A question asked frequently is whether preventing ignition sources can be relied upon as the only means of protection against dust explosions. The general consensus is that this may be possible in certain cases, but not in general. Relying on preventing ignition sources only is definitely inadequate if the minimum electric spark ignition energy of the dust is in the region of vapors and gases, ie, <10 mJ. However, for dusts of very high MIEs, it may be argued that several types of process plants could be satisfactorily protected against dust explosions solely by eliminating ignition sources.

According to Scholl (1989), one may distinguish between two categories of ignition sources. Organizational ignition sources, which can largely be prevented by enforcing adequate working routines, include:

- Smoking
- Open flames
- Open light (bulbs)
- Welding (gas/electric)
- Cutting (gas/rotating disc)
- Grinding

Operational ignition sources arise within the process itself and include:

- Open flames
- Hot surfaces
- Self-heating and smoldering nests
- Exothermic decomposition
- Heat from mechanical impact between solid bodies (metal sparks/hot-spots)
- Exothermic decomposition of dust via mechanical impact
- Electric sparks/arcs, electrostatic discharges

7.5.3.2 Preventing Self-Heating, Smoldering, and Burning in Large Dust Deposits

The risk of self-heating in powder/dust deposits depends on the inherent properties of the material. Therefore, possible disposition to self-heat should be known or assessed for any material before admitting it to storage silos or other part of the plant in which conditions are favorable for self-heating and further temperature rise to smoldering and burning.

Possible means of preventing self-heating and self-ignition in powders with a disposition to self-heat include:

- Control of temperature, moisture content, and other important powder/dust properties before admitting powder/dust to, eg, storage silos. Adjustment of powder/dust properties to acceptable levels by cooling, drying, etc., whenever required
- Ensuring that hot solid bodies do not become embedded in the powder/dust mass
- Continuous monitoring of temperature in powder mass at several points by thermometer chains
- Rolling of bulk material from one silo to another, or within the same silo, whenever onset of self-heating is detected, or as a routine after certain periods of storage, depending on the powder/dust type
- Inerting of bulk material in silo by suitable inert gas, eg, nitrogen
- Monitoring of possible development of gaseous decomposition/oxidation products, eg, CO and methane, for early detection of self-heating.

Means for preventing and controlling self-heating/self-ignition, and methods for extinction of smoldering combustion inside large dust deposits, eg, in silos, are discussed in Chapter 6.

Some synthetic organic chemicals, in particular, cyclic compounds, can decompose exothermally and become ignited by a hot surface, a smoldering nest, frictional heat, or other ignition source. Such decomposition does not require oxygen, and therefore inerting has no effect. Adiabatic exothermal decomposition of bulk powder at constant volume can, due to the very high powder concentration, generate much higher pressures than a dust explosion in air. Zwahlen (1989) suggested the following alternative safety measures:

- Process the hazardous powder in the wet state, as a slurry or suspension.
- If wet processing is impossible, avoid processes involving moving mechanical parts in contact with the powder that can give rise to ignition.
- Keep the processed batches of the powder as small as feasible.
- Keep strict control to prevent foreign bodies from entering the process.
- Detectors for observing early temperature and pressure rise and sprinkler systems must be provided.
- Use of additives that suppress the decomposition tendency may be helpful in some cases.

7.5.3.3 Preventing Ignition by Open Flames/Hot Gases

Most potential ignition sources of the open-flame type can be avoided by enforcing adequate organizational procedures and routines. This in particular applies to prohibition of smoking and other use of lighters and matches, and to enforcement of strict rules for performing hot work. Hot work must not be carried out unless the entire area that can come in contact with the heat from the work, indirectly as well as directly, is free of dust, and hazardous connections through which the explosion may transmit to other areas, have been blocked. It is important to note that hot work also includes disc-cutting and grinding operations. Gas cutting torches are particularly hazardous because they work with excess oxygen. This gives rise to ignition and primary explosion development in which explosions in air would be unlikely. Factory inspectorates in most industrialized countries have issued detailed regulations for hot work in factories containing combustible powders or dusts.

In certain situations in the process industry, hot gaseous reaction products may entrain combustible dust and initiate dust explosions. Each such case has to be investigated separately and the required set of precautions tailored to serve the purpose in question.

7.5.3.4 Preventing Ignition by Hot Surfaces

Hot surfaces may occur in industrial plants both intentionally and unintentionally. The first category includes external surfaces of hot process equipment, heaters, dryers, steam pipes, and electrical equipment. The equipment on which hot surfaces may be generated unintentionally include engines, blowers, fans, mechanical conveyors, mills, mixers, bearings, and unprotected light bulbs.

A further category of hot surfaces arises from hot work. During grinding and disc-cutting, glowing hot surfaces are often generated, which may be even more effective as initiators of dust explosions than the luminous spark showers typical of these operations. A hot surface may ignite an explosive dust cloud directly, or via ignition of a dust layer that subsequently ignites the dust cloud. Parts of glowing or burning dust layers may loosen and be conveyed to other parts of the process where they may initiate explosions.

The hot surface temperatures of various apparatuses can, if the surface is covered by a dust layer, be significantly higher than it would be without dust, due to thermal insulation by the dust. This both increases the ignition hazard and may cause failure of the apparatus due to increased working temperature. The measures taken to prevent ignition by hot surfaces must cover both modes of ignition, ie, both direct ignition of the explosive dust cloud, or via ignition of a dust layer that subsequently ignites the cloud. The recommended measures include:

- Removal of all combustible dust before performing hot work
- Prevention/removal of dust accumulations on hot surfaces (electrical apparatuses etc.)
- Isolation or shielding of hot surfaces
- Use of electrical apparatus approved for use in the presence of combustible dust
- Use of equipment with minimal risk of overheating
- Inspection and maintenance procedures that minimize the risk of overheating

7.5.3.5 Preventing Ignition of Dust Clouds by Smoldering Nests

As discussed in Section 7.3.2, Pinkwasser (1985, 1986) studied the possibility of dust explosions being initiated by smoldering lumps ("nests") of powdered material that were conveyed through a process system.

Infrared radiation detection and subsequent extinction of smoldering nests and their fragments during pneumatic transport in dust-extraction ducts can be an effective means of preventing fire and explosions in downstream equipment, for example dust filters. Normally the transport velocity in the duct is known, and this allows effective extinction by synchronized injection of a small amount of extinguishing agent just when the smoldering/burning nest or fragment passes the injection nozzles located at the optimal distance downstream of the detection point. Most often, water is used as an extinguishing agent, and it is applied as a fine mist. Such extinction systems are quite common in the wood industries, but also to some extent in the food, feed, and some other industries.

7.5.3.6 Preventing Ignition by Heat From Accidental Mechanical Impacts

Mechanical impacts can in principle produce two different kinds of potential ignition sources. The first is small flying fragments/particles of burning solid material, most often metal, which have become torn away and from the bulk and ignited by the impact. The second is hot-spots in which the impacting bodies touch. Sometimes, eg, in rotating machinery, impacts may occur repeatedly at the same points on one or both of the impacting bodies and this may give rise to hot-spots of appreciable size and temperature. In the latter case the hazardous source of ignition will then be a hot surface, and what has been said in Section 7.5.3.4 applies (see also Section 2.2.5 of Chapter 2 and Section 7.3.5).

With regard to single accidental impacts, research has revealed that, in general, the ignition hazard associated with single accidental impacts is considerably smaller than often believed in the past. This, in particular, applies to dusts of natural organic materials such as grain and feedstuffs, when being exposed to accidental sparking from impacts between steel hand tools like spades or scrapers, and other steel objects or concrete. In such cases, the ignition hazard is probably nonexistent.

However, if metals that are more sophisticated are involved, such as titanium or hard aluminum alloys, energetic spark showers can be generated. In the presence of rust, luminous, incendiary thermite flashes can result. Thermite flashes may also result if a rusty steel surface covered with aluminum paint or a thin smear of aluminum is struck with a hammer or another hard object. However, impact of ordinary soft unalloyed aluminum on rust seldom results in thermite flashes, but just in a smear of aluminum on the rust. For a given combination of impacting materials, the incendivity of the resulting sparks or flash depend on the sliding velocity and contact pressure between the colliding bodies.

Although the risk of initiation of dust explosions by accidental single impacts is probably smaller than believed by many in the past, there are special situations in which the ignition hazard is real. It would in any case seem to be good engineering practice to:

- Remove foreign objects from the process stream as early as possible
- Avoid construction materials that can give incendiary metal sparks or thermite flashes (titanium, magnesium, aluminum, etc.)
- Whenever special noise signals indicating accidental impact(s) in process stream are observed, inspect process and remove cause(s) of impact immediately in a safe way

7.5.3.7 Electric Sparks and Arcs and Electrostatic Discharges

The various types of electric sparks and arcs and electrostatic discharges are described in Section 2.2.6 of Chapter 10 and Section 7.3.6. Sparks or arcs due to breakage of live circuits can occur when fuses blow, in rotating electric machinery, and when live leads are accidentally broken. The main rule for minimizing the risk of dust explosions due to such sparks and arcs is to:

- Obey regulations for electrical installations in areas containing combustible dust (see Chapter 10)

The electrostatic hazard is more complex and it has not always been straightforward to specify clearly defined design guidelines. However, Glor (1988), having contributed substantially to developing a unified approach, recommended the following measures:

- Use conductive materials or materials of low dielectric strength, including coatings, (breakdown voltage across dielectric layer or wall <4 kV) for all plant items that may accumulate very high charge densities (pneumatic transport pipes, dust deflector plates, and walls of large containers that may become charged due to ionization during gravitational compaction of powders). This prevents propagating brush discharges.
- Earth all conductive parts of equipment that may become charged. This prevents capacitive spark discharges from equipment.
- Earth personnel if powders of minimum ignition energies (MIE) < 100 mJ are handled. This prevents capacitive spark discharges from humans.
- Earth electrically conductive powders (metals, etc.) by using earthed conductive equipment without nonconductive coatings. This prevents capacitive discharges from conductive powder.
- If highly insulating material (resistivity of powder in bulk $> 10^{10}$ Ωm) in the form of coarse particles (particle diameter > 1 mm) is accumulated in large volumes in silos, containers, hoppers, etc., electrostatic discharges from the material in bulk may occur. These discharges can be hazardous when a fine combustible dust fraction of minimum ignition energy < 10 mJ is present simultaneously. So far, no reliable measure is known to avoid this type of discharge in all cases, but an earthed metallic rod introduced into the bulk powder will most probably drain away the charges safely. It is, however, not yet clear whether this measure will always be successful. Therefore, under these circumstances, the use of inerting (see Section 7.5.2.1) or explosion venting or explosion suppression (see Sections 7.5.4.4 and 7.5.4.6), should be considered.
- If highly insulating, fine powders (resistivity of powder in bulk $> 10^{10}$ Ωm) with a minimum ignition energy ≤10 mJ as determined with a low-inductance capacitive discharge circuit, has accumulated in large volumes in silos, containers, hoppers, etc., measures of explosion mitigation should be considered (see following Section 7.5.4). There is no experimental evidence that fine powders without any coarse particles will generate discharges from powder heaps, but several explosions have been reported with such powders in situations in which all possible ignition sources, with the exception of electrostatics, have been effectively eliminated.

If combustible powders are handled or processed in the presence of a flammable gas or vapor (*hybrid mixtures*), the use of electrically conductive and earthed equipment is absolutely essential. Insulating coatings on earthed metallic surfaces may be tolerated provided that the thickness is less than 2 mm, the breakdown voltage less than 4 kV at locations where high surface charge densities have to be expected, and conductive powder cannot become isolated from earth by the coating. If the powder is nonconducting (resistivity of the powder in bulk $> 10^6$ Ωm), measures of explosion prevention (eg, inert gas blanketing) are strongly recommended. If the resistivity of the powder in bulk is less than 10^6 Ωm, brush discharges, which would be incendiary for flammable gases or vapors, can also be excluded.

However, experience has shown that even in the case of powders of resistivities in bulk $< 10^6$ Ωm, it is very difficult in practice to exclude all kinds of effective ignition sources when flammable gases or vapors are present. In such cases, large amounts of powders should only be handled and processed in closed systems blanketed with an inert gas.

Glor also emphasized that, due to increasing use of nonconducting construction parts in modern industrial plants, the chance of overlooking unearthed conducting items is high. Therefore, the effort to ensure proper earthing of all conducting parts must be maintained, in particular in plants handling dusts of low MIE. Adequate earthing is maintained as long as the leak resistance to earth does not exceed 10^6 Ω for process equipment and 10^8 Ω for personnel. However, in practice, one should aim for considerably lower resistances to earth.

7.5.4 Mitigating Dust Explosions That Are Initiated Despite Preventive Measures

7.5.4.1 Reducing Sizes of Explosive Dust Clouds by Good Process Design

7.5.4.1.1 Minimizing Volumes of Process Equipment

This measure can be considered part of developing inherently safer plants (see Section 7.5.4.9). Then the volumes of process equipment should not be larger than the volumes required by the process. Nevertheless, one often finds industrial plants with eg, silos that are considerably larger than required by the process. This can either be due to inadequate design in the first place, or due to the plant being used for another purpose than originally designed for.

7.5.4.1.2 Minimizing Volumes of Dust Clouds Generated at Transfer Points

This measure can also be considered as part of applying the concept of inherently safer plants (see subsequent Section 7.5.4.9). Undesired dust clouds are practically always generated when powder/dust/pelletized material, etc. is falling freely under gravity. Whenever possible, therefore, efforts should be made to design transfer points in such a way that the material is flowing smoothly in bulk, rather than being dispersed as a cloud. For example, by having an inclined chute at transfer points between chain or belt conveyors, dusting can be reduced considerably. Another example is the very smooth discharge of material from a silo on to a chain/belt conveyor, which can be obtained if the silo hopper is designed to produce mass flow rather than funnel flow.

7.5.4.2 Partial Inerting by Inert Gas

In Table 7.2, partial inerting, as opposed to complete inerting discussed so far, has been included as a possible means of mitigating dust explosions. The concept, discussed by Eckhoff (2004), implies that a smaller fraction of inert gas than that required for complete inerting is added to the air. In this way, the ignition sensitivity, the explosion violence, and the maximum constant-volume explosion pressure will be reduced, in some cases appreciably. This offers a new possibility for applying mitigatory measures such as explosion venting or automatic explosion suppression in situations in which the explosion violence of the dust in air only is too severe to permit the use of such techniques. More research is needed to establish correlations between the oxygen content in the gas phase and various ignitability and explosibility parameters of various dusts.

7.5.4.3 Isolation (Sectioning)

CEN (2004b) issued a standard for design of explosion isolation systems. In Section 2.4.3.1 of Chapter 2 three main reasons are given for trying to prevent a gas explosion in one process unit from spreading to others via pipes and ducts. These reasons also apply to dust explosions. First, there is always a desire to limit the extent of the explosion as far as possible. Second, a dust flame propagating in a duct between two process units can give rise to violent flame jet ignition of the dust cloud in the second volume. The third main reason is pressure piling, which implies the increase of the pressure of the unburnt dust cloud in the downstream process unit(s) to above atmospheric pressure, due to compression caused by the expansion of the hot combustion gases in the unit in which the explosion starts, as well as in the connecting duct(s) (see Section 2.4.5.2.1 in Chapter 2).

As for gases, two categories of methods are used for obtaining explosion isolation, viz. passive methods activated by the propagating explosion itself, and active ones, which require a separate flame/pressure sensor system, which triggers a separately powered system for operating the isolation mechanism. For obvious reasons, the passive systems are generally preferable, as long as they function as intended and are otherwise suitable for the actual purpose.

Passive isolation systems include the concept of flame propagation interruption in ducts by providing a vented 180° bend system, as illustrated in Fig. 7.62. This concept is used quite frequently to interrupt dust explosions in pipes and ducts.

The basic principle is that the explosion is vented at a point at which the flow direction is changed by 180°. Due to the inertia of the fast flow caused by the explosion, the flow will tend to maintain its direction rather than making a 180° turn. However, the boundaries for the applicability of the principle have not been fully explored. Parameters that may influence performance include explosion properties of dusts, velocity of flame entering the device, direction of flame propagation, and direction, velocity, and pressure of initial flow in duct. The removal of part of the screw in screw conveyors will ensure that a plug of bulk powder/dust will always remain as a choke. Therefore, the modified conveyor will prevent transmission of a dust explosion through the screw. It has in fact been demonstrated experimentally that in some cases even

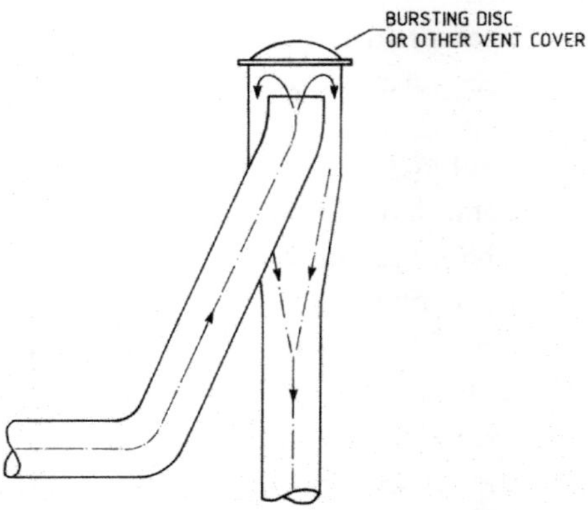

Figure 7.62 Passive device for interrupting dust explosions in pipes and ducts by combining change of flow direction and venting. Flow direction may also be opposite to that indicated by the arrows.
From Eckhoff, R.K., 2003. Dust Explosions in the Process Industries, third ed. Gulf Professional Publishing/Elsevier. (first ed. published in 1991, second ed. in 1997). ISBN:0-7506-7602-7.

unmodified screw conveyors can effectively stop the propagation of a dust explosion. Specially designed rotary locks are also used for preventing explosion transfer between process units or a process unit and a duct.

Active isolation methods include various kinds of fast-response mechanical valves. The required closing time of an automatic explosion isolation valve depends on the distance between the remote pressure or flame sensor and the valve, and on the type of dust. Often closing times as short as 50 ms, or even shorter, are required. Often this is obtained by using an electrically triggered explosive charge for releasing the compressed air or nitrogen that operates the valve. The slide valve must be sufficiently strong to resist the high pressures of 5–10 bar (g) that can occur on the explosion side after valve closure (in the case of pressure piling effects and detonation, the pressures may transiently be even higher than this).

Another active explosion isolation method is flame interruption by fast automatic injection of extinguishing chemicals ahead of the flame in pipes connecting process units. This is a special application of automatic explosion suppression, which will be described in Section 7.5.4.5. Important design parameters are type of dust, initial turbulence in primary explosion, duct diameter, distance from vessel in which primary explosion occurs, method used for detecting onset of primary explosion, and type, quantity, and rate of release of extinguishing agent.

7.5.4.4 Dust Explosion Venting

7.5.4.4.1 Main Principle

CEN (2002a) issued a draft standard for design of dust explosion venting systems. CEN (2003) also produced a draft standard for design of specific explosion venting devices. The main principle of dust explosion venting is the same as for venting of gas explosions and outlined in Section 2.4.5.6 of Chapter 2 and illustrated in Fig. 2.75.

7.5.4.4.2 Sizing of Dust Explosion Vents

Several parameters have an influence on the required area for venting of dust explosions:

- Enclosure volume
- Length/diameter ratio of enclosure
- Maximum overpressure P_{red} that the enclosure can withstand
- Static opening overpressure P_{stat} of vent cover
- Mass of vent cover
- Burning rate of dust cloud

For some time it was thought by many that the burning rate of a cloud of dust in air was a constant property of a given dust, which could be determined once and for all, eg, in the standard 1 m^3 closed vessel test (see Fig. 7.18). However, a cloud in air of a given dust can burn with widely different combustion rates, depending on the dust concentration, the turbulence, and the degree of dust dispersion in the actual industrial situation. This means that the required vent area also depends markedly on the specific industrial situation of dust cloud generation and flame propagation. During the last few decades, further experimental evidence in support of this fact has been produced. As a result, a differentiated view on dust explosion vent sizing has gradually evolved, which has also been taken into account in the latest dust explosion-venting standard issued by CEN (2002a). Experimental evidence supporting a differentiated vent-sizing approach is given in Chapter 6 of Eckhoff (2003). It is foreseen that the use of CFD-based numerical codes for simulating turbulent dust explosions in complex geometries will expand in the years to come. This also applies to venting of dust explosions.

7.5.4.4.3 Vent Covers

A wide range of vent cover designs are in use. One classical and simple type of vent cover is a light but rigid panel, eg, an aluminum plate, held in position by a rubber clamping profile as used for mounting fixed windows in cars. Other methods for keeping the vent cover in place include various types of clips. When choosing a method for securing the panel, it is important to make sure that the pressure, P_{stat}, needed to release the vent panel is small compared with the maximum tolerable explosion pressure, P_{red}. It is further important to anchor the vent panel to the enclosure to be vented, eg, by means of a wire or a chain. Otherwise, the panel may become a hazardous projectile in the event of an explosion. Finally, it is also important to make sure that rust formation or other processes do not increase the static opening pressure of the vent cover over time.

Bursting panels constitute a second type of vent covers. In the past, such panels were often "home made," and adequate data for the performance of the panels were lacking. A primary requirement is that P_{stat}, the static bursting pressure of the panel, is considerably lower than the maximum permissible explosion pressure, P_{red}. Today, high-quality bursting panels are manufactured in several countries. Fig. 7.63 shows one example.

Figure 7.63 Reinforced 6 m^3 vented bag filter enclosure fitted with a modern 0.85 m^2, three-layer bursting panel. P_{red} is 0.4 bar (g).
From Eckhoff, R.K., 2003. Dust Explosions in the Process Industries, third ed. Gulf Professional Publishing/Elsevier. (first ed. published in 1991, second ed. in 1997). ISBN:0-7506-7602-7.

Modern explosion vent panels burst reliably at the P_{stat} values for which they are certified, and are manufactured in a wide range of sizes and shapes, and coatings may be provided that allow permanent contact with various types of chemically aggressive atmospheres. Often a backing film of Teflon is used as an environmental protection to prevent the vent panel from contaminating the product inside the enclosure that is equipped with the vent. However, the upper working temperature limit of Teflon is about 230°C.

Hinged explosion doors constitute a third category of vent covers. Such doors may take a variety of different forms, depending on the equipment to be vented and other circumstances. Various kinds of calibrated locking mechanisms to ensure release at the predetermined P_{stat} have been developed. Hinged doors may be preferable if explosions are relatively frequent.

The final category of vent covers to be mentioned are the reversible ones, ie, covers that close as soon as the pressure has been relieved. The purpose of such covers is to prevent secondary air being sucked into the enclosure after the primary explosion has terminated, and giving rise to secondary explosions and fires. The reversible vent covers include counter-balanced hinged doors and spring-loaded, axially traversing vent covers.

7.5.4.4.4 Potential Hazards Caused by Venting

Explosion venting prevents rupture of the enclosure in which the explosion takes place. However, significant hazards still remain. These include:

- Ejection of strong flame jets and flame balls from the vent opening
- Emission of blast waves from the vent opening
- Reaction forces on the equipment caused by the venting process
- Emission of solid objects (vent panels and other possible objects)
- Emission of toxic combustion products

In general, flame ejection will be more hazardous the larger the vent and lower the static opening pressure of the vent cover. This is because with a large vent and a weak cover, efficient venting will start at an early stage of the combustion process inside the enclosure. Then, large clouds of unburnt explosive mixture may be pushed out through the vent and subsequently ignited when the flame passes through the vent. The resulting, secondary fire-ball outside the vent opening can present a substantial hazard. If, on the other hand, the enclosure is strong, allowing the use of a small vent and a high P_{red}, mainly combustion products are vented, and the flame outside the vent is considerably smaller.

Reaction forces from explosion venting can significantly increase both the material damage and the extent of the explosion. Process equipment can tilt, ducts can become torn off, and secondary explosive clouds can be formed and ignited. Whenever an explosion vent is installed, it is therefore important to make an assessment of whether the equipment to be vented is able to withstand the reaction forces from explosion venting. A simple first-order quasistatic consideration says that the maximum reaction force equals the maximum pressure difference between the interior of the vessel being vented and the outside atmosphere, times the vent area. Experiments have confirmed that this simplified model in fact predicts reaction forces in quite good agreement with the forces actually measured, as long as the duration of the pressure peak is not too short. However, for very fast explosions, dynamic (impulse) considerations may be required.

7.5.4.4.5 Vent Ducts

One traditional solution to the flame jet/flame ball problem is to use vent ducts. As illustrated in Fig. 7.64, this means that a duct of cross-sectional area at least equal to the vent area is mounted between the vent and a place where a strong flame jet/flame ball will not present any hazard. However, vent ducts will generally increase the flow resistance somewhat and therefore also the pressure drop between the interior of the enclosure and the ambient

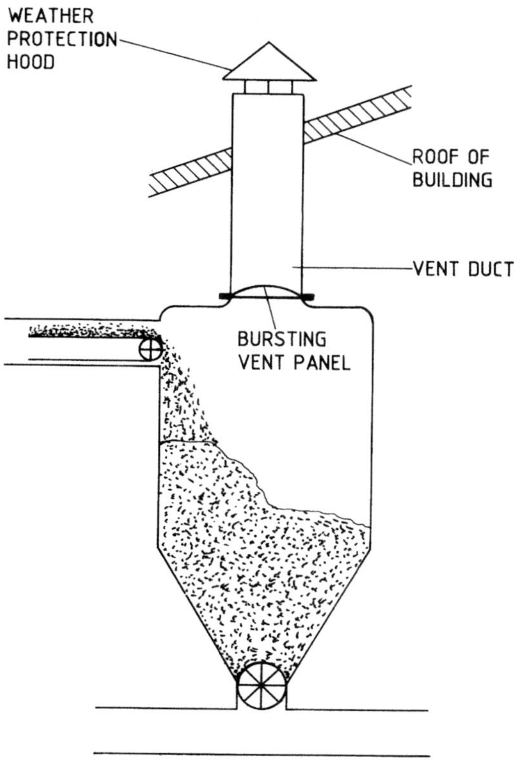

Figure 7.64 Illustration of the principle of vent ducts.
From Eckhoff, R.K., 2003. Dust Explosions in the Process Industries, third ed. Gulf Professional Publishing/Elsevier. (first ed. published in 1991, second ed. in 1997). ISBN:0-7506-7602-7.

atmosphere. Consequently, adding a vent duct inevitably increases the maximum explosion pressure in the vented vessel somewhat. In general, this increase grows with increasing duct length, increasing number of sharp bends, and decreasing duct diameter.

7.5.4.4.6 The Quenching Tube

In some applications in which vent ducts are difficult to implement, the quenching tube, invented by Alfert and Fuhre (1989) may provide a good solution. The principle of this device is illustrated in Fig. 7.65.

If a dust explosion occurs in the enclosure to be vented, and the bursting panel, which constitutes an integral part of the quenching tube assembly, bursts, the explosion is vented through the comparatively large specially designed wall of the quenching tube. The wall is designed to yield

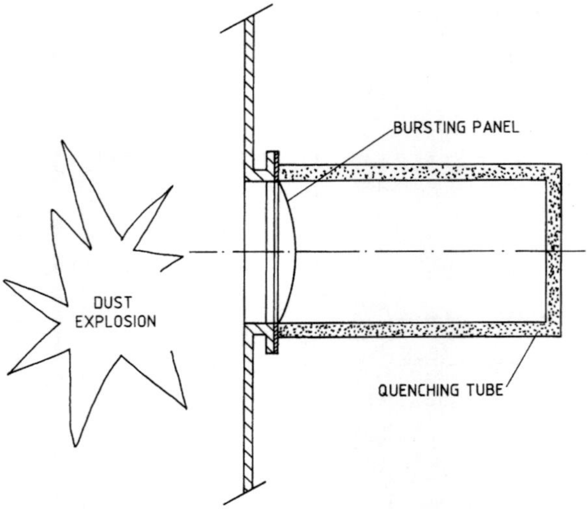

Figure 7.65 Illustration of the principle of the quenching tube for flame and dust-free venting of dust explosions.
From Eckhoff, R.K., 2003. Dust Explosions in the Process Industries, third ed. Gulf Professional Publishing/Elsevier. (first ed. published in 1991, second ed. in 1997). ISBN:0-7506-7602-7.

low-pressure drop, but high retention efficiency for dust particles and efficient cooling of combustion gases. This means that flame ejection from the vent is effectively prevented and the blast effects and reaction forces significantly reduced. Furthermore, burning lumps of powder and other smaller objects that could be ejected through an open vent, are retained inside the quenching tube. However, any toxic gaseous combustion products, eg, carbon monoxide, will escape to the atmosphere. The increase of the maximum explosion pressure in the vented enclosure due to the flow resistance through the quenching tube wall is mostly moderate, and can normally be compensated for by a moderate increase of the vent area.

7.5.4.5 Explosion-Pressure-Resistant Design

In some special cases, eg, when producing and handling highly toxic materials, full confinement of potential accidental dust explosions is required. Then all process equipment containing such materials must be able to withstand an internal dust explosion. In many cases, one can assume that the maximum internal pressure exerted by the explosion is approximately static, but not always. The strength of some materials, including structural steels, is highly strain-rate sensitive. This means that the stress at which plastic deformation starts, depends on the rate of loading. On the other hand, the damage to a structure also depends on how quickly the structure responds to the pressure loading. The natural period of vibration of the mechanical structure is normally used as a measure of the response time. If the duration of the pressure peak is long compared with the natural period of vibration, the loading can be considered as being essentially a static load. If, on the other hand, the pressure pulse is short compared with the response time of the structure, the damage is determined by the impulse, ie, the time integral of pressure.

7.5.4.6 Automatic Explosion Suppression

7.5.4.6.1 Applicability in General

CEN (2002b) drafted a standard for explosion suppression systems. The draft does not, however, give any quantitative guidelines for design of such systems. The basic principle of automatic explosion suppression is described in Section 2.4.5.8 in Chapter 2 and illustrated in Figure 2.77. Automatic suppression of dust explosions has been found to be feasible for organic dusts of maximum rate of pressure rise in the standard 1 m^3 closed ISO-vessel of up to 300 bar/s, ie, $K_{St} = 300$ bar m/s. It is somewhat unclear, however, whether the method can also be used for aluminum dusts of K_{St} in the range 300–600 bar m/s. Moore and Cooke (1988) found that, for aluminum flake of $K_{St} = 320$ bar m/s, it was difficult to ensure lower suppressed explosion pressures than about 2 bar (g), even under optimum conditions for suppression. In the case of dusts of natural organic materials and plastics of K_{St} up to 300 bar m/s, the corresponding suppressed explosion pressures would typically be 0.2–0.4 bar (g). It was therefore concluded that reliable suppression of the very violent aluminum flake explosions is difficult. However, they showed that a combination of explosion suppression and venting can reduce the maximum explosion pressure to a level significantly lower than the level for venting only. An alternative approach is to reduce K_{St} by partial inerting (see Section 7.5.4.1).

7.5.4.6.2 Applicability to Large Systems

Moore and Bartknecht (1987) conducted dust explosion suppression experiments in large vessels of volumes up to 250 m^3 and were able to show that successful suppression of explosions in clouds of organic dusts is possible even in such large volumes. However, as the vessel volume increases, more suppressant and faster injection are required for successful suppression. The actual design of suppression systems depends very much on the specific design of the suppressors, and other details which vary somewhat from supplier to supplier. Therefore, it is difficult to specify generally applicable quantitative design criteria.

Moore and Bartknecht employed three standardized types of suppressors. The smallest type, of volume 5.4 L, was used for vessel volumes up to 5 m^3, whereas suppressors of 20 L were used in the vessel volume range of 5–30 m^3. The largest suppressor type of 45 L was used for the larger process volumes. The performance of the suppression system in large process volumes was verified experimentally in vessels of up to 250 m^3, for which 10 of the 45 L suppressors were required for successful suppression of St 2 dust explosions (organic dusts). For St 1 dusts, seven such suppressors were sufficient. However, these results refer to dust clouds of very high turbulence and homogeneity, and later investigations have shown that considerably smaller total suppressor volumes are required if the dust cloud is less turbulent and less homogeneous, which is often the case in industrial practice.

Experiments have shown that the two explosion protection methods venting and suppression are to a great extent complementary. In practice, cost-effective safety is achieved by using either one of the two methods, or a combination of both, depending on the circumstances.

7.5.4.6.3 Influence of Type of Suppressant (Extinguishing Agent)

Traditionally, halogenated hydrocarbons (halons) were used as suppressants in automatic dust explosion suppression systems. However, long before the environmental problems caused by these chemicals became a major issue, Bartknecht (1978) showed that powder suppressants, such as $NH_4H_2PO_4$, were in general much more effective for suppressing dust explosions than halons. Therefore, powder suppressants have been used for suppressing dust explosions for many years. But powders differ in their suppressive power, and efforts have been made to identify the most effective ones. For example, addition of only 30 wt% of $NH_4H_2PO_4$ powder is required to prevent flame propagation in dust clouds in air of Pittsburgh bituminous coal, whereas with $CaCO_3$ dust (limestone) 70 wt% is needed. $NaHCO_3$ has proved an effective agent for suppressing aluminum dust explosions. This material can also in some cases be used even in the food industry. It is soluble in water and can therefore be removed effectively by water only. Superheated steam (water at $>180°C$) has also been used as a nonpolluting suppressant.

7.5.4.7 Flexible Options for Explosion Prevention and Mitigation

Figs. 7.66 and 7.67, based on an analysis by Farber, illustrate how a given process plant can be protected against hazardous dust explosions by choosing quite different overall strategies. In Fig. 7.66 the main strategy is explosion prevention by inerting using CO_2, whereas in the strategy adopted in Fig. 7.67 explosion mitigation/control by venting and isolation plays a central role. Whenever a solution is developed for a given process plant, cost effectiveness will be a major concern.

7.5.4.8 Good Housekeeping (Dust Removal/Cleaning)

7.5.4.8.1 General Outline

The main prerequisite for disastrous secondary explosions in factories is that sufficient quantities of combustible dust have accumulated outside the process equipment to permit development of large secondary dust clouds (see Section 7.2.7). In other words, the possibility of extensive secondary explosions can be eliminated if the outside of process equipment, and shelves, beams, walls, and floors of workrooms are kept free of dust.

Significant quantities of dust may accumulate accidentally outside process equipment due to discrete events such as bursting of sacks or bags or erratic discharge from silos or filters. In such cases it is important that

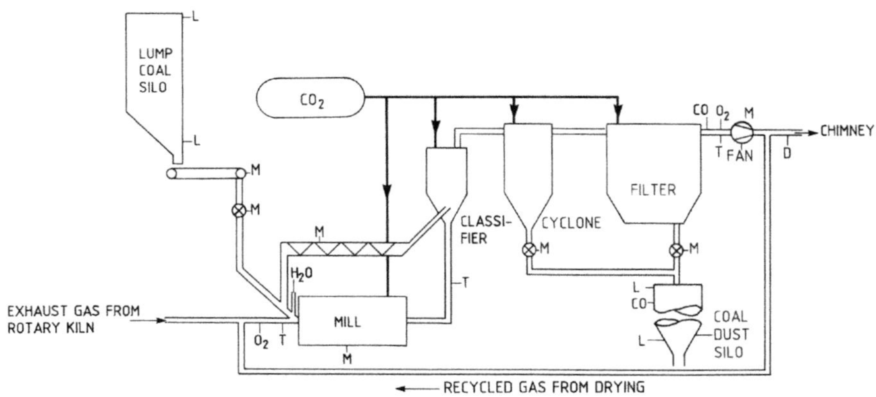

Figure 7.66 Comprehensive sensor system for monitoring, controlling, and interlocking of a process for milling and drying of coal. Explosion protection based on inerting with CO_2. *CO*, Carbon monoxide concentration sensors; *D*, Dust concentration sensor, *L*, Level sensors for coal and coal dust in silos; *M*, Movement sensors for mechanical components; *O_2*, Oxygen concentration sensors; *T*, Temperature sensors.
From Eckhoff, R.K., 2003. Dust Explosions in the Process Industries, third ed. Gulf Professional Publishing/Elsevier. (first ed. published in 1991, second ed. in 1997). ISBN:0-7506-7602-7.

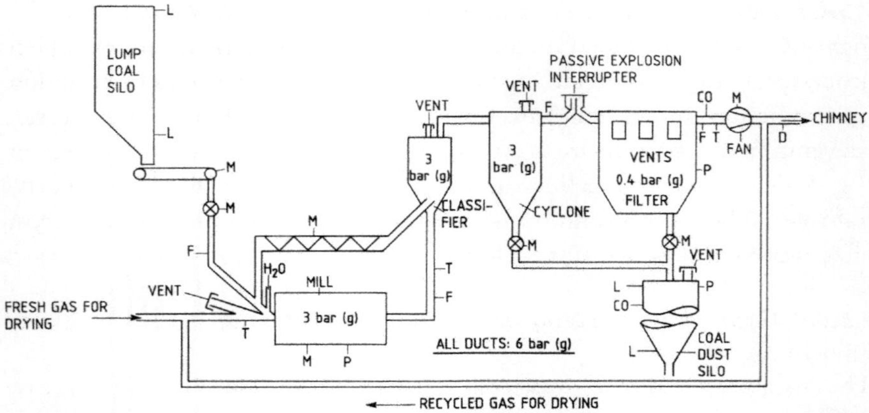

Figure 7.67 Comprehensive sensor system for monitoring, controlling, and interlocking of a process for milling and drying of coal. Explosion protection based on venting and explosion shock resistant design. *CO*, Carbon monoxide concentration sensors; *D*, Dust concentration sensor; *F*, Flame sensor; *L*, Level sensors for coal and coal dust in silos; *M*, Movement sensors for mechanical components; *P*, Pressure sensors; *T*, Temperature sensors.
From Eckhoff, R.K., 2003. Dust Explosions in the Process Industries, third ed. Gulf Professional Publishing/Elsevier. (first ed. published in 1991, second ed. in 1997). ISBN:0-7506-7602-7.

the spilt dust be removed immediately. In case of large dust quantities, the main bulk may be sacked by hand using spades or shovels, whereas industrial, explosion-proof vacuum cleaners should be used for the final cleaning. In the case of moderate spills, dust removal may be accomplished by vacuum cleaning only. Effective dust extraction should be provided in areas where dusting occurs as part of normal operation, eg, at bagging machines.

Considerable quantities of dust can accumulate outside process equipment over a long time due to minor but steady leaks from process equipment. The risk of such leaks is comparatively large if the working pressure inside the process equipment is higher than ambient pressure, whereas running the process at slightly lower than ambient pressure reduces the leaks.

It is important that process equipment be inspected regularly for discovery and sealing of obvious accidental leak points as early as possible. However, often one has to accept a certain unavoidable level of dust leaks from process equipment. It is then important to enforce good housekeeping routines by which accumulations of explosible dust outside process equipment are removed at regular intervals, preferably by explosion-proof vacuum cleaning.

Use of compressed air for blowing spilled dust away should be prohibited. By this method dust is not removed, but only transferred to another location in the same room. Besides, dust explosions can result if the dust concentration in the cloud that is generated is in the explosive range and an ignition source exists in the same location.

7.5.4.8.2 Industrial Explosion-Proof Vacuum Cleaners

The subject was discussed by Kühnen (1978), Wibbelhoff (1984), and Beck and Jeske (1989). Beck and Jeske listed the requirements to mobile type 1 vacuum cleaners recommended in F. R. Germany for removal of combustible dusts:

- The fan must be on the clean side and protected against impacts by foreign bodies.
- The electric motor and other electric components must satisfy the general requirements to such components that are to be used in areas containing explosible dusts. Motors must be protected against short-circuits and overheating.
- The exhaust from the vacuum cleaner must be guided in such a way that it does not hit dust deposits and generate dust clouds.
- All electrically conducting parts of the equipment, including hose and mouthpiece, must be earthed with a resistance to earth of less than 1 MΩ.
- Vacuum cleaner housings must be constructed of materials that are practically nonflammable. Aluminum and aluminum paints must not be used.
- A clearly visible sign saying "No suction of ignition sources" should be fitted to the housing of the vacuum cleaner.

Fig. 7.68 shows an example of a large mobile vacuum cleaner for combustible dusts in industry.

Sometimes it is useful to install stationary vacuum cleaning systems rather than having mobile ones. Then a central dust-collecting station with suction fan is connected to a permanent tube system with a number of plug-in points for vacuum cleaning hoses at strategic locations. Good housekeeping is essential because clean workrooms exclude the possibility of extensive secondary explosions. Cleanliness also improves the quality of the working environment in general.

7.5.4.9 Application of the Concept of "Inherently Safer Plant Design" to Dust Explosion Prevention and Mitgation

7.5.4.9.1 Outline of Basic Concept

The following introductory paragraph is just a copy of the first paragraph of Section 2.4.1.1 in Chapter 2. According to Kletz (2003), Khan and Amyotte

Figure 7.68 Large mobile vacuum cleaner for collecting combustible/explosible dusts in industry. Both the main vessel and the connecting ducts are designed to withstand internal explosion pressures of 9 bar (g). Power requirement 45–55 kW. *From Eckhoff, R.K., 2003. Dust Explosions in the Process Industries, third ed. Gulf Professional Publishing/Elsevier. (first ed. published in 1991, second ed. in 1997). ISBN: 0-7506-7602-7.*

(2003), Hendershot (2006), and Amyotte et al. (2007), inherently safer plant design is based on four main principles, viz. *minimization, substitution, moderation,* and *simplification.*

- *Minimization* implies using so little of the hazardous material that there is no significant risk of hazardous effects.
- *Substitution* implies using a material or a process, which is less likely to give rise to hazardous situations.
- *Moderation* implies using the hazardous material in the least hazardous form.
- *Simplification* implies elimination of unnecessary complexity, and design of "user friendly" plants.

Amyotte et al. (2009) specifically discussed the application of inherent safety principles to dust explosion prevention and mitigation.

7.5.4.9.2 Two Examples of Application in a Dust Explosion Context

7.5.4.9.2.1 Design of Inherently Safer Silos Silos of various sizes for storage of combustible materials are often of the "funnel flow" type illustrated in Figs. 7.69 and 7.70.

This type of silo is characterized by the flow out of the silo taking place in a limited zone round the silo axis. The material will enter and leave the silo

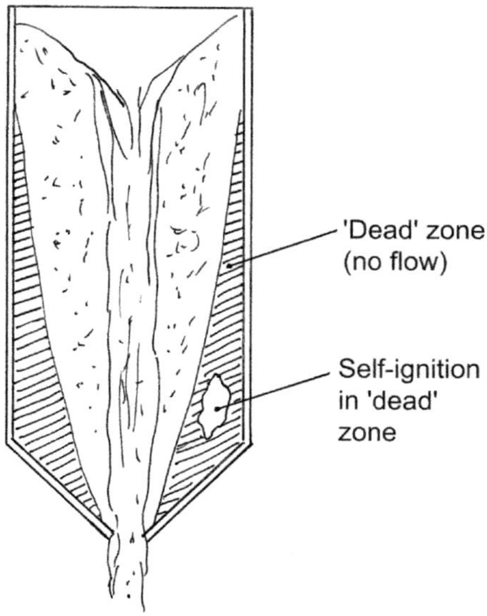

Figure 7.69 Illustration of a funnelflow silo with "dead" zones in which self-ignition can develop despite recycling of the rest of the material in the silo.

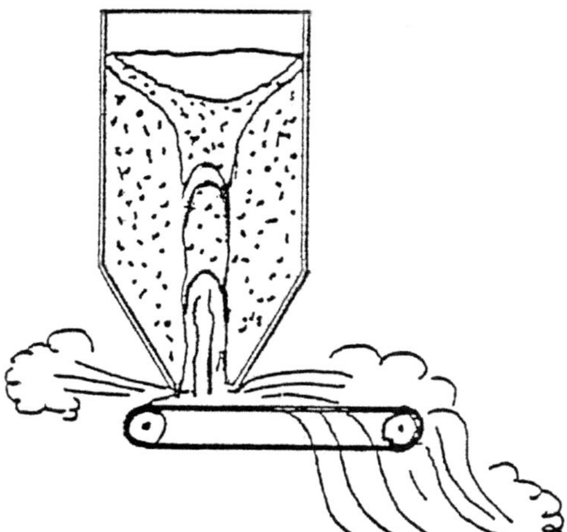

Figure 7.70 Illustration of a funnel-flow silo with uncontrolled "flooding" of the material, which can cause excessive spillage of combustible dust as both clouds and layers in the workrooms.

in opposite order ("last in/first out"). The flow is often unreliable and in some cases unaided gravity flow will not empty the silo completely, leaving behind "dead" zones in which self-heating and self-ignition may develop. If this happens, recycling of the material in the silo will not remove the self-heating zone, and a hazardous situation may arise (see Chapter 6, in particular Section 6.4.4). The situation is illustrated in Fig. 7.69.

Fig. 7.70 illustrated another hazard associated with funnel-flow silos. Sometimes the flow of material out of the silo can become fluidized ("flooding") on its way downward in the central core of the silo, and explosive dust clouds may be generated and dust layers accumulate in the area outside the silo.

An inherently much safer solution is given in Fig. 7.71.

Here, the principle of *substitution* is adopted, which implies "using a material or a process, which is less likely to give rise to hazardous situations". In this case, the silo has been specifically designed to give "mass" flow instead of "funnel" flow during discharge of the stored material from the silo. In this way the risk of "flooding" is virtually eliminated. A mass-flow silo also to some extent remixes/homogenizes the material in the silo horizontally

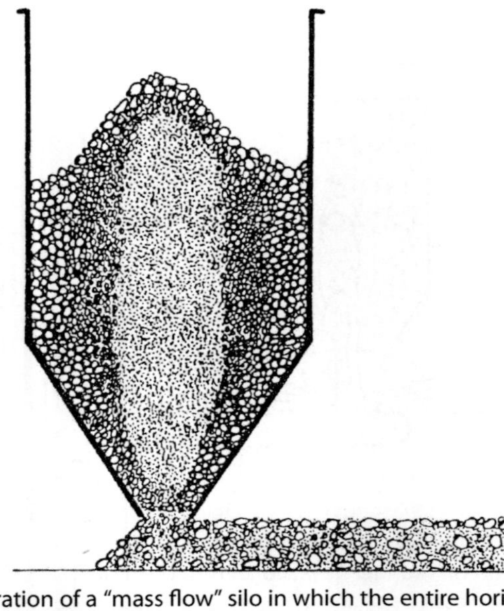

Figure 7.71 Illustration of a "mass flow" silo in which the entire horizontal cross section of material flows downward during the discharge process.

during material discharge. In cases in which the material stored in the silo has become horizontally segregated during silo filling from the top, this will contribute to a more homogeneous material flowing out of the silo at the bottom.

7.5.4.9.2.2 Production of Magnesium Ferrosilicon (MgFeSi) Alloys According to Traditional Method In a Norwegian plant for producing Magnesium Ferrosilicon (MgFeSi) alloys (an additive in steel production), *substitution* of an existing process by a new one gave a product that was both less liable to be the source of dust explosions, and more homogeneous in quality. The substitution therefore yielded both a safer process and a better product.

The entire production process just prior to the explosion was outlined by Eckhoff (1994). In the traditional process, the molten MgFeSi alloy was poured from the melting pot onto a concrete floor where it was left for solidification and cooling. The resulting solidified and cooled alloy flake of several m diameter was then broken up into smaller pieces and sent through the system for crushing, screening, and packing illustrated schematically in Fig. 7.72.

During these mechanical processes, considerable amounts of fine dust were produced, which was removed by an integrated dust-extraction system also indicated in Fig. 7.72. It was well known that the fine alloy dust represented a significant dust explosion hazard, in particular because of the Mg content.

7.5.4.9.2.3 Dust Explosion in 1984 In 1984, the plant experienced a rather extensive dust explosion that swept through the system represented by Fig. 7.72. The alloy quality being crushed and screened on the day of the explosion contained an average of about 10% Mg. It was known even before the explosion that the minimum ignition energy (MIE) of clouds of MgFeSi dusts in air decreased with increasing Mg content. However, laboratory tests conducted after the explosion revealed that the fine dust involved in the explosion in fact had an Mg content of nearly 20%, rather than the average of 10% in the alloy as a whole.

7.5.4.9.2.4 Metal Segregation During Solidification of Poured Flake Further investigations disclosed that when the molten alloy, after pouring from the melting pot onto the concrete floor, solidified as a large flake, significant selective migration of Mg occurred, which caused

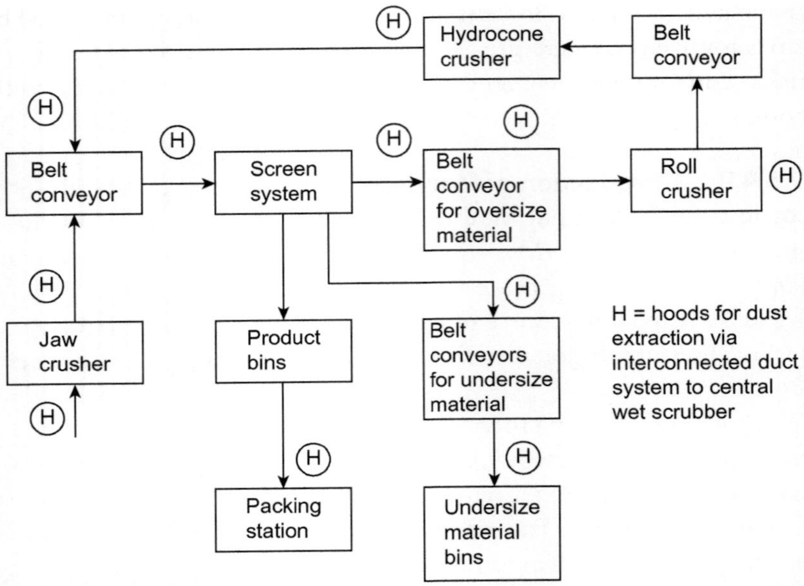

Figure 7.72 Block diagram of plant for crushing, screening, and packing of MgFeSi at Aalvik, Norway, which was made inherently safer by introducing a new pouring procedure for molten MgFeSi alloy.
From Eckhoff, R.K., 1994. Dust Explosion Hazards in the Ferro-Alloys Industry. Proceedings of the 52th Electric Furnace Conference, Nashville, TN, USA, November 13–16, 1994. Iron & Steel Soc. Inc., Warrendale, PA, USA, pp. 129–146.

significant enrichment of the metal at one end of the flake. Furthermore, because the Mg-enriched alloy tended to be more brittle and therefore created more fine dust than the bulk of the alloy, the fine dust that was produced during crushing and screening (Fig. 7.72) turned out to contain a much higher percentage of Mg than the nominal average for the total charge of molten alloy in the melting pot.

7.5.4.9.2.5 New Pouring/Solidification Method Improved Product Quality and Reduced Explosion Risk After several years of development a new pouring method for molten alloy was introduced in 1989. Instead of pouring the entire melting pot content onto one single big concrete floor, it was poured into a number of quite small (on the order of 1 m × 0.5 m) rectangular trays. This made the distribution of the various phases, eg, silicides, very fine-grained, and greatly improved the product homogeneity also with respect to the content of Mg. In addition, the excessive accumulation of Mg in the fine dust fraction, which was experienced when crushing and grinding solidified alloy flakes from the old pouring method, was practically eliminated.

As a result, this rather simple change of production method yielded both a more homogeneous and hence better product, and a reduced risk of dust explosions.

CHAPTER EIGHT

Explosives, Pyrotechnics, and Propellants

8.1 CHEMICAL COMPOSITION OF SOME EXPLOSIVES

An explosive material is a solid or liquid material or a mixture of such materials, which after activation liberates gases and heat due to a fast exothermal chemical reaction. A common feature of these materials, which makes them differ from explosive gases, sprays/mists, and dust clouds, is that the oxygen of the atmosphere (normally air) is not an essential participant in the heat-liberating chemical reaction. The different materials of this type used in common explosives may be grouped in the following categories:

(a) Dry mixtures of powders of fuels, ie, materials that can burn (eg, metals, carbon, sulfur, starch, or wood) and materials that can supply oxygen (eg, nitrates)

(b) Water-based suspensions (slurries) or gels of powders of fuels in liquid solutions of solid materials that can supply oxygen (eg, ammonium nitrate)

(c) Nitrated organic compounds (eg, nitroglycol, glycerine, dinitrotoluene, trinitrotoluene, trinitro-phenylmethyl-nitramine). Such materials contain the fuel (carbon, C, and hydrogen, H) and the source of oxygen for combustion (nitro-groups, NO_2) in the same molecule.

(d) Primary explosives (eg, lead azide), which, when exposed to mechanical impact above a certain strength, will decompose spontaneously by detonation. Such primary explosives are therefore used to initiate detonation in other less reactive explosives, which would just deflagrate when exposed to less vigorous ignition sources.

Commercial explosives used in practices are often mixtures/combinations of materials belonging to the categories (a), (b), (c), and (d). One example is given in Table 8.1.

Another example is ammonium nitrate/fuel oil (AN/FO), which is simply a mixture of solid crystalline ammonium nitrate moistened with diesel oil. In practice, categories (a) and (b) may interchange, depending on the amount of water added. For example, a saturated solution of ammonium nitrate in water may exist side by side with solid crystalline ammonium nitrate.

Explosion Hazards in the Process Industries
ISBN: 9780128032732
http://dx.doi.org/10.1016/B978-0-12-803273-2.00008-6

Table 8.1 Composition of a typical gelatine dynamite

Chemical component	% By mass
Ammonium nitrate	60–65
Nitroglycol	20–21
Nitroglycerine	5–6
Dinitrotoluene	6–8
Nitrocellulose (providing gelatine)	1
Additives to produce desired consistency (eg, starch, "kiselguhr," wood powder)	Balance to 100%

European Committee for Electrotechnical Standardization (CENELEC, 1997) defines explosive substances as solid, liquid, pasty or gelatinous substances, and preparations which are liable to react exothermally and under rapid gas generation even without the participation of atmospheric oxygen, and which, under specified test conditions under partial confinement, will detonate or deflagrate rapidly during heating. It is important to note that this definition not only comprise the genuine explosives, but also pyrotechnics and propellants. Equivalent to explosive substances are:

(a) substances or preparations capable of exploding under certain realistic conditions, and intended for use as blasting explosives, but not generally defined as explosives

(b) initiation and igniting devices and pyrotechnics

(c) other products in which explosive substances or substances or preparations in accordance with (a) are totally or partially confined and in which an explosion is initiated

8.2 COMBUSTION PROPAGATION IN EXPLOSIVES/PYROTECHNICS/PROPELLANTS

8.2.1 Deflagration and Detonation

As in the case of premixed gases, mists/sprays and dust clouds, explosives, and many pyrotechnics and propellants also possess two distinctly different modes of exothermal reaction, viz. deflagration and detonation. In fact, with these materials the difference between the two modes is generally much more dramatic than with gases, mists/sprays, and dusts.

In the case of deflagration, the velocity of propagation of the reaction zone is much lower than the velocity of sound in the material. In a deflagration wave, the direction of flow of the reaction products is opposite to the

Table 8.2 Comparison of deflagration and detonation properties of a nitroglycol explosive of density 1.5 g/cm^3

Property	Deflagration	Detonation
Flame front propagation velocity	0.3 mm/s	7.3 km/s
Thickness of reaction zone	10 mm	1 mm
Liberated energy in reaction zone	1.9 kJ/g	6.7 kJ/g
Mass conversion rate	0.45 kg/m^2 s	~10^7 kg/m^2 s
Energy production rate	85 MJ/m^2 s	7.4×10^{10} J/m^2 s

The data for deflagration were obtained at fully unconfined conditions.

direction of propagation of the reaction. The burning of a gunpowder or a rocket-propellant charge is a deflagration process.

The mode of exothermic reaction of an explosive material once ignited, whether a deflagration or a detonation, depends on the mode of initiation of the reaction. Deflagration results if the ignition source is not generating a shock wave of the strength required for initiation of detonation. However, if the reactive material is confined, the liberated heat in a deflagration will generate a pressure rise in the still unreacted material, and substantial acceleration of the exothermal chemical reaction can result.

Table 8.2 illustrates the dramatic difference between the features of detonation and unconfined deflagration for one specific explosive material. The ratio of the two mass conversion rates is more than 10^7 and of the two energy production rates 10^9.

8.2.2 Experimental Methods for Measuring Detonation Properties

8.2.2.1 Dautriche Method

The Dautriche method for determination of detonation rates of explosives, as described by Meyer (1987), is illustrated in Fig. 8.1.

The sample of the explosive to be tested is placed in a confinement tube, which may or may not be surrounded by a steel sleeve. The part of the sample along which the detonation velocity is to be measured, is defined by two blasting caps. Each end of a piece of detonating cord of known detonation velocity is connected to a blasting cap, and the cord is placed on a lead plate as indicated in the figure. The detonation wave in the test sample is initiated by the detonator/booster at the upstream end of the test tube, and a steady detonation wave front starts to travel along the confinement tube. As soon as this wave hits the first blasting cap, a second detonation wave is instantaneously initiated at the end of the detonating cord connected to that cap.

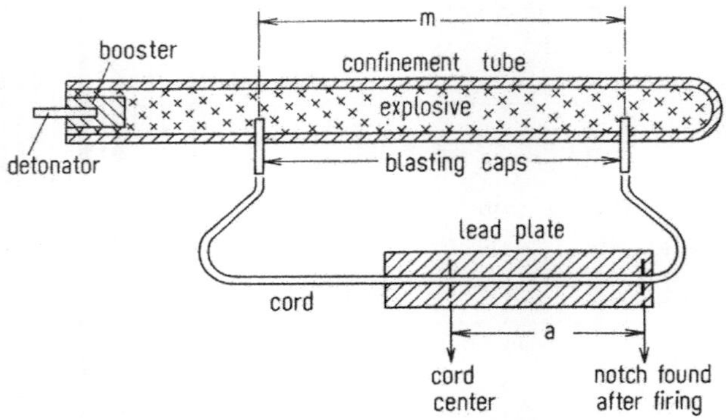

Figure 8.1 Measurement of detonation velocities in explosives.
From Meyer, R., 1987. Explosives, third ed. VCH Verlagsgesellschaft, Weinheim, Germany. ISBN:3-527-26599-6/0-89573-600-4.

At the same time, the detonation wave in the test sample continues down the confinement tube and eventually reaches the second blasting cap. This then initiates a third detonation wave at the other end of the detonating cord. Eventually the two detonation wave fronts in the cord, traveling at equal but opposite directions collide, which produced a notch in the lead plate at the meeting point. Clearly, the time required for the two opposite detonation waves to travel from the first blasting cap to the collision point at the lead plate is the same. Hence, for the wave going directly into the detonating cord (total length L), the time is $(½L + a)/D_{cord}$, whereas for the other wave it is $m/D_{sample} + (½L - a)/D_{cord}$. By equating the two expressions one gets

$$m/D_{sample} = 2a/D_{cord} \quad [8.1]$$

or

$$D_{sample} = D_{cord} \cdot m/2a. \quad [8.2]$$

8.2.2.2 The Detonating Cap Sensitivity Test

In this case the objective of the test is to determine the response of a sample of the explosive to a standardized detonating cap. The results are used to determine the classification of the explosive as a transport hazard. According to Meyer (1987), the test method illustrated in Fig. 8.2 has been used in Germany. The explosive sample is placed into a 200-mm-long cardboard sleeve of 80-mm internal diameter and wall thickness between

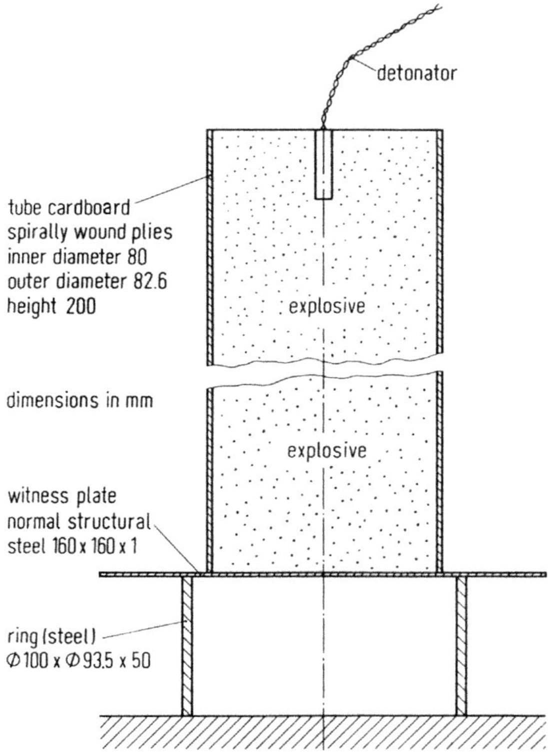

Figure 8.2 The cap sensitivity test for classifying explosives with regard to the explosion hazard that they represent when being transported.
From Meyer, R., 1987. Explosives, third ed. VCH Verlagsgesellschaft, Weinheim, Germany. ISBN:3-527-26599-6/0-89573-600-4.

1.3 and 1.4 mm. One end of the sleeve is sealed by a thin cardboard disk, which is glued into position. Because the cap sensitivity can be influenced by the density of the sample under test, the density is determined by weighing.

The cardboard sleeve containing the test sample is placed upright onto a steel plate of 1 mm thickness, resting on a steel ring of 50 mm height, inside diameter 100 mm, and wall thickness 3.5 mm. A standardized detonating test fuse is inserted at the top of the sample and initiated. If a proper detonation wave is then being initiated in and propagates downward through the sample, a sharp circular hole will be cut out of the steel plate. However, if the steel plate is not affected, or only suffers denting with or without fissure, this is regarded as a nondetonation.

8.2.2.3 The Trauzl Lead Block Test

The Trauzl lead block test is a comparative method for the determination of the strength of an explosive. The principle of the test is illustrated in Fig. 8.3.

A sample of 10 g of the test material, wrapped in tin foil, is introduced into a central bore hole (125 mm deep, 25 mm diameter) in a massive cylinder of soft lead, of height and diameter equal to 200 mm. A standardized blasting cap with an electric primer is introduced into the center of the explosive charge, and the remaining free space is filled with quartz sand of standard grain size. After the explosion, the volume of the resulting bulge is determined by filling it with water. The volume of the original cavity is deducted from the result obtained. Ten grams of picric acid may be used as a standard explosive for comparison.

In a modified lead block test, recommended by Federal Institute for Materials Testing (BAM) in Germany, the test sample is prepared as follows: A special instrument wraps the sample in a piece of tinfoil and molds it into a cylinder of 11 cm^3 capacity. The resulting bulk density should be only slightly higher than the bulk density by pouring. Liquids are filled into thin-walled cylindrical glass ampoules or, in special cases, directly into the cavity of the lead block. The initiation is effected by means of an electric blasting cap containing lead azide as the initiating charge and PETN as the secondary charge. The empty space above the test sample is filled with dried quartz sand as in the original method.

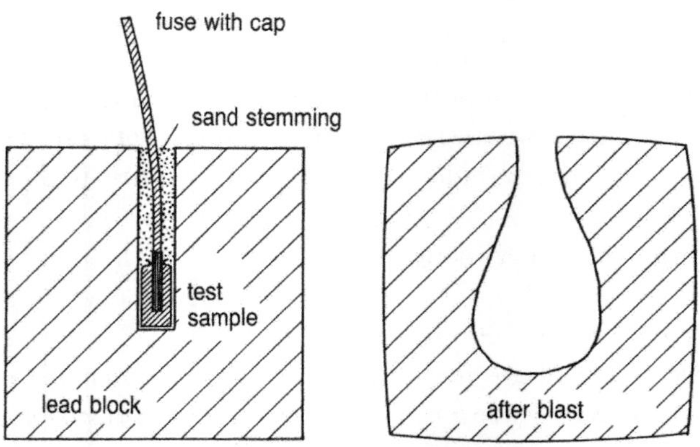

Figure 8.3 The Trauzl lead block test for comparing the strength of explosives. *From Meyer, R., 1987. Explosives, third ed. VCH Verlagsgesellschaft, Weinheim, Germany. ISBN:3-527-26599-6/0-89573-600-4.*

8.3 IGNITION/INITIATION OF EXPLOSIVES/ PYROTECHNICS/PROPELLANTS

8.3.1 Decomposition Temperature

The decomposition temperature of explosive materials, propellants, and pyrotechnics is the minimum temperature at which a small sample of the material, placed in an externally heated test tube, bursts into flame, decomposes rapidly or detonates violently.

Meyer (1987) describes the following method for determining the decomposition temperature: 0.5 g (0.01 g in the case of initiating explosives!) is placed in a test tube which is then immersed in a liquid metal bath (preferably Wood's metal) at 100°C. The temperature of the bath is then raised at a rate of 20°C/min. until deflagration/decomposition takes place.

CENELEC (1997) defines the decomposition temperature as the lowest temperature at which a given quantity of the material tested ignites in a carefully standardized differential thermal analysis (DTA) test. The basic principle of DTA is that the sample under test and a sample of the same size of an inert material are heated in parallel and the temperature difference between the two samples recorded continuously. The sample size is in the range 0.1—0.5 g, whereas the heating rate shall not exceed 5°C/min. Fig. 8.4 shows an example of a DTA test. In this particular case, the first-recorded difference between the temperature of the test sample and the inert reference sample is a modest negative peak due to endothermic melting of the test sample. However, as the temperature rises further, the major exothermal peak of the decomposition reaction appears. The decomposition temperature in this test is defined as the "extrapolated peak initial temperature," ie, the reference sample temperature at which the steepest tangent to the rising part of the decomposition peak intersects with the base line, as indicated in Fig. 8.4.

Table 8.3 gives some results for a range of materials obtained by the differential thermal analysis method.

8.3.2 Drop-Hammer Sensitivity Test

The sensitivity to impact of solid, liquid or gelatinous explosives is tested by the drop-hammer method. The principle of this method is illustrated in Fig. 8.5.

The samples of the explosives are subjected to the action of failing weights of different masses. The parameter to be determined is the height of fall, for one specific weight mass, at which a sufficient amount of impact

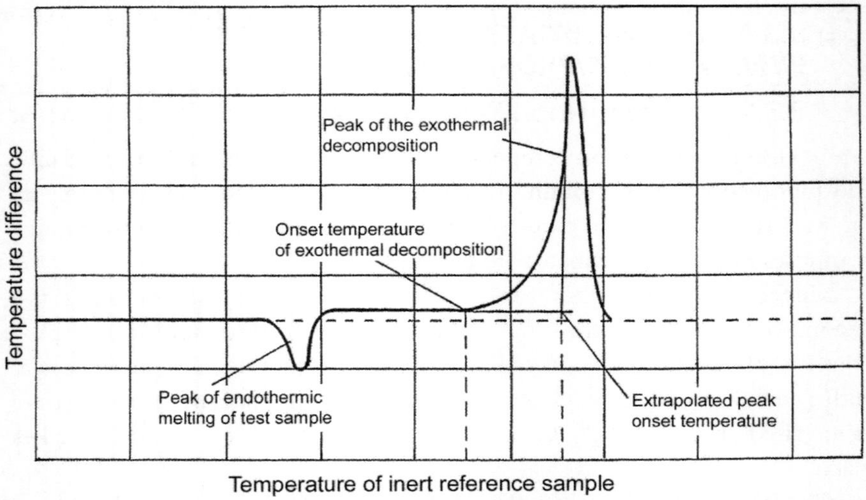

Figure 8.4 Example of result of differential thermal analysis test for the decomposition temperature of explosives, propellants, and pyrotechnics.
From CENELEC, March1997. Installations of electrical apparatus on aread endangered by explosive substances. Draft Europ. standard prEN 50273, CENELEC Central Office, Brussels.

energy is transmitted to the sample for it to decompose or to explode. According to Meyer (1987), US Bureau of Mines developed two somewhat different standard procedures, depending on whether the sample to be tested is a solid or a liquid:

8.3.2.1 Solids, US Bureau of Mines

A sample of 0.020 g of the explosive is tested, using a fixed dropping mass, usually 2 kg. The result of the test is the minimum height at which at least one of 10 repeated trials results in explosion. The sample of the explosive being tested is held between two flat, parallel, hardened steel surfaces, as illustrated in Fig. 8.5.

8.3.2.2 Liquids, US Bureau of Mines

In the case of liquids, the same basic procedure is used as with solids (Fig. 8.5) but with the following modifications:

- The mass of explosives tested is 0.007 g
- A disk of desiccated filter paper saturated with the liquid to be tested is placed on the anvil, whereafter the plunger is lowered onto the sample filter paper

Table 8.3 Decomposition temperatures of explosive substances used as blasting explosives, propellants, ignition substances or pyrotechnics

Chemically uniform substances

No.	Chemical name	CAS-no.[b]	EINECS[c]- no.	Decomposition temperature °C
1	Mannitol hexanitrate	15825-70-4	239-924-6	140[d]
2	Mercury difulminate	628-86-4	211-057-8	151[d]
3	Glycerol trinitrate	55-63-0	200-240-8	155[e]
4	Oxydiethylene dinitrate	693-21-0	211-745-8	170[f]
5	Pentaerithrityl tetranitrate	78-11-5	201-084-3	164[d]
6	*N*-methyl-N,2,4,6-tetranitroaniline	479-45-8	207-063-0	171
7	Ethylene dinitrate	628-96-6	211-063-0	172[f]
8	Cellulosenitrate (nitrocellulose)	9004-70-0		176[d]
9	Perhydro-1,3,5-trinitro-1,3,5-triazine	121-82-4	204-500-1	208[d]
10	Picric acid	88-89-1	201-865-9	214[d]
11	Styphnic acid	5704-04-1	227-193-6	224[d]
12	Dipicrylamine	131-73-7	205-037-8	244[f]
13	Lead 2,4,6-trinitro-*m*-phenylene dioxide	15245-44-0	239-290-0	256[d]
14	Lead diazide	13424-46-9	236-542-1	268[d]
15	2,4,6-Trinitrotoluene	118-96-7	204-289-6	271[d]
16	Dipricrylamine	131-73-7	205-037-8	301[f]
17	1,3,5-Trichlorotrinitrobenzene	2631-68-7	220-115-1	305[f]
18	1,3,5-Trinitrobenzene	99-35-4	202-752-7	327[f]
19	2-Chloro-1,3,5-trinitrobenzene	88-88-0	201-864-3	330[f]
20	2,4,6-Trinitrobenzoic acid	129-66-8	204-958-2	337[f,a]
21	Ammonium perchlorate	7790-98-9	232-235-1	341[f]
22	5-ISMN (isosorbide-5-mononitrate)			183–193
23	ISDN (isosorbide dinitrate)			175–200

[a]At 190°C exothermal prereaction (decarboxylization).
[b]Chemical abstract service number.
[c]European inventory of existing commercial chemical substances.
[d]DTA open; 0.1 g sample mass.
[e]DTA open; 0.5 g sample mass.
[f]DTA pressure-tight 0.5 g sample mass.
From CENELEC, March1997. Installations of electrical apparatus on aread endangered by explosive substances. Draft Europ. standard prEN 50273, CENELEC Central Office, Brussels.

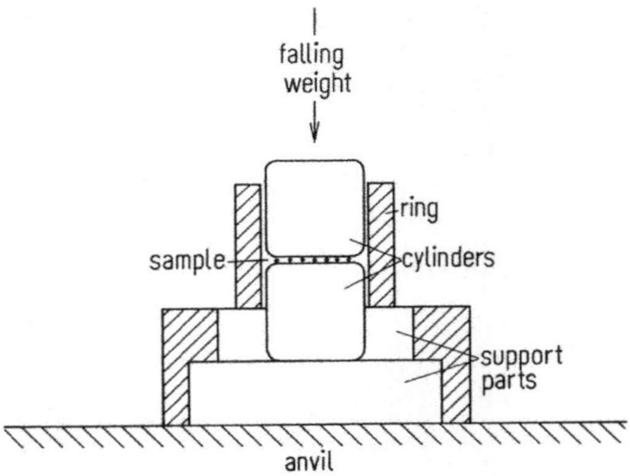

Figure 8.5 Illustration of principle of drop-hammer impact sensitivity tests.
From Meyer, R., 1987. Explosives, third ed. VCH Verlagsgesellschaft, Weinheim, Germany. ISBN:3-527-26599-6/0-89573-600-4.

The US Bureau of Mines' drop-hammer method has been modified by BAM, to improve the reproducibility of the data obtained. In the modified test, the sample is placed in a confinement device consisting of two coaxial cylinders placed one on top of the other and guided by a ring. All parts have the same hardness. Cylinders and rings are renewed after each test causing failure. If the sample is a powder or a paste, the upper cylinder is slightly pressed into the charged confinement device as far as it will go without flattening the sample. If liquids are tested, the distance between the cylinders is 1 mm. The charged device is put on the anvil of the drop-hammer apparatus, and the drop weight, guided by two steel rods, is unlocked. For sensitive explosives such as primary explosives, a small drop hammer is used; for insensitive explosives a larger hammer. The small hammer involves the use of drop weights of up to 1000 g, whereas the drop weights used with the larger hammer are 1, 5, and 10 kg. Then the drop heights are 10—50 cm for the 1-kg weight, 15—50 cm for the 5-kg weight and 35—50 cm for the 10-kg weight.

8.3.3 Electric Spark Sensitivity Test

8.3.3.1 Original US Bureau of Mines Test

In this method, described by Dorsett et al. (1960), the standard dust-layer thickness tested was 1.6 mm. The layer was resting on a 25-mm-diameter steel plate that also served as the negative electrode. The positive needle point electrode, connected to a capacitor bank charged to 400 V, was lowered by hand toward the surface of the dust layer until a spark discharge occurred. After an ignition occurred, the steel plate was cleaned, a new dust layer formed, and the process repeated at progressively lower capacitance values until the lowest that gave at least one ignition in 20 trials was identified. The minimum ignition energy was defined as $\frac{1}{2}CU^2$, in which C is the capacitance and U the charging potential of 400 V.

8.3.3.2 "Nordtest Fire 016"

This method, described by Nordtest (1982), is primarily intended to be used for pyrotechnics and explosives in pulverized form. It may, however, also be applied to man-made and natural combustible materials, which, when distributed as a thin layer resting on a flat metal surface, are able to propagate self-sustained combustion.

The test apparatus is illustrated in Fig. 8.6.

The dust/powder is poured gently into the disc-shaped cavities formed by the slidable supporting plate/hole plate assembly and excess dust is removed by a scraper. Planar, circular dust/powder samples of thickness

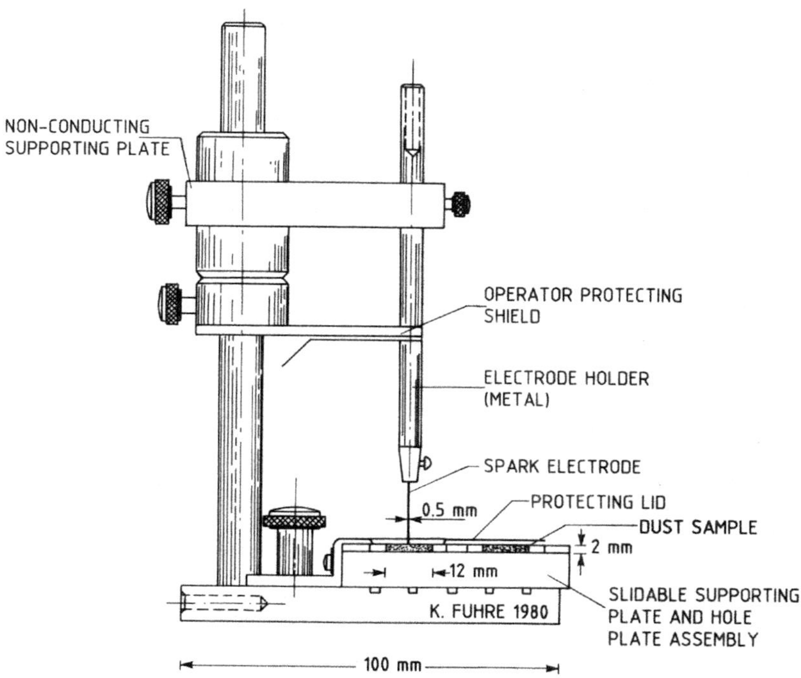

Figure 8.6 Cross-section of Nordtest (1982) apparatus for determination of electric spark sensitivity of layers of explosives, pyrotechnics, and propellants.

2 mm and diameter 12 mm are thus obtained. The metal bottom of the cavities acts as one of the two electrodes forming the spark gap. A thin tungsten wire pointing downward toward the dust/powder layer and with its tip just above the dust/powder surface, acts as the second electrode.

An electric spark of the desired net energy and discharge time is passed through the sample, and it is observed whether or not ignition occurs. An electric spark generator that permits independent variation of spark energy and spark duration is required. Twenty apparently identical tests are carried out at each combination of net spark energy and discharge time, yielding a frequency of ignition in the range 0—100%. If the first spark discharge at a given set of conditions does not give ignition of the dust sample exposed to the spark discharge, the same sample is just shifted horizontally a short distance to allow the next spark to pass through fresh dust/powder layer within the same sample. However, if ignition occurs, the whole sample tested is discarded and the test is continued with a new sample.

The minimum electric spark ignition energy, defined as the net spark energy yielding an ignition frequency of about 5%, is determined for various spark discharge durations Δt. The ultimate result of the test is an electric spark ignition sensitivity profile $E_{min}(\Delta t)$, as illustrated in Fig. 8.7.

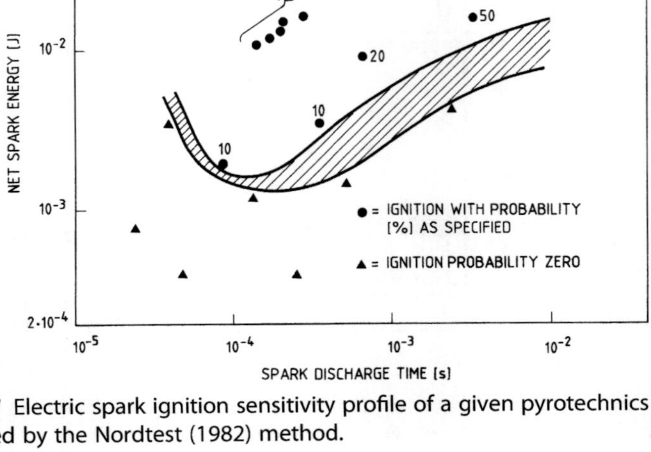

Figure 8.7 Electric spark ignition sensitivity profile of a given pyrotechnics product as determined by the Nordtest (1982) method.

8.4 CASE HISTORIES OF ACCIDENTAL EXPLOSIVES/ PYROTECHNICS/PROPELLANTS EXPLOSIONS

8.4.1 Ammonium Nitrate Explosion Catastrophe in Ludwigshafen, Germany (1921)

At 0730 on September 21, 1921, a major explosion occurred at Oppau near Ludwigshafen, in the works of the Badische Anilin und Soda Fabrik (BASF). The account given here is taken from Marshall (1987). The explosion killed 561 people including 4 who stayed in Mannheim 7 km away from the explosion center. Fifteen hundred people were injured, and in the town of Oppau 75% of the houses were destroyed. The total number of destroyed houses amounted to around 1000.

In the explosion about 4500 tonnes of a mixture of ammonium nitrate and ammonium sulfate blew up. The material was most probably detonated by explosive charges that were customarily used to break up caked material. According to one source, the material was a cocrystallized mixture of ammonium nitrate with ammonium sulfate, in the molecular proportions 2:1, which would correspond to about 58% ammonium nitrate by weight. Another source claimed that the mixture which blew up was 50% by weight of nitrate and sulfate. The same source claimed that it was proved impossible to detonate any mixture containing less than 60% nitrate, but that the mixture was not homogeneous and may have contained up to 70% nitrate in some places. If, however, as pointed out by Marshall (1987), the mean content was 50% nitrate, some places must have been considerably leaner than this, and it is difficult therefore to see why the whole heap blew up. The decision to use explosives to break up caked material was based on exhaustive tests, and many thousands of actual cake-breakup blastings had been conducted. According to BASF, the crater generated by the explosion was 10 m deep, 75 m wide and 115 m long. Within a radius of 250–300 m, all buildings were totally destroyed or so badly damaged that they had to be demolished.

8.4.2 Ammonium Nitrate Explosion Catastrophe in Texas City, USA (1947)

This account is also based on Marshall (1987). The disaster occurred at 0912 on April 16th and 17th, 1947 in the ships "Grandcamp" and "Highflyer." Grandcamp was carrying about 2300 tonnes of ammonium nitrate in paper sacks. According to one source, a fire had started among the sacks about an hour before the explosion. It was decided, to avoid damaging the nitrate, not to use water for extinguishing the fire. Instead, one tried to put the fire out by battening down the hatches and admitting steam. As pointed out by Marshall, admitting hot steam probably just worsened the situation by heating up the nitrate.

The explosion killed over 200 people who had crowded to the dockside to watch the fire, the ship's crew of 32 people, and 4 people in two aircraft circling overhead. Severe damage occurred and new widespread fires were initiated in the nearby Monsanto complex, as well as in the tank parks of a number of oil companies in the neighborhood. Pieces of wreckage traveled up to 1 km, a 1 tonne piece of propeller shaft traveled 4 km.

The day after, at 1310 on April 17th, the second ship, "Highflyer," which had caught fire the previous day as a consequence of the explosion on "Grandcamp," also blew up. This ship was also carrying ammonium nitrate and, according to some reports, sulfur.

The total death toll due to the two explosions was estimated at 552. In addition, 200 people were missing and 3000 injured. The material damage was estimated to be in excess of $100 million (1947 prices).

This disaster shows that, in some circumstances, ammonium nitrate anaerobic fires may escalate to a detonation even when barely confined. According to Marshall (1987), the hatches of the hold must already have been destroyed or displaced by the fire. The quantity of ammonium nitrate that actually exploded cannot be stated with certainty. Marshall (1987) suggests that about 2000 tonnes of ammonium nitrate may have detonated in each of the two explosions, and that this would have been equivalent to about 1000 tonnes of 2,4,6 trinitrotoluene (TNT) in each case. The explosions led to pronounced "domino" effects by starting fires in nearby chemical plant oil storages and dry goods warehousing.

From the actions taken, it seems clear that the crew on "Grandcamp" had no training in how to deal with anaerobic fires. Furthermore, the harbor was inadequately equipped to deal with major fires. The death toll was undoubtedly considerably aggravated by the phenomenon of "negative evacuation," ie, people flocked to the scene to witness the fire. Had the authorities acted (or been able to act) to evacuate the area, the death toll could have been far less. According to Marshall (1987), the disaster in Texas City is not the only example of people being killed while being spectators at the scene of a disaster.

8.4.3 Explosives-Mix Explosion in Mustang, Nevada, USA (1998)

8.4.3.1 Introduction

On January 7, 1998, at 0754, two explosions in rapid succession destroyed the Sierra Chemical Company (Sierra) Kean Canyon plant east of Reno near Mustang, Nevada, killing four workers and injuring six others. Because of the loss of life and extensive damage, the US Chemical Safety and Hazard Investigation Board (CSB) sent a team to investigate the explosion in an attempt to understand the causes of this accident. The present account is a brief summary of the comprehensive CSB (1998) report.

8.4.3.2 The Kean Canyon Plant and Its Production

The Sierra Kean Canyon plant manufactured explosive boosters, mixed custom flux for gold-smelting operations, and repackaged bulk soda ash for sale to the mining industry. When initiated by a blasting cap or detonation cord, boosters provide the added energy necessary to detonate less-sensitive blasting agents or other high explosives. The boosters manufactured at the Kean Canyon plant consisted of a base mix and a second explosive mix, called Pentolite, both of which were poured into cardboard cylinders. The operators working in the plant were responsible for the preparation of the explosive mixtures, the operation of the mixing pots, and the pouring of the mixtures into the booster cylinders. Fig. 8.8 gives an approximate outline of the part of the plant that was demolished by the accidental explosion.

The boosters were filled in the two Booster Rooms indicated in the figure. The primary explosives used in the base mix were 2,4,6 trinitrotoluene (TNT), pentaerythritol tetranitrate (PETN), and a mixture of TNT and hexahydro 1,3,5 trinitro 1,3,5 triazine (RDX). Pentolite was a mix of TNT and PETN.

8.4.3.3 The Explosions

CSB's investigation team identified four possible explosion scenarios that could lead to the explosions that occurred, but based on the seismic data, interviews of workers, and the physical evidence observed during the investigation, they considered one of these as being more likely than the others. They concluded that the first explosion occurred in the plant's Booster Room 2, indicated in Fig. 8.8, whereas the second, larger explosion followed 3.5 s later in the PETN drying building, which is also indicated in Fig. 8.8. The two explosions were recorded by the Seismology Laboratory at the University of Nevada, Reno, and they were also able to conclude that

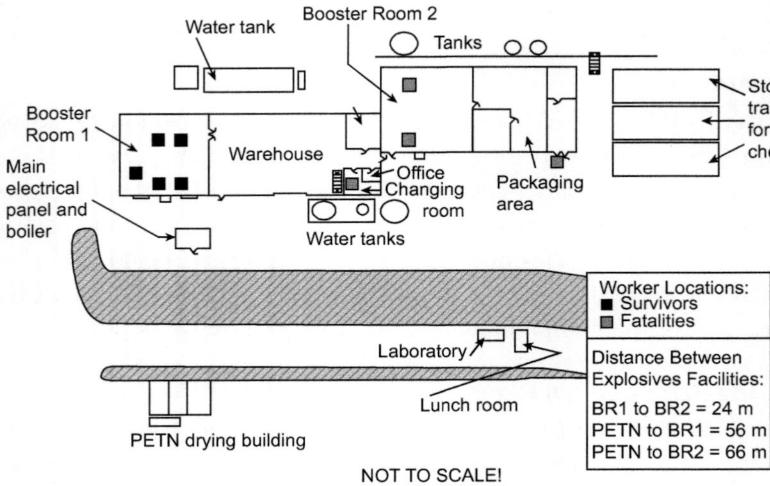

Figure 8.8 Approximate plan of the Sierra Chemical Company Kean Canyon plant east of Reno near Mustang, Nevada, United States.
From CSB, 1998. Explosives manufacturing incident, Sierra Chemical Company, Mustang, Nevada, 1998. Report No. 98-001-I-NV, U.S. Chemical Safety and Hazard Investigation Board (CSB), Washington, DC, USA.

the site of the first explosion was somewhat further to the north than that of the second, larger explosion, which in accordance with Booster Room 2 being located further to the north than the PETN building. The interval of 3.5 s between the explosions was estimated by the laboratory to be accurate to ± 0.2 s.

Most probably, the first explosion was initiated in a mixing pot in Booster Room 2 (see Fig. 8.8), when an operator who had left about 25–50 kg of base mix in his mixing pot the day before, turned on the mixing stirrer motor of the pot the next morning. During the evening and night the mix had stratified and solidified, and the explosion initiation occurred when the bottom of the mixer blade, which was embedded in the solidified explosives mix in the pot, forced the explosive material by impact, shearing or friction against the pot wall. Alternatively, explosive material was pinched between the mixer blade and the pot wall, causing the initial detonation in the pot. The shock wave from this initial detonation then instantaneously detonated the other several thousand kilograms of explosives in the Booster Room 2, leading to the first main explosion that demolished Booster Room 2, and also the rest of the building. Only three walls were left of Booster Room 1, as indicated in Fig. 8.9, whereas Booster Room 2 was completely leveled with the earth, as can be seen in Fig. 8.10.

Figure 8.9 Remains of Booster Room 1 of the Sierra Chemical Company Kean Canyon plant east of Reno near Mustang, Nevada, USA, after explosives mix explosion in 1998. *From CSB, 1998. Explosives manufacturing incident, Sierra Chemical Company, Mustang, Nevada, 1998. Report No. 98-001-I-NV, U.S. Chemical Safety and Hazard Investigation Board (CSB), Washington, DC, USA.*

Figure 8.10 Remains of Booster Room 2 of the Sierra Chemical Company Kean Canyon plant east of Reno near Mustang, Nevada, USA, after explosives mix explosion in 1998. *From CSB, 1998. Explosives manufacturing incident, Sierra Chemical Company, Mustang, Nevada, 1998. Report No. 98-001-I-NV, U.S. Chemical Safety and Hazard Investigation Board (CSB), Washington, DC, USA.*

When the first explosion occurred, a worker in Booster Room 1 saw a huge fireball engulf a truck, which was parked immediately outside the building. This worker was thrown against one of the walls in Booster Room 1, as the ceiling and another wall of the room collapsed on top of him and the four other workers. Seconds later a second, the same worker, heard the even more powerful (louder) second explosion in the PETN building. CSB concluded this explosion was most likely initiated by a heavy piece of equipment or burning debris from the first explosion falling through the reinforced concrete roof or the skylight of the building. The site of the PETN building and an adjacent magazine was now transformed to a 13 m by 20 and 2-m-deep crater. The explosions were felt as far as 35 km away.

A total of 11 employees were at the site at the time of the explosions. After the explosions, five workers in Booster Room 1 were trapped temporarily under the collapsed building, but were able to crawl out within a few minutes. Three of the five workers were seriously injured, whereas two only received minor injuries. Concerned about possible additional explosions, the workers from Booster Room 1, after calling for other possible survivors, went to the entrance to the facility. There they met two other workers who had been in the gravel pit below the site, approximately 100 m southwest of the PETN building. The other four workers who were in or near Booster Room 2 were killed by the explosions.

The blast effects of the explosions leveled the site and threw structural materials, manufacturing equipment, raw materials from the booster and flux operations, and other fragments up to 1 km away. The legs and cross bracing from an empty tank, which previously stood at the corner of the changing room, were imbedded in a motor home located 300 m from the production building. The doors of one of the large magazines and a portable magazine located west of the production facility were sprung open by the negative pressure pulse; however, large quantities of explosive materials that were stored inside did not detonate. Many undetonated boosters had been scattered throughout the site by the explosion. Other hazards at this time included fires, toxic chemicals, and potential detonation of the explosives in Booster Room 1 as the fire progressed.

8.4.3.4 Events Leading to the Explosion

The CSB investigation concluded that the following sequence of events led to the explosion catastrophe. The day before the accidental explosion, one melt/pour operator working in Booster Room 2 (see Fig. 8.8) needed to leave work early, at 1500. When he left, there were between 25 and

50 kg of base mix left in one of his mixing pots (No. 5). He mentioned that he had left some explosive mix in the pot to another operator in the room, who later confirmed this by visual inspection. This other worker, from having looked into the mixer, indicated that the depth of explosive mix left in pot 5 was about 100 mm, which matched his estimate of there being 25—50 of explosives in the mixer. The mixer blades extended about 50 mm down into the mix. Metallurgical analysis of mixer parts retrieved after the explosion provided further evidence supporting the conclusion that explosive material was left in pot 5, because the analysis showed that damage to the hub of the mixing blade was consistent with it having been in contact with detonating explosives.

An overcurrent protection device on the electrical mix motors in Booster Room 2 would stop the motor if the blade was unable to break up the explosives, but not before the start-up torque was applied to the explosives. Due to the solidified material in the explosives draw-off line on the bottom of the pot, it would be impossible for the explosives to simply break free of the pot without causing friction with the interior of the pot and shearing a portion of the explosives in the draw-off line.

Without continuous agitation, the different explosives and binders of the mix tended to stratify due to their different densities. This stratification would increase the sensitivity of portions of the explosive material left in the pot. Turning on an agitator immersed 50 mm into a solidified mass of stratified explosives would present a high risk of explosion from the impact.

Leaving material in the mixing pots overnight was not according to common practice in the plant, but this was not for safety reasons. Several months before the explosion, when material had been left overnight in the Comp B mixing pot in Booster Room 1, management made it clear that this was an unacceptable practice because it delayed the operation of the day-shift workers.

The operator who left the material in his pot had been working in Booster Room 2 for 8 weeks prior to the incident. His normal practice was to leave both of his mixing pots empty in agreement with common practice. Because he was the only person working his production line, he would normally know whether his pot was empty when he started work the next day. Because the pots in Booster Room 2 heated material much faster than the pots in Booster Room 1 could do, it is possible that he on the day before the explosion thought that leaving material in the pot would not delay production the next day. It was acceptable practice at the Kean Canyon facility to alter normal processes without discussion or management approval.

At the end of each day, operators were instructed to leave a steam-line valve to each pot partially open to keep the boiler cycling to prevent freezing of condensate in the lines. This amount of steam would be insufficient, however, to maintain any quantity of explosive mix in the pots above its melting point if outside temperatures were below the freezing point. During the night before the explosion, the outside temperature dropped to between −4 and −7°C. The operator who had left explosives in his pot offered the remaining material to the operator on the other production line in Booster Room 2. Because the operator who was leaving did not reach a firm agreement on whether the second operator would use the residual explosives, it is possible that no steam valves were left open that afternoon because leaving the valves open would make the remaining base mix too runny to pour. The operator who left early may have mistakenly thought that his remaining base mix would be used that afternoon and, thus, he failed to look in the pot the next morning before turning on the steam and mixer motor. The investigation team concluded that this was the most likely scenario.

8.4.3.5 Key Findings and Root/Contributing Causes

- US standards require that companies using highly hazardous materials have in place an integrated safety management system. CSB's investigation of the Kean Canyon explosion catastrophe revealed that many essential elements of process safety management were missing or deficient. The process hazard analysis (PHA) of the facility and operations was inadequate, and supervisors and workers from the plant had not been involved. The PHA for Booster Room 1 was conducted by company personnel from other locations and did neither consider safe location of buildings nor human factors issues. These deficiencies in the PHA program allowed unsafe conditions and practices to remain unrecognized and uncorrected. No PHA was conducted for Booster Room 2.
- There was no safety inspection or safety auditing program. Safety "walk-through" inspections were unfocused and did not examine process safety management (PSM) program effectiveness. The result was that management was generally unaware of unsafe practices and conditions.
- The training programs for personnel working in the plant were inadequate. Worker training was conducted primarily in an ineffective, informal manner that overrelied on the use of on-the-job training. Poor management and worker training led to a lack of knowledge of the hazards involved in manufacturing explosives.

- Written operating procedures were either nonexisting or not available to the working personnel. Managers believed that, without using a blasting cap, it was almost impossible to detonate the explosive materials they used or produced. Personnel primarily relied on experience to perform their jobs. Procedures and other safety information were not available in the language spoken by most workers. Operators routinely made changes in the steps they took in manufacturing explosives. This resulted in the use of inconsistent and hazardous work practices. There were no written procedures for Booster Room 2.
- The facility was built with insufficient separation distances between different operations, and the construction of buildings was inadequate. Because unrelated chemical operations were located in the same building as Booster Room 2, an additional fatality and extensive property damage resulted. Close proximity of structures allowed the explosion to spread to a second building.
- Safety inspections by regulatory organizations were conducted infrequently and inspectors generally did not have expertise in explosives manufacturing safety. This allowed unsafe conditions in the plant to remain uncorrected.
- Employees had not been involved in developing or conducting process safety activities. This resulted in a lack of understanding of process hazards and controls by workers. It also resulted in management not benefiting from the experience and insights of workers.
- Reclaimed, demilitarized explosive materials purchased by Sierra sometimes contained foreign objects. The risk of using contaminated explosive materials had not been adequately examined.

8.4.4 Pyrotechnics Disaster in Enschede, The Netherlands (2000)

8.4.4.1 Introduction

The following short summary is based on the accounts by Bruyn et al. (2002) and Weerheijm et al. (2002), which are in turn based on the results of the extensive forensic and technical investigations of this catastrophic accidental fire/explosion undertaken by Netherlands Forensic Science Institute (NFI) and TNO Prins Maurits Laboratory (TNO-PML).

8.4.4.2 Overall Development and Consequences of the Disaster

The disastrous fire/explosion occurred on Saturday afternoon May 13, 2000 on the premises of the company S.E. Fireworks (SEF) in the city of

Enschede in the Netherlands. SEF imported and traded with fireworks, and also performed firework displays and shows. Within an hour, the incident developed from a moderate fire and some initiated fireworks in one of the buildings into a series of three explosions of increasing violence. The first was a relatively small explosion in a container. Within a minute, seven garage boxes (prefabricated concrete storage facilities) exploded. This was followed about 1 min later by a further explosion of storage cells in the central storage building, whereby the most violent explosion occurred in storage cell C11. The resulting blast wave was comparable to that from explosion of 4000—5000 kg of TNT. Most of the evidence for quantifying the final explosion strength was obtained from the window breakage and the observed damage of about 700 houses in the surroundings.

Twenty-two people were killed and 947 were injured. A complete residential area was destroyed with about 500 houses being completely demolished and about 1350 partly damaged. The main cause of house damage was massive conflagration of typical old houses with wooden floors, initiated by the throw out of fireworks.

8.4.4.3 The Situation Prior to the Fire/Explosion

The layout of the facility prior to the fire/explosion is illustrated in Fig. 8.11.

The facility consisted of a central storage building containing 14 storage cells (C2—C15), 7 prefabricated garage boxes (M1—M7) and 14 ISO containers (E1—E14).

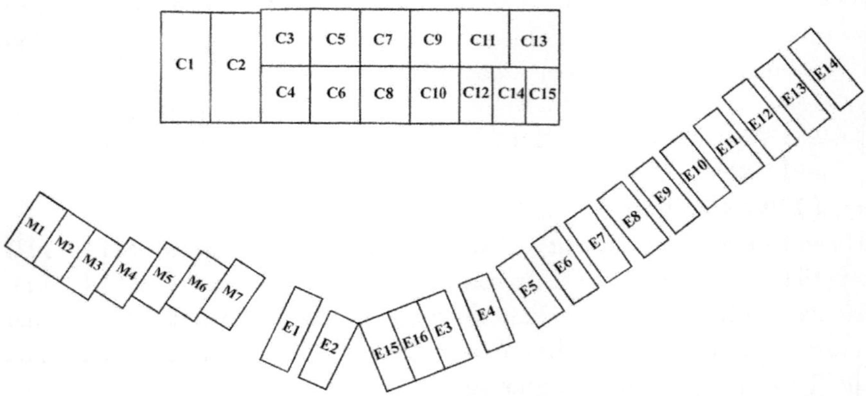

Figure 8.11 Layout of the pyrotechnics storage premises of the company S.E. Fireworks (SEF) in the city of Enschede in the Netherlands in 2000.
Based on figure from Weerheijm, J., Bruyn, P.C.A.M. de, Wees, R.M.M. van, Karelse, J.W., 2002. The fireworks disaster in Enschede. Part 21: Safety analysis. Proceedings of the 6th International Conference Probabilistic Safety Assessment and Management, 23—28 June 2002, San Juan Porto Rico, USA.

The central storage building had wooden doors and was constructed in cast reinforced concrete with a wall and roof thickness of 20 cm. Cell 2 was used for fireworks preparation and workshops. The majority of the cells in the central building (C3–C11 and C13) were 4 m wide, 4 m deep, and 2.8 m high, whereas three smaller cells (C12, C14, and C15) were only 2.5 m wide. The walls and roof of the prefabricated concrete boxes had a thickness of 5 cm. Six of these boxes (M1–M6) were 2.6 m wide, 5.2 m deep, and 2.3 m high (volume 31 m^3), and fitted with the original thin, corrugated steel sheet doors. M7 was slightly larger and had a wooden door. To prevent direct flame contact between cell doors in case of fire, the walls in the central building and the garage boxes were extended externally by 0.5 m. The International Organization for Standardization (ISO) containers were standard 6-m-long (20 ft) transport containers, with wooden floors. They were not equipped with any additional fire protective measures.

The total licensed quantities for storage were 158.500 kg (gross mass) of 1.4S or 1.4G fireworks. In some cells, 3G fireworks could be stored, the maximum permissible stored quantity for the whole facility being 2000 kg.

8.4.4.4 Sequence of Main Events

Due to the fact that many people that were attracted to the accident, several private video recordings taken from various angles were made available to the investigators. These were helpful in the reconstruction of the disaster. The reconstructed sequence and times of the main events were:

1445: First observation of fire in C2 (ejection of fireworks)
1500: Fire reported to fire brigade
1508: Preliminary inspection of the site by the fire brigade
1516: Fire in C2 under control
1524: Observed smoke and fire effects from C4
1533: Observation of smoke from between containers E1 and E2
1534: Massive deflagration and minor explosion of contents in container E2
1540: Massive explosion of garage storage boxes M1–M7
1546: Explosion in C11 in the central building followed almost simultaneously by explosions in the other cells in the main building, and in a number of the transport containers (E).

Fig. 8.12 shows the fireball from the final explosion from distance of about 600 m, whereas Fig. 8.13 shows a top view of the completely demolished explosion area, with search sections taped out for the forensic investigation.

Figure 8.12 Photograph, taken at a distance of 600 m, of fireball created by the final major explosion on the pyrotechnics storage premises of the company S.E. Fireworks (SEF) in the city of Enschede, the Netherlands, in May 2000.
From Bruyn, P.C.A.M. de, Weerheijm, J., Karelse, J.W., Wees, R.M.M. van. 2002. The fireworks disaster in Enschede. Part 1: overview and reconstruction. Proceedings of the 6th International Conference Probabilistic Safety Assessment and Management, 23—28 June 2002, San Juan Porto Rico, USA.

Figure 8.13 Top view photograph of the explosion area after the final major explosion on the pyrotechnics storage premises of the company S.E. Fireworks (SEF) in the city of Enschede in the Netherlands, in May 2000.
From Bruyn, P.C.A.M. de, Weerheijm, J., Karelse, J.W., Wees, R.M.M. van. 2002. The fireworks disaster in Enschede. Part 1: overview and reconstruction. Proceedings of the 6th International Conference Probabilistic Safety Assessment and Management, 23—28 June 2002, San Juan Porto Rico, USA.

8.4.4.5 *Initial Fire in Cell C2*

The fire in C2 was first noticed at about 1445 h when activated firework was ejected and landed outside the SEF premises. A small fire in a garden was reported. When the fire brigade arrived, the doors of cell C2 were open at both sides and the plastic skylights were gone. Some fast pressure buildup must have occurred because the doors on both sides were blown out. No external blast damage was noticed. Possible mechanisms for initiation of the fire/explosion in C2 included:

- ejected fireworks
- sabotage, arson or improvised explosive devices
- malfunctioning or defects of the electrical and gas installation or other equipment
- fire caused by human activities on the SEF area
- (self-)ignition of firework articles, pyrotechnics or other fuels caused by instability or external effects

The effects of the fire in C2 were:

- ejected fireworks, with the possibility of fire initiation outside the main building
- fire jets and heat radiation directed toward the garage boxes (M) and containers (E)
- fire heating of the walls to adjacent C-cells

The fire in C2 was attacked from both sides of the central building. However, the firemen on the side facing the garage boxes (M) were killed in the subsequent explosion, and hence no direct witness reports of the fire development on that side of the building were available.

8.4.4.6 *Initiation and Possible Consequences Fire/Explosion in E2*

The following possible initiation mechanisms for the fire/explosion in E2 were examined:

- external fire
- fireworks on and/or under the container
- burning magnesium (ejected from the workshop) on top of container
- fireworks before door slit

Considering the timeframe, the very limited fire resistance of the container, the presence of an old small trailer with wooden floor between E1 and E2, a small fire would be sufficient to initiate a fire and activate the fireworks in E2. The effects following ignition in E2 were:

- smoke from door slit
- strong smoke development followed by intensive firework reactions

- flame jet (in two pulses)
- ejected fireworks and very severe external massive reaction of fireworks

For the possible consequences of the E2 reaction for the further escalation of the disaster, the following effects were examined:

- ejection of debris
- failure of container (fragments and blast)
- flame jet (and possible external fireball)
- ejected fireworks

8.4.4.7 Initiation and Possible Consequences of Fires/Explosions in the Garage Boxes

The damage analysis strongly indicates that the most severe explosion of those in the garage boxes occurred in M7. The observed debris velocity of up to 200 m/s was most probably caused by the failure of the 5-cm-thick walls of the prefabricated garage box M7 The resulting severe crushing of the fireworks in M6 and the thermal loading caused the exothermal chain reaction of the fireworks in M6, and subsequently in the other cells. The reactions in the garage boxes occurred within the timeframe of less than 0.5 s. The resulting explosion blast was comparable to that of the explosion of 800 kg of TNT. A fireball of diameter 170 m was formed. The garage boxes were completely destroyed and fragmented into small debris. The combined effect of blast, flame, and debris was sufficient to initiate the final major explosions in the central building and the remaining E containers.

8.4.4.8 Initiation and Explosion Effects of Explosion in Central Storage Building

Considering the central building, the strength of the explosion was far sufficient to blow the wooden doors into the cells and the fireball engulfed the whole building. The contents of all cells could have been ignited. The local damage, however show clearly that the explosion in storage cell C11 was most severe and dominant. A single explosion in C11 and the sequential sympathetic reactions in the other cells can explain the total damage. In analogy with the garage box analysis, the required local pressure and gas pressure were related theoretically to the required reacted mass of pyrotechnics per second to explain the observed damage and the sympathetic reactions in the adjacent cells. Crucial in the explanation is the reaction velocity of the fireworks in C11. Suggested root causes of the devastating mass explosion in C11 included:

- storage of firework of the transport class 1.1
- combined storage of 1.3G and 1.1 class fireworks. Fireworks of the class 1.3G were stored, but due to door impact the packages were severely damaged and the firework obtained the 1.1 characteristics

- after initiation of the stored 1.3G class fireworks, temperature and confinement conditions accelerated the deflagration process toward a detonation-like reaction flame jet and possible external fireball

8.4.4.9 Observed Explosion Effects

The major explosion effects were craters, fireballs, blasts, and debris. The video recordings and the damage at the explosion area showed that the sequence of three main explosions, ie, in container E2, in the garage boxes (M1—M7) and the central building had successively increasing strengths. For this reason the final disastrous explosion destroyed evidence and traces of the preceding explosions, which hampered detailed analysis. Nevertheless, the following conclusions could be drawn from remaining evidence.

No crater or evidence for blast damage due to the E2 explosion was found. A very severe fireworks fire and the ejection of firework articles from the container into the surroundings characterized the "explosion" in E2. The observed effects correspond to fireworks of transport classification 1.3G.

As regards the second main explosion in the garage boxes M in Fig. 8.11, video recordings showed that the fireball of the explosion increased to a diameter of about 85 m in 0.3 s. The explosion appeared to be due to an exothermal reaction of the contents of the garage boxes, which started in M7 and terminated in M1. All the boxes were completely destroyed, and no remaining debris could be recollected. However, the video recordings showed debris being ejected at velocities on the order of 200 m/s. The exothermal reactions were violent, but a detonation did definitely not occur. The concrete floors showed in the garage boxes were severely cracked, and the floor slab of M7 was moved more than 1 m horizontally, and a large depression of the soil at the original location was found. However, no crater was formed. The walls of the boxes were clearly sheared off and the direction of the deformed reinforcement bars formed clear evidence for the propagation direction of the chemical reactions in the cells (from M7 toward MI). In most cases, the blast strength of an explosion can be quantified from the building damage in the surroundings and especially from the window breakage. The final explosion destroyed most of this evidence as regards the explosions in the garage containers. In one video recording of the second explosion, however, window and roof tile damage is visible. The radius of the observed fireball corresponded to exothermal reaction of about 17.000 kg of propellant. Further valuable information about the strength of the explosion in the garage boxes was obtained from the seismic signals that were recorded of both of the two last major explosions. The ratio of the signals, the local damage, and the blast damage to the surrounding area suggested that the explosion in the garage containers produced a blast wave of the same strength as that produced by about 800 kg of TNT. The strength of the explosion in the garage boxes was by far sufficient to blow the wooden doors into the cells of the central building. Furthermore, the fireball engulfed the whole central building, and the contents of all the C-cells were probably ignited. The central building was completely destroyed (see Fig. 8.13). Violent exothermal reactions occurred in many of the cells in the central building, but the final, major explosion was clearly dominated by the reaction in C11. This emerges from the facts that a large crater of depth 1.3 m was formed just under cell C11. Although the crater extended to the areas below some other C-cells, it was clearly the result of a single major explosion event. The floors of the other cells showed no damage of the type to be expected from independent, severe explosion reactions. Only the edges of the cell floors adjacent to C11 were severely damaged and contributed to the crater.

Debris from the central building caused many casualties and severe damage to buildings. Debris was found up to a distance of 580 m from the explosion center. From the angle of impact and the traveling distance of the largest debris, ejection velocities could be estimated. Because the angle of ejection was unknown, only an approximate range of ejection velocities could be determined. For most of the collected debris, the estimated velocities were in the range 30–100 m/s. The highest estimated velocity was 150 m/s.

The diameter of the final major fireball was 135 m, corresponding to exothermal reaction of about 86.500 kg of pyrotechnics/propellants. It is evident that this fireball was not caused by the contents of cell C11 only. Exothermal reactions of the pyrotechnic material stored in other cells and also the storage containers must have contributed. On the video recordings, the shape of the fireball shows quite clearly some additional "sub sources" which may be caused by container reactions.

8.4.4.10 Conclusions

- Despite extensive investigation, no definite evidence for the initial cause of the chain of events had been found by the time of the publication of the papers by Bruyn et al. (2002) and Weerheijm et al. (2002). There was no indication of sabotage, and no traces of high explosives had been detected, only of fireworks-related substances.
- In addition to the initial fire in the central building, three other crucial events promoted the escalation of the accident. These were the severe firework reactions in container E2, followed by the explosions in the

garage boxes and finally by the massive explosion in the storage cell C11 of the central building, accompanied by the reactions in the other storage cells and containers.

- Window breakage, structural damage, crater dimensions, debris, and the seismic signals enabled quantification of two major explosions. The first occurred in the garage boxes and had a blast strength equivalent to about 800 kg TNT. The blast produced by the final, much more powerful explosion had a strength equivalent to about 4000−5000 kg TNT.
- Despite the violent character of the explosions, they were all subsonic deflagrations. This is in accordance with the nature of the exothermal reactions being the cause of the explosions. No evidence of detonations was traced.
- The types of stored firework articles, the excessive quantities stored, and partly inadequate storage conditions, probably made it possible for the event to escalate from the initial moderate fire to the final disastrous explosion.
- Much larger quantities of some types of fireworks were stored than licensed, which contributed to making the storage unsafe. If the stored quantities had been in agreement with the premises of the licenses granted for storage, the fire in the workshop of the central storage building could never have escalated to the disaster that actually occurred.
- The minimal fire resistance of the storage containers, and inadequate safety distances between the various storage units, also contributed to the escalation.
- Weerheijm et al. (2002) concluded that further research would be needed to enable definite conclusions to be drawn on the origin and stepwise escalation of the catastrophic explosions in Enschede, and also for enabling specification of improved requirements for safe storage and transportation of the types of pyrotechnics that gave rise to the catastrophe.

8.5 MEASURES FOR PREVENTING AND MITIGATING ACCIDENTAL EXPLOSIONS OF EXPLOSIVES, PROPELLANTS, AND PYROTECHNICS

8.5.1 General Safety Considerations for Facilities Producing Explosives According to Sejlitz (1987)

8.5.1.1 Basic Rule

The planning of a process and/or a work place should always be based on the following three-point basic rule:

- a ***minimum number of persons*** for handling
- the ***smallest possible quantity*** of explosive material in
- the ***shortest possible time***

It is interesting to note that the basic rule is in a way an adaption of the principles of design of "inherently safer plants" (see Chapters 2 and 7) to the explosives, pyrotechnics, and propellants industries.

8.5.1.2 Characteristics of the Explosive Material

Before conducting any work with explosives, propellants, and pyrotechnics, all their basic safety properties must be known. These data make it possible to evaluate both the various risks of ignition/initiation associated with handling and processing the material and the expected consequences of such ignition/initiation. The most important basic safety properties are (see Sections 8.2 and 8.3):

- ignition/initiation sensitivity (eg, impact, friction, heat, and electric/electrostatic sparks)
- chemical stability (eg, possibility of exothermal decomposition during storage)
- effect of ignition/initiation (eg, when the material is unconfined, enclosed, or in the processing vessels/containers
- health risks (eg, acute and long-term exposure)

8.5.1.3 Explosion Protection

On the basis of the basic safety data of the material(s), one can decide whether the planned work:

- can be performed in an open work room, or whether it
- must be performed under shielded conditions

The degree of rigidity of shielding depends on the severity of the effects of ignition/initiation of the material in question, and may take the form of:

- simple protection screens inside the work room
- smaller, laboratory-scale bunkers inside the work room
- special full-scale bunkers (often with pressure-resistant observation windows)

8.5.1.4 Automation and Remote Process Operation

In the case of dangerous operations, it is often preferable to automate the process and run and control it remotely. However, this may also sometimes introduce additional risk elements, such as:

- Reduced operator attention may result in operational errors being detected too late.

- Automated, remotely operated processes may prevent the operator from acquiring the knowledge of the basic processes that he may need in the case of irregularities and errors.
- The possibilities of efficient interaction with the process, in case of irregularities and errors, are reduced
- The work may appear less interesting and challenging to the workers

8.5.1.5 Quantities of Explosive Material

As already mentioned, the quantities of explosive material being handled at any time at each work place should be as small as practically feasible. The following principles should be taken into account:

- For a given production volume per day, the mass of material handled at any work station at any time gets smaller if the transfer rate of explosive material through the process (mass per unit time) increases
- A well-designed material flow system permits frequent transports of explosive material to and from the work rooms, which reduces the amount of explosive material stored in the work rooms at any time. This makes it possible to keep the buffer stores of explosive material in the work rooms at a minimum

8.5.1.6 Avoidance of Accidental Ignition/Initiation

The equipment in which explosive materials are processed/treated/handled shall be designed to ensure that the materials are not exposed to:

Mechanical Impact

Only approved equipment and tools shall be available at the work place. Experience has shown that if nonapproved equipment such as hammers, rubber mallets, crowbars etc. have been available, operators have used these to solve problems that have arisen, and by doing sc initiated explosion accidents.

Friction

Equipment and tools shall be designed/constructed in such a way that dust cannot enter threaded couplings, bearings etc. Furthermore, foreign bodies and impurities (hand tools, bolts, and nuts, gravel from transportation boxes etc.) must be prevented from entering apparatuses/equipment containing explosive material, thereby giving rise to explosion initiation by friction.

Exothermal Chemical Decomposition

Accidental explosions have occurred in cases in which explosive materials that have first been made chemically unstable by impurities or heating, have been enclosed in eg, a ball valve or between the walls of a double-wall pipe. Whenever such possibilities exist, the equipment/apparatus must be controlled and cleaned regularly.

Adiabatic Compression

This can occur during pressing and other treatment of gun powder, if the air in the pressing tool has not been evacuated before pressing, or if the air trapped in the gun powder is being subjected to a pressure rise by other means.

Heating/Fire

All equipment/apparatus in which explosive materials are heated electrically shall be equipped with several independent protection systems against overheating. Hot work (welding etc.) always presents a significant fire/explosion risk factor if carried out on premises in which the rooms and the equipment/apparatuses have not been adequately cleaned before the hot work is initiated.

Electrostatic Discharges

Static electricity is always a significant risk factor when producing/treating/handling very sensitive explosive materials. The possibility of ignition/initiation by electrostatic sparks (discharges between two electrically conducting electrodes) can be effectively eliminated by connecting all conducting components/apparatuses/equipment/tools, as well as the operator, to earth. For the operator, using electrically conducting shoes is only effective if the floor is also conducting and connected to earth. If this is not the case, the operator must be equipped with a separate cable to earth. If the material can be ignited/initiated by brush discharges, special precautions must be taken to avoid such discharges.

8.5.1.7 Good Practices for Apparatus Design

Quite often, accidental initiation/ignition of explosive materials occurs because they become exposed to excessive mechanical forces in contact with parts of apparatuses or tools, or to heat. Such accidents can occur even under normal operating conditions, but they are more common under circumstances deviating from normal operation, eg, startup, shut-down, cleaning, maintenance/overhaul, repair, and irregularities in the performance of equipment/apparatus.

It is important to consider the possibilities of irregular performance of equipment/apparatus, and the ease by which it can be cleaned, maintained, overhauled, and repaired, at the earliest stage of selecting/designing new equipment/apparatus. Careful consideration should be given to the following aspects:

Selection of Construction Materials

The construction materials used in equipment/apparatus must not be able to react chemically with the explosive materials produced/treated/handled in the equipment apparatus. They should also be able to withstand mechanical stress and wear due to rubbing etc. On the other hand, the use of mechanically strong materials may increase the risk of ignition/initiation by impact, rubbing etc., and in such cases softer and less mechanically strong construction materials will have to be used. However, all types of construction materials can represent an ignition/initiation risk if dimensioned/used in an incorrect way. There are no materials that are either fully adequate or fully inadequate in any kind of practical situation.

Free Space Around Moving Parts of Equipment/Apparatus

Hazardous frictional stresses may be generated if an explosive material gets squeezed between moving parts of equipment/apparatus and another object, eg, a tool. To prevent other objects from creating such situations, the free space around moving parts should either be so large that the introduction of another object will not be able to create hazardous stresses, or so small the another object will not be able to enter the space.

Design and Construction of Equipment/Apparatus

It is essential that the design and construction of equipment/apparatus facilitates easy cleaning. It is a good working principle that equipment/apparatus should be designed and constructed in such a way that complete cleaning or inspection can be performed daily in a safe and efficient way. This requires that:

- All surfaces are smooth, preferably honed or polished.
- There are no narrow gaps, grooves, holes or cracks in which explosive material can escape the cleaning operation.

- There are easy-to-handle and well-dimensioned couplings and other fixing elements, and that bolts and nuts are secured in an adequate way.
- Pipelines are easy to dismantle and inspect.
- There are effective arrangements and procedures to prevent spill of explosive material, and accumulation of dust from such materials

8.5.1.8 Written Work Instructions

In addition to a good process design, it is important that the operator has the knowledge, skills, and motivation to run the process in a safe way. Hence, well-written, easy-to-understand work instructions shall always be readily available. These may include information on the following aspects:

- Properties of the explosive material(s)
- The work operations to be undertaken
- Risks associated with the work
- Guidelines/regulations to ensure safe work procedures
- Measures to be taken in case of deviation from normal material and process conditions
- Routines for cleaning/housekeeping
- Control and maintenance of production equipment
- Special work instructions for more comprehensive and difficult installation and maintenance work

8.5.1.9 Ten Mnemonic Rules

- Work with the smallest practically possible quantity of explosive material at any time
- Be well informed about methods of work and safety instructions
- Make sure that there are adequate escape routes
- Use only approved tools and other apparatus/equipment
- Keep the work place in order. Materials, tools, and apparatus/equipment, which are not used for the time being shall not be stored in the work rooms
- Give notice of/correct any errors/undesired deviations from normality
- Make sure that all explosive materials that are not being processed are well packed and clearly marked
- Make sure that all trash/refuse explosive materials are well packed and clearly marked
- Never conduct production work with explosive material in a store room for explosive materials
- Keep the work place clean and tidy at any time. Spill of explosive material should be removed immediately (by sweeping)

8.5.2 General Safety Recommendations for Facilities Producing Explosives Suggested by CSB (1998)

Partly based on the experience gained from investigating the Sierra's Kean Canyon plant explosion catastrophe in 1998, CSB (1998) recommended that the following aspects be considered:

- Adequate process safety management (PSM) in explosives manufacturing facilities requires both careful planning and careful implementation.
- Prevention of primary explosions, as well as further propagation of explosions, requires clear understanding of explosives safety principles and safe practices.
- Process hazard analyses (PHAs) shall include examination of safety-distance requirements, building design, human factors, past incident reports, and lessons learned from explosives manufacturers.
- Written operating procedures shall be specific to the actual process of concern and address all phases of the operation.
- Procedures, chemical hazards, and process safety information shall be communicated in the language(s) understood by personnel involved in manufacturing or handling of explosives.
- Explosives training and certification programs for workers and line managers shall provide and require demonstration of a basic understanding of explosives safety principles and job-specific knowledge.
- Process changes, such as the construction or modification of buildings, or changes in explosive ingredients, equipment, or procedures are analyzed and PSM elements shall be updated to address these changes.
- Prestartup safety reviews shall be performed to verify operational readiness whenever changes/modifications have been made.
- Adequate implementation of all elements of mandatory official process safety management standards shall be verified by performing periodic assessments and audits of safety programs.
- Employee-participation programs shall effectively include workers and address and resolve their safety issues.
- Explosives safety programs shall provide an understanding of the hazards and control of detonation initiation sources. These include:
  - foreign objects in raw materials
  - impact by tools or equipment
  - impingement
  - friction
  - electrical/electrostatic sparks
  - other electrostatic discharges

Specific handling requirements for raw materials shall be followed strictly. Substitute raw materials may introduce additional ignition risks requiring special concern.

CSB (1998) further suggests that the following issues be addressed in plant design or plant modification:

- Unrelated chemical or industrial operations or facilities shall be separated from explosives facilities using quantity—distance guidelines
- Operations in explosives manufacturing plants shall be separated by adequate intraplant distances to reduce the risk of propagation
- Facilities shall be designed to reduce secondary fragmentation that could result in the propagation of explosions.
- Process and safety training guidelines for personnel involved in the manufacture of explosives shall be developed and disseminated. Such guidelines shall include methods for the demonstration and maintenance of proficiency.
- Safety guidelines for screening of reclaimed explosive materials shall be developed and disseminated.
- Adequate frequencies of safety inspections of explosives manufacturing facilities shall be maintained due to their potential for catastrophic accidental explosions.

CHAPTER NINE

Destructive and Harmful Effects by Pressure, Solid Fragments, and Heat From Accidental Explosions

9.1 EFFECTS OF EXTERNAL BLAST WAVES FROM EXPLOSIONS

9.1.1 Basic Blast Wave Characteristics

The problem addressed is the harmful/damaging effects of *external* blast waves from the various kinds of explosions discussed in previous chapters. Pressure development inside fully or partly confined areas where explosions take place are not discussed here, but in the various previous Chapters (2, 5 and 7). Predicting strengths of blast waves from gas/vapor, spray/mist, and dust explosions is a very complex matter. Fig. 9.1 may serve as a simple, qualitative illustration of the problem. Consider a given volume of just air that is at the outset fully enclosed and compressed to an overpressure P_e on the order of, say, 10 bars (g).

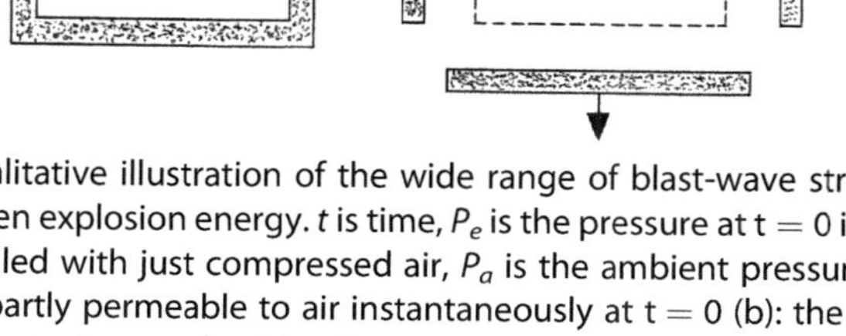

Figure 9.1 Qualitative illustration of the wide range of blast-wave strengths that can arise from a given explosion energy. *t* is time, P_e is the pressure at t = 0 inside an initially closed vessel filled with just compressed air, P_a is the ambient pressure (a): the vessel walls become partly permeable to air instantaneously at t = 0 (b): the vessel walls are fully removed instantaneously at t = 0.

Explosion Hazards in the Process Industries
ISBN: 9780128032732
http://dx.doi.org/10.1016/B978-0-12-803273-2.00009-8

Fig. 9.1a illustrates the situation in which the solid wall is at some moment made instantaneously partly permeable to airflow. The wall permeability may then be defined as the mass of air that flows through the wall per unit of time at some specified pressure difference between the inside and the outside. It is assumed that the wall permeability can vary over a wide range to illustrate the wide range of explosion violence than can be experienced in the surroundings from a *deflagration* of a given explosive cloud (gas/vapor, spray/mist, or dust). This variation, in turn, is due to the wide range of possible inherent burning rates of a combustible cloud of a given chemical composition, and the degree of confinement of the exploding cloud. With low wall permeability in Fig. 9.1a, the high initial pressure, eg, 10 bars (g), will be relieved very gently into the ambient atmosphere, and the resulting blast wave moving into the surrounding atmosphere at the velocity of sound will resemble a gentle "gush" rather than a violent abrupt "bang." With increasing permeability of the vessel walls, the strength of the blast wave experienced in the surroundings will increase, which will also the destructive effects of the blast wave in the surroundings.

The explosion experiments with premixed methane/air in a 50-m^3 horizontal explosion tube discussed in Chapter 2 illustrate this important point. The experiments are described in Section 2.1.6 of Chapter 2, and some relevant experimental results are given in Fig. 2.10. Quite substantial differences in the violence with which the gas mixture exploded inside the 50-m^3 tube were found, due to differences in the internal mechanical inventory of turbulence-generating obstacles in the tube.

As would be expected, also the peak values of the blast waves emitted into the surroundings from the various types of explosions at various distances away from the fully open downstream end of the explosion tube varied considerably, depending on the violence of the explosion inside the tube. These are shown in Fig. 9.2.

As can be seen the "side-on" blast-wave overpressures (see Section 9.1.2 for definition) at a given distance varied substantially with the estimated average flame speed in the exploding gas mixture inside the explosion tube. For example, at 10 m away from the exit of the tube, an estimated flame speed of 170 m/s gave a peak overpressure of only about 0.03 bar(g), whereas a flame speed of 300 m/s gave about 0.5 bar(g), ie, more than 10 times higher.

Reverting to Fig. 9.1, case (b) illustrates the extreme singularity of *detonation* of the explosive cloud (gas/vapor, spray/mist, or dust). At t = 0, as soon as the walls have been instantaneously removed, the compressed air

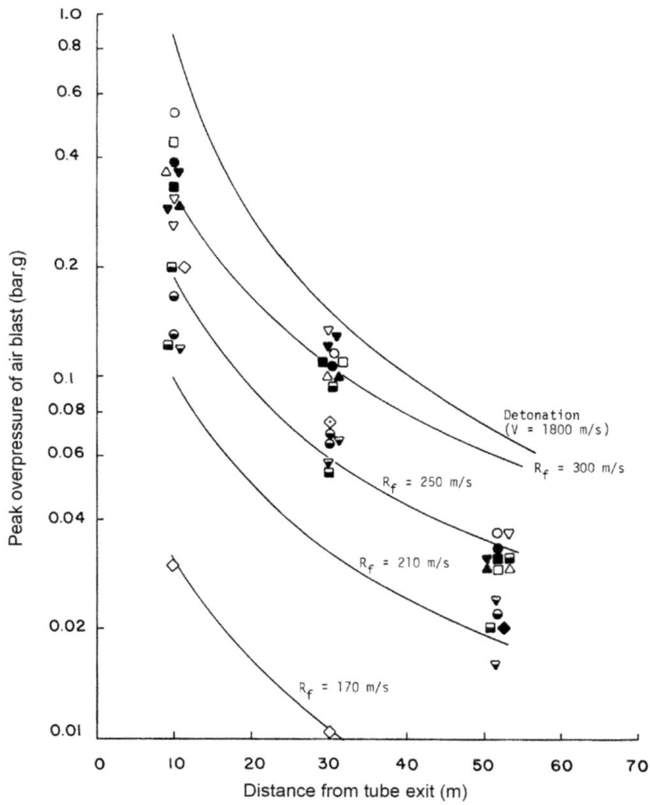

Figure 9.2 "Side-on" blast overpressures from methane/air deflagrations of various strengths produced by the same methane/air volume in the same vented vessel. The various symbols represent various arrangements of turbulence-producing baffle rings inside the explosion tube (see Fig. 2.10). R_f is an estimated average flame speed in the burning fuel/air mixture inside the explosion tube.
From Moen, I.O., Lee, J.H.S., Hjertager, B.H., Fuhre, K., Eckhoff, R.K., 1982. Pressure development due to turbulentflame propagation in large-scale methanelair explosions. Combustion and Flame 47, 31–52.

appears as a wall-free pressurized air pocket surrounded by ambient air at atmospheric pressure. This is an unstable situation, and a shock wave moving away from the high-pressure zone at supersonic velocity will be generated, followed by airflow away from the compressed pocket.

Fig. 9.2 also gives the computed blast-wave decay curve from a proper detonation of a fuel gas/air mixture of detonation velocity 1800 m/s, corresponding to that the methane/air mixture used in the experiments in the 50 m^3 explosion tube. As can be seen the blast wave from the detonation is not much stronger than those generated by the most violent turbulent deflagrations.

For the sake of completeness, it should be briefly mentioned here that detonations in explosive clouds of gas/vapor, spray/mist, or dust do not necessarily have to be initiated by some specific detonating charge. They can also develop directly from highly turbulent deflagrations. This phenomenon is called deflagration-to-detonation transition (see Section 2.1.7 in Chapter 2).

An important point in the present context is that blast waves from *detonations* are much simpler to predict than the wide spectrum of blast waves from all possible *deflagration* modes that can occur with a given quantity of a given explosive cloud. Therefore, much of the existing knowledge about blast-wave effects of various types of explosions is based on detonation experiments with genuine explosives (see Chapter 8). This is reflected in the following sections of the present chapter dealing with the effects of external blast waves from explosions.

9.1.2 Basic Blast-Wave Characteristics

The profile of the pressure as a function of time of an *ideal blast wave* (as from a detonation) passing a pressure sensor placed at a given distance from the explosion source is illustrated in Fig. 9.3.

The pressures measured are the so-called "side-on" pressures, ie, actual static pressures in the various parts of the wave as the wave passes the pressure sensor. This is achieved by using a pressure sensor having the small pressure-sensing face parallel with the direction of blast-wave propagation. This is contrary to measurements of "head-on" pressures using sensors having their small pressure-sensing areas at 90° to the direction of blast-wave propagation. In the latter case, neglecting any complicating interference of reflected waves, the pressure measured will be the sum of the static pressure and the dynamic pressure.

A soon as the blast-wave front illustrated in Fig. 9.3 reaches the observer at some supersonic velocity, the pressure increases practically instantaneously from the ambient value P_o to a peak value $P_o + \Delta P_+$. Then it decreases comparatively slowly down to a negative minimum value, whereafter it finally rises again to the ambient pressure P_o.

Because deflagrations are generally slower combustion processes than detonations, they also generally produce weaker blast waves than detonations of similar energy release. A typical weak deflagration blast wave moving at a velocity only slightly higher than the sound velocity into the ambient air is illustrated in Fig. 9.4. Such waves have fronts of considerably

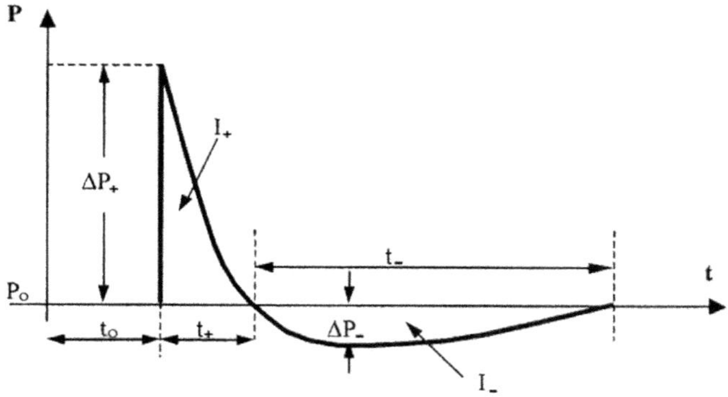

Figure 9.3 Characteristic pressure profile of a blast wave emitted by a detonation. P_o is ambient atmospheric pressure (abs); ΔP_+ is the peak magnitude of the first positive-pressure phase; ΔP_- is the peak magnitude of the second negative-pressure phase; t_o is the time from the explosion instant to the arrival of the blast-wave front at the point of measurement; t_+ is the duration of the first positive-pressure phase; t_- is the duration of the second negative-pressure phase; I_+ is the positive impulse (integral of pressure vs. time); and I_- is the negative impulse (integral of pressure vs. time).
From Sochet, I., Gardebas, D., Calderara, S., Marchal, Y., Longuet, B., 2011. Blast wave parameters for spherical explosives detonation in free air. Open Journal of Safety Science and Technology 1, 31—42.

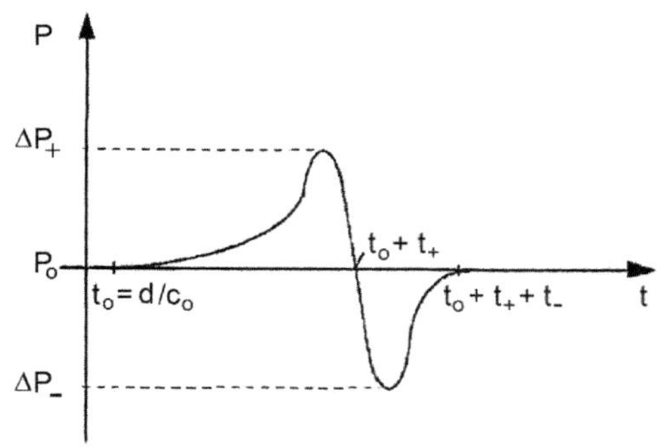

Figure 9.4 Typical pressure—distance profile of a blast wave from a slow deflagration when passing the measurement point at the sound velocity. The same symbol explanations as given for Fig. 9.3 apply. In addition: d is the distance from the explosion site to the point of pressure measurement, and c_o is the velocity of sound in the ambient atmosphere.
From Sochet, I., September 5—10, 2010. Blast effects of external explosions. Plenary Lecture at the 8th ISHPMIE, Yokohama, Japan. Preliminary manuscript received from the author.

slower pressure-rise rates than those of extremely steep detonation wave fronts illustrated in Fig. 9.3.

In addition, the maximum positive pressure rise ΔP_+ is typically much lower than for detonation waves. The propagation speeds also are typically considerably lower than those of detonations, often not much higher than the velocity of sound in the ambient atmosphere.

In addition, as opposed to detonations, the magnitude of the negative-pressure phase ΔP_- (the "suction phase") is often on the same order as that of the preceding positive peak ΔP_+. Therefore, in the case of deflagrations the negative-pressure phase can present a hazard potential on the same order as the positive phase.

9.1.3 TNT Equivalents

Marshall (1987) simply defined the TNT-equivalent concept as "the amount of TNT that would produce the same damage effects as the actual explosion." Sochet (2010) and Sochet et al. (2011) presented a more detailed and precise definition: The ratio of the mass of TNT to the actual mass of another explosive for which detonation of the TNT produces the same peak blast-wave overpressure at a given distance from the explosion source as detonation of the actual mass of the other explosive. The concept of "another explosive" not only comprises genuine explosive materials (see Chapter 8), but also other combustible systems including explosive premixed clouds in air of gas/vapor, spray/mist, and dust (see Chapters 2, 5 and 7). As pointed out by Sochet (2010), the main reason for using TNT as the reference material is twofold. First, there is large amount of experimental data on the blast-wave characteristics produced by this particular explosive material. Second, there is a large amount of published data on harm and damage caused by blast waves from TNT detonations.

According to Sochet (2010), Lannoy (1984) analyzed 23 explosion accidents that had occurred in the gas, oil, and chemical process industries. He then tried to estimate the TNT equivalent of the material that gave rise to the actual explosion under the prevailing circumstances, in terms of the mass of detonating TNT that would cause the same blast peak overpressure as 1 kg of that material under those circumstances. As pointed out by Sochet (2010), a hypothetical theoretical upper limit of this ratio would be the ratio of the two respective heats of combustion and decomposition. If the actual material is a hydrocarbon, this ratio is about 10. For example, the heat of combustion of propane is about 50 kJ/g, whereas the heat of decomposition of TNT is only about 5 kJ/g. This would imply that, at this theoretical upper limit, detonation of 10 kg of TNT would produce the same peak blast-wave overpressure as a hydrocarbon gas/vapor explosion involving only 1 kg of the actual fuel. However, on the basis of his analysis, Lannoy concluded that, on the whole, the "efficiencies," in terms of peak blast overpressures, of the energies released in the 23 hydrocarbon gas/vapor explosions reviewed, were considerably less than this simple consideration would indicate. Lannoy's analysis in fact suggested that for the majority (about 80%) of the 23 cases investigated the TNT equivalent was only about 2, ie, detonation of 2 kg of TNT would yield about the same peak blast overpressure as 1 kg of the hydrocarbon under the actual explosion circumstances.

Fig. 9.5 gives the classical relationship between the "side on" peak blast overpressure from TNT detonations as a function of the "scaled" distance between the point of pressure measurement and the explosion point. According to Marshall (1987), this relationship was first presented by Hopkinson (1915). It says that the blast-wave peak pressure at any "scaled" distance from the explosion source is the same and independent of the actual size of the explosive charge. The "scaled" distance is then defined as the ratio of the actual distance (m) and the cube root of the mass of the explosive TNT charge ($kg^{1/3}$). For this reason, the "scaled" distance has the dimensions length/$mass^{1/3}$ ($m/kg^{1/3}$).

As pointed out by Sochet (2010), Sochet et al. (2011), and others, the applicability of the TNT-equivalent concept for assessing harmful/destructive blast-wave effects from gas/vapor, spray/mist, and dust explosions, has considerable limitations. However, because of its simplicity the method is nevertheless widely used. Fig. 9.2 illustrates that there is not just one single TNT equivalent for one given fuel/air cloud. Instead there will be a wide range of potential TNT equivalents, depending on the actual circumstances under which the cloud explodes. The influences of the degree of confinement of the exploding cloud, and of obstacle configurations inside the confinement that promote turbulence generation in the cloud, and hence enhanced combustion rate, are of particular importance. It is beyond the scope of the present text to discuss this complex area any further. More details, including the so-called multienergy method developed by Berg (1985), can be found in other texts, eg, Baker and Tang (1991).

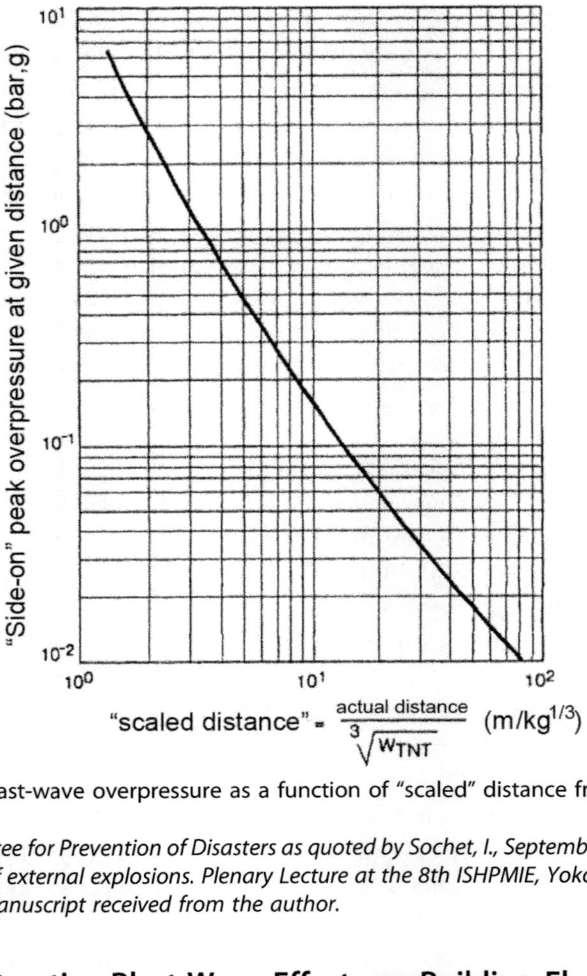

Figure 9.5 Blast-wave overpressure as a function of "scaled" distance from explosion source.
From Committee for Prevention of Disasters as quoted by Sochet, I., September 5–10, 2010. Blast effects of external explosions. Plenary Lecture at the 8th ISHPMIE, Yokohama, Japan. Preliminary manuscript received from the author.

9.1.4 Destructive Blast Wave Effects on Building Elements and Buildings

Table 9.1 shows frequently used estimates of potential external damage from explosions on the basis of estimates of peak blast-wave side-on overpressures at the distances from the explosion center of concern.

Tables 9.2–9.7 give similar data from other sources.

The damage levels typical of the four zones in Table 9.5 are as follows: In zone A, buildings may be damaged beyond economical repair. In zone B, buildings suffer partial collapse and/or failure of some structural members. In zone C, buildings are still usable, but some structural repairs are required. In zone D, damages are less severe, eg, broken windows, light cracks in walls,

Table 9.1 Structural damage caused at various overpressure levels

Overpressure (bar, g)	Type of damage
0.001	Noise of low frequency (10−15 Hz) up to 137 dB
0.002	Breaking of large glass windows already under strain
0.003	Noise 143 dB glass breakage due to a sonic boom from a jet fighter plane
0.007	Breakage of small windows already under strain
0.01	Typical pressure for window breakage
0.02	Safe limit: 95% Probability for no serious damage for pressure less than 0.02 bar. Small damage to house roofs (roof tiles and gutter) 1−10% of window glass broken
0.03	Limited minor structural damage
0.035−0.07	Generally large- and small-window breaks, occasional damage to window frames
0.05	Small damage to house structure
0.06	1% Structural damage, 99% of window glass shattered
0.07	Partial destruction of houses, made uninhabitable (danger of collapsing)
0.07−0.14	Corrugated asbestos panel shattered, failure of fasteners for corrugated steel and aluminum panels
0.09	Minor damage to steel frames of building (distortion)
0.10	Light structure collapse, pressure vessels intact
0.14−0.21	Nonreinforced-concrete or cinder-block walls destroyed
0.16	Lower limit for serious structural damage
0.17	50% Destruction of brickwork of houses
0.21	Heavy machinery (1.5 ton) within the building suffer little damage; steel frames of building distorted and separated from foundation; 50% structural damage
0.21−0.28	Light structures made out of self-supporting steel plates (without frame) demolished; rupture of empty oil storage tanks
0.28	Cladding of light industrial buildings ruptures
0.30	Major building damage (collapse)
0.34	Wooden utility poles snapped; minor damage to machinery (20 tons) inside buildings
0.34−0.41	Total destructions of houses; reinforced walls break; 99% structural damage to buildings
0.48	Loaded freight cars overturned
0.48−0.55	Failure of nonreinforced 20−30-cm thick brick walls due to shearing or fixtures; failure of pipe trestles (pipe rupture)
0.62	Complete destruction of loaded freight cars
0.7	Destruction of industrial buildings; machinery (3.5 tons) within building heavily damaged due to dislodging
0.8−1.5	Destruction of earthquake-proof concrete and steel-framed concrete buildings

Source unknown.

Table 9.2 Damage effects of blast waves on structures and equipment for various peak side-on overpressures

Structural element	Failure	Peak side-on overpressure (approximate) (psi)
Glass windows, large and small	Shattering usually, occasionally frame failure	0.5–1.0
Corrugated-asbestos siding	Shattering	1.0–2.0
Corrugated-steel or asbestos paneling	Connection failure followed by buckling	1.0–2.0
Brick-wall panel, 8 or 12 in. thick, not reinforced	Shearing and flexure failures	3.0–10.0 (7–8)
Wood-siding panels, standard house construction	Usually failure occurs at the main connections, allowing a whole panel to be blown in	1.0–2.0
Concrete or cinder- block wall panels, 8 or 12 in. thick, not reinforced	Shattering of the wall	1.5–5.5 (2–3)

1 psi = 0.069 bar.
From Glasstone and Dolan, and Brasie and Simpson as reproduced in Table 17.42A in Lees, F.P., 1996. Loss prevention in the process industries. Hazard, Identification, Assessment and Control, second ed. Butterworth-Heinemann, Oxford, UK, ISBN:0-7506-1547-8.

Table 9.3 Damage effects of blast waves on structures and equipment for various peak side-on overpressures

Structural element	Failure	Peak side-on overpressure (approximate) (psi)
Self-framing steel panel building	Collapse	3–4
Oil storage tanks	Rupture	3–4
Wooden utility poles	Snapping failure	5
Loaded rail cars	Overturning	7

1 psi = 0.069 bar.
Data from Brasie and Simpson as reproduced in Table 17.42B in Lees, F.P., 1996. Loss prevention in the process industries. Hazard, Identification, Assessment and Control, second ed. Butterworth-Heinemann, Oxford, UK, ISBN:0-7506-1547-8.

and damage to wall panels and roofs. It should be pointed out here that the buildings referred to in Table 9.7 may largely have been simple brick buildings, which may have been more easily damaged by a given blast-wave strength than more modern buildings. This may also apply to window damage.

Table 9.4 Various damage effects produced by blast waves for various peak side-on overpressures

	Peak side-on overpressure (psi)
Annoying noise (137 dB), if of low frequency (1−15 Hz)	0.02
Occasional breaking of large glass windows already under strain	0.03
Loud noise (143 dB). Sonic boom glass failure	0.04
Breakage of windows, small, under strain	0.1
Typical pressure for glass failure	0.15
'Safe distance' (probability 0.95, no serious damage beyond this value)	0.3
Missile limit	
Some damage to house ceilings; 10% window glass broken	
Limited minor structural damage	0.4
Large and small windows usually shattered; occasional damage to window frames	0.5−1.0
Minor damage to house structures	0.7
Partial demolition of houses, made uninhabitable	1.0
Corrugated asbestos shattered	1−2
Corrugated steel or aluminum panels, fastenings fail, followed by buckling	
Wood panels (standard housing), fastenings fail, panels blown in	
Steel frame of clad building slightly distorted	1.3
Partial collapse of walls and roofs of houses	2
Concrete- or cinder-block walls, not reinforced, shattered	2−3
Lower limit of serious structural damage	2.3
50% Destruction of brickwork of house	2.5
Heavy machines (3000 lb) in industrial building suffer little damage	3
Steel-frame building distorted and pulled away from foundations	
Frameless, self-framing steel-panel building demolished	3−4
Rupture of oil storage tanks	
Cladding of light industrial buildings ruptured	4
Wooden utilities poles (telegraph, etc.) snapped	5
Tall hydraulic press (40,000 lb) in building slightly damaged	
Nearly complete destruction of houses	5−7
Loaded train wagons overturned	7
Brick panels, 8−12 in. thick, not reinforced, fail by shearing or flexure	7−8
Loaded train boxcars completely demolished	9
Probable total destruction of buildings	10
Heavy (7000 lb) machine tools moved and badly damaged	
Very heavy (12,000 lb) machine tools survived	

1 psi = 0.069 bar.

From Clancey as reproduced in Table 17.43 in Lees, F.P., 1996. Loss prevention in the process industries. Hazard, Identification, Assessment and Control, second ed. Butterworth-Heinemann, Oxford, UK, ISBN:0-7506-1547-8.

Table 9.5 Classification categories for blast-overpressure damage to buildings proposed by Stephens

Zone	Damage level	Peak side-on overpressure (kPa)
A	Total destruction	>83
B	Severe damage	>35
C	Moderate damage	>17
D	Light damage	<3.5

1 kPa = 0.01 bar.
From p. 17/201 in Lees, F.P., 1996. Loss prevention in the process industries. Hazard, Identification, Assessment and Control, second ed. Butterworth-Heinemann, Oxford, UK, ISBN:0-7506-1547-8.

Table 9.6 Explosion damage to three types of process storage facilities

	Peak overpressure	
Damage level (%)	**(psi)**	**(10^{-5} bar)**
Steel floating-roof petroleum tank		
20	3.5	24,132
99	20.0	137,895
Vertical cylindrical steel pressure vessel		
20	12.0	82,737
99	14.0	96,527
Spherical steel petroleum tank		
20	8.0	55,158
99	16.0	110,316

1 psi = 0.069 bar.
From Rausch, Tsao, and Rowley based on data from Pickering and Bockholt. Reproduced from p. 17/202 in Lees, F.P., 1996. Loss prevention in the process industries. Hazard, Identification, Assessment and Control, second ed. Butterworth-Heinemann, Oxford, UK, ISBN:0-7506-1547-8.

Table 9.8 and Fig. 9.6 provide some more detailed information about the critical peak blast-wave overpressures at which window-glass panes of various thicknesses, areas, and height-to-breadth ratios will break.

9.1.5 Harmful Blast Effects on Humans

9.1.5.1 Harmful Effects due to Overpressure

As pointed out by L'Abbé and Crupi (1991a), the ear is the most sensitive human organ with regard to injury from overpressure. The threshold overpressure for eardrum perforation is between 0.35 and 0.5 bar. Persons being subjected to overpressures of less than 0.35 bar may still feel pain and/or

Table 9.7 Some outer-limit distances for various effects of detonation of 45 tons of a high explosive

Effect	Limit of effect (m)	Scaled distance $R/W^{\frac{1}{3}}$ (m/kg$^{1/3}$)
Minor damage, inner zone	305	8.8
Average serious damage	360	10.4
Earthquake damage, normal ground	640	18.5
Average missile flight	1530	44
Minor damage, outer zone	3050	88
Average glass-breakage limit	30,500	880

From Robinson as reproduced in Table 17.45 in Lees, F.P., 1996. Loss prevention in the process industries. Hazard, Identification, Assessment and Control, second ed. Butterworth-Heinemann, Oxford, UK, ISBN:0-7506-1547-8.

Table 9.8 Breaking pressure of plain glass windows as function of window area, aspect ratio, and glass thickness

Window thickness (mm)	Window size (m × m)	Pane area (m^2)	Breaking, range (kPa)	Pressure average (kPa)	Ratio of thickness to area (1/m)
5	1.0 × 1.0	1.03	6.2–7.6	6.9	0.00471
	1.0 × 0.5	0.515	8.3–11.7	10.0	0.0189
	0.48 × 0.48	0.233	15.2–20.7	18.0	0.0921
3	1.0 × 1.0	1.03	3.5–5.2	4.35	0.00283
	1.0 × 0.5	0.515	6.9–8.3	7.6	0.0113
	0.48 × 0.48	0.233	10.3–12.4	11.35	0.553
4 (32 oz)	1.0 × 1.0	1.03	4.8	4.8	0.00377

1 kPa = 0.01 bar.
From the analysis of results from Marshall, Harris, and Moppett by Gilbert, as reproduced on p. 17/252 in Lees, F.P., 1996. Loss prevention in the process industries. Hazard, Identification, Assessment and Control, second ed. Butterworth-Heinemann, Oxford, UK, ISBN:0-7506-1547-8.

temporary deafness and "ringing" in the ears, but this normally persists for less than 36 h.

However, overpressures of 1 and 2 bars cause a 50% and 85% probability of eardrum perforation, respectively. Eardrum perforation is not a permanent injury, and in the majority of cases it will heal by itself. However, there are considerable individual variations in the damaging effects experienced at a given overpressure. Some of these are medical, others are related to the geometrical orientation of the person in relation to the location of the exploding charge and environmental conditions.

Damage to the chain of small bones in within the middle ear can to some extent be repaired by special surgery. Damage to the inner ear will normally result in some degree of irreversible loss of hearing.

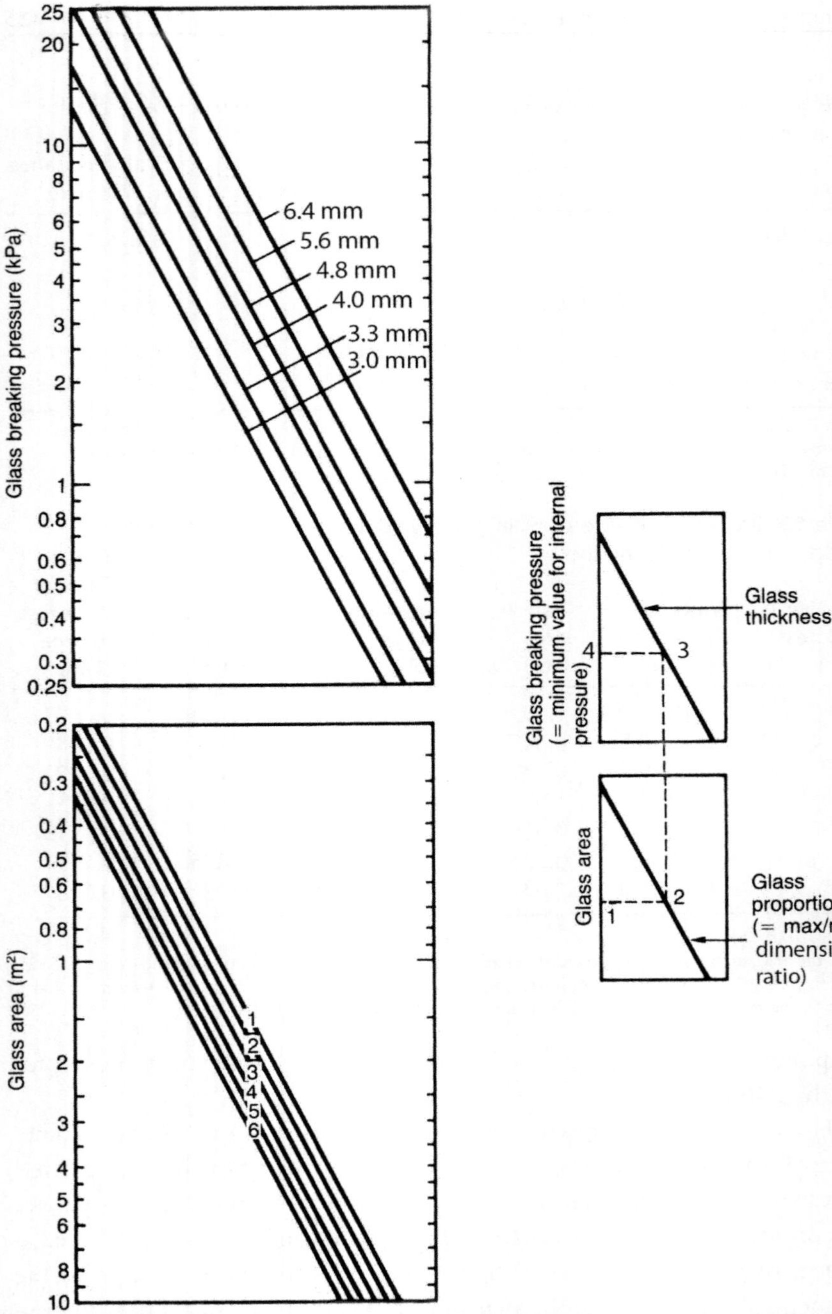

Figure 9.6 Nomogram giving the critical peak-blast overpressure at which window panes will break as a function of the window-pane area, the height-to-breadth ratio of the pane, and the thickness of the glass.

Based on figure from Building Research Establishment, United Kingdom, as reproduced in Fig. 17.110 in Lees, F.P., 1996. Loss prevention in the process industries. Hazard, Identification, Assessment and Control, second ed. Butterworth-Heinemann, Oxford, UK, ISBN:0-7506-1547-8.

At blast overpressures >3 bars internal bleeding can result within the nose, mouth, and larynx. However, this is not regarded as a serious injury.

Injury to the eyes is usually associated with fragmentation, rarely with overpressure. Injury by fragments will be discussed in the following. Also, injury to the brain is usually associated with impact (acceleration) either by the pressure wave itself or by a solid surface.

The areas within the thoracic region that are most susceptible to injury from overpressure are interfaces between parts of different densities.

A pressure wave travels through water faster and with greater intensity than through air. Another well-known difference between air and water is that air is compressible whereas water is not. When blast wave impacts a human body, a small part of the wave is reflected, whereas the remainder travels through the body. As the wave travels through the mass of water-containing organs within the thoracic cavity, it also hits the delicate lung alveoli. When pressure builds up around the thin soft walls of the alveoli, the gas inside gets compressed. This may cause the delicate alveoli walls to rupture and internal bleeding may result. Excessive bleeding can seriously impede breathing and lead to death.

Fig. 9.7 is a simplified summary of the effect of blast waves on the human body. The graph has been oversimplified to make it easy to use, and should therefore only be considered as a preliminary guide. TNT was used as the reference explosive material throughout. All the thresholds for the various injure categories were determined for blast waves of constant positive-phase duration of 1.5 ms.

L'Abbé and Crupi (1991a) provided the following explanation of how to use Fig. 9.7. Consider detonation of an explosive charge equivalent to 4.0 kg of TNT. A person 5 m away from the point of detonation will then be hit by a blast-wave peak of 0.75 bar overpressure. This can be read off Fig. 9.7 by drawing a vertical line upward from 4.0 kg to intersection with the 5-m distance line, and then drawing a horizontal line from the intersection to the left, and reading off the resulting pressure on the vertical axis. In this case, there would be a probability of eardrum perforation of approximately 25%. On the other hand, if the person was only 1.5 m away from the same 4.0 kg of detonating TNT, the resulting experienced overpressure would be approximately 10 bars. This would mean eardrum perforation and loss of hearing, and only approximately 15% probability of survival.

Fig. 9.7 only applies to outdoor situations. Inside buildings/structures, reflections of blast waves may create situations that are more hazardous. Fig. 9.8 gives some further data on rupture of human air drums by blast waves from detonations.

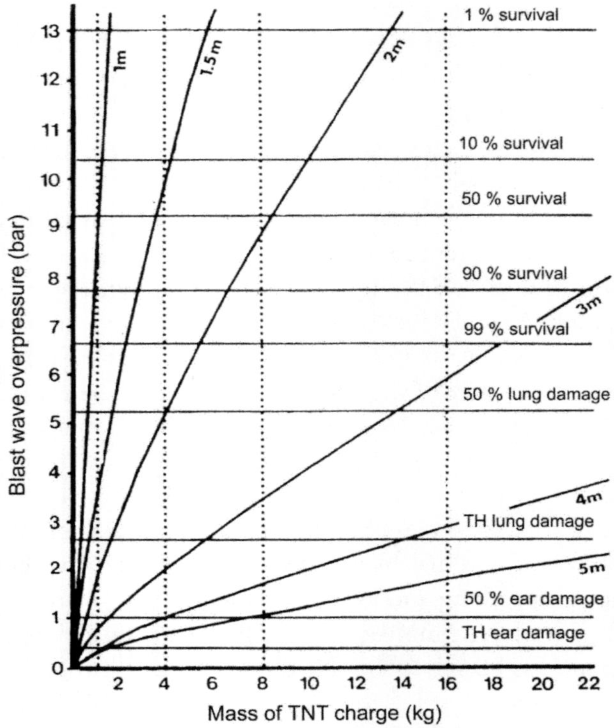

Figure 9.7 Effect on a person of the blast wave from detonation of a TNT charge, as a function of the mass of the charge, and the distance between the charge and the person. Assumption: The duration of the positive phase of the blast wave is 1.5 ms throughout.

From L'Abbé, R.L., Crupi, V.G., April 1991a. Blast effects. Part I: overpressure injury, fifteenth ed. Europex Newsletter, 4–6.

According to Lees (1996), a normal healthy human ear responds to sound waves within the frequency range 20–20,000 Hz. A blast wave from a detonation is a very short sequence of an amplified sound wave. However, the eardrum is unable to respond normally to pulses of durations shorter than about 0.3 ms. When being exposed to strong pulses of shorter durations than this, it ruptures. There is a direct relationship between the peak overpressure of the blast-wave pulse, and the probability of air drum breakage in humans, as shown in Fig. 9.8. Fig. 9.9 gives correlations between the maximum incident peak overpressure of the blast wave, the duration of the positive phase of the pressure pulse, and the probability of survival due to lung damage.

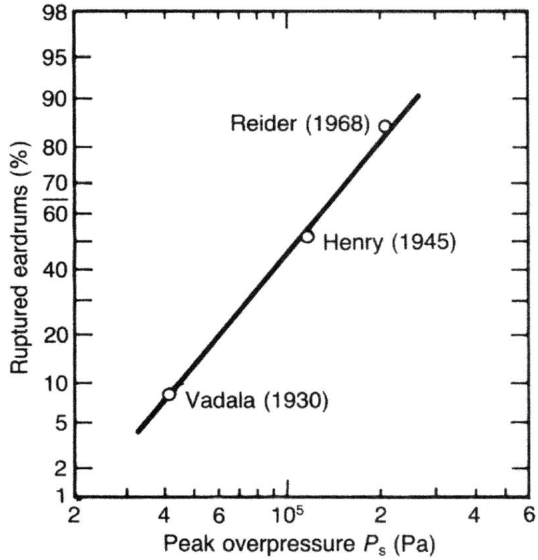

Figure 9.8 Percentage eardrum ruptures in humans as a function of peak overpressure of blast wave.
Fig. 17.104 in Lees, F.P., 1996. Loss prevention in the process industries. Hazard, Identification, Assessment and Control, second ed. Butterworth-Heinemann, Oxford, UK, ISBN: 0-7506-1547-8.

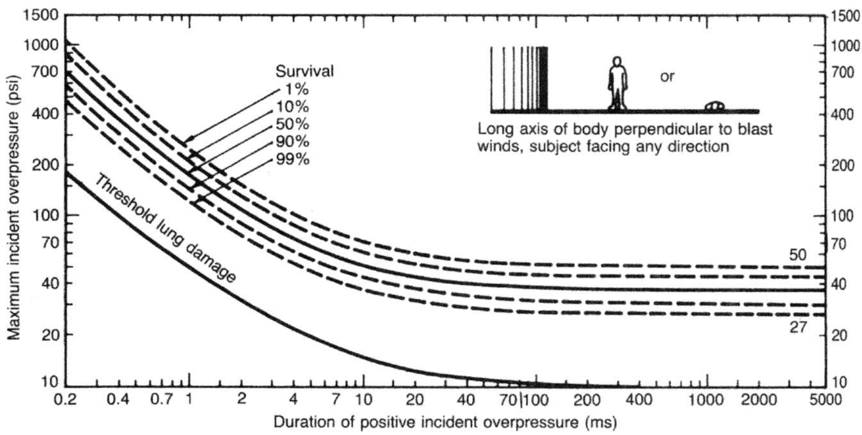

Figure 9.9 Lung injury survival curves for a 70-kg male person exposed to a blast wave, as functions of the maximum incident overpressure and the duration of the main positive pressure pulse. 1 psi = 0.069 bar.
From Bowen et al. as reproduced in Fig. 17.106 in Lees, F.P., 1996. Loss prevention in the process industries. Hazard, Identification, Assessment and Control, second ed. Butterworth-Heinemann, Oxford, UK, ISBN:0-7506-1547-8.

Once a blast wave has passed through the lung, the compressed gases in the alveoli expand and in some cases this can force gases into the bloodstream. For TNT charges of >20 kg, the threshold of significant lung damage is estimated at approximately 2.7 bars, whereas at 50% probability of severe lung damage the expected threshold is 5.5 bars.

Possible overpressure damage to the stomach and bowel includes bleeding in the tissues surrounding delicate blood vessels, and also severe weakening of the bowel walls resulting in perforations and infections. These injuries can be difficult to detect and can cause long-term problems.

Serious injury to the hands, arms, legs, and feet occurs when exposed to very high overpressure exceeding 15 bars. Because these body parts do not contain gases, they are essentially noncompressible.

According to Lees (1996), Bowen et al. in their analysis of lung damage by blast waves distinguished between three different body orientation cases as illustrated in Fig. 9.10.

In case (A), the applicable peak overpressure for lung injury is just the peak incident pressure p. In case (B), the applicable peak overpressure for lung injury is (p + q) in which q is the peak dynamic pressure. In case (C), with the body in an arbitrary position in front of a surface that reflects the incident blast wave, the applicable peak overpressure for lung injury is p_r, p_r being the peak-reflected overpressure. Lees (1996) gives equations by which both q and p_r can be calculated.

However, lung damage is not the only possible harm that blast waves can do to people. Table 9.9 shows the injuries that humans can suffer when being blown against a solid wall by a blast wave. The degree of injury for a given blast overpressure depends on whether the person is standing stiff legged or is seated. Being seated is somewhat safer than standing stiff legged.

9.1.5.2 Harmful Effects due to Acceleration of the Body Caused by Blast Waves

The following is based on the account by L'Abbé and Crupi (1991c). They were primarily addressing acceleration/deceleration-type injuries in humans in connection with accidental explosions during disposal of explosive materials, but their considerations are equally relevant in the context of accidental industrial explosions.

Injuries caused by positive or negative acceleration can be produced in two ways. The first is when a victim is hit by a projectile or fragment; the second is a when the victim is hit by a blast wave. In the latter case, the acceleration of various body parts depends on the size, shape, and mass of the person and the characteristics of the blast wave.

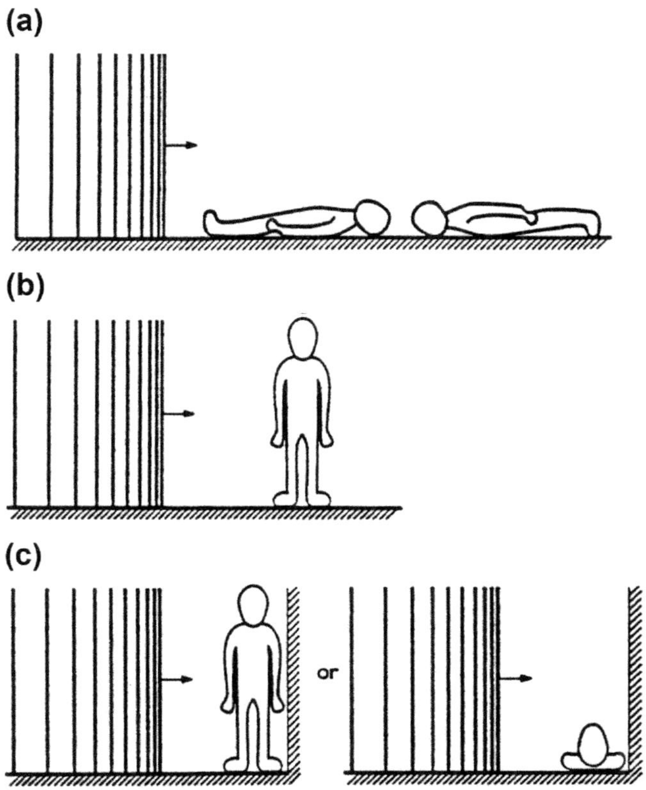

Figure 9.10 Influence of body orientation on blast-wave effects on humans: (a) long axis of body parallel to the direction of the blast wave, (b) long axis of body perpendicular to the direction of the blast wave, (c) thorax near a surface against which the blast wave reflects following perpendicular approach.
From Bowen et al. as reproduced in Fig. 17.105 in Lees, F.P., 1996. Loss prevention in the process industries. Hazard, Identification, Assessment and Control, second ed. Butterworth-Heinemann, Oxford, UK, ISBN:0-7506-1547-8.

Acceleration injury can range from quite minor ones to brain bleeding and limb amputation. According to L'Abbé and Crupi (1991c), tolerance levels when a human body is exposed to acceleration were scarce. Tentative degrees of injury could only be estimated by extrapolating from animal test data, tests on dead bodies, biomechanical modeling, and noninjurious testing on humans. Acceleration-type injuries caused by blasts, or impacts onto a rigid surface, can cause severe and fatal injuries. However, it has been shown that protective suits and helmets greatly reduce head, chest, and neck injuries.

In the case of excessive acceleration of the head and by both the initial impact of the blast wave and the subsequent impact if the body falls over

Table 9.9 Injury to a person displaced by a blast wave and impacting with a solid wall, as a function of the impacting velocity

	Impact velocity (m/s)
Standing stiff legged impact	
Mostly "safe"	
No significant effect	<2.44
Severe discomfort	2.44–3.05
Injury	
Threshold	3.05–3.7
Fracture threshold (heels, feet, legs)	4.0–4.9
Seated impact	
Mostly "safe"	
No significant effect	<2.44
Severe discomfort	2.44–4.3
Injury	
Threshold	4.6–7.9
Skull fracture	
Mostly "safe"	3.0
Threshold	4.0
50%	5.5
Near 100%	7.0
Total body impact	
Mostly "safe"	3.0
Lethality threshold	6.4
Lethality 50%	16.5
Lethality near 100%	42

From Glasstone as reproduced in Table 17.54 in Lees, F.P., 1996. Loss prevention in the process industries. Hazard, Identification, Assessment and Control, second ed. Butterworth-Heinemann, Oxford, UK, ISBN:0-7506-1547-8.

and hits the floor, an unprotected person is unlikely to survive. However, according to L'Abbé and Crupi (1991c), when the person is adequately protected, the acceleration is dramatically reduced, and the person will survive. This also applies if the protected person is hit by the blast wave from the back.

In 1989, a series of acceleration experiments were conducted in Canada using a test manikin of the type used in car collision experiments. The objective was to measure head and chest accelerations for a protected and nonprotected manikin when exposed to a blast wave from an explosion. The manikin had the same body size, shape, mass, and articulation characteristics as an average human being. An accelerometer that could measure accelerations in three dimensions was mounted at the center of gravity of the head, another one at the center of gravity of the chest. The experiments produced accurate head and chest acceleration data, which could be related to existing tolerance data. Fig. 9.11 gives the head and chest acceleration data measured in a protected and unprotected manikin standing 3 m away from detonation of 4 kg of 75% forcite dynamite.

Head acceleration for the unprotected manikin reached 291 g's, whereas the acceleration for the protected manikin reached only 33 g's. The helmet therefore reduced acceleration by 89%. In the chest region, little difference was observed between the protected and unprotected manikin. The readings were 25 and 27 g, respectively.

Deceleration injury occurs when a victim impacts a rigid surface following an explosion. Injuries can range from minor lacerations and bruises to ruptured organs, bone fractures, and massive brain injury. The spine and spinal cord are very delicate structures, the injuries of which can lead to long-term disabilities including paralysis.

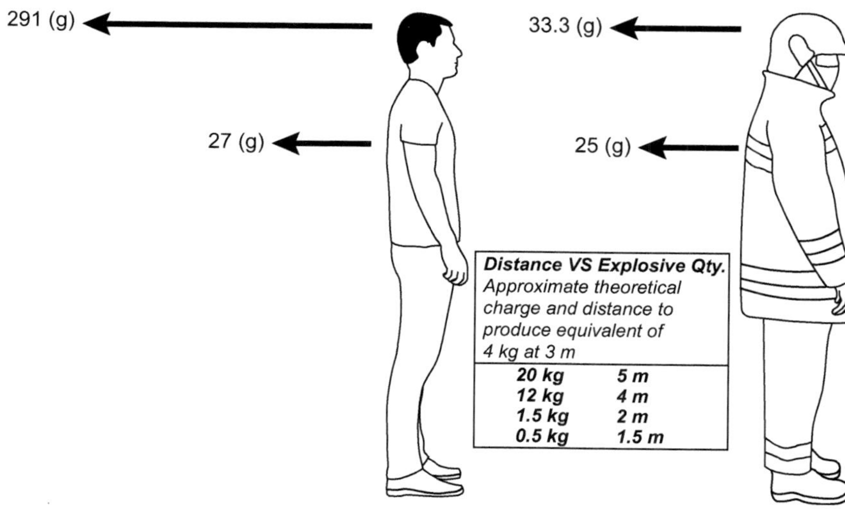

Figure 9.11 Initial body acceleration of an unprotected and protected person when exposed to the blast wave from detonation of 4 kg of dynamite (75% forcite) at a distance of 3 m away from the person. The *black area* of the back head indicates a fatal acceleration level. The figure also gives a table of combinations of other quantities of dynamite and corresponding distances that give blast waves of the same strength impacting on the person, as 4 kg detonated 3 m away.
From L'Abbé, R.L., Crupi, V.G., December 1991c. Blast effects. Part III: acceleration/deceleration type injuries, seventeenth ed. Europex Newsletter, 8–12.

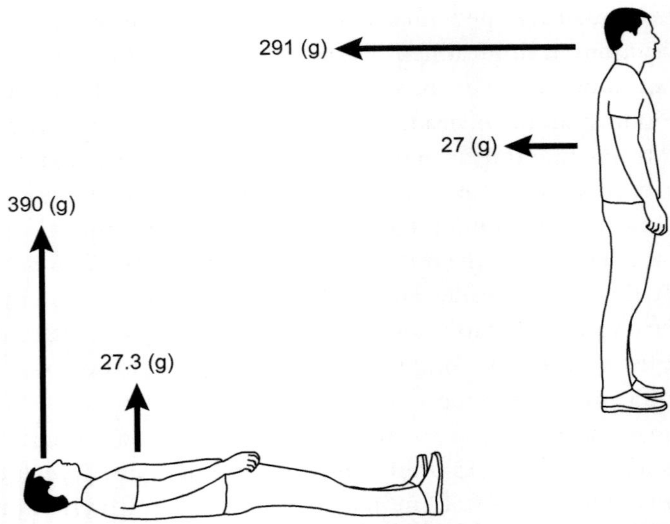

Figure 9.12 Initial body acceleration of an unprotected person when exposed to the blast wave from detonation of 4 kg of dynamite (75% forcite) at a distance of 3 m away from the person, as in Fig. 9.11, and body acceleration when the body hits the floor. The *black area* of the back head indicates a fatal level of acceleration.
From L'Abbé, R.L., Crupi, V.G., December 1991c. Blast effects. Part III: acceleration/deceleration type injuries, seventeenth ed. Europex Newsletter, 8–12.

Fig. 9.12 illustrates the decelerations measured in the head and chest regions when the manikin impacted the ground, following the detonation of 4 kg of 75% forcite dynamite at 3 m away from the unprotected manikin.

In this case the acceleration of the head when hitting the floor was 390 g, which a human cannot survive. Due to a somewhat softer surface, head accelerations in the case of a real human being would have been somewhat lower, but in any case serious-to-severe injuries would be expected.

Another concern is the fact that both accelerations and decelerations occur within a very short duration without allowing for the brain to fully recover. This sudden change in "acceleration" direction compounds the severity of the problem.

Based on accepted impact tolerance levels, an unprotected person would suffer a critical to unsurvivable head injury, severe spinal injury, and/or serious injury to the organs within the thoracic/abdominal region. A state-of-the-art bomb-disposal suit and helmet should reduce these to the moderate injury level in the head and chest/abdomen region.

Another significant point is the chest impact velocity when a person is blown onto his back. Until several years ago, manufacturers of bomb suits typically mounted the wireless radio on the rear of the suit, directly in line with the spinal column. Being blown back onto a radio could, in some cases, lead to severe spinal injury and paralysis. It was the opinion of the Canadian R&D Group that nothing should be mounted on the outside rear of the bomb suit, because of the injurious potential in the case of a fall. It was also considered that the suits should incorporate some type of spinal protection. The head should be protected with a maximum amount of impact protection in view of the very high head-impact velocities observed (up to 7.4 m/s).

According to L'Abbé and Crupi (1991c), two very distinct body trajectories were observed in some manikin tests. The first resulted in impacts on the back of the body. The second resulted in an impact on the face/front. Impact velocities in the latter case were generally lower, with resulting lower expected levels of chest injury. However, even at impact velocities of about 3 m/s, head injury would be considered critical for an unprotected individual hitting a hard surface, whereas a moderate-level injury would be expected for an individual equipped with a good protective helmet.

9.2 HARMFUL EFFECTS ON HUMANS OF MECHANICAL FRAGMENTS EJECTED BY EXPLOSIONS

In Chapter 3, the role of fragments (missiles) ejected from shattered process equipment in escalating Boiling Liquid Expanding Vapor Explosions (BLEVEs) is discussed. In such cases, the fragments ejected from the first vessel that BLEVEd may puncture other neighboring vessels and give rise to secondary BLEVEs, explosions, and fires. In the present section, however, focus will primarily be on the harm that flying fragments from explosions may cause to humans.

9.2.1 Properties of Fragments Ejected From Various Explosion Types

Because of the damaging potential of solid objects ejected by accidental explosions, it is important to know expected traveling distances and mass distributions of such objects. Accidental explosions in the process industries can produce various types of ejected objects. In the case of process vessel explosions, the vessel can be torn apart in a number of fragments of various masses. Lees (1996) gives a wealth of data on properties of fragments from various types of explosions. Fig. 9.13 gives distributions of traveling distances of fragments from accidental vessel explosions for some of the event groups defined in Table 9.10.

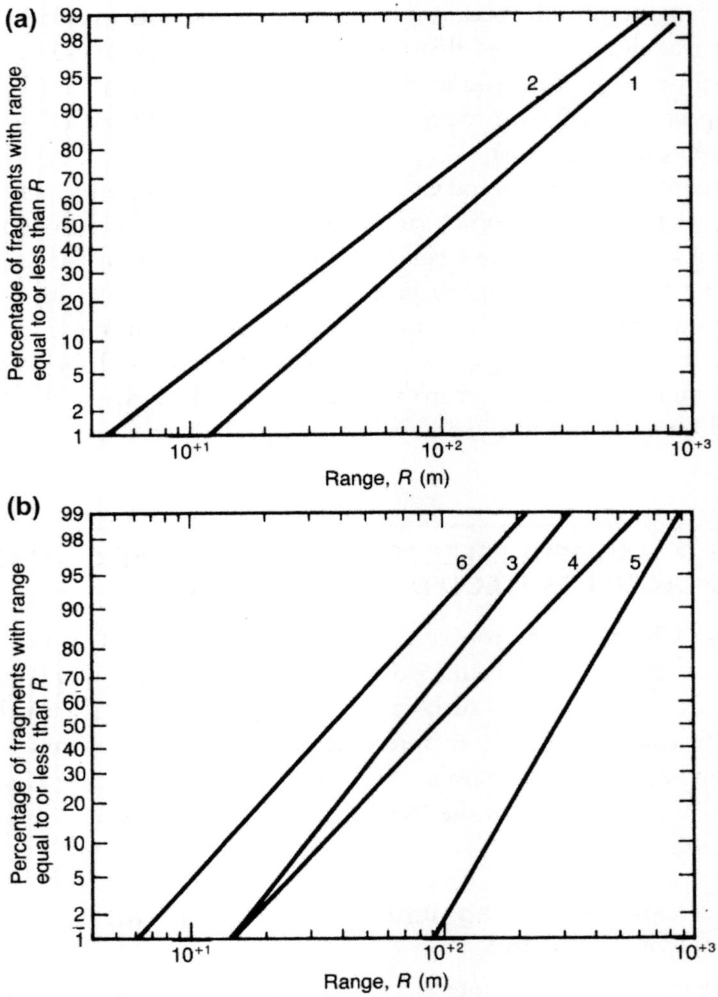

Figure 9.13 Distributions of traveling distances of fragments from accidental vessel explosions. The numbers marking the lines refer to the event groups in Table 9.10. *From Baker et al. (1978) as reproduced in Fig. 17.100 in Lees, F.P., 1996. Loss prevention in the process industries. Hazard, Identification, Assessment and Control, second ed. Butterworth-Heinemann, Oxford, UK, ISBN:0-7506-1547-8.*

Table 9.10 Number of fragments produced in a selection of 20 pressure vessel explosions

Event group	No. of events	Material	Source energy (J)	Vessel details		No. of fragments
				Shape	Mass (kg)	
1	4	Propane, anhydrous ammonia	$1.49–5.95 \times 10^5$	Rail tank car	25,542–83,900	14
2	9	LPG	3814–3921	Rail tank car	25,464	28
3	1	Air	5.2×10^{11}	Cylinder pipe and spheres	145,842	35
4	2	LPG, propylene	550	Semitrailer (cylinder)	6343–7840	31
5	3	Argon	$244–1133 \times 10^{10}$	Sphere	48.3–187	14
6	1	Propane	24.8	Cylinder	512	11

From Baker et al. as reproduced in Table 17.50 in Lees, F.P., 1996. Loss prevention in the process industries. Hazard, Identification, Assessment and Control, second ed. Butterworth-Heinemann, Oxford, UK, ISBN:0-7506-1547-8.

Fig. 9.14 gives estimated mass distributions of fragments from explosions within three of the event groups listed in Table 9.10.

More recently, Gubinelli and Cozzani (2009) investigated a selection of industrial accidents involving fragment ejection. They were able to identify fragmentation patterns typical of various types of explosion accidents. A selection of reference fragmentation patterns was defined on the basis of geometrical characteristics of the categories of process vessels that are most

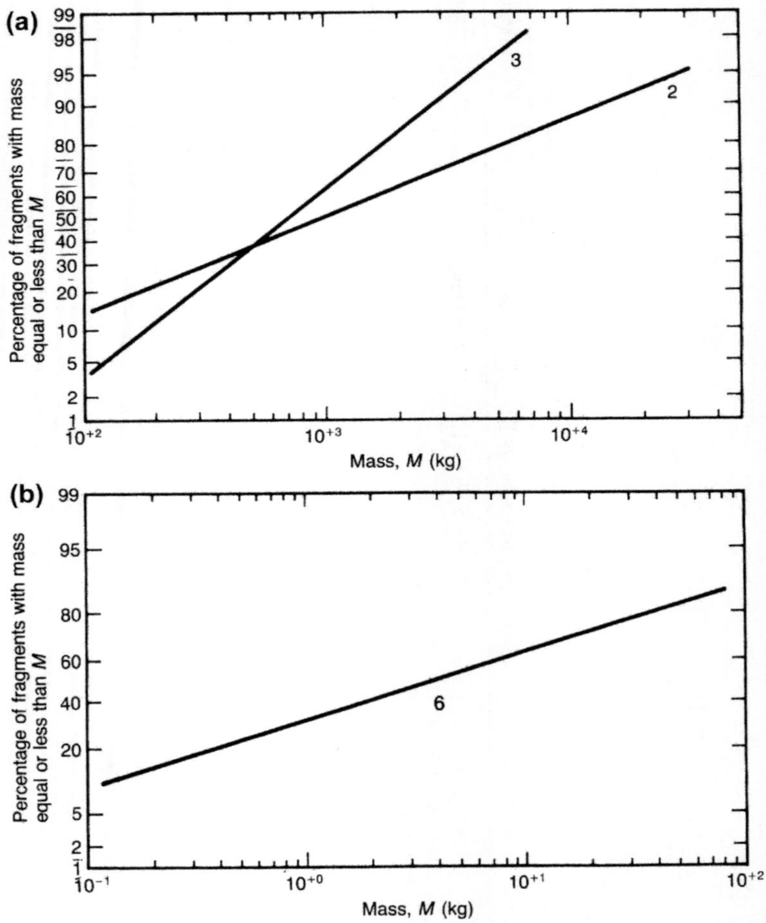

Figure 9.14 Mass distributions of fragments from accidental vessel explosions. *From Baker et al. as reproduced in Fig. 17.95 in Lees, F.P., 1996. Loss prevention in the process industries. Hazard, Identification, Assessment and Control, second ed. Butterworth-Heinemann, Oxford, UK, ISBN:0-7506-1547-8. The numbers marking the lines correspond to the event groups in Table 9.10.*

frequently involved in fragmentation accidents. Gubinelli and Cozzani used a database from analyses of more than 140 vessel-fragmentation accidents for validating their approach.

Tugnoli et al. (2014) proposed a procedure for calculating probability distributions for the initial directions of fragments ejected from shattered process vessels. A detailed review of a large number of past accidents provided the basis for validating the procedure.

9.2.2 Penetration of Fragments

9.2.2.1 Penetration of Fragments into Various Materials

Although Section 9.2 primarily deals with harm to humans by fragments from explosions, it may be useful just to mention that much work has also been done on penetration of ejected fragments into various materials. According to Lees (1996), the depth of penetration *t*, ie, the minimum thickness of the material to prevent the fragment from going through it, is given by Eq. [9.1].

$$t = \mathrm{K}M^{n_1} \cdot V^{n_2} \qquad [9.1]$$

Here, *t* is the minimum thickness (m), K is a constant, *M* is the mass (kg) of the fragment, *V* is the velocity of the fragment (m/s), and n_1 and n_2 are exponents. Values of K and n_1 and n_2 for various target materials are given in Table 9.11.

Eq. [9.1] is valid for compact blunt steel fragments such as solid cylinders of length-to-diameter ratios approximately unity and of masses up 1 kg.

Also according to Lees (1996), the depth of penetration t of large blunt fragments of masses >1 kg into various materials may be estimated by applying Eq. [9.2].

$$t = (CM/A)\cdot log_{10}\left(1 + 5\cdot 10^{-5}\cdot V^2\right) \qquad [9.2]$$

Table 9.11 Values of constant K and indices n_1 and n_2 in Eq. [9.1]

	K	n_1	n_2
Concrete (crushing strength 35 MN/m^2)	18×10^{-6}	0.4	1.5
Brickwork	23×10^{-6}	0.4	1.5
Mild steel	6×10^{-5}	0.33	1.0

From the 'High Pressure Safety Code' as quoted on p. 17.224 in Lees, F.P., 1996. Loss prevention in the process industries. Hazard, Identification, Assessment and Control, second ed. Butterworth-Heinemann, Oxford, UK, ISBN:0-7506-1547-8.

Table 9.12 Values of constant C in Eq. [9.2]

	C
Concrete (unreinforced, crushing strength 15 MN/m^2)	10×10^{-4}
Concrete (1.4% reinforcement, crushing strength 40 MN/m^2)	3.5×10^{-4}
Brickwork	25×10^{-4}
Mild steel	0.5×10^{-4}
Alloy steel	0.3×10^{-4}

As quoted on p. 17.225 in Lees, F.P., 1996. Loss prevention in the process industries. Hazard, Identification, Assessment and Control, second ed. Butterworth-Heinemann, Oxford, UK, ISBN: 0-7506-1547-8.

in which t (m) is the depth of penetration, C is the constant given for various materials in Table 9.12, A (m^2) is the fragment area facing the target, and M (kg) and V (m/s) mass and velocity of the fragment.

9.2.2.2 Penetration of Ejected Fragments into Human Bodies

The following is based on the account by L'Abbé and Crupi (1991b).

The injurious potential of a fragment penetration into a human body is dependent upon the fragment energy (function of mass and impact velocity), the fragment shape, the material density, the probability of fragment breakup, the probability of fragment tumbling, and the properties of the body tissue and location of impact. Irregularly shaped fragments tend to deposit a great deal of their preimpact kinetic energy rapidly along the wound tract, producing tissue destruction.

In situations of war, one distinguishes between primary and secondary fragments. Primary fragments are parts of the explosive device's casing, whereas secondary fragments can be parts of a structure (building), furniture, etc. ejected by the blast. Secondary fragments tend to be irregular in shape and have masses ranging from a fraction of a gram to several tons. Though their velocity may be relatively low, their range and injurious potential can be substantial. In accidental explosions in the process industries, all fragments will normally be of the secondary type.

9.2.2.2.1 Injuries to Head and Neck

The head and neck constitute only about 10% of the projected body area. However, experience from wars show that the percentage of serious to fatal fragment injuries due to impact on head and neck represent a considerably higher percentage. The brain, brain stem, and eyes are among the most sensitive areas. For a 10-g glass fragment, moderate injury to the eye can occur at fragment velocities of as low as 15 m/s. Serious injury to the eye can occur at much lower impact energies than impacts against the skin, and with substantially greater long-term effects.

9.2.2.2.2 Injuries to Chest and Abdomen

Fragment penetration and/or impact in the chest and abdomen regions can have very serious consequences. In the chest, the most sensitive areas include the heart, major blood vessels, the spinal cord, the windpipe, the food pipe, and the lungs. In the abdominal region, sensitive areas include the solid organs, the spinal cord, major blood vessels, and the gastrointestinal tract. Experience from war indicates that penetrating wounds in the spinal cord in the chest, and in the abdominal cavities are among the most serious ones.

9.2.2.2.3 Injuries to Arms and Legs

Serious fragment injuries in arms and legs can cause traumatic amputations. Amputations can be classified according to their life-threatening nature. For example, amputation of up to three fingers or toes is classified as a moderate injury, amputation above the knee is classified severe. Depending on the degree of severity amputations can cause long-term disability and/or psychological trauma.

Fig. 9.15 shows the body regions in which fragment penetration can cause critical injury.

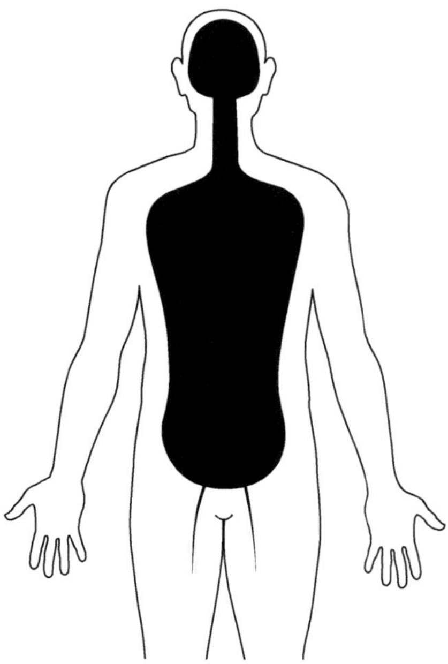

Figure 9.15 *Black area* denotes body regions in which penetration of fragments from explosions can cause critical injury.
From L'Abbé, R.L., Crupi, V.G., September 1991b. Blast effects. Part II: fragmentation injury, Europex Newsletter, sixteenth ed. 11–12.

Table 9.13 Probabilities of flying glass fragments penetrating the human abdomen wall for various fragment masses and velocities

A. Penetration of abdomen wall: impact velocity (ft/s)

Mass of glass fragment (g)	Probability of penetration (%)		
	1	**50**	**99**
0.1	235	410	730
0.5	160	275	485
1.0	140	245	430
10.0	115	180	355

B. Injury from penetrating 10-g fragments

Effect	Impact velocity (ft/s)
Skin laceration:	
Threshold	50
Serious wounds:	
Threshold	100
50%	180
Nearly 100%	300

1 ft = 0.30 m.

From Glasstone as reproduced in Table 17.58 in Lees, F.P., 1996. Loss prevention in the process industries. Hazard, Identification, Assessment and Control, second ed. Butterworth-Heinemann, Oxford, UK, ISBN:0-7506-1547-8.

Glasstone's data for injuries caused by glass fragments of different masses and impact velocities are given in Table 9.13.

Lees (1996) also presented results from work by Eisenberg, Lynch, and Breeding, some of which are reproduced in Table 9.14. In this case the mass of the glass fragment is constant and equal to 10 g.

Table 9.14 Some data for the degree of injury from a 10 g glass fragment impacting on a human body at various velocities

Injury	Impact velocity (m/s)
Skin laceration:	
Threshold	15
Serious wound:	
Threshold	30
50%	55
Near 100%	90

Data extracted from table reproduced on p.17/255 in Section 17.40 in Lees, F.P., 1996. Loss prevention in the process industries. Hazard, Identification, Assessment and Control, second ed. Butterworth-Heinemann, Oxford, UK, ISBN:0-7506-1547-8.

9.3 HAZARDOUS EFFECTS ON HUMANS OF DIRECT CONTACT WITH, AND THERMAL RADIATION FROM, FLAMES/FIREBALLS

This section is based on L'Abbé and Crupi (1992). They point out that burn injuries can be divided into two main categories. The first is caused by short-duration contact with, or radiation flashes from, fireballs; the second is caused by longer-duration contact with flames. In the first case, exposure durations are normally fractions of a second, whereas in the second case, the exposure times vary from some seconds to several minutes.

It appears that harmful exposure to short-duration radiant heat flashes can in most cases be prevented by normal clothing. However, prevention of harmful effects from long-duration contacting flames requires special protective clothing. Fig. 9.16 gives the approximate percentages

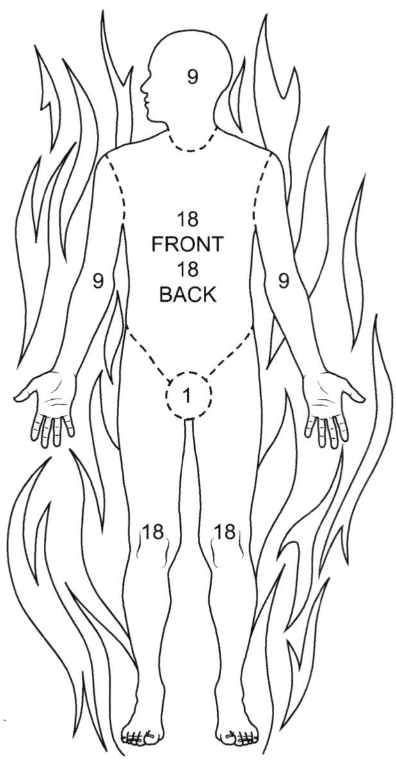

Figure 9.16 Illustration of the percentages of the total human body area represented by the various body parts.
From L'Abbé, R.L., Crupi, V.G., April 1992. Blast effects. Part IV: heat and heat stress, eighteenth ed. Europex Newsletter, 8–10.

of the total surface area of a human body that are represented by the various parts of the body.

According to L'Abbé and Crupi (1992), Fig. 9.16 is called the "Diagram of Nines," and is used to estimate the severity of burn injuries. In this diagram, the total surface area of the human body has been divided into parts representing 9%, or multiples of 9%, of the total area. As can be seen, the head and the arms represent 9% each, whereas the legs and front and back of the torso represent 18% each.

Table 9.15 gives severity of burns as a function degree of burn and percentage of the body affected. If the fraction of the body area that is burnt is known, the seriousness of injury can be estimated. For example, if a person has suffered a second- or third-degree burn to the front of one leg (half of 18%), this is classified as a minor injury.

A second example according to L'Abbé and Crupi (1992) is a person carrying an explosive device in both hands. If the device detonates, the following could be expected: Second- or third-degree burns to the front of both arms (9% of total area), to the torso (18%), to the face and neck (1/2 of 9%), to the upper half of both legs and to the genitalia (10%). This adds up to >40% of the body area, and would be classified as a critical injury.

Table 9.15 Severity of burns as a function degree of burn and percentage of the body affected

Degree of burn	Percentage of total body surface affected	Severity of injury
1 or 2	<10%	Minor
3	<10%	Moderate
3	<10% face, hand, and genitalia involvement	Serious
2 or 3	10–19%	Moderate
2 or 3	10–19% face, hand, and genitalia involvement	Serious
2 or 3	20–29%	Serious
2 or 3	20–29% face, hand, and genitalia involvement	Severe
2 or 3	30–39%	Severe
2 or 3	30–39% face, hand, and genitalia involvement	Critical
2 or 3	40–89%	Critical
2 or 3	90%+	Usually fatal

From L'Abbé, R.L., Crupi, V.G., April 1992. Blast effects. Part IV: Heat and heat stress, eighteenth ed. Europex Newsletter, 8–10.

It is important to note that inhalation of hot combustion gases can cause serious burns inside the mouth and even critical burns in the breathing organs. Therefore, protective suits and helmets must not only possess an adequate degree of fire resistance and provide adequate coverage of the body; in addition, they must provide adequate ventilation to reduce the risk of injuries due to inhalation of hot combustion gases.

CHAPTER TEN

Design of Electrical Apparatuses for Hazardous Areas

10.1 CLASSIFICATION OF HAZARDOUS AREAS

10.1.1 The Overall Objective of Area Classification

The overall objective of area classification is to *minimize the probability of accidental ignition* of explosive atmospheres in, eg, the process industries. The area classification forms the traditional basis for design of electrical apparatus. The basic philosophy is that requirements that are more strict are to be enforced in the design of electrical apparatuses to be used in areas where the probability of occurrence of explosive atmospheres is high, than in equipment to be used in areas where this probability is low. The prime objective of area classification is to assign different "zone" classes to the various parts of the industrial plant of concern according to the likelihood that explosive atmospheres may occur in the various parts.

Area classification is concerned with minimizing the probability of ignition of an explosive atmosphere. This is not at all the same as *minimizing the explosion risk*. The basic difference is that the explosion risk concept not only addresses the probability of ignition, but indeed also the consequence of an ignition. Explosion *risk* is often defined as the *product* of the *probability of ignition* and the *expected consequence* of an ignition. In other words, a minute ignition probability combined with a catastrophic explosion resulting from an ignition can easily give rise to an unacceptably high explosion risk. A survey of a selection of methods of risk and hazard evaluation/analysis is given in Chapter 11.

10.1.2 Gases and Vapors

10.1.2.1 Definition of Area Classification

The Institute of Petroleum (2002) is a well-established source of information on classification of areas containing combustible gases/vapors. They define area classification as the division of a facility into hazardous areas and nonhazardous areas, and the subdivision of the hazardous areas into zones (see definitions in Section 10.1.2.7). A hazardous area is then defined as a three-dimensional space in which a flammable gas/vapor atmosphere may

Explosion Hazards in the Process Industries
ISBN: 9780128032732
http://dx.doi.org/10.1016/B978-0-12-803273-2.00010-4

be expected to be present at such frequencies as to require special precautions for the control of potential ignition sources including fixed electrical equipment. All other areas are nonhazardous in this context, though they may, in part or whole, form part of a wider restricted area within the facility, in which all work is carried out under special controls. Examples include petroleum distribution installations and offshore production installations.

The International Electrotechnical Commission (IEC) has defined a hazardous area as an area within which an explosive atmosphere is present, or may be expected to be present, in quantities such as to require special precautions for the construction, installation, and use of electrical and other apparatus to prevent such apparatus from becoming an ignition source. Area classification is then the effort to divide a given process environment (called "area" despite being three-dimensional) into different parts (zones) characterized by different probabilities of occurrence of explosive atmospheres. This subdivision has to be documented on maps showing the extent of the various zones in all three dimensions.

10.1.2.2 Purpose of Area Classification

The purpose of area classification is to avoid ignition of releases, intentional as well accidental, that may occur in the operation of facilities handling flammable gases, liquids, and vapors. The approach is to reduce to an acceptable minimum level the probability of coincidence of a flammable atmosphere and a source of ignition.

As pointed out previously, it is not the aim of area classification to guard against the ignition of major catastrophic releases of flammable gases or vapors, eg, due to rupture of a pressure vessel or a pipeline. In properly run facilities, such events have a very low probability of occurrence. The probability of such releases must be kept below acceptable limits by correct design, construction, maintenance, and operation procedures of facilities.

10.1.2.3 Limitations of Area Classification

As already pointed out, the traditional area classification system was developed at a time when more comprehensive risk analysis methods were not available, and does neither address catastrophic failure, nor account for the consequences of explosions. It is now realized that conventional area classification is not an adequate tool to serve as a basis for estimating explosion risks. The lack of acceptance criteria, in terms of maximum acceptable risk (frequency $\times$ consequence), is a major problem in applying conventional area classification as the sole basis for selection of equipment that reduces the risk of ignition. In comprehensive risk analyses, the probabilities of both the occurrence of explosive atmospheres and the ignition of such atmospheres, and the potential consequences of explosions of such atmospheres in the actual process situation, have to be quantified.

More recently, conventional area classification procedures have often been supplemented or even replaced by thorough quantitative risk-assessment procedures that also account for the quantitative consequences of explosions and fires. If the result of such an analysis is that the explosion/fire risk is unacceptable, then appropriate measures for reduction or, if possible, elimination of the risk have to be taken. This may be accomplished by one or a combination of several of the measures indicated in Fig. 2.66 in Chapter 2. Careful consideration must be given to both technical risks during normal operating conditions and periods of limited technical malfunction during normal operation and to risks during maintenance and repair work. Possible human errors must also be considered. In the context of preventing the occurrence of effective ignition sources, risks associated with the use of tools in maintenance and repair work are of prime concern.

The area classification approach has traditionally been used as a basis for specifying requirements for electrical equipment in potentially explosive atmospheres. It neither considers the consequences of possible explosions, nor does it cover catastrophic events. Therefore, traditional area classification often has to be supplemented and sometimes even replaced by more comprehensive risk analysis, which not only accounts for the likelihood of ignition but also for the consequences, should ignition occur. In any case, the overall aim must be to ensure the required level of safety by means of cost-effective solutions.

10.1.2.4 Small-Scale Operations

Certain locations handling only small quantities of flammable fluids can, in the context of area classification, be classified as "nonhazardous." For example, this may apply to laboratories for testing small petroleum fluid samples. It is not possible to specify a given cut-off point, as this must be judged according to the circumstances. For instance, when draining gasoline from a vehicle fuel tank in an enclosed garage or a below-ground inspection pit, due precautions must be taken to prevent ignition. Such areas shall be classified as Zone 1, and only Zone 1 (or Zone 0) electrical equipment is permitted (see definition of zones in Section 10.1.2.7 in the following). In addition, due precautions must be taken to prevent ignition from any other type of ignition source.

Table 10.1 Capacity thresholds of process equipment located inside and outside buildings below which area classification may not be required

	Gas (volume corrected to 1 bar (abs) pressure) (L)	Liquefied flammable gas (L)	Flammable liquid at a temperature above its flash point (L)
Inside	50	5	25
Outside	1000	100	200

From The Institute of Petroleum, August 2002. Area Classification Code for Installations Handling Flammable Fluids, second ed. The Institute of Petroleum, London.

Each vessel containing flammable fluids should be treated individually, by considering the nature of its surroundings and the extent to which people need to be present. As a rough guide given by the Institute of Petroleum, hazardous area classification may not be needed if the maximum amounts of material that could be released are below the quantities given in Table 10.1.

However, in certain circumstances, ignition of quite small quantities of flammable gas/vapor mixed with air can cause danger to persons in the immediate vicinity. When this is the case, as in a relatively confined location from which rapid escape would be difficult, area classification may be needed down to quite small quantities of fluid.

10.1.2.5 Good Standard of Design and Operation

When applying the area classification technique described here, it is assumed that the facilities to which it is applied are designed, constructed, maintained, and operated in accordance with good industry practice to reduce releases of combustible gases/vapors to a minimum. Equipment and piping should be designed to international standards or national equivalents. The recommendations of the Institute of Petroleum (IP) Model Codes (see The Institute of Petroleum, 2002) or their equivalents, regarding good operational and maintenance practice, should also be followed in accordance with good industry practice to reduce releases to a minimum.

10.1.2.6 Management of Area Classification

As a rule, area classification should be carried out before the design and layout of equipment are finalized. At this stage, it may be possible to make considerable improvements at little cost. The area classification should always be reviewed and drawings modified, if necessary, on completion of design and before any change is made to existing plants handling flammable fluids.

Area classification should be incorporated into a company's Health, Safety, and Environmental Management System. The person responsible for the coordination of the area classification should be identified and be competent in this field. The work, which requires an interdisciplinary approach, should be carried out by persons who have full knowledge of the process systems and equipment, in consultation with safety, loss prevention, and electrical engineering personnel, as appropriate. Agreements reached on the area classification should be formally recorded, continually reviewed, and kept updated. Records, such as drawings and/or tabulated data sheets, should include details as to the type of protection selected to meet the zone requirements and the apparatus subgroup and temperature class.

In principle, the classification of an area entails consideration of all the actual and potential sources of release of flammable fluid present. In practice, the procedure can be simplified by adopting a standardized area classification diagram. In other cases, a procedure of considering individual point sources will be required. A detailed procedure for this is given by The Institute of Petroleum (2002).

10.1.2.7 Definition of Zones

Areas are subdivided into zones based on the likelihood of occurrence and duration of a flammable atmosphere, as follows:

- ***Zone 0***
  This is the part of a hazardous area in which a flammable atmosphere is present continuously or for long periods.
- ***Zone 1***
  This is the part of a hazardous area in which a flammable atmosphere is likely to occur for shorter periods in normal operation.
- ***Zone 2***
  This is the part of a hazardous area in which a flammable atmosphere is not likely to occur in normal operation and, if it occurs, will exist only for a short period.
- ***Nonhazardous area***
  This is an area in which a flammable atmosphere is not likely to occur at all.

10.1.2.8 Sources and Grades of Release

For the purpose of area classification, a point source of release is defined as a point from which a flammable gas, vapor, or liquid may be released into the atmosphere. The following three grades of release are defined in terms of their likely frequency and duration:

- ***Continuous grade release***
  This is a release that is continuous or nearly so, or that occurs frequently and for short periods.
- ***Primary grade release***
  This is a release that is likely to occur periodically or occasionally in normal operation.
- ***Secondary grade release***
  This release is unlikely to occur in normal operation, but which may occur infrequently and for short periods in nonnormal situations.

The grade of release is dependent solely on the frequency and duration of the release. It is completely independent of the rate and quantity of the release, the degree of ventilation, or the characteristics of the fluid, although these factors determine the extent of vapor travel and in consequence the dimensional limits of the hazardous zone.

To assist understanding of the boundaries of the definitions of the different grades of release, the following quantities are suggested. A release should be regarded as continuous grade if it is likely to be present for more than 1000 h/year and primary grade if it is likely to be present for between 10 and 1000 h/year. A release likely to be present for less than 10 h/year and for short periods only should be regarded as secondary grade. This assessment should take account of any likelihood of leaks remaining undetected. When releases are likely to be present for less than 10 h/year but are anticipated in normal operation (eg, routine sampling points) they should be regarded as primary grade releases.

The allocation of the grade of release should be reviewed in the course of the design stages to determine if practicable and economical design or engineering improvements can be made to reduce the number of continuous and primary grade releases. Assessment of the grade of release is not always obvious and will require experienced engineering and operational judgement. Releases that occur regularly but with short duration should generally be classified as primary grade sources giving rise to a Zone 1 area.

10.1.2.9 Relationship Between Grade of Release and Class of Zone

There is, in most cases, under unrestricted "open air" conditions a direct relationship between the grade of release and the type of zone to which it gives rise; ie,

- Continuous grade normally leads to Zone 0
- Primary grade normally leads to Zone 1
- Secondary grade normally leads to Zone 2

However, this may not always be true. For example, poor ventilation may result in a more stringent zone whereas, with generous ventilation, the opposite will be true. In addition, some sources may be considered to have a dual grade of release with a small continuous or primary grade and a larger secondary grade. Examples include vents with dual-purpose process requirement, and pump seals.

10.1.2.10 Area Classification Drawings

Area classification records can comprise detailed drawings with notes and/or can be in the form of tabulations. The area classification drawings should be in sufficient scale to show all the main items of equipment and all the buildings in both plan and elevation. The positions of all openings such as doors, windows, and ventilation inlets and outlets, and utility entries, if not sealed vapor-tight, should be included. This is because the positioning of these openings can affect the sizing of related external hazard zones.

Records should be marked up to show the boundaries of all hazardous areas and zones present using the shading convention adopted in current standards as indicated in Fig. 10.1. It is acceptable to indicate any requirement for small local zones/areas, eg, around pumps and control valves, in a note on the drawing.

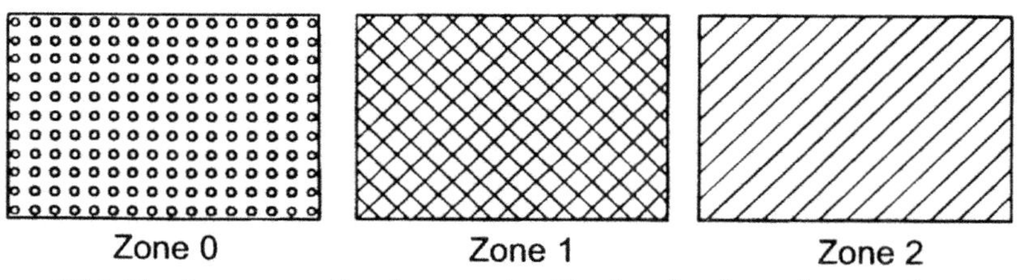

Figure 10.1 Shading convention for area classification drawings of areas where explosive clouds of gases and vapors can occur.
From The Institute of Petroleum, August 2002. Area Classification Code for Installations Handling Flammable Fluids, second ed. The Institute of Petroleum, London.

The final area classification should include a record of all additional supporting details. It is necessary to clearly distinguish regions on the drawing in which different gas properties prevail, eg, hydrogen with a Gas Group IIC on part of a drawing in which mainly hydrocarbons are present. This may be illustrated using half-width hatching for the "hydrogen" region.

10.1.2.11 Combustible-Fluid Properties and Area Classification

10.1.2.11.1 Class I Fluids

Class I fluids (see Tables 2.3 and 2.4 in Chapter 2) have very low flash points, and will always produce a vapor in the flammable range in air even at temperatures far below ambient. Facilities handling Class I fluids must always be area classified.

10.1.2.11.2 Class II and III Petroleum With Subdivisions (1) and (2)

Materials with flash points according to Class II or III (see Tables 2.3 and 2.4 in Chapter 2) will often be stored or handled at temperatures below their flash points, ie, at temperatures at which the vapor pressure is too low to give explosive fuel/air mixtures. In such cases, and when the possibility of releases in the form of a flammable mist or spray can also be excluded, the surrounding area does not have to be classified. However, when a material is held under pressure and there is a possibility of mist or spray formation on release, that material may produce a flammable atmosphere regardless of the storage conditions and flash point. In such circumstances, therefore, those materials classified as Class II(1) or III(1) (ie, nonhazardous) should be classed as II(2) or III(2), respectively. In such a case, it must be confirmed that the liquid temperature cannot be raised by any means in the event of release, eg, by contact with a hot surface or proximity to an adjacent nonelectrical source of ignition. As a typical example, within the United Kingdom mainland a maximum ambient temperature of 30°C can be assumed, whereas typically offshore in the North Sea a maximum would be 24°C. Aviation fuels of the kerosene type (flash point > 38°C) may, therefore, under these conditions, be classed as nonhazardous when stored away from processing areas and hot lines and vessels. Because of the greater variability in temperature in typical process areas, Class II liquids should, however, be regarded as hazardous, ie, Class II(2), in such areas. Special assessment may be required in climate regions of high normal ambient temperatures. When petroleum materials are to be stored or handled that are in the Class II(2) or III(2) condition, or are likely to be exposed to conditions of temperature above the flash point, the facilities should be classified as Class I.

10.1.2.11.3 Unclassified

Petroleum materials having a flash point above 100°C can be given the subdivision accorded to Class III petroleum and should be regarded as Class III(2) when handled at or above their flash points. A distinction, however, should be made between distillates in this unclassified range and products such as bitumens. When such materials are stored in heated fixed roof tankage (wherein the ullage space is essentially unventilated), the flash point as sampled or recorded will not be indicative of the presence or absence of a flammable atmosphere that may have accumulated in the tillage space. In common with the ullage space of fixed roof tanks or road or rail tank vehicles containing all classes of petroleum, the ullage space should be classified Zone 0, with a 1.5 m Zone 1 surrounding vents and other roof openings. It should also be recognized in respect of heated petroleum of high flash point that, although flash point and ignition temperature are different characteristics, at very high temperatures the effects can converge. That is, the high temperatures, for example, of a hot surface can create a flammable condition locally which may be ignited by autoignition by that hot surface, as well as by an alternative ignition source.

10.1.2.11.4 Fluid Categories

In some situations and processes, flammable fluids may have quite extreme and varied temperatures and pressures. To facilitate area classification by the point-source method developed by The Institute of Petroleum (2002) in such cases, the concept of fluid categories has been introduced. This categorizes the fluids according to their potential for rapid production of flammable vapor when released to the atmosphere. The fluid categories used are defined in Table 10.2.

10.1.2.11.5 Relationship Between Petroleum Class and Fluid Category for Simple Situations

For simple situations, Table 10.3 may be used to convert petroleum classification to fluid category. The degree of vaporization that will occur on release to the atmosphere is reduced systematically when going from Category A via Category B to Category C.

From the definitions in Table 10.3 it follows that for the same flammable fluid, at various stages of its processing or handling, there can be different fluid categories, depending upon the temperatures and pressures at the specific points of release. Therefore, the situation at any specific point of release must be evaluated separately.

Table 10.2 Fluid categories

Fluid category	Description
A	A flammable liquid that, on release, would vaporize rapidly and substantially. This includes: **1.** Any liquefied petroleum gas or lighter flammable liquid. **2.** Any flammable liquid at a temperature sufficient to produce, on release, more than about 40% vol. vaporization with no heat input other than from the surroundings.
B	A flammable liquid, not of Category A, but at a temperature sufficient for boiling to occur on release.
C	A flammable liquid, not of Categories A or B, but which can, on release, be at a temperature above its flash point, or form a flammable mist or spray.
G(i)	A typical methane-rich natural gas.
G(U)	Refinery hydrogen.

From The Institute of Petroleum, August 2002. Area Classification Code for Installations Handling Flammable Fluids, second ed. The Institute of Petroleum, London.

10.1.2.12 Specific Examples of Area Classification

The Institute of Petroleum (2002) has produced a set of examples of area classification of simple industrial operations in the open. Often a specific case requiring classification is close to one of the examples provided, and this may justify area classification by just copying this specific example, either without any changes or by appropriate adjustment to suit the case to be addressed. The Institute of Petroleum (2002) advises on modifications that may be made.

Figs. 10.2–10.4 present samples of the comprehensive collection of examples provided by The Institute of Petroleum (2002).

10.1.2.13 Area Classification by Point-Source Method Developed by The Institute of Petroleum (2002)

Generally, a process plant will have a Zone 2 area inside the plant boundaries, within which there may be local Zone 1 and even Zone 0 areas.

Some release sources are true point sources, eg, vents, drains, and sample points. Other equipment, eg, pump units, are assemblies of several individual point sources. Determination of hazard radii using the point-source methodology entails consideration of each identifiable potential release point in such assemblies. The point-source method comprises the following steps:

- identify point sources
- determine grades of release
- determine fluid categories

Table 10.3 Relationship between petroleum class and fluid category for simple situations

Classification of petroleum by the based (except for LPG) on closed-cup flash points			**Fluid category**		
Class	**Description**	**Handled above flash point**	**Handled above boiling point**	**Can be released as mist**	**Handled below boiling point and cannot he released as mist**
Class 0	Liquefied petroleum gases (LPGs)	Yes	A	A	A[2]
Class I	Flash point less than 21°C	Yes	B	C	C
Class II (1)	Flash point 21–55°C	No	N/A[1]	C	N/A[1]
Class II (2)	Flash point 21–55°C	Yes	B	C	C
Class III (1)	Flash point 55–100°C	No	N/A[1]	C	N/A[1]
Class III (2)	Flash point 55–100°C	Yes	B	C	C
Unclassified (1)	Flash point greater than 100°C	No	N/A[1]	C	N/A[1]
Unclassified (2)	Flash point greater than 100°C	Yes	B	C	C

Note 1: Not applicable (N/A) because liquids not handled above their flash point cannot be above their boiling point.
Note 2: Cryogenic fluids need special consideration.
From The Institute of Petroleum, August 2002. Area Classification Code for Installations Handling Flammable Fluids, second ed. The Institute of Petroleum, London.

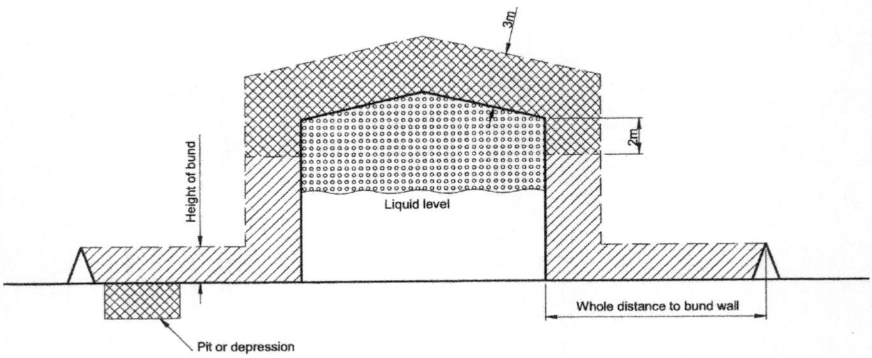

Figure 10.2 Area classification of a bunded (or embankment-enclosed) tank with cone or dome roof.
From The Institute of Petroleum, August 2002. Area Classification Code for Installations Handling Flammable Fluids, second ed. The Institute of Petroleum, London.

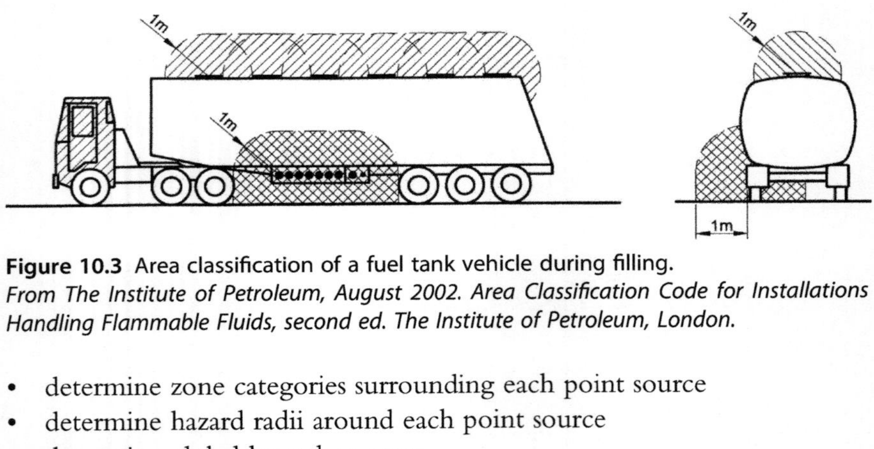

Figure 10.3 Area classification of a fuel tank vehicle during filling.
From The Institute of Petroleum, August 2002. Area Classification Code for Installations Handling Flammable Fluids, second ed. The Institute of Petroleum, London.

- determine zone categories surrounding each point source
- determine hazard radii around each point source
- determine global hazardous areas

10.1.2.13.1 Identify Point Sources

The first stage in the area classification of a plant or facility is the identification of the possible point sources of release. Usually these are small, and the associated equipment typically includes valves, flanges, vents, sampling, and drainage points, instrument connections, rotating machinery such as pumps and compressors, and any areas where spillage from these sources could collect. The Institute of Petroleum (2002) provides the following list of equipment configurations that often give rise to leak point sources:

- pumps
- equipment drains and liquid sampling points

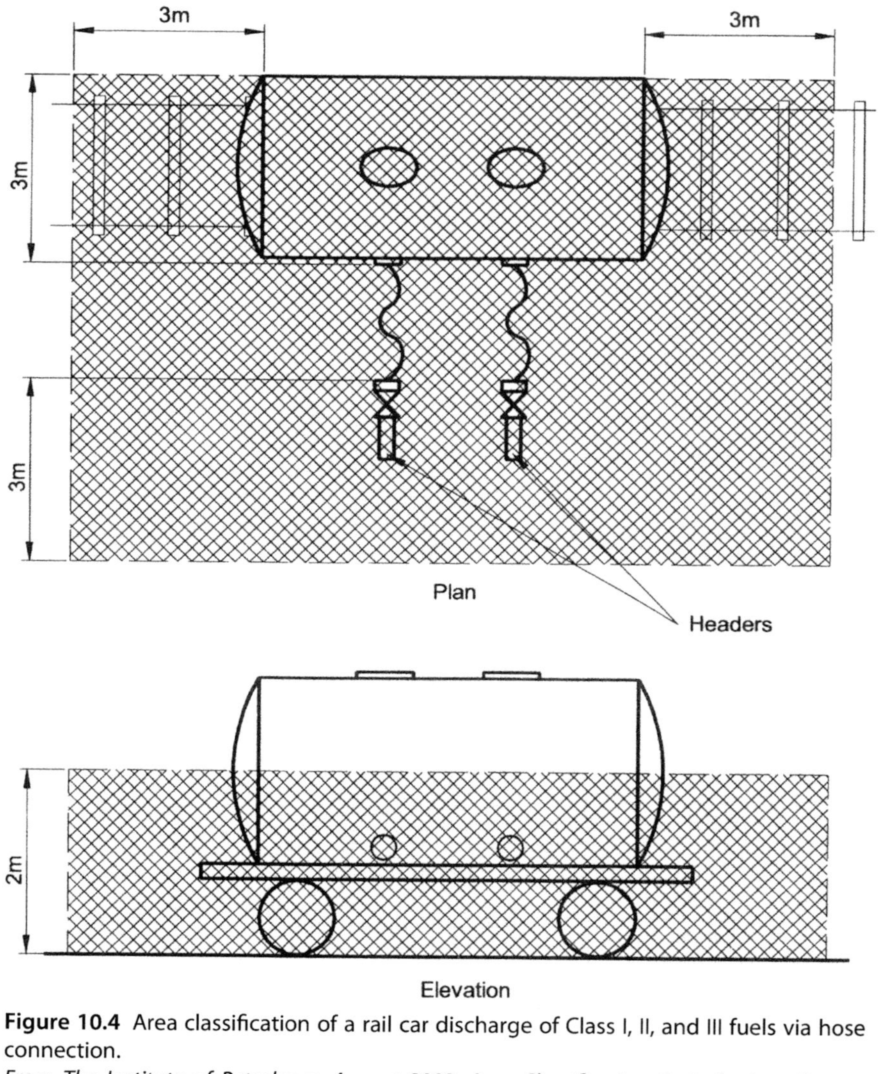

Figure 10.4 Area classification of a rail car discharge of Class I, II, and III fuels via hose connection.
From The Institute of Petroleum, August 2002. Area Classification Code for Installations Handling Flammable Fluids, second ed. The Institute of Petroleum, London.

- compressors
- vents
- piping systems
- pig receivers and launchers
- liquid pools during spillage

- sumps, interceptors, and separators
- pits and depressions
- surface-water draining systems

10.1.2.13.2 Determine Grades of Release

In principle, the area classification procedure entails the consideration of all actual and potential sources of release of flammable gases or vapors. All continuous and primary grade sources of release should be identified and assessed to determine the extent of the resulting Zone 0 and Zone 1 hazardous areas for each source. For secondary grade sources of release, it is often sufficient to consider only the sources located toward the periphery of a plant, which will be the point sources defining the outer boundary of the overall Zone 2 area. The extent of vapor travel, and hence the hazard radius for each point of release to be assessed, will be a function of the fluid characteristics and vapor-forming conditions during release, including mass or mass rate and the rate of vaporization. In assessing the grade of a release the following two points are central:

1. In the case of operationally controlled releases, eg, sampling and draining points, vents, filter cleaning, and pig-receiving operations, no single grade of release is generally applicable. This is because the operational frequency chosen will determine whether the equipment release should be graded as primary, secondary, or continuous.
2. For uncontrolled releases, eg, from pump seals, glands, and pipe flanges, The Institute of Petroleum (2002) has provided release-hole sizes assessed and compiled for various situations on the basis of experience. Alternatively, the manufacturers of various types of equipment may have provided typical specific release rates, or actual failure rates have been determined through observation for specific release situations.

10.1.2.13.3 Determine Fluid Categories

The fluid category for each point has to be determined, eg, by using tables provided by The Institute of Petroleum (2002). It is important to consider that the fluid state under shutdown conditions may give rise to a different vapor-producing potential than the state under normal operating conditions. An example would be the fluids in a fractionating column or receiver, which at shutdown might be less well separated from lighter fractions.

10.1.2.13.4 Determine Zone Category for Each Point Source

The zone category around a point of release is a function of the grade of release and the type of ventilation in the area of the release.

10.1.2.13.5 Determine Hazard Radii for the Zones Around Each Point Source

The methodology for determining the hazard radius is dependent on the grade of release (primary, continuous, or secondary). The Institute of Petroleum (2002) has produced a system of block diagrams and tables by which hazard radii may be estimated. However, if the case considered is not covered by the standard fluid categories and tabulated values, dispersion modeling using an adequate numerical code will have to be carried out.

10.1.2.13.6 Determine Global Hazardous Area Drawing

The global area classification drawing is obtained by constructing the envelopes of the various zone categories on the drawing containing all the individual hazardous zones for the point sources.

10.1.3 Clouds of Liquid Droplets at Temperatures Below Flash Point of Liquid

As pointed out by The Institute of Petroleum (2002), and indicated in Tables 10.2 and 10.3, flammable atmospheres may also be formed when flammable liquids being handled below their flash points are released in the form of a mist or spray. Such materials are normally regarded as nonhazardous from a flammability-of-liquids point of view.

However, they should be treated as hazardous when being pumped or stored under pressure, because they may then be capable of producing an explosive mist or spray if being released from a narrow hole or flange leak. As Tables 10.2 and 10.3 indicate, they should then be regarded as a Category C fluid generating a hazardous area as appropriate.

There is little knowledge on the formation of flammable mists and the appropriate extents of associated hazardous areas. Some of the issues related to the formation and the hazard of flammable mists are discussed by Bowen and Shirvill (1994). Pressure differentials of a few bars are sufficient to atomize commonly encountered liquids. Generation of mists created by the impact of liquid streams on a surface close to the point of release also appears possible. However, Bowen and Shirvill did not make any recommendations for the assessment of the extent of flammable clouds of mists and sprays, and further research is needed. They do suggest, however, that porous spray guards can be used around flanges and known potential leakage points, causing the material to coalesce back to a liquid below its flash point, rendering it nonhazardous. Nevertheless, this is only practicable in a few specialized applications, and requires rigorous control over maintenance activities to ensure that removed guards are replaced after work completion.

As a preliminary solution, The Institute of Petroleum (2002) suggested as follows: When the fluid temperature is at least 5°C below its flash point, and the pressure is atmospheric or only a few m liquid head in nonpressurized storage tanks, the fluid can be treated as nonhazardous. However, if being pumped and/or stored under overpressure, it should be regarded as a Category C fluid (see Table 10.2), generating a hazardous area because of the possibility of mist or spray being generated by leaks from a small hole or a flange.

10.1.4 Dust Clouds

10.1.4.1 Basic Similarities and Differences Between Dusts and Gases/Vapors

As discussed both in Chapter 1 and in Chapter 7, explosive gas/vapor mixtures and explosive dust clouds, once existing, exhibit very similar ignition and combustion properties, such as

- flammability/explosivity limits
- laminar burning velocities and quenching distances
- the response of the burning velocity to cloud turbulence
- detonation phenomena
- adiabatic constant-volume explosion pressures of similar magnitudes
- well-defined minimum ignition energies, and
- minimum ignition temperatures for given experimental conditions

Recognition of these similarities may have contributed to the development of a gas/dust "harmonization" concept. However, there are two fundamental ways in which dusts differ from gases, and both should have a major impact on the ways that electrical apparatuses are constructed to prevent them from becoming potential initiators of dust explosions and fires.

The first fundamental difference is in the ranges of hazardous fuel concentrations. For combustible gases/vapors, flame propagation is only possible when the fuel/air mixing ratios lie between the lower and the upper flammability limits. Dust flame propagation, however, is not limited only to the flammable dust concentration range of clouds. The state of settled layers/deposits constitutes an additional singular regime of flame propagation. This is because, contrary to combustible gases and liquids, settled powders/dusts in air will always have some air trapped in the voids between the particles, which facilitates sustained, although often very slow, combustion propagation throughout the deposit.

The second basic difference between dusts and gases is in the generation and sustainment of explosive clouds. The paramount question is whether there will be an explosive dust cloud in the first place. The physics of generation and sustainment of dust clouds and premixed gas clouds are so substantially different that explosive dust clouds are highly unlikely to be generated in a variety of situations in which explosive gas clouds may form quite readily, eg, by release from process equipment to the open atmosphere. The important consequence is that in practice the explosive dust clouds that give rise to primary dust explosions are always found inside process equipment, such as mills, mixers, ducts, buckets elevators, cyclones, filters, and silos. This, of course, will also be reflected in area classification.

The National Fire Protection Association (1997) contains a paragraph that in an excellent way, using practical terms, clarifies the basic difference between gases and dusts with regard to their abilities to migrate through narrow passages. In a slightly modified form this paragraph says as follows:

> *Walls are much more important in separating hazardous and non-hazardous zones in the case of combustible dusts than in the case of combustible gases. Only completely non-perforated solid walls make satisfactory barriers in the case of gases, whereas closed doors, light-weight partitions, and even partial partitions could make satisfactory barriers between hazardous and non-hazardous zones in the case of dusts.*

10.1.4.2 Definition of Zones

Many countries including the United States, Germany, and Norway traditionally used a dust zone classification system based on two hazardous zones only. However, in many countries this system has gradually been replaced by a common three-zone concept. Explosive Atmospheres (ATEX) Directive (1999) defines the three hazardous zone categories that have to be identified in areas containing combustible dusts, in the context of European standardization. These definitions, also shared with IEC, are:

Zone 20

A place in which an explosive atmosphere in the form of a cloud of combustible dust in air is present continuously, or for long periods, or frequently.

Zone 21

A place in which an explosive atmosphere in the form of a cloud of combustible dust in air is likely to occur occasionally in normal operation.

Zone 22

A place in which an explosive atmosphere in the form of a cloud of combustible dust in air is not likely to occur in normal operation but, if it does occur, will persist only for a short period.

In the context of the philosophy of the ATEX Directive (1999), layers/deposits of combustible dusts are not considered as hazardous in themselves. They are only taken into account if they are considered to be possible sources of generation of explosive dust clouds, or if they can give rise to dust fires that can ignite dust clouds. The symbols used for identifying the various zone categories are shown in Fig. 10.5.

10.1.4.3 Examples of Area Classification

Figs. 10.6 and 10.7 illustrate the application of area classification in two practical cases.

The interior of the hopper in Fig. 10.6 is classified as Zone 20 because an explosive dust cloud will be generated there during the quite frequent emptying of bags into the hopper. Because of the dust-extraction system, explosive dust clouds are not expected to extend outside the hopper in normal operation. However, abnormal situations may arise in which this may occur, eg, if a bag bursts during being emptied, or if the dust-extraction system fails. Therefore, a Zone 22 is assigned to a limited area just outside the hopper opening.

The interior of the cyclone in Fig. 10.7 is classified as Zone 20 because of the frequent presence of an explosive dust cloud, or even continuously, for long periods. However, the dust concentration in the gas leaving the

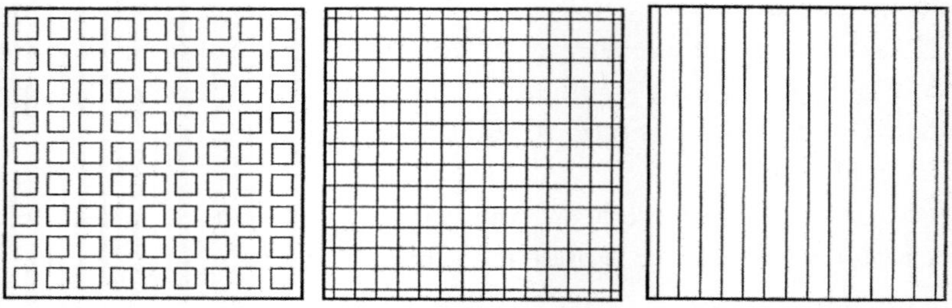

Figure 10.5 Standardized system for marking zones 20, 21, and 22 on hazardous area classification maps of industrial plant.

From CENELEC, September 2002. Electrical Apparatus for Use in the Presence of Combustible Dust. Part 3: Classification of Areas Where Combustible Dusts Are or May Be Present. European Standard EN 50281-3. CENELEC Central Secretariat, Brussels.

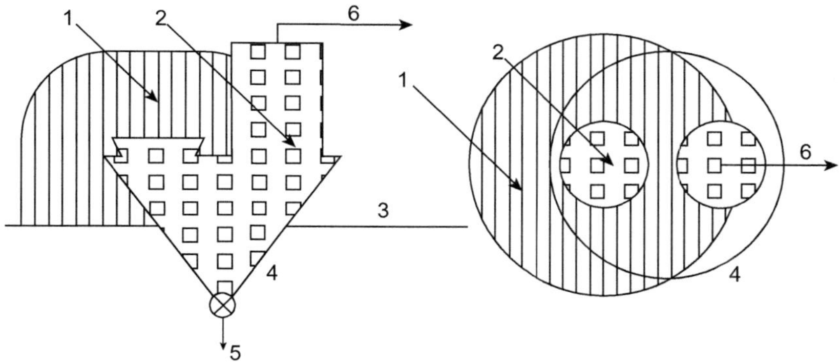

Legend: 1 Zone 22 usually I m radius
2 Zone 20
3 Floor
4 Bag discharge hopper
5 To process
6 To extract

Figure 10.6 Marking of zones 20 and 22 on a hazardous area classification map of a station for emptying bags containing combustible powder into a hopper equipped with dust extraction.

From CENELEC, September 2002. Electrical Apparatus for Use in the Presence of Combustible Dust. Part 3: Classification of Areas Where Combustible Dusts Are or May Be Present. European Standard EN 50281-3. CENELEC Central Secretariat, Brussels.

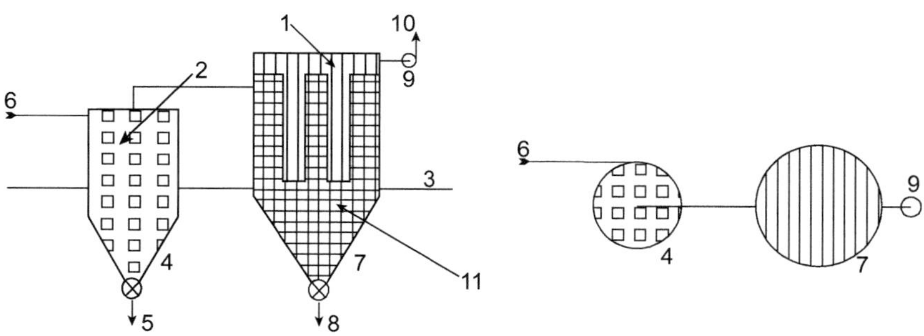

Legend: 1	Zone 22	6	Inlet	11	Zone 21
2	Zone 20	7	Filter		
3	Floor	8	To fines bin		
4	Cyclone	9	Extract fan		
5	To product silo	10	To outlet		

Figure 10.7 Marking of zones 20, 21, and 22 on hazardous area classification map of dust-collection plant comprising a cyclone and a filter in series.

From CENELEC, September 2002. Electrical Apparatus for Use in the Presence of Combustible Dust. Part 3: Classification of Areas Where Combustible Dusts Are or May Be Present. European Standard EN 50281-3. CENELEC Central Secretariat, Brussels.

cyclone and entering the dusty side of the filter is normally below the minimum explosive concentration, except for short periods when the bags are blown or shaken to release accumulated dust. Hence, the dusty side of the filter is classified as Zone 21. Normally, only negligible quantities of dust will exist on the clean side of the filter. However, filter bags may burst, or other abnormal situations may arise causing explosive dust clouds to be generated on the clean side of the filter, which is therefore classified as Zone 22.

10.1.5 Explosives, Pyrotechnics, and Propellants

10.1.5.1 Introduction

The European Committee for Electrotechnical Standardization (CENELEC) (1997) defines explosive substances as solid, liquid, pasty, or gelatinous substances, and preparations which are liable to react exothermally and under rapid gas generation even without the participation of atmospheric oxygen, and which, under specified test conditions under partial confinement, will detonate or deflagrate rapidly during heating. It is important to note that *this definition not only comprise*[sic] *the genuine explosives, but also pyrotechnics and propellants.*

10.1.5.2 Definitions of Hazardous Zones According to CENELEC

According to CENELEC (1997), hazardous zones in the context of explosives, pyrotechnics, and propellants are areas where, during manufacture, production, processing, or storage of such substances, there may be a finite risk of ignition due to electrical equipment. CENELEC (1997) introduced the following zone definitions:

Zone E1

Areas where explosive substances

- come in contact with electrical equipment as a function of design and/or process, or
- may to a significant extent appear as dust, vapor, liquid, sublimate, or another state.

Zone E2

Areas where explosive substances

- do not come in contact with electrical equipment as a function of design and/or process, but
- may occasionally occur in the form of dust, vapor, liquid, sublimate, or another state.

Zone E3

Areas where explosive substances

- do neither come into contact with electrical equipment as a function of design and/or process, nor
- occur in the form of dust, vapor, liquid, sublimate, or another state, neither as a function of design nor of process.

10.1.5.3 Definitions of Hazardous Zones According to Swedish Standards

The Swedish Standard (SS) 421 08 24 (1988) defines the various zones as follow:

Zone E1

Areas where

- explosives are handled, and where
- considerable quantities of dust, vapor, condensate and/or sublimate from explosive substances can normally occur.

All of the interiors of containers, vessels, piping, etc. for explosive substances are normally classified as Zone E1, irrespective of whether there is a risk of dust generation.

Zone E2

Areas where

- explosives are handled (not stored), and where
- considerable quantities of dust, vapor, condensate, and/or sublimate from explosive substances occur only rarely, or where
- only nondusting handling of explosive substances is taking place occasionally, but where specific requirements as to electrical apparatuses are nevertheless considered necessary.

Zone E3

Areas where

- explosive substances are only stored in packed form so that spreading of the explosive substances in the area is effectively prevented.

10.1.5.4 Temperature Classes

SS 421 08 24 (1988) specifies the following temperature classes for explosive substances:

- ET 1: Apparatuses suitable for substances having a minimum ignition temperature $\geq 180°C$
- ET 2: Apparatuses suitable for substances having a minimum ignition temperature $< 180°C$

SS 421 08 24 (1988) also gives minimum ignition temperatures for a range of explosive substances. For 2,4,6 trinitrotoluene (TNT), the value is 290–310°C, ie, the temperature class is ET 1.

10.1.5.5 Example of Area Classification According to the Swedish Standard

The example is a casting plant for TNT as illustrated in Fig. 10.8.

The process taking place in this example is as follows: At first the drums containing the TNT chips to be cast are preheated in the preheating room. No significant dust generation is expected, and the entire room is classified as E2.

The drums are then transferred to the melting room in which the TNT chips are poured from the drums into a melting pot. This process creates considerable amounts of TNT dust. Despite local dust extraction just above the pouring point, the dust cloud may at times extend across the entire room. The molten TNT is poured from the melting pot into intermediate containers. These are transported from the melting room to the casting room by an elevated transportation track running outside the building, from which the containers are hanging freely while being transported.

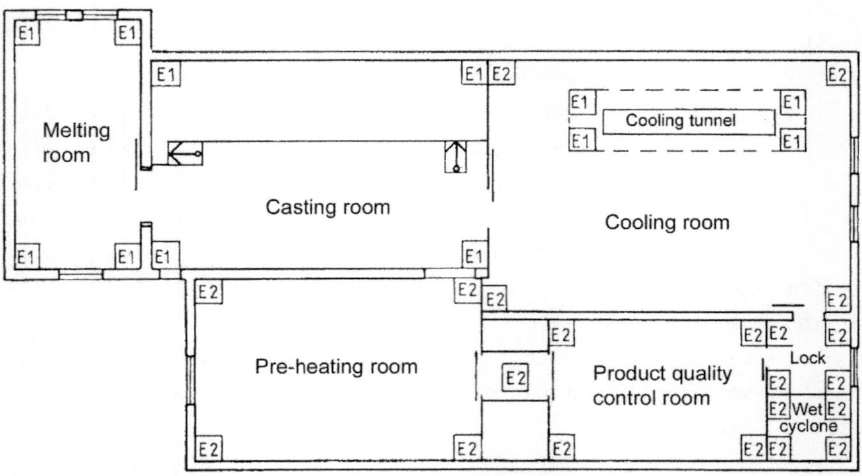

Figure 10.8 Area classification of plant for casting of TNT.
From Swedish Standard SS 421 08 24, 1988. Classification of Hazardous Areas Endangered by Explosives (in Swedish). Produced by SEK (The Swedish Electrotechnical Commission). SIS (The Swedish Standardization Commission).

The entire melting room is classified as E1, whereas the transportation area outside the melting room to the casting room, including a 0.5-m wide zone on each side of, as well as above and below, the suspended containers, is classified as E2.

In the casting room, the molten TNT is poured from the transportation container into molding pot. Further TNT chips are added to the molten material while this is being stirred and shaken. Even during this process there will be some dust generation, most of which is removed by local dust extraction.

When the TNT has reached the desired slurry-like consistency, it is poured into high-explosive shells, mines, etc. according to the production plan. During this process, there will be some generation of TNT vapor, which solidifies as soon as it makes contact with cold surfaces. Furthermore, some spill of molten TNT during casting is unavoidable. Therefore, the floor in the casting room is to be kept wet by applying water, to both collect settling TNT dust and prevent spilled TNT from the casting operations from becoming firmly stuck to the floor.

The conditions in the casting room are generally such that the entire room is to be classified as E1. The outdoor transportation track from the casting room to the cooling room is classified as E2 in the same way as the track from the melting room to the casting room.

In the cooling room the cast products are cooled down, normally in special cooling cupboards or cooling tunnels, which are usually equipped with local dust extraction. The interior of these tunnels and cupboards is to be classified as E1, whereas the cooling room in general is E1. However, often there is a need to remove excess TNT that has become attached to the outside of high-explosive shells etc. during the casting process. In such operations TNT dust is likely to be generated, and a local spherical E1 Zone with a radius of 2 m around the point of this kind of material removal has to be established.

The room for the wet cyclone that collects the dust from the various local dust extraction stations is classified as E2. Finally, the quality of the cast products has to be controlled in a special product quality control room, which is normally classified as E2.

The temperature class for electrical and other equipment requiring consideration is ET 1, which means that the maximum permissible temperature of any hot surface in all classified areas is 180°C.

10.2 BASIC DESIGN CONCEPTS FOR ELECTRICAL APPARATUSES FOR USE IN CONTACT WITH SUBSTANCES THAT CAN CAUSE EXPLOSIONS

10.2.1 Gases and Vapors

A series of standardized basic design concepts for electrical apparatuses intended for use in explosive gas atmospheres have been available for a long time. The details are described in a corresponding series of international and national standards (IEC, European Committee for Standardization (CEN), CENELEC, American Society for Testing and Materials (ASTM), etc.). The following summary is mainly based on British Broadcasting Corporation (BBC) (1983). Most of the figures are taken from Eckhoff (1996b).

10.2.1.1 Intrinsic Safety (Ex i)

This design concept can be used for apparatuses to be used in all three zones (0, 1, and 2). An intrinsically safe circuit is a circuit in which no spark or any other thermal effect is capable of causing ignition of a given explosive atmosphere. The basic principles of intrinsically safe design are given in Fig. 10.9.

To prevent hot-surface ignition of an explosive gas/vapor atmosphere by electrical apparatuses, the apparatuses must be designed to ensure that the temperatures of all surfaces that can make contact with the explosive atmosphere are below the minimum ignition temperature of the explosive atmosphere (see Section 2.2.4.4 in Chapter 2). For practical reasons it has been agreed internationally to standardize a limited number of temperature classes for electrical apparatuses, and these are given in Table 10.4. Table 2.2 in Chapter 2

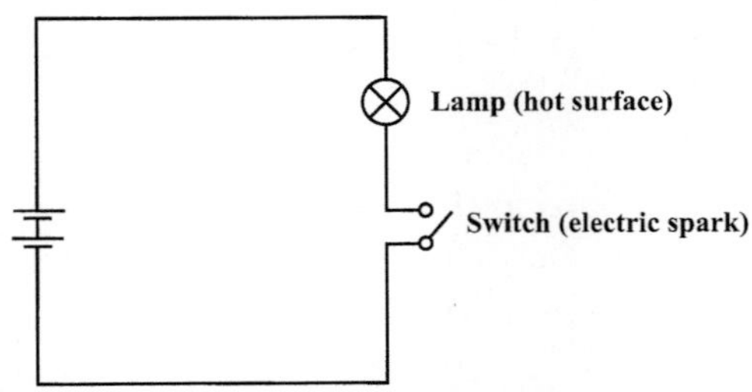

Figure 10.9 Illustration of two types of ignition sources that are to be prevented by intrinsically safe (Ex i) systems.
- Incendiary electric spark/arcs generated by closing or breaking electrical circuits (switch in figure)
- Incendiary hot surfaces (lamp in figure).

Table 10.4 Temperature class symbols indicating internationally standardized maximum surface temperatures of electrical apparatus

T1	450°C
T2	300°C
T3	200°C
T4	135°C
T5	100°C
T6	85°C

gives the minimum ignition temperatures and the corresponding temperature classes for a range of combustible gases and vapors in air.

The intrinsic safety concept originated in the United Kingdom about 100 years ago through the pioneering work by Wheeler and others. British Standard 1259 from 1945 contains detailed specifications for intrinsically safe circuits for mining underground. Another British Standard from 1949, dealing with *Intrinsically Safe Transformers primarily for Bell Signalling Circuits*, covered electrical devices for mining facilities underground. These systems had to satisfy the following conditions:

- the secondary voltage of the supply transformer of these circuits was limited to 15 V
- the current was limited (by a resistor) to a maximum of 1.5 A
- the signal was to be given by a bell
- the local cables in the mine were two bare wires laid in parallel

An intrinsically safe apparatus is electrical apparatus in which all the circuits are intrinsically safe. Intrinsically safe electrical apparatuses are produced to different standards depending on the ignition sensitivity of the explosive atmosphere in which the apparatuses are to be used (gas group I for coal mines only and groups IIA, IIB, and IIC for other activities). The reference gases are methane for group I, and propane, ethylene, and hydrogen for groups IIA, IIB, and IIC respectively.

In addition, intrinsically safe electrical apparatuses are divided into the categories Ex ia and Ex ib. Ex ia comprises electrical apparatus that shall not be capable of causing ignition neither in normal operation, nor with one fault or any combination of two faults. Category Ex ib comprises electrical apparatuses incapable of causing ignition in normal operation and with one single fault. The traditional approach has been that only Ex ia apparatus can be used in all the three zones 0, 1, and 2, whereas Ex ib can be used only in zones 1 and 2. However, in recent years, IEC has opened up for several other solutions acceptable in Zone 0. Most of these are special combinations of at least two parallel independent protection methods that are acceptable in Zone 1.

In intrinsically safe circuits a distinction is made between intrinsically safe electrical apparatus and *associated* electrical apparatus. An electrical apparatus is intrinsically safe when all its circuits are intrinsically safe. In an associated electrical apparatus not all circuits will be intrinsically safe but contain circuits that can affect the intrinsic safety.

Typical examples of intrinsically safe electrical apparatus are passive two- or four-terminal networks, limit switches, and measuring instruments containing a coil. Associated electrical apparatus is, for example, power supply units for intrinsically safe circuits with intrinsically safe output. Associated electrical apparatus must either be mounted outside the hazardous area or be fitted with an additional type of protection.

Safety barriers are isolating elements generally accommodated outside hazardous areas to isolate intrinsically safe circuits from nonintrinsically safe circuits so reliably that any chance of mutual influence is excluded. Safety barriers with galvanic isolation are used, ie, transformers, direct-current transformers and relays, and those without galvanic isolation, ie, safety barriers containing diodes or Zener diodes.

In principle, the intrinsic safety of, eg, simple inductive or capacitive circuits can be assessed theoretically by means of the appropriate curves in Figs. 2.30 and 2.31 in Section 2.2.6 of Chapter 2, based on the minimum ignition energy (MIE) of the explosive gas atmosphere in which the apparatus is to be located. However, often direct testing of the electrical apparatus of concern, using the standard spark test apparatus shown in Fig. 2.29 in Section 2.2.6 of Chapter 2 is required. This apparatus is designed to produce both closing and breaking contacts in the actual explosive gas mixture. The contact arrangement consists of a rotating disc carrying two tungsten wires, and another rotating cadmium disc with two slots in it, over which the tungsten wires slide. The two discs rotate in opposite directions, the contact holder with 80 rev/min, the disc with about 20 rev/min. During each revolution of the contact holder an average of eight contact closures and eight contact breaks occur.

The "minimum ignition current" (MIC) is used as the basic criterion for assigning the appropriate gas group to any given gas (see Section 2.2.6.1 in Chapter 2). The reference gas mixtures used for representing each of the four gas groups are:

- Gas group I: (8.3 ± 0.3) volume% methane in air
- Gas group IIA: (5.25 ± 0.25) volume% propane in air

- Gas group IIB: (7.8 ± 0.5) volume% ethylene in air
- Gas group IIC: (21 ± 2) volume% hydrogen in air

10.2.1.2 Flame-Proof Enclosures (Ex d)

This design concept can be used for apparatuses to be used in zones 1 and 2. Flameproof design implies that electric parts which can ignite an explosive atmosphere are placed in an enclosure which can withstand the pressure developed during an internal explosion of an explosive mixture and which prevents the transmission of the explosion to any explosive atmosphere surrounding the enclosure. The principle of flameproofing is illustrated in Fig. 10.10.

The following aspects are central in the design of flameproof enclosures

- For enclosures in which the contents are essential in service, the volume to be considered is the free volume. For luminaires, the volume is determined without the lamps fitted.
- The flameproof joint is the place at which corresponding surfaces of two parts of an enclosure come together and prevent the transmission of an internal explosion to the explosive atmosphere surrounding the enclosure.

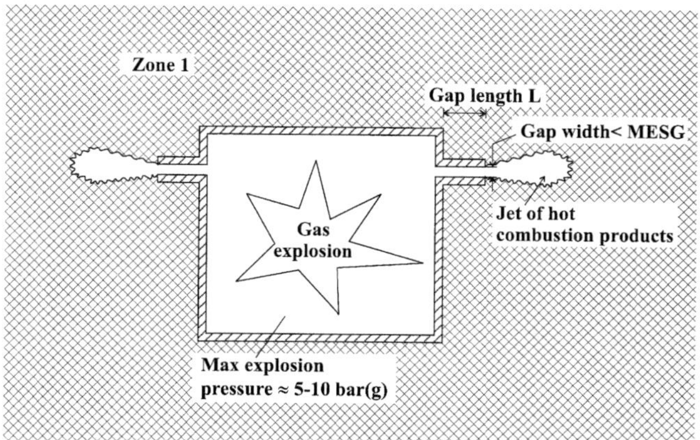

Figure 10.10 Illustration of a flameproof enclosure (Ex d). The enclosure must satisfy three main requirements:
- Gap widths < MESG at actual conditions
- Enclosure to withstand the maximum internal overpressure that an internal gas explosion can produce at actual conditions
- Temperature of external enclosure surface < minimum ignition temperature (min. ign. temp.) at actual conditions.

- The width of flameproof joint is the shortest path through a flameproof joint from the inside to the outside of an enclosure.
- The gap of flameproof joint is the distance between the corresponding surfaces of a flameproof joint when the electrical apparatus has been assembled. For cylindrical surfaces, the gap is the difference between the diameters of the bore and the cylindrical component.
- The maximum experimental safe gap for an explosive mixture is the maximum gap of a joint of 25 mm width which prevents any transmission of an explosion in 10 tests made under standard test conditions, using the apparatus illustrated in Fig. 2.44 in Section 2.2.8 of Chapter 2.
- Pressure piling is the increase in the explosion pressure above the normally expected pressure, eg, because of subdivision of the flameproof enclosure. See Fig. 2.60 in Section 2.4.5.2 of Chapter 2.
- A quick-acting door or cover is a door or cover provided with a device which permits opening or closing by a simple operation, such as the movement of a lever or the rotation of a wheel.
- A door or cover fixed by screws is a door or cover the opening or closing of which requires the manipulation of several screws or nuts.

The design of flameproof joints is specified in detail in international standards. The flameproof joints can be flanged, cylindrical, spigot, and threaded joints. The design of nonthreaded gaps shall withstand the mechanical stresses arising. The surfaces of joints shall be machined so that their average roughness does not exceed certain limits. The joint surfaces should be protected against corrosion, by grease, electroplating, chemical treatment, etc. Painting is not acceptable. The width of a joint shall be at least the minimum values specified in the standards, and interconnecting compartments shall be precluded as far as possible.

Eckhoff et al. (2013) and Arntzen at al. (2014) studied experimentally the extent to which mechanical damage (eg, scratches) and rusting of flame-gap surfaces will deteriorate the protective performance of flame gaps. It was found that the damaging effect, if any, was less than often anticipated.

As for intrinsically safe apparatus, the design of flameproof enclosures also depends on the ignition properties of the explosive gas atmosphere in which the enclosure is to be used. If the standard maximum experimental safe gap (MESG) (see Table 2.2 in Chapter 2) of the gas atmosphere is known, the gas group to which the gas belongs is also known. International standards then specify the design requirements to be satisfied. Nevertheless, flameproof enclosures shall also be tested for their ability to withstand the pressure of an explosion inside the enclosure, and to prevent transmission of an internal to an explosive gas cloud on the outside.

The tests of the ability of the enclosure to withstand the pressure of an internal explosion are made in two stages. The maximum explosion pressure, or reference pressure, is first measured by igniting a relevant explosive gas mixture inside the enclosure to be tested. The number of tests to be made and the explosive mixture considered depends on the gas group with which the enclosure has to comply. When the reference pressure has been determined, a proper pressure test of the enclosure is carried out, either as a static or as a dynamic test. In a static test, a test pressure exceeding the reference pressure by a factor of typically 1.5 shall be maintained inside the enclosure for a specified time. In a dynamic test, the enclosure is also subjected to 1.5 times the reference pressure, and the rate of rise of pressure shall be similar to that prevailing in the explosion experiments for determination of the reference pressure. The tests are considered satisfactory if the enclosure has not suffered any damage or permanent deformation liable to weaken any of its parts.

For the test for nontransmission of an internal explosion, the enclosure is placed in a test chamber and the same explosive mixture is introduced inside both the enclosure and the test chamber in which the enclosure is placed. The initial pressure before igniting the inside mixture is normally atmospheric both inside and outside the enclosure.

10.2.1.3 Increased Safety (Ex e)

This design concept can be used for apparatuses to be used in zones 1 and 2. Increased safety is a type of protection by which measures are applied to prevent of excessive temperatures and occurrence of arcs and sparks in the interior and on the external parts of electrical apparatus in normal service. The principles of the concept of increased safety are illustrated in Fig. 10.11. The following objectives are central in the design of increased-safety enclosures:

- To ensure that temperatures on the outside and inside surfaces of the apparatus do not exceed the maximum permissible temperature. The maximum permissible temperature in turn depends on the minimum ignition temperature of the combustible substance with which these surfaces may make contact.
- To ensure that the minimum clearances between conducting parts at different potential satisfy specific requirements stated in international standards

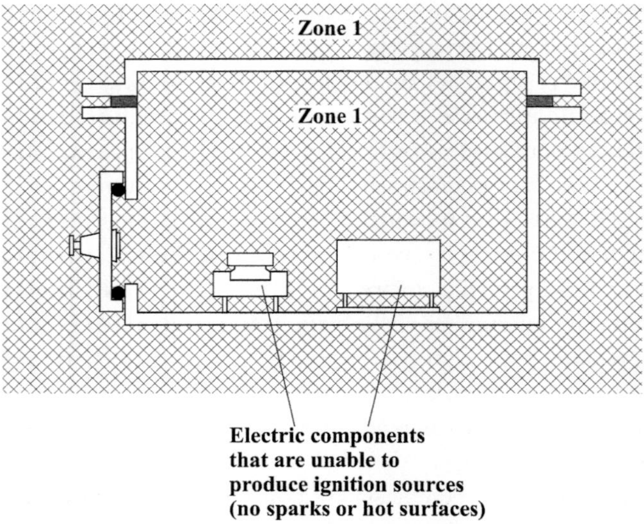

Figure 10.11 Illustration of increased safety enclosure (Ex e). The enclosure must satisfy the following main requirements.
- No ignition sources permitted inside enclosure
- Satisfactory sealing (IP-protection) against ingress of water and dust
- Enclosure body made of earthed antistatic material
- Temperature of external enclosure surface $<$ minimum ignition temperature (min. ign. temp.) at actual conditions.

- To ensure that the creeping distances between conducting parts at different potential shall meet specific requirements stated in international standards. The creeping distance is the shortest distance between two conducting parts along the surface of the insulating parts. To increase creeping distances, insulating surfaces of insulating parts can be provided with ribs or notches.

Notches in the surface of insulating parts may be taken into consideration for calculating the creeping distance only if the notches are at least 3 mm deep and 3 mm wide. Ribs are only permitted if they are at least 3 mm high and their width is appropriate to the mechanical strength of the material, the minimum being 1 mm.

For the increased-safety type of protection, special attention is paid to the design of terminals for external connections and the internal connections. As these parts can be ignition sources, it is required that each connection shall be made such that the contact practically does not deteriorate due to heating in service, alteration of the insulating material, or vibration, ie, its contact resistance remains constant even during a protracted period in service. Terminals for external connections shall be generously dimensioned to permit the effective connection of conductors of a cross section at least corresponding to the rated current of the electrical apparatus. The terminals shall be

- fixed in their mountings without possibility of self-loosening
- constructed in such a way that the conductors cannot slip out from their intended location
- designed to assure proper contact without deterioration of the conductors
- designed/machined without sharp edges
- designed so that they cannot be twisted or permanently deformed during normal tightening
- designed to assure that the contact that they are providing are not appreciably impaired by temperature changes in normal service.

The size of the terminal compartment should enable perfect connections to be made using common tools. In the interior of electrical apparatus, connections shall not be subject to undue mechanical stress. Solid insulating materials shall have mechanical characteristics which are suitable at temperatures at least 20K above the temperature attained in continuous rated service.

Normally, enclosures containing electrically live bare parts shall as a minimum provide a degree of protection of IP 54 (for definition of IP, see Section 10.2.3.2.1), whereas for enclosures containing electrically insulated parts only, a degree of protection of IP 44 is sufficient.

10.2.1.4 Pressurized Apparatus (Ex p)

This design concept can be used for apparatuses to be used in zones 1 and 2. Pressurized apparatus Ex p is a type of protection by which the entry of a surrounding atmosphere into the enclosure of the electrical apparatus is prevented by maintaining, inside the said enclosure, a protective gas at a higher pressure than that of the surrounding atmosphere. The principle is illustrated in Fig. 10.12.

The overpressure is maintained either with or without a continuous flow of the protective gas. It is then assumed that no flammable gas or vapor is introduced into the electrical apparatus, and that it contains no internal source of flammable gas or vapor. The protective gas can be can be air, inert gas, or another suitable gas.

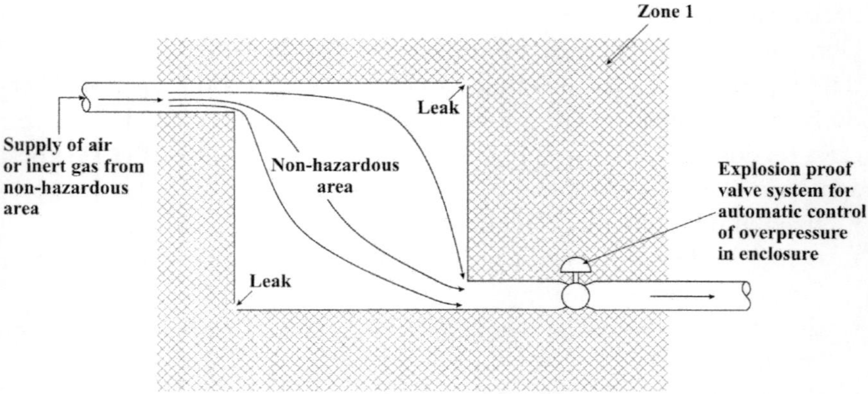

Figure 10.12 Illustration of pressurized enclosure (Ex p).

Before the application of voltage to the electrical apparatus, the enclosure housing the apparatus has to be purged with a protective gas. This implies that a sufficient quantity of the protective gas is passed through the apparatus enclosure and any connected ducts, until the concentration of any explosive gas at any point within the enclosure has been reduced to a value sufficiently lower than the lower explosive limit. Then the protective gas is pressurized to maintain a slight overpressure within the enclosure while the circulation is maintained. A minimum overpressure of 0.5 mbar (50 Pa) shall be maintained relative to the external (atmospheric) pressure at any point within the enclosure. Typically, the quantity of protective gas required for purging will be at least five times the volume of the free space in the enclosure and its associated ducts.

This type of continuously purged pressurized enclosure, besides providing protection against explosion, also aids in dissipating any heat developed by the electrical apparatus inside the enclosure.

10.2.1.5 Oil-Filled Enclosures (Ex o)

This design concept can be applied to apparatuses for use in zones 1 and 2. Oil immersion is a type of protection in which the electrical apparatus or parts of the electrical apparatus are immersed in a specified-quality mineral oil in such a way that an explosive atmosphere which may be above the oil or outside the enclosure cannot be ignited. The principle is illustrated in Fig. 10.13.

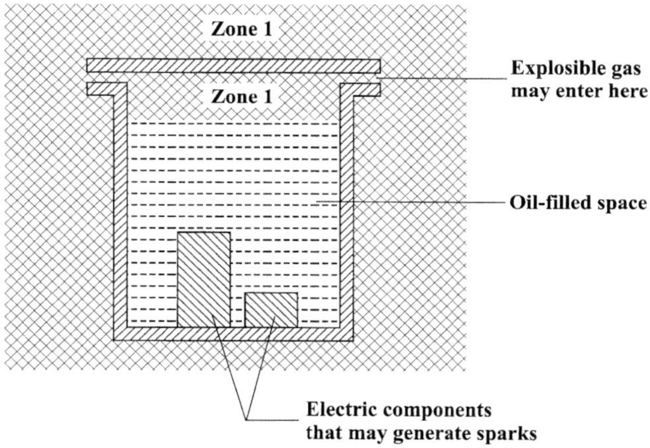

Figure 10.13 Illustration of oil-filled enclosure (Ex o).

Current standards assume that the electrical apparatus immersed in the oil is fixed in its operating position in accordance with specific installation instructions. All parts capable of producing arcs or sparks in normal service shall be immersed in the oil at a minimum depth. Devices shall be provided so that the oil level can be easily checked when the electrical apparatus is in service. The highest and lowest oil levels permissible in normal service should be clearly marked. To ensure that electrical apparatuses satisfy the requirements for this type of protection, certain tests shall be carried out with the highest and the lowest oil level corresponding to a fault in the level gauge, and with an explosive mixture of air and hydrogen above the oil.

10.2.1.6 Sand-Filled Enclosures (Ex q)

This design concept, illustrated in Fig. 10.14, can be applied to apparatuses for use in zones 1 and 2. The following is based on BBC (1983). Sand filling is a type of protection in which the enclosure of electrical apparatus is filled with a material in a finely granulated state so that, in the intended conditions of service, any arc occurring within the enclosure of an electrical apparatus will not ignite the surrounding atmosphere.

No ignition shall be caused either by flame or by excessive temperature of the surfaces of the enclosure. The arcing current is the average value of the root mean square current in the arc caused by a short circuit during its

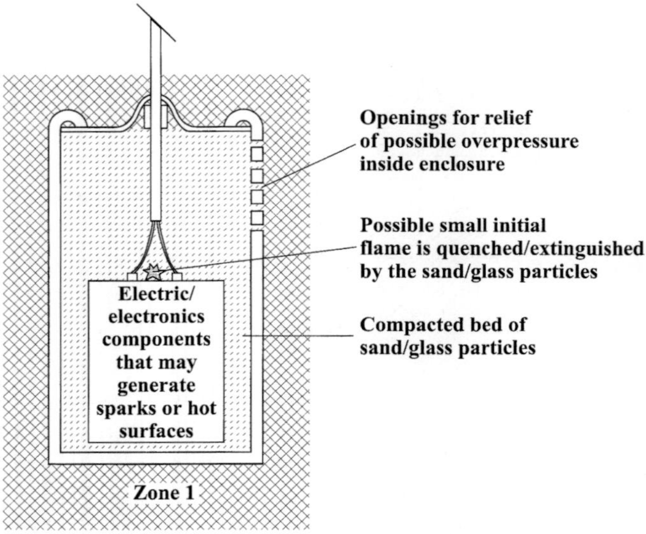

Figure 10.14 Illustration of sand-filled enclosure (Ex q).

development within the filling material of the electrical apparatus. The short-circuit current is value calculated for the point in a network at which the powder-filled electrical apparatus is connected.

Short-circuit time or arcing time is the time during which the arcing current flows through the electrical apparatus, counted from its initiation to its final extinction.

The screen is a perforated-metal sheet fixed inside the enclosure, within the mass of the filling material, in such a way as to cover all the live parts of the electrical apparatus inside the enclosure. The minimum safe height is the shortest vertical distance between the free surface of the filling material, after it is suitably shaken down, and the nearest live part, which prevents transmission of ignition by an electric arc of such current and duration as has been specified for the construction of the electrical apparatus. When the electrical apparatus is provided with a screen, the minimum safe height is the sum of the protection height and the height of the reserve layer. The protection height is the distance between the screen and the nearest live part of the electrical apparatus inside the enclosure. The height of the reserve layer is the thickness of the filling material above the screen, designed to fill up any accidentally formed voids in the safety layer.

The enclosure of powder-filled electrical apparatus shall generally be of metal. Some other materials may be permitted, provided their mechanical and thermal properties are satisfactory. The enclosure in its normal service condition, with all openings closed as in normal use, shall comply with the degree of protection IP 54, and adequate mechanical strength shall be confirmed in a hydraulic test.

The quartz sand used as filling material or a filling material of similar quality shall not contain more than 0.1% by weight of water at the time of filling and shall be suitably shaken down to prevent the formation of voids. No part made of any organic material shall be used above the live parts, or between them and the walls of the enclosure.

Bare live parts shall be suitably spaced from each other and from the walls of the enclosure. For electrical apparatus with nominal voltage less than or equal to 500 V and factory enclosed so as to prevent dismantling without destruction, a distance of 4 mm between live parts, and a distance of 5 mm between live parts and the enclosure are sufficient.

Irrespective of its volume, the enclosure shall be subjected to a hydraulic type test with a pressure of 0.5 bar without the occurrence of permanent deformation exceeding 0.5 mm in any of its dimensions. The pressure shall be applied for at least 1 min.

10.2.1.7 Encapsulation by Molding (Ex m)

This design concept, illustrated in Fig. 10.15, can be applied to apparatuses for use in zones 1 and 2. Encapsulation (Ex m) is a type of protection in which the parts that could ignite an explosive gas atmosphere are enclosed by a polymer compound seal in such a way that the explosive atmosphere cannot come in contact with any sparking or heating which may occur within the encapsulation. The polymer compound can be thermosetting,

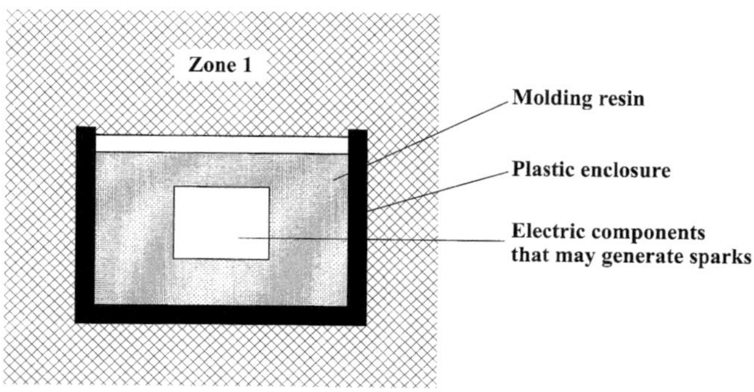

Figure 10.15 Illustration of encapsulated/molded enclosure (Ex m).

thermoplastic, and elastomer materials with and without fillers and/or other additives. The temperature range of the polymer is the temperature range within which its characteristics satisfy the standard both in service and during storage. The continuous service temperature of the polymer is the maximum temperature to which it can be continuously exposed. The choice of polymer for a particular application depends on the duty that the polymer has to perform in the apparatus to be encapsulated.

The suitability of the polymer and the encapsulation in which it is used are verified by tests of eg,

- dielectric strength of the polymer
- water absorption of the polymer
- thermal endurance of polymer and other parts of the component
- maximum surface temperature of the component

10.2.1.8 Design Concepts for Zone 0 Other Than Ex ia

During recent decades, it has been agreed internationally that there is a need for additional design concepts than just Ex ia for electrical apparatuses to be used in Zone 0. The new Zone 0 concepts are typically combinations of at least two independent applications of design concepts for Zone 1 apparatuses.

10.2.1.9 Special Design for Use in Zone 2 Only (Ex n)

Since 1995, special concepts for design of special apparatuses for use in Zone 2 only have been developed. These concepts are essentially just more liberal versions of the design concepts adopted for Zone 1 (and 0) apparatuses. Up to now, all the Zone 2 design requirements have been collected in a single independent Ex n standard. In the future, specific minimum Zone 2 requirements may be integrated in the various separate standards describing the design concepts outlined in previous Sections 10.2.1.1–10.2.1.7.

10.2.2 Clouds of Liquid Droplets

So far no specific approved standards or guidelines exist for designing electrical apparatuses for preventing ignition of explosive clouds of liquid droplets by such apparatuses. Some work to develop such standards has been initiated. In the case of sprays of liquids of low boiling points, in which one can assume immediate complete evaporation as soon as the drops have been generated, the protection methods for gases described in Section 10.2.1 would apply.

10.2.3 Dust Clouds

10.2.3.1 Introduction

The paragraph from NFPA (1997) quoted in Section 10.1.4.1 also indeed applies to design of electrical apparatuses. In this context, the paragraph may be slightly rephrased by replacing the initial word "walls" by "simple enclosures." Hence:

> *Simple enclosures are much more important in separating hazardous and non-hazardous zones in the case of combustible dusts than in the case of combustible gases. Only completely non-perforated solid walls make satisfactory barriers in the case of gases, whereas closed doors, light-weight partitions, and even partial partitions could make satisfactory barriers between hazardous and non-hazardous zones in the case of dusts.*

Eckhoff (2006, 2009) pointed out some important basic differences between gas/vapor and dust explosions and expressed some critical views on the lack of consideration of these differences in European directives and standards specifying electrical apparatuses to be used in areas containing explosive dust clouds.

10.2.3.2 Preventing Ignition of Combustible Dusts by Keeping Potential Ignition Sources Inside Enclosures That Prevent Hazardous Ingress of Dust

10.2.3.2.1 The "IP" Code for Prevention of Dust Ingress

The use of suitable enclosures to keep dust away from delicate electrical and mechanical components has long traditions. Irrespective of specific hazardous effects, the presence of dusts is generally incompatible with delicate equipment and components, just from the point of view of cleanliness and tidiness. Specific reasons for applying this concept include:

- Combustible dust can form an explosive cloud inside the enclosure and cause a dust explosion there
- Combustible dust can form dust layers inside the closure and cause dust fire there
- Electrically conductive dust can cause short-circuiting inside the enclosure
- Abrasive and/or corrosive dusts can damage delicate mechanical components inside the enclosure

In the context of preventing ignition of combustible dusts, only the first and second reasons are relevant. Furthermore, when considering that formation of explosive dust clouds inside enclosures of a reasonable standard, by ingress of dust from the outside, is highly unlikely the possibility of dust fire is in fact the only hazard that has some relevance in the present context.

IEC (2001) produced a standard, named the "IP" code, which defines various "degrees of protection" against ingress of solid objects including dust particles and ingress of water. The initials "IP" just means "International Protection." According to IEC (2001), the "degrees of protection" offered by a given enclosure are to be specified by two digits, the first referring to ingress of solid objects, the second to ingress of water. For solid objects, six levels of protection are defined, ranging from objects larger than 50 mm (digit 1) to dusts (digits 5 and 6). For water, the corresponding range is from gentle dripping (digit 1) to continuous complete immersion (digit 8). The code also specifies specific test methods by which enclosures can be tested for compliance with the requirements of the various degrees of protection.

It is important to note that so far the IEC "IP" code, IEC (2001), does not cover protection against ingress of explosive gases. Satisfactory protection against ingress of gas used to be very difficult to achieve using simple enclosure technology, eg, flanges with gaskets, because gas molecules will migrate though even very tiny openings, in particular if there is a pressure drop across the opening. For this reason, it was necessary to introduce all the additional technologies to prevent ingress of explosive gases into enclosures (oil-filled, pressurized, and molded enclosures) described in Section 10.2.1, ignition of explosive gas that has entered the enclosure (enclosures filled with glass beads or sand, elimination of potential ignition sources inside enclosure), or transmission of an explosion inside the enclosure to a possible external explosive atmosphere ("flame proof[sic]" enclosures), also described in Section 10.2.1. Perhaps IEC in a future-revised version of its IP code may consider including even protection against ingress of gases.

However, in the case of dusts, the additional protection methods developed for gases are not required to prevent formation of explosive dust clouds inside enclosures. Commonplace enclosure technology, eg, flanges with suitable gaskets, provides the protection required. The IEC IP code specifies two levels of prevention of ingress of dust into enclosures, viz. "dust protected" (digit 5) and "dust-tight" (digit 6), which are defined as follows:

Dust protected

A limited quantity of dust is allowed to penetrate into the enclosure under certain conditions (IP 5X).

Dust tight

No dust is allowed to penetrate into the enclosure (IP 6X).

IEC (2001) also specifies the tests to which enclosures of categories IP 5X and IP 6X have to be subjected. The enclosure to be tested is placed inside a test chamber in which a very dense cloud of fine talcum powder is maintained continuously during the test period either by a powder-circulation pump, as illustrated in Fig. 10.16, or by some other means. Depending on the practical circumstances in industry in which the enclosure to be tested is to operate, tests can be conducted either with a slight negative pressure inside the enclosure to be tested, as illustrated in Fig. 10.16, or with no pressure difference across the enclosure wall.

Fig. 10.17 shows an actual test cabinet according to IEC (2001) during dust-ingress testing of a three-phased geared electric motor.

CENELEC (1998) specified the IP requirements for combustible dusts as follows:

Zone 20 and Zone 21: IP 6X

Zone 22: IP 6X for electrically conductive dusts

IP 5X for electrically nonconductive dusts

CENELEC (1998) also specified marking codes to be used to identify the degree of protection offered by a given enclosure.

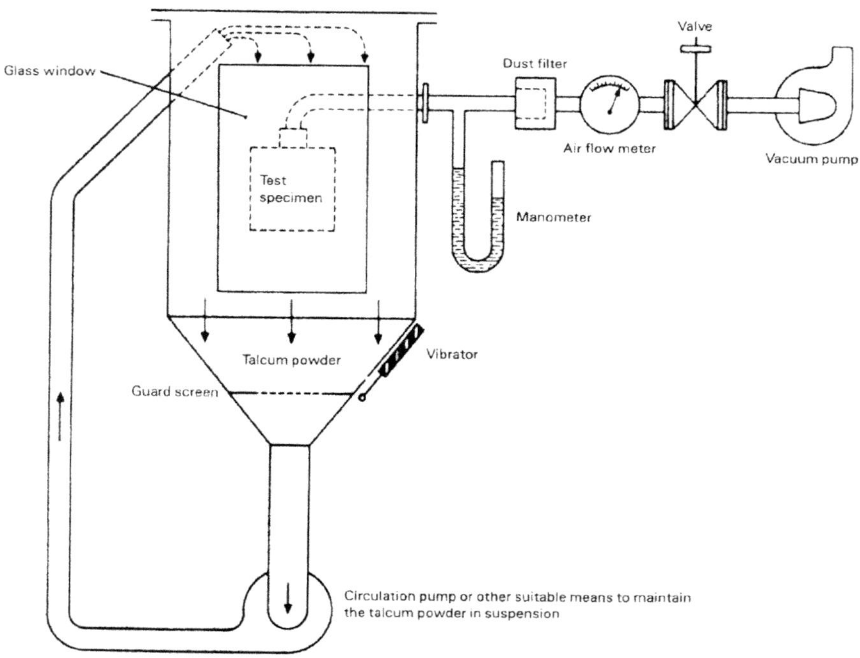

Figure 10.16 Illustration of method used for testing enclosures for their ability to prevent ingress of dust.
From IEC, 2001. Degrees of Protection Provided by Enclosures (IP Code), second ed. of IEC International Standard 60529. Central Office of International Electrotechnical Commission, Geneva, Switzerland.

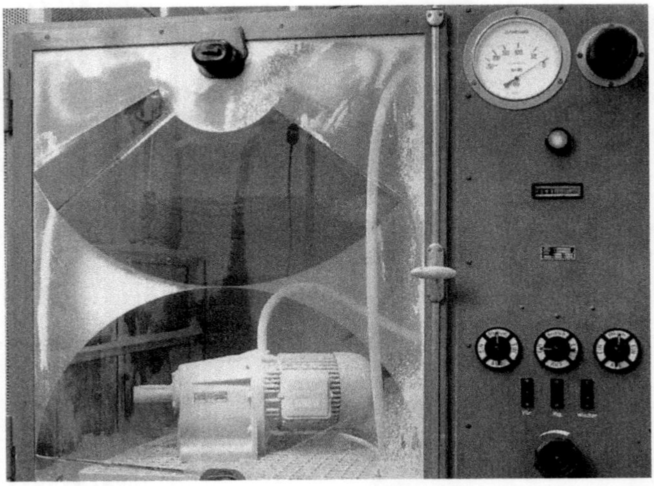

Figure 10.17 Photograph of an actual test cabinet according to IEC (2001) during dust ingress testing of a three-phased geared electric motor.
From Greiner, H., 2001. IP Degrees of Protection, Report SD 101 E issued by Danfoss Bauer GmbH, Esslingen, Germany.

10.2.3.2.2 Design for Preventing Ignition of Dust Clouds and Dust Layers by Hot External Surfaces of Enclosures

Ignition of dust clouds and dust layers by hot surfaces is discussed in Chapter 5. Although it is known that the minimum ignition temperature of a given dust cloud is not a true inherent constant of the cloud, results from the laboratory-scale tests illustrated in Fig. 7.25 are regarded as representative of the dust tested. It is customary to require that the maximum temperature of the dust-free enclosure surface does not exceed 2/3 of the minimum ignition temperature for dust clouds, in °C, as measured in this standard test.

In the case of ignition of dust layers by hot surfaces, the test method illustrated in Fig. 6.4 in Chapter 6 is used. Tests with a given dust, varying the thickness of the dust layer, shows that the minimum ignition temperature decreases systematically with increasing layer thickness. It is customary to require that the maximum temperature of the enclosure surface be at least 75°C lower than the minimum ignition temperature determined in the test. Fig. 10.18 indicates how the maximum-permissible enclosure surface temperature decreases systematically with increasing dust-layer thickness, for three different dusts having minimum hot-surface ignition temperatures of 250, 320, and 400°C, respectively for 5-mm layer thickness. However, if a large part of

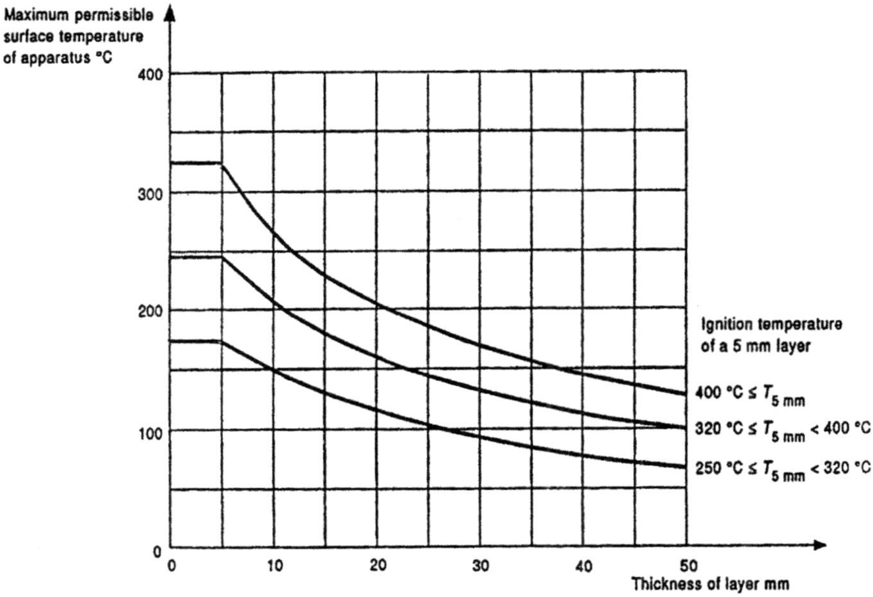

Figure 10.18 Graphs for maximum permissible temperatures of enclosure surfaces as functions of the thickness of a layer of combustible dust on the surface. It is assumed that the dust layer on enclosure surface does not significantly increase the temperature of the surface as compared to that attained without a dust layer.
From CENELEC, September 1998. Electrical Apparatus for Use in the Presence of Combustible Dust. Part 1—2: Selection, Installation and Maintenance. European Standard EN 50281-1-2. CENELEC Central Office, Brussels.

the hot surface is covered by a comparatively thick dust layer, the surface temperature of the enclosure may increase to a value significantly higher than that attained in the absence of dust. In that case, Fig. 10.18 does not apply, and special assessment will be required. This may imply both special tests and mathematical model simulations.

10.2.3.2.3 Other Requirements to IP Enclosures

- Enclosures made of plastic materials must be able to withstand certain specified thermal load tests, in the range of both low and high temperatures. This also applies to any plastic materials used for cementing
- Enclosures must be able to withstand relevant tests for mechanical strength
- Earthing and bonding of metallic parts. All metal parts, which by becoming electrostatically charged can give rise to electrostatic discharges that can ignite clouds or layers of the dust in question, must be properly earthed and bonded

- To avoid propagating brush discharges (see Chapters 2 and 4), enclosures made of plastic materials must satisfy certain requirements to the maximum-permissible insulation resistance to earth, the maximum permissible breakdown voltage across the thickness of the plastic wall, or the minimum permissible thickness of external plastic insulation on metal.
- Exposed parts of enclosures must not contain metals that are able to generate impact sparks that can ignite clouds or layers of the actual dust

10.2.3.3 Other Proposed Methods for Design of Enclosures for Electrical Apparatus to Be Used in Areas Containing Combustible Dusts

10.2.3.3.1 Background

For a long time, design of equipment for areas containing combustible dusts to prevent ignition of the dust has essentially been based on two basic principles. The first is isolation of potential ignition sources by means of enclosures that keep the dust out to the required extent (IP 5X or 6X), the second prevention of ignition of the actual dust (layer or cloud) by the enclosure surface. This philosophy was until recently governing the standardization work in both the IEC and Europe. There does not seem to be any valid reason for departing from this simple, sound philosophy as the basis in international standardization work also in the future. Nevertheless, a new series of dust standards for electrical apparatus has been produced in an attempt to align standards for dusts with those for gases.

10.2.3.3.2 Pressurized Enclosures (Ex p)

The basic idea of the Ex p standard for gases has been outlined in Section 10.2.1. The controversy represented by this standard in the context of combustible dusts is discussed by Eckhoff (2003). The dust standard is both superfluous and inherently self-contradictory.

If the interior of electrical apparatus enclosures has to be kept entirely free of dust particles (dusts of abrasive or corrosive materials may damage delicate moving components, or accumulation of layers of electrically conductive dusts may short-circuit electric/electronic circuitry), it does not represent substantial difficulties to design dust-tight enclosures satisfying the requirements of the enclosure standard IP 6X. In the case of comparatively large apparatus enclosures, eg, instrument cabinets, with doors and windows fitted with rubber gaskets and locking arrangements, it may be difficult to completely prevent ingress of dust. However, transmission of explosive dust clouds from the outside of the enclosure to its interior through possible narrow gaps and holes can be entirely excluded. Hence, a pressurization standard for combustible dusts to prevent dust explosions inside enclosures is superfluous.

The self-contradiction of the standard lies in the following erratic assumption: If particles in a dust cloud embracing an enclosure should enter it though narrow holes and gaps, the particles can stay and accumulate suspended in the atmosphere inside the enclosure, and eventually form an explosive dust cloud inside the enclosure, as if the dust particles were gas molecules. This assumption is basically wrong, If significant quantities of dust particles do, over some time, enter an enclosure at all, they will not accumulate as cloud, but as a layer. For a more extensive discussion, see Eckhoff (2003, 2006, 2009).

10.2.3.3.3 Encapsulation by Molding (Ex m)

As part of the effort to align dust standards with gas standards, a new standard for encapsulation of electrical equipment for combustible dust atmospheres by molding has been produced. With this type of protection electrical parts that can ignite an explosive gas atmosphere are moulded into a compound material to eliminate the possibility of contact between the electrical parts and the explosive atmosphere parts that can ignite an explosive atmosphere are molded into a compound material excluding atmospheric contact with these parts. The compounds can be thermosetting, thermoplastic, epoxy resins, elastomers, etc. with or without fillers. It is difficult to see that this comprehensive standard for dusts, which is to a large extent an edited copy of the corresponding gas standard, is very helpful. As outlined in Section 10.2.1.7 the basic issue of the molding concept is to prevent an explosive gas/vapor from entering an enclosure containing ignition sources. However, this kind of sophisticated molding solutions are superfluous with combustible dusts. As discussed in Section 10.2.3.2 simple gasket sealing solutions that satisfy the relevant IP requirements, are fully adequate. A relevant specific issue with dusts would rather be to make sure that molded components embedded in dust deposits do not give rise to self-heating/self-ignition of the dust layer/deposit.

10.2.3.3.4 Intrinsically Safe Electrical Apparatus (Ex i)

In practice, electrical circuits, switches, etc. to be used in areas containing combustible and/or electrically conductive dusts, should always be kept inside enclosures. This will prevent significant quantities of dust from making contact with electrical components in general, including components that may generate electric sparks and/or hot surfaces. Hence, direct adoption of the entire concept of "intrinsically safe design" from the gas/vapor domain into the domain of combustible dusts does not seem to be an optimal approach.

However, in some special situations there can be a genuine need for intrinsically safe apparatus even in environments containing combustible powders/dusts in some highly specialized applications. One example is capacitive level indicators for solid bulk materials stored in silos and bins. In this case, a live capacitor "plate," in the form of a bare-metal rope, carrying a voltage with respect to earth, is directly exposed to the combustible powder/dust inside the silo or bin. An ignition risk could arise from electrical sparks generated by direct contact between the energized bare-metal rope and any grounded-metal part of the silo.

Fig. 10.19 illustrates the application of this type of level indicator. The basic principle of measurement is as follows: A short electric pulse of a few ns (nano seconds) duration is emitted from the sensor head at the top of the silo and travels down the vertical metal rod/rope. At the point at which the rod/rope becomes immersed in the powder, its impedance changes abruptly, which causes a partial reflection of the electrical pulse from this point and backward to the sensor head at the top. The distance d from the sensor head at the silo top to the powder/liquid surface is then

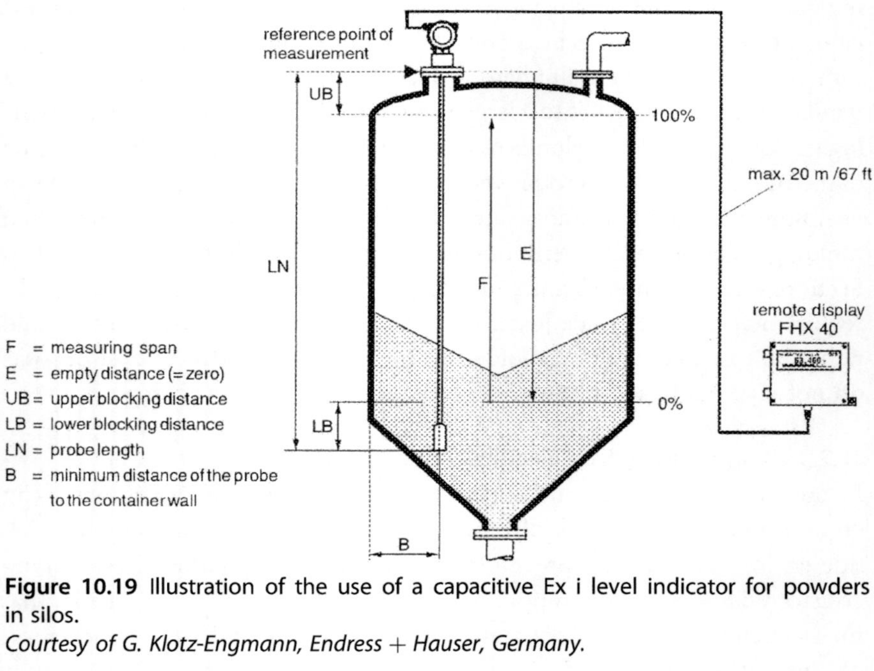

Figure 10.19 Illustration of the use of a capacitive Ex i level indicator for powders in silos.
Courtesy of G. Klotz-Engmann, Endress + Hauser, Germany.

d = ½ ct. Here, c is the speed of the electromagnetic wave pulse passing along the metal rope/rod, and t is the time from pulse emission at the sensor head till the return of the reflected pulse from the point along the rod/rope at which the surrounding medium changes from gas to bulk powder.

10.2.4 Explosives/Pyrotechnics

CENELEC (1997) does not specify any specific standards for design of electrical equipment to be used in areas where explosives, propellants, or pyrotechnics are produced and/or handled. Instead, the standard specifies conditions under which electrical apparatus standards for gases and vapors (Section 10.2.1) or for dusts (Section 10.2.3) apply. The following general guidelines are given:

- In areas endangered by explosive substances, only electrical apparatus being absolutely necessary for the operation of the electrical installations shall be used. When this is not practicable, apparatus should be installed in the area with the lowest risk.
- Electrical apparatus shall be selected on the basis of its design or protected by additional structural measures and installed in such a way that in prescribed use the protection against the penetration of dust and water and against electrical, chemical, thermal, or mechanical influences required for operational safety is maintained.
- Cable glands should be either made from a suitable noncorrodible material or electroplated with, eg, chromium or nickel.
- The design of enclosures for electrical apparatuses should minimize the possibilities of accumulation of dust layers, and facilitate cleaning operations.
- In cases in which electrical apparatus in an area E1 is subjected only to accumulation of explosive substances as layers, at the same time as the apparatus is provided with additional devices for preventing accumulation of such substances on and inside the electrical apparatus (eg, dust covers or additional enclosures), the requirements for area E2 apply.
- Electrical apparatus—with the exception of cables and cords—may be used in general at ambient temperatures not exceeding 40°C. The influence of adjacent heat sources shall be taken into account.
- In cases in which ambient temperatures exceed 40°C, electrical apparatus that has been specially designed for these conditions shall be used.
- When explosion-proof electrical apparatus is used, it shall be ensured that the explosion protection as in prescribed use is retained.

CHAPTER ELEVEN

Review of Some Methods of Hazard and Risk Analysis

11.1 INTRODUCTION

Hazards and risk analysis is a very large subject in itself, covered by a substantial amount of published information, including a number textbooks. Khan and Abbasi (1998) made the important point that the terms *hazard* and *risk* have different meanings and should therefore not be used interchangeably, as is often done in practice. They suggest that *hazard* be related to the *source of the harm* under consideration, whereas *risk* is the expected impact of this harm (eg, probability times consequence). The term "hazard analysis" comprises a number of different systematic methods for identifying the hazards to be associated with a given process or plant. Such analyses can be used as a basis for optimal selection of means for preventing and mitigating accidental explosions.

The Center for Chemical Process Safety (CCPS) (1985), Cox (1986, 1987), Khan and Abbasi (1998), and Gould et al. (2000) presented informative reviews of various techniques used for hazard/risk analyses in the process industries. In 1992, CCPS (1992) published a second edition of CCPS (1985). Only the first edition (1985) was available when the present chapter was written. In the first edition the following list of common hazard/risk assessment methods were presented:

- Preliminary hazard inventories
- Process/systems checklists
- Safety audits
- Relative ranking based on Dow and Mond hazard indices
- "What-if?" analysis
- Failure modes and effect analysis (FMEA)
- Hazard and operability study (HAZOP)
- Fault tree analysis (FTA)
- Event tree analysis (ETA)
- Cause–Consequence analysis (CCA)
- Human-error analysis (HEA)

Explosion Hazards in the Process Industries
ISBN: 9780128032732
http://dx.doi.org/10.1016/B978-0-12-803273-2.00011-6

Gould et al. (2000) grouped a much larger number of 37 identified methods in four main categories, viz. methods for *process hazards identification*, methods for *hardware hazards identification*, methods for *control hazards identification*, and methods for *human hazards identification*.

From the various reviews of hazard/risk evaluation methods just cited, it appears that disagreement has arisen on both the number of existing methods and the way in which they should be organized and presented. For example, Khan and Abbasi (1998) presented FTA, FMEA, What-if analysis, and hazard indices analysis (see following sections) as sub-sections in a main HAZOP section. CCPS (1985) and Gould et al. (2000), on the other hand, presented all these techniques as separate methods on the same level as HAZOP. For the purpose of the present limited review, a slightly modified version of the comparatively short list from CCPS (1985), given above, will be followed.

11.2 PRELIMINARY HAZARD INVENTORIES

A preliminary hazard inventory can provide an important input to many safety studies. CCPS (1985) uses the term "preliminary hazard analysis" to characterize this kind of data collection. A main objective is to record all stocks of hazardous material or energy, and note relevant details of storage conditions. When carried out at the conceptual stage of a process plant project, such an inventory may contribute to plant layout optimization and suggest process changes eg, to reduce stored quantities of hazardous material. Generally, the collected information will be useful input to more comprehensive risk assessment activities. However, the inventory itself is little more than a "screening" designed to identify problem areas.

More details are given by CCPS (1985, 1992), Cox (1986, 1987), Khan and Abbasi (1998), and Gould et al. (2000).

11.3 PROCESS/SYSTEMS CHECKLISTS

Checklists are lists of specific safety-related requirements that have to be fulfilled by both hardware and operators in process plants. The purpose is to ensure compliance with standard procedures. The use of checklists ensures a minimum-acceptable level of process safety. They should be easy to use, and, if they are, inexperienced employees in particular can benefit greatly from going systematically through the various points of safety issues related to their job. However, checklists also offer a common basis for management reviews of individual employees' work.

Checklists should be as detailed as required for ensuring that standard procedures are being followed, but they also offer opportunities for identifying problems that require further attention.

Checklists are not only useful for up-and-running activities. The method is also used for controlling the development of new projects from the initial design stage right through to plant shutdown. Then, such lists are also frequently used by various staff and management functions as a basis for formal approval before a project can move from one stage to the next.

However, the value of checklists are necessarily limited to the experience base of the checklist author(s), and should be audited and updated regularly.

More details are given by CCPS (1985).

11.4 SAFETY AUDITS

In general, a safety audit is an on-site walk-through inspection in a process plant by qualified people with the purpose to ensure that the operating and maintenance procedures are in accordance with existing standards. The rigorousness of such an audit can vary from a very informal spontaneous visual inspection of the quality of housekeeping, to a very formal week-long examination by a team with adequate backgrounds and responsibilities. What is said in the following relates to the latter, more comprehensive form of safety audit. A complete audit of this kind will normally require a team of two to five people for at least 1 week. Personnel assigned to safety audit inspections need to be very familiar with relevant safety standards and procedures. Special technical skills are helpful for evaluating instrumentation, electrical systems, pressure vessels, process materials, and chemistry, and other matters requiring special attention.

The auditing team should look for major risk situations. Reviewing general housekeeping and attitudes of personnel should not be primary objectives, although they can be significant indicators of where to find problems and places where real improvements are needed. Various hazard evaluation techniques, such as checklists, "what-if" questions, and raw-material evaluations, may be used during the review.

To be able to conduct a complete comprehensive safety audit, the auditing team must have access to relevant codes and standards, detailed plant descriptions (eg, piping and instrumentation drawings and flowcharts), and written procedures for plant startup and shutdown in normal operation and emergencies. In addition, they must have access to personnel injury reports, hazardous incidents reports, maintenance records (eg, checks of critical instruments, pressure-relief valve tests, pressure-vessel inspections), and process-material characteristics (eg, toxicity and reactivity information).

One main aim of a safety audit is to keep operating personnel alert to process hazards that have already been identified. However, comprehensive audits may also review whether currently adopted operating procedures need to be revised, and they may seek to identify any performed changes of equipment/processes/procedures that may have introduced new hazards. They may also identify places where new safety methods/procedures should be introduced for fighting existing hazards in existing production and maintenance activities. The report of the auditing team should include observed deviations from designed and planned procedures and notification of any new safety items discovered during the audit. However, the responsibility for implementing any corrective actions remains with the plant management.

Whereas safety audits are most commonly conducted in operating process plants, they are also applicable to pilot plants, laboratories, and storage facilities. A comprehensive safety audit is intended to complement other current efforts to maintain and improve plant safety, including regular informal visual inspections.

A comprehensive safety audit should be regarded as a cooperative effort to improve the overall safety and performance of the plant, rather than as an untimely and unpleasant interference with normal operations. Cooperation is essential. Workers are likely to become defensive unless considerable efforts are made to present the review as a common benefit. The audit often includes interviews with many people in the plant, such as operators, maintenance staff, engineers, management, and safety staff, depending on the plant organization. Having the support and involvement of all these groups is an important condition for fulfillment of the aims of the audit.

At the end of a safety audit, recommendations should be given for specific actions that should be taken. The recommendations should include responsibilities and completion dates. A followup evaluation or reinspection should be planned for verifying whether the corrective actions taken are adequate and sufficient.

Safety audits are usually conducted on a regularly scheduled basis. Special-emphasis reviews or followup/resurvey inspections can be scheduled intermittently. Some "high-risk" plants are audited as often as every 2 or 3 years. For "low-risk" plants, the intervals may be as long as 5—10 years.

More details are given by CCPS (1985), Cox (1986, 1987), Khan and Abbasi (1998), and Gould et al. (2000).

11.5 RELATIVE RISK RANKING USING DOW AND MOND INDICES

The Dow and Mond indices constitute a direct and relatively easy route toward obtaining estimates of the various risks in a process plant. The methods assign penalties and credits based on plant features. Penalties are assigned to process materials and process conditions that can contribute to an accident. Credits are assigned to plant safety features that can prevent and/or mitigate the effects of an accident. The penalties and credits are combined to derive an index that is a relative ranking of the risk.

Both the Dow and the Mond risk-ranking methods address risks in individual process units. The two methods are closely related as the Mond index was developed as an extension of the Dow index. The primary difference between the two is that the Mond index not only takes into account flammability and chemical reactivity but also the toxicity of materials in the assessment of material factors for the various process units.

The Dow index is easier to apply than the Mond index because of the extensive use of tables and graphs in place of traditional equations, but both methods employ the same basic calculation method. The Dow guide contains methods for estimating outage time and business interruption costs, which are compatible with either index. The two methods comprise the following seven general steps:

- Identify on the plot plan the process units that are expected to have the greatest effect or contribute the most to a fire, explosion, or release of toxic material
- Determine the material factor for each process unit
- Evaluate the appropriate contributing hazard factors, considering fire, explosion, and toxicity
- Calculate the unit hazard factor and damage factor for each process unit
- Determine the fire-and-explosion index and area of exposure for each unit
- Calculate maximum probable property damage
- Evaluate the costs of maximum probable days of outage and business interruption

Further details are given by CCPS (1985), Cox (1986, 1987), Khan and Abbasi (1998), and Gould et al. (2000).

11.6 "WHAT-IF" ANALYSIS

The basic feature of a "What-if" analysis is to conduct a systematic examination of a process or operation by asking questions that begin with "What if...." The process/operation can include buildings, power systems, raw materials, products, storage, material handling, operating procedures, work practices, in-plant environment, plant security, management practices, etc. A "What-if?" analysis is a powerful tool, but the people conducting it must be experienced.

A "What-if" analysis includes the following steps:

- Define study boundaries
- Gather needed information
- Define the team
- Conduct the reviews
- Record results

Two types of boundaries limit what any given "What-If" analysis will include. The first is the limited selection of consequence categories that should/can be covered by the study. The second is the limitation(s) of the actual physical system(s) to be investigated. The two boundaries are closely related. Often, the consequences to be considered (public safety, plant damage, etc.) will define what portions of the plant should be included in the study. In general, the consequences are divided into three main categories, viz. *public risk*, *worker risk*, and *economic risk*. These can be further divided into more detailed risk categories. The economic risk could, for example, be defined for one specific plant section as "explosions/fires leading to loss of production for more than 3 days." However, considerable care has to be exercised when going into detail.

The formulation of the exact questions to be asked is left to the team conducting the examination, unless there is already a list of adequate questions from reviews. The questioning usually starts at the upstream end of the process and follows the process flow to the defined end boundary of the review. Alternatively, the questioning can focus on one particular consequence category at a time, eg, personnel safety or public safety, and continue right through the part of the plant included in the review. Usually a small group of two or three experts conducts the examination and reports the findings. The findings are usually accident-event sequences that result from the "What-If" questions. The questions essentially suggest an initiating event, and perhaps a failure from which an undesirable event sequence could occur.

For example, a question might be "What if the raw material is the wrong concentration?" The team would then attempt to determine how the process would respond, eg, "If the acid concentration had been doubled, the reaction could not have been controlled, and a rapid exothermal reaction would result."

The questions and answers (including hazards, consequences, and suggested solutions) are all recorded. The study has ended when the team has completed its recordings and has reported the results to the responsible authorities. The recordings and report have then to be reviewed at the appropriate level of management.

More details are given by CCPS (1985), Cox (1986, 1987), Khan and Abbasi (1998), and Gould et al. (2000).

11.7 HAZARD AND OPERABILITY STUDIES (HAZOP)

11.7.1 Outline

According to Wikipedia (2015), a "hazard and operability study" (HAZOP) is a structured and systematic examination of a planned or existing process or operation to identify and evaluate problems that may represent risks to personnel or equipment, or prevent efficient operation. The study is carried out by a suitably experienced multidisciplinary team (HAZOP team) during a set of meetings. The HAZOP technique is essentially *qualitative*, and aims to stimulate the imagination of participants to identify potential hazards and operability problems. The HAZOP technique was initially developed to analyze chemical process systems and mining operation processes but has later been extended to other types of systems and also to complex systems, such as nuclear power plants, and to use software to record the deviation and consequence.

The overall purpose of HAZOP is the same as of FMEA, ie, to identify systematically all the possible ways in which the system investigated could fail, evaluate these, and formulate recommendations for preventive and mitigating measures.

However, as opposed to FMEA, HAZOP contains a systematic method for identifying failure modes. This involves scrutiny of a large number of possible deviations from normal operating conditions, which are generated by applying guide words, such as more, less, reverse, etc., to each of the parameters describing process conditions in each component, plant item, or line in the plant. However, HAZOP in its original form has disadvantages, and some industrial companies have modified the way in which the results of the study are handled. Instead of "recommendations," the output is "identified problems," which leaves more room for a coordinated rational design revision, which is not only cheaper, but probably also safer.

More details are given by CCPS (1985, 1992). Khan and Abbasi (1998) discussed some inherent limitations in the original HAZOP method.

11.7.2 Some Details of the HAZOP Method

According to Wikipedia (2015), the HAZOP method is typically applied to processes (planned or existing) for which design information is already available. For continuous processes, the HAZOP procedure commonly starts with a piping and instrumentation diagram and a process–flow diagram, which are examined in sections. The sections are chosen so that, for each section, meaningful *design intent* (the desired, or specified range of performance for that item, not just its design duty point) can be specified.

Consider as an example a pipe in a chemical plant that is intended to transport 2.3 kg/s of 96% sulfuric acid at 20°C and a pressure of 2 bars from a pump to a heat exchanger. A careful designer will be able to foresee possible deviations, eg, a hotter or stronger acid or reduced or lack of flow in a line, even before the design process reaches the HAZOP stage. The intended duty of the heat exchanger may be to heat 2.3 kg/s of 96% sulfuric acid from 20°C to 80°C, but its full design intent will also include "glimpse of the obvious" functions, such as maintaining containment of hot acid and the heating fluid, and preventing leakage of one into the other.

When defining the various sections to be analyzed, attention must be paid to the complexity of the whole system and the magnitude of the hazards it might pose. The HAZOP team then determines what are the possible significant *deviations* from each intended performance, feasible *causes* of the deviations, and likely *consequences* of them. It can then be decided, during either the HAZOP session or subsequent analysis, whether existing/designed safeguards are adequate, or whether additional actions are required to reduce the risks to acceptable levels.

For batch and other sequential operations, a logical flow diagram for the HAZOP study should be available. This is because various types of equipment may have different design intents at different points in the operation, and hazards may arise from performing operations out of scheduled sequence.

HAZOP meetings recorded by hand generally take 3 to 4 h per day. It is then possible to analyze about 30 specific items per meeting. In a medium-sized chemical plant, the total number of items to be considered may be about 1200. Then, about 40 such meetings would be needed. Various software programs are now available to assist in recording the analyses performed in the meetings.

11.7.3 Guide Words and Parameters

To identify deviations, the team applies (systematically, in order) a set of *Guide Words* to each section of the process. To prompt discussion, or to ensure completeness, it may also be helpful to explicitly consider appropriate *parameters* which apply to the design intent. These are general words such as Flow, Temperature, Pressure, and/or Composition. The current standard notes that guide words should be chosen which are appropriate to the study and neither too specific (limiting ideas and discussion) nor too general (allowing loss of focus). A typical set of guide words are given in Table 11.1. The last four guide words apply to batch or sequential operations. When a guide word is meaningfully applicable to a parameter eg, NO FLOW, MORE TEMPERATURE, their combination should be recorded as a credible potential deviation. The distinction between some guide words may not always be remembered by the team and should be specified. For example, "LESS COMPOSITION" should suggest less than 96% of normal sulfuric acid, AS WELL AS COMPOSITION should suggest a

Table 11.1 Examples of guide words used in HAZOP analyses and their meanings

Guide word	Meaning
NO OR NOT	Complete negation of the design intent
MORE	Quantitative increase
LESS	Quantitative decrease
AS WELL AS	Qualitative modification/increase
PART OF	Qualitative modification/decrease
REVERSE	Logical opposite of the design intent
OTHER THAN	Complete substitution
EARLY	Relative to the clock time
LATE	Relative to the clock time
BEFORE	Relating to order or sequence
AFTER	Relating to order or sequence

From Wikipedia, 2015. Internet Publication Found March 2015 by Seeking Hazop on Google.

contaminant, whereas OTHER THAN COMPOSITION should suggest something else, such as oil. Deviations should be well observed by the plant. If, for example, a 60% sulfuric/15% nitric acid mixture could be fed instead of just sulfuric acid, the possibility could be flagged up against LESS, AS WELL AS, OTHER THAN.

HAZOP-type studies may also be carried out by considering applicable guide words and identifying elements to which they are applicable or by considering the parameters associated with plant elements and systematically applying guide words to them. Table 11.2 gives an overview of commonly used guide words, parameter pairs, and common interpretations of them.

Once the causes and effects of any potential hazards have been established, the system being studied can then be modified to improve its safety. The modified design should then be subject to another HAZOP, to ensure that no new problems have been added.

11.7.4 HAZOP Team

Any HAZOP study is conducted by a team of selected people. The team should be as small as possible, but it is essential that it comprises relevant skills and experience. Team members should be able to display "intuition and good judgment," and be able to ensure that meetings are held in "a climate of positive thinking and frank discussion." The number of team members should preferably be at least 4–5. In a comprehensive HAZOP analysis, there will be many successive HAZOP meetings, and new team members may be co-opted and old members replaced as the analysis develops. However, normally the Study Leader and Recorder will remain the same persons throughout. Earlier it was sometimes suggested that the Study Leader could also be the Recorder, but it is now generally recommended that these positions be held by two persons. Software is now available from several suppliers to aid the Study Leader and the Recorder. As many as 20 individuals may be involved in a comprehensive HAZOP analysis, but is recommended that no more than 7–8 be in the team at any time, as a larger team is likely to make slower progress. Each team member should have a definite role as indicated in Table 11.3.

Further information on the HAZOP method is provided by CCPS (1985), Cox (1986, 1987), Khan and Abbasi (1998), and Gould et al. (2000).

Table 11.2 Examples of commonly used HAZOP guide words in relation to certain process elements of a plant

Parameter/guide word	More	Less	None	Reverse	As well as	Part of	Other than
Flow	High flow	Low flow	No flow	Reverse flow	Deviating concentration	Contamination	Deviating material
Pressure	High pressure	Low pressure	Vacuum		Delta-p		Explosion
Temperature	High temperature	Low temperature					
Level	High level	Low level	No level		Different level		
Time	Too long/too late	Too short/too soon	Sequence step skipped	Backward	Missing actions	Extra actions	Wrong time
Agitation	Fast mixing	Slow mixing	No mixing				
Reaction	Fast reaction/ runaway	Slow reaction	No reaction				Unwanted reaction
Startup/shutdown	Too fast	Too slow			Actions missed		Wrong recipe
Draining/Venting	Too long	Too short	None		Deviating pressure	Wrong timing	
Rendering inert	High pressure	Low pressure	None			Contamination	Wrong material
Utility failure (instrument air, power)			Failure				
Distributed control system failure			Failure				
Maintenance			None				
Vibrations	Too low	Too high	None				Wrong frequency

From Wikipedia, 2015. Internet Publication Found March 2015 by Seeking Hazop on Google.

Table 11.3 Typical titles, alternative titles, and roles of members of HAZOP teams

Name	Alternative	Role
Study leader	Chairman	A person who is experienced in HAZOP but not directly involved in the design of the actual plant/process. The role is to ensure that the HAZOP procedure is followed carefully. He/she is responsible for ensuring that discussions lead to definite conclusions and are adequately recorded, problems are documented, and recommendations passed on.
Recorder	Secretary or scribe	A person to record discussions accurately but comprehensively, and to document problems and recommendations.
Designer	Representative of the team which has designed the process	A person being able to explain any design details and/or provide further information.
User	Representative of those who will use it	A person who can evaluate the usefulness of the plant and pose questions on its operability, and questions on the effect of deviations.
Specialist	According to specialty, eg, chemist or human factors specialist	A person with relevant technical knowledge and training in a specific field, eg, the effect of varying reaction conditions, human reliability analysis, and human-error identification.
Maintainer	If appropriate	A person concerned with maintenance of the plant.

From Wikipedia, 2015. Internet Publication Found March 2015 by Seeking Hazop on Google.

11.8 FAILURE MODES, EFFECT, AND CRITICALITY ANALYSIS (FMECA)

In literature other than CCPS (1985), eg, in the review paper by Khan and Abbasi (1998), assessment of "criticality" is not considered part of this method, which is then called "failure mode and effect analysis (FMEA)." The present account, however, is essentially based on CCPS (1985).

The purpose of FMECA is to identify systematically all of the possible ways in which a process plant can fail, to evaluate these, and to formulate recommendations for preventive and mitigatory measures. The results are recorded in a standard format in which recommendations for action can be included. The weakness of FMECA is that there is no specified method for identifying the failure modes, their effects, and criticalities. The engineer is expected to do this from either first principles or experience, and the only formal requirement imposed on him/her is the reporting format.

As defined by CCPS (1985), FMECA is concerned almost entirely with possible failure of equipment/apparatuses. The three central data categories collected are data on the ways in which equipment/apparatuses can fail, data on the destructive effects that can result from various modes of failure, and data for the estimated probabilities of failure. These data in turn provide the basis for determining the changes should be made for reducing the risk of such failures in the future. Each failure type is considered individually as an independent occurrence, with no relation to other failure types in the system, except for any secondary effects that the failures considered may produce.

Generally, FMECA is first conducted on a qualitative basis. Quantitative data may then subsequently be included to establish a criticality ranking for the system being evaluated. Criticality rankings are generally expressed as probabilities but may also be indicated in other ways. In some instances, they are designated in categories from 1 to 10 (or from A to Z) to show the principal items that could generate problems. These categories are often not based on probabilities but reflect experience. The FMECA formal setup comprises the following five steps:

- Determine the level of resolution
- Develop a consistent format for the analysis
- Define the problem to be addressed and its boundary conditions
- Complete the FMECA table
- Report the results

An FMECA table comprising all data from the analysis should be produced. The table should be organized in a deliberate but systematic manner to reduce the possibility of omissions and to ensure the completeness of the FMECA. One systematic approach will be to start somewhere at the system boundary, as defined on a reference drawing, and systematically evaluate the items in order as they appear in the process flow path. Each equipment item can then be checked off or "red-lined" on the reference drawing when its failure modes have been evaluated completely. All entries for each item or system being addressed in the FMECA should be completed before proceeding to the next item.

The following items should be standard entries in the FMECA table:
- Equipment identification
- Complete equipment description
- Possible failure modes, effects of failures
- Criticality ranking

The last item means that the severity of any potential failure be assigned a given criticality ranking characteristic/number as for example:
- No process upsets (best)
- Minor process upset, small hazard to facilities and personnel, process shutdown not required
- Major process upset, significant hazard to facilities and personnel, orderly process shutdown required
- Immediate hazard to facility (worst) and personnel, emergency shutdown required

Further details on FMECA are given by CCPS (1985). Khan and Abbasi (1998), and Gould et al. (2000) give detailed reviews of the slightly simpler failure mode and effect analysis method (FMEA).

11.9 FAULT TREE ANALYSIS (FTA)

The method is applied to complex systems, whether the complexity is due to the nature of the process itself, or to the instrumentation required for running the process. In the basic technique, the "Fault Tree Analysis," the failure modes must first be identified, eg, by HAZOP (see Section 11.11 following). These failure modes are named "top events." Examples of "top events" would be a fire around a distillation tower, and a dust explosion in a milling plant.

For each "top event," the fault tree analyst(s) must identify all those events or combinations of events that could lead directly to the top event. The precise logical relationships between causes and effects are expressed by AND or OR gates and are usually presented in diagrammatic form. The immediate causes of the top event have their own contributory causes, and these can be presented in a similar way, so that a complete fault tree is built up. This process ceases when all of the causative factors at the bottom of the tree are of a simple kind for which frequencies of occurrence or probabilities can be estimated.

FTA is a widely used tool for system safety analysis. One of the primary strengths of the method is the systematic, logical development of the many contributing failures that might result in an accident. This type of development requires that the analyst have a complete understanding of the system/plant operation and the various equipment failure modes. FTA breaks down an accident into its contributing equipment failures and human errors. The method, therefore, is a "reverse-thinking" technique; that is, the analyst begins with an accident or undesirable event that is to be avoided and identifies the immediate causes of that event. Each of the immediate causes is examined in turn until the analyst has identified the basic causes of each event. The fault tree is a diagram that displays the logical interrelationships between these basic causes and the accident.

The result of the FTA is a list of combinations of equipment and human failures that are sufficient to result in the accident. These combinations of failures are known as minimal cut sets. Each minimal cut set is the smallest set of equipment and human failures that are sufficient to cause the accident if all the failures in that minimal cut set occur simultaneously. Thus, a minimal cut set is logically equivalent to the undesired accident stated in terms of equipment failures and human errors. A fault tree is a graphical representation of the interrelationships between equipment failures and a specific accident. The following symbols are used to display these relationships:

OR Gate

This gate indicates that the output event occurs if any of the input events occur.

AND Gate

This gate indicates that the output event occurs only when all the input events occur.

INHIBIT Gate

This gate indicates that the output event occurs when the input event occurs and the inhibit condition is satisfied.

DELAY Gate

This gate indicates that the output event occurs when the input event has occurred and the specified delay time has expired.

BASIC Event

This event represents a basic equipment fault or failure that requires no further development into more basic faults or failures.

INTERMEDIATE Event

This event represents a fault event that results from the interactions of other fault events that are developed through logic gates such as those defined above.

UNDEVELOPED Event
This event represents a fault event that is not examined further because information is unavailable or because its consequence is insignificant.
EXTERNAL or HOUSE Event
This event represents a condition or an event that is assumed to exist as a boundary condition for the fault tree.

The synthesis of fault tree is best done by engineers with good experience of the type of process system under consideration. It is much easier to teach a technically qualified and experienced person how to construct a fault tree than to teach a reliability specialist all the technical details of the system. On the other hand, the quantitative analysis of a fault tree is a separate activity in which the reliability specialist will play the dominant role. Further details are given by CCPS (1985).

11.10 EVENT TREE ANALYSIS (ETA)

11.10.1 Outline

An event tree can be looked at as the reverse of a fault tree, starting with the initial or "bottom events" and exploring all possible "top events" that can result from it. Each outcome has further outcomes and all of these can be related by means of decision gates. At each gate, the conditional probabilities must be estimated for each of the alternative branches. On this basis the probabilities of the final hazard, or "top event," can be calculated.

The horizontal core of Fig. 2.66 in Chapter 2 is a typical event tree leading to gas explosions and fires in offshore and onshore process industries. Here, the initial event is leakage of a combustible gas, the intermediate event ignition of the gas, and the final, or "top," event an explosion, a fire, or a combination of both. The various boxes in Fig. 2.66 do not constitute part of the basic event tree, but indicate actions that can be taken to prevent the various events in the chain leading to the top event.

ETA evaluates potential accident outcomes that might result following, eg, an equipment failure or a process upset constituting the initiating event. As opposed to the FTA, the ETA starts with the analyst identifying a possible initiating event and then develops sequences of secondary events that can lead to potential accidents. Both successes and the failures of any safety measures along the event chain should be taken into account.

ETAs are well suited for assessing possible consequences of a given initiating event. The general procedure for ETA contains four steps:

- Identification of the initiating event of concern
- Identification of the safety functions designed to deal with that initiating event
- Construction of the event tree
- Description of the various sequences of secondary events and their results

11.10.2 Identification of the Initiating Event

Selecting the appropriate initiating event is a crucial part of any ETA. This event can be a system or equipment failure, a human error, or a process upset that can result in a variety of effects depending on how well hardware and operators respond to the event. If the selected event leads directly to a specific well-defined accident, an FTA for determining its basic cause(s) would be more appropriate than an ETA. In most applications of ETA, the initiating event is anticipated, and the plant design includes systems, barriers, or procedures intended to mitigate the effects of the initiating event.

11.10.3 Identification of Safety Measures/Functions Designed to Deal With the Initiating Event

Such safety functions usually include:

- Safety systems responding automatically to counteract undesired effects of the initiating event (eg, automatic shutdown systems)
- Alarms that alert the operator when the initiating event occurs (eg, when pressures and/or temperatures exceed maximum permissible levels)
- Prescribed operator actions in response to alarms
- Barriers (eg, release containment methods) for limiting the effects of initiating events

Clearly, such safety functions influence the ultimate effects of any accident resulting from an initiating event. The event tree analyst should identify all safety functions that can change the result of the initiating event in the chronological order in which they are expected to respond. The descriptions of these safety functions should state their intended purpose. The successes and failures of the safety functions are accounted for in the event tree.

11.10.4 Construction of the Event Tree

The first step in the construction of an event tree is to enter the initiating event and the safety functions that should be taken into account in the analysis.

The next step is to evaluate the efficiency of the various safety functions. Only two possibilities are normally considered, viz. successful action and complete failure of the safety function. It is then assumed that the initiating event has occurred. The analyst must evaluate whether the success or failure of the safety function affects the course of the accident. If the accident is affected, a branch point is inserted in the event tree to distinguish between the success and the failure of the safety function. If a given safety function does not affect the course of the accident, the accident path proceeds, with no branch point, to the next safety function.

Every branch point developed in the event tree creates additional accident paths that must be evaluated individually for each of the subsequent safety systems.

11.10.5 Description of Event Sequences Leading to Accidents

The last step of the ETA is to describe the sequences of events leading to accidents. The sequences of events following an initial event can have a variety of outcomes. Some of these may be safe recovery and return to normal operation, or an orderly shutdown. However, the sequences of significance from a safety point of view are the ones resulting in accidents.

Once the relevant event sequences have been described, they can be ranked according to the severity of their outcomes. The structure of the event tree should show the progression of the chain of events leading up to the accident in a clear way. This can facilitate the specification of additional/modified procedures/safety systems that will yield the most cost-effective protection against the identified potential accidents.

11.11 CAUSE–CONSEQUENCE ANALYSIS (CCA)

11.11.1 Outline

The purpose of CCA is to enable identification of both the basic causes and the consequences of potential accidents. This is achieved by incorporating both FTA and ETA in the method. A major strength of CCA is that it facilitates communication between involved experts, not least because the cause–consequence diagram displays the interrelationships between the basic causes of accidents and the consequences. If appropriate data are available, CCA can also be used to quantify the expected frequencies of occurrence of various consequences.

11.11.2 Typical Applications and Results

CCA can be used in both the design phase and the operating phase of a process/plant to identify potential accidents, assess their expected consequences, and identify basic causes of the potential accidents. In addition, if quantification is desired and adequate data are available, probabilities of each type of accident may be worked out.

11.11.3 Required Input

Two categories of data input are required for executing a CCA. The first is adequate knowledge of the type of hardware/component failures or process upsets that could cause accidents in the actual plant/process. The second is adequate knowledge of any existing safety systems or emergency procedures that could influence the outcome of an accident.

11.11.4 Required Staff

A CCA is best performed by a small team of two to four people possessing the variety of experience and expertise required for that particular analysis. At least one team member should be experienced in CCA (or FTA and ETA). The other members should have adequate knowledge of and experience with the various operations/systems to be analyzed, and of/with possible interactions between them.

11.11.5 Time and Cost Requirements

The time and cost requirements for a CCA are highly dependent on the number, complexity, and level of resolution of the events included in the analysis. Scoping-type CCAs for several initiating events can usually be accomplished within 1 week or less. Detailed CCAs may require 2–6 weeks, depending on the complexity of the supporting FTAs.

More details on CCA are given by CCPS (1985) and Gould et al. (2000).

11.12 QUANTITATIVE RISK ANALYSIS (QRA)

QRA consists of four major steps:

- Identification of the types of potential failures
- Estimation of consequences of these failures
- Estimation of failure probabilities
- Assessment of overall risks of the various failure types (failure risk = failure probability · failure consequence)

Failure modes are identified first by establishing the location of the main inventories of hazardous material and then by detailed scrutiny of the diagrams of process flow and instrumentation using checklist methods or HAZOP. Once the list of failure cases has been identified, the consequence of each failure type must be calculated.

QRA has been criticized by pointing at

- inaccurate mathematical models
- incomplete analysis of actual practical problem
- inaccurate primary-failure probability data
- inadequate acceptability criteria
- difficulty of checking final result
- complexity and laboriousness of method

Khan and Abbasi (1998) pointed out that, even at that time, QRA had been in existence for many years. Before being introduced in the chemical process industries (CPI), it was used extensively in the nuclear industries. Unfortunately, the application of QRA in the process industries is much more difficult than in the nuclear industries. This is because of the greater diversity of processes, hazardous materials, equipment types, and control schemes in the process industries. This diversity requires continuous addition of new capabilities in QRA. A typical QRA comprises the following four steps.

- Hazards identification;
- Estimation of frequencies of occurrence of the identified hazards;
- Estimation of the consequences of the various identified hazards; and
- Quantitative assessment of the risks to be associated with the various identified hazards.

The first step seeks an answer to the first basic question: What can go wrong? This is the most important step because hazards that are not identified cannot and will not be quantified, which in turn inevitably leads to underestimation of the total risk. The techniques used for hazard identification include Checklists, HAZOP, FMEA, and "What-If" analysis.

After the various hazards have been identified, the scope of a QRA has to be defined. This second step involves answering another key question: How likely are the various accidents? Answering this question requires quantification of the probability of each accident scenario. FTA may be used for a third purpose.

The third step consists of quantitative assessment of the consequences/negative impacts of the various identified hazards. CCPS (1999) issued a set of guidelines for consequence analysis of chemical releases. The consequences are often measured in terms of the number of expected fatalities, but they may also include measurements of number of injuries of various categories, and value of the property lost. This assessment of consequences can be very demanding due to the great variety of materials, processes, and technologies involved. According to Khan and Abbasi (1998), consequence analysis was the component of QRA that was growing most rapidly at that time.

One major challenge is to estimate the consequence of a given failure. Inherent in this challenge is the often-unforeseeable scale of *accident escalation* (see also Section 1.6 in Chapter 1). The well-known Piper Alpha catastrophe in the North Sea in 1988 is a dramatic example. The initiating event was a comparatively limited leak of propane (due to lack of conveyance of essential information from one working team to the next). The propane mixed with the ambient air and formed an explosive vapor cloud that ignited and exploded. The explosion destroyed production equipment, piping, etc., and this caused more leaks of oil and gas, and after some time the entire platform was on fire. Due to the weakening of the steel structures of the platform by the heat from the flames, the entire platform eventually tipped over and sank in the sea. One hundred and sixty-seven people lost their lives. Only 61 survived. In insurance loss, the figure was 3.4 billion US$.

Table 11.4 gives examples of escalation effects in some other accidental explosion cases.

The last step of a QRA is to calculate the estimated resulting risk. Despite the considerable resources and effort required for the execution of a comprehensive QRA, it still appears to be the most favored and most frequently used scheme for risk analysis in the process industries. For this reason, one seeks, to the extent possible, to optimize and minimize the resources required for executing the various steps involved in the analysis.

11.13 HUMAN-ERROR ANALYSIS (HEA)

11.13.1 Overview

According to CCPS (1985), the focus on HEA in the field of process safety was inspired by industrial engineering efficiency analyses. Systematic methods for human-performance and human-error evaluations were developed during and after World War II. In the context of safety studies, the first forms of HEA were conducted for the military, the chemical industry, and transportation systems. This was followed by quantitative analysis of human errors in the weapons and nuclear power industries.

These techniques are most successfully applied by professionals trained in human-performance analysis. However, it is also possible to perform HEA using operations-oriented staff members, eg, training specialists.

HEA most often constitutes part of a wider spectrum of methods of hazard evaluation. This can be accomplished by either considering the

Table 11.4 Escalation effects from accidental explosions

Marietta, Ohio, 1962

Benzene vapors from a 6 in. relief valve on a phenol plant ignited and an explosion occurred. Flying debris ruptured pipework, releasing flammable liquids.

Louisville, Kentucky, 1965

A compressor circulating gaseous monovinylacetylene exploded. Fragments from the explosion, together with jet flames and transmission through piping, led to a series of further explosions.

Feyzin, France, 1966

A refinery storage sphere suffered a boiling liquid expanding vapor explosion (BLEVE), throwing pieces of steel up to 100 ton some three-quarters of a mile. One fragment cut the legs from under an adjacent sphere, toppling it, and breaking its 8 in. connecting pipe. Another landed on a pipeway and severed 40 lines.

Texas City, Texas, 1969

A 100-ft high refinery column disintegrated, apparently due to detonation of vinyl acetylene. Pieces were distributed uniformly over a radius of 1500 ft. One section weighing some 800 lb traveled 3000 ft.

Romeoville, Illinois, 1977

Roof fragments from an exploding cone roof tank in a refinery struck a covered floating roof tank and a floating roof tank, both of which ignited.

Abqaiq, Saudi Arabia, 1978

Jetting effects of gas escaping from a pipeline threw a 22 ft pipe section 400 ft, where it struck the vapor space of a storage spheroid and released further flammable vapor.

Texas City, Texas, 1978

Rupture of a refinery storage sphere due to overpressure led to a large vapor cloud fire. This set off a series of BLEVEs which generated numerous missiles, causing further vessel failures. In addition to the original sphere, two further spheres, five horizontal bullets, and four vertical bullets were destroyed.

Bantry Bay, Eire, 1979

Explosions on an oil tanker at the jetty generated numerous missiles. One fragment weighing 1000 lb was found at the foot of a large crude oil storage tank 1800 ft from the ship.

Linden, New Jersey, 1979

A vapor cloud entered an unused control room. When the cloud exploded, debris from this building severed pipes, releasing more hydrocarbons.

Ponce, Puerto Rico, 1979

A 13-ft diameter dimerizer vessel suffered massive failure. The 15 ton steel end traveled 1900 ft and struck an adjacent paraxylene plant, setting it on fire.

Mexico City, Mexico, 1984 see Appendix 4

Romeoville, Illinois, 1984

A massive failure occurred of a 55-ft high × 8-ft diameter refinery absorption column. Most of the vessel was propelled 3500 ft where it struck and toppled a 138 kV power transmission tower. Later in the incident, a piece of a vessel which had suffered a BLEVE traveled 500 ft, shearing off pipes.

Table 11.4 Escalation effects from accidental explosions—cont'd

Pascagoula, Mississippi, 1986
Fragments, from an explosion in the area of a batch still, punctured two atmospheric storage tanks and a pressure storage vessel, releasing flammable material.

Grangemouth, United Kingdom, 1987
Overpressure of a low-pressure separator due to gas "breakthrough" caused the 30-ft high × 10-ft diameter vessel to disintegrate. One piece weighing 3 tons was thrown 3300 ft.

Torrance, California, 1987
A propane treater vessel failed and rocketed into a central pipe rack, cutting several pipes, including the refinery flare lines.

Antwerp, Belgium, 1989
An explosion on an ethylene oxide distillation column caused a large fire and generated missiles which did extensive damage throughout the plant.

Brief summaries of case histories as presented in first and second part of Table 17.49 in Lees, F.P., 1996. Loss prevention in the process industries. Hazard, Identification, Assessment and Control, second ed. Butterworth-Heinemann, Oxford, UK, ISBN:0-7506-1547-8.

human-performance characterization of system operation as in integral part of the wider hazard evaluation, or performing an HEA separately and integrating the results into the wider hazard evaluation afterward. Generally, specific techniques for HEA constitute a specialized sub-discipline of hazard evaluation at large.

However, there are situations in which HEA is best regarded as an independent exercise in its own right. This would typically be when foreseen accidents are either known or expected, to be initiated essentially by human error.

Kirytopoulosa et al. (2014) were concerned with the problem of road tunnel safety, which may appear quite different from process safety. However, these workers specifically addressed a problem that is both central and common in a wide range of safety assessment situations, viz. the role of possible errors due to organizational and human factors. Based on the Cognitive Reliability and Error Analysis Method (CREAM) for human reliability analysis, they developed a fuzzy logic system for obtaining quantitative estimates of the (tunnel) operator's performance in safety-critical situations.

In any case, it is clear from what has been said above that the buildup and maintenance of an integrated system for preventing and mitigating explosions in the process industries also must contain a strong element related to human relations and human behavior.

11.13.2 Personnel Categories Involved

A number of different personnel categories may be involved, including:

- Workers in the plant.
- Foremen in the plant.
- Workers from the maintenance department.
- Plant engineers.
- Safety engineers.
- Purchasing department officers.
- Safety manager.
- Middle management.
- Top management.
- Suppliers of equipment.
- Explosion experts/consultants.

Adequate prevention and mitigation of accidental explosions cannot be realized unless there is meaningful communication between the various categories of personnel involved. If such communication is lacking, the result can easily become both unsatisfactory and confusing.

In general, meaningful communication may be defined as conveyance and proper receipt and appreciation of adequate information whenever required.

11.13.3 Requirements for Adequate Human Performance

However, to receive, appreciate, and use information obtained/received in a proper way, any involved person from any of the categories mentioned in Section 11.13.2 must have:

- Adequate knowledge.
- Adequate motivation.
- Adequate resources and deciding power.

Knowledge about accidental explosions can be acquired by reading, listening to lectures, talking to experts, etc., although experience from actual explosion prevention and mitigation work is perhaps the best form of knowledge.

Genuine motivation is perhaps more difficult to achieve indirectly. It seems to be a law of life that people who have themselves experienced serious explosion accidents possess the highest level of motivation, in particular if the accident caused injuries and perhaps even loss of life. This applies to workers as well as top management. However, high levels of motivation can also result from good demonstrations of real explosions, including their initiation by various ignition sources, as well as their propagation and damaging effects. Video and film can also be a good help, if used properly.

11.13.4 The Role of Top Management

The final element in the foregoing Section 11.13.3, viz. adequate resources and the authority to put plans into practice, is in reality controlled by the top management. Verhaegen (1989) concluded from this that the real responsibility for establishing and running a proper safety assurance system must always lie with the top management.

11.13.5 Ten Essential Elements for Proper Safety Management

Summarizing the experience of a large, multinational chemical company, Verhaegen suggested that the following 10 essential elements be involved to ensure proper safety management:

- Top management responsibility.
- Safety statement (explicit commitment from top management).
- Objectives and goals (specification of long- and short-term expectations).
- Stated standards (written guidelines and rules).
- Safety committees (a dedicated organization for handling safety issues at all levels).
- Safety audits (regular reexamination of work practices).
- Accident records (written analyses of accidents. Why did they happen? How can similar future accidents be prevented?).
- Safety personnel (qualified specialists essential as advisers, but overall responsibility remains with top management).
- Motivation (by information, involvement, etc.).
- Training (a continual process; courses essential; the message must get through!).

Verhaegen emphasized the problem that a good safety organization is in reality often kept active by one or two dedicated individuals. If they change position within the company, or even leave, the safety organization may suffer. Management should foresee this problem and provide a workable solution.

REFERENCES

Abbasi, T., Abbasi, S.A., 2007. The boiling liquid expanding vapour explosion (BLEVE): mechanism, consequence assessment, management. Journal of Hazardous Materials 141, 489–519.

Abe, Y., Nariai, H., Hamada, Y., 2002. The trigger mechanism of vapour explosions. Journal of Nuclear Science Technology 39, 845–853.

Addai, E.K., Gabel, D., Krause, U., 2015. Experimental investigation of the minimum ignition temperature and the lower explosible/flammable fuel concentration, of clouds of hybrid mixtures of a combustible dust and a combustible gas or solvent vapour. Journal of Hazardous Materials (Paper being reviewed for publication).

Alfert, F., Fuhre, K., 1988. Ignition of Gas Clouds by Hot Surfaces. Which Maximum Temperature Is Relevant and Acceptable for Equipment in Classified Areas? CM1 Report No. 873301-1 Christian Michelsen Research, Bergen, Norway (from 2001: GexCon AS).

Alfert, F., Fuhre, K., 1989. Flame and Dust Free Venting of Dust Explosions by Means of a Quenching Pipe. Report No. 89/25820-1, January. Chr. Michelsen Institute, Bergen, Norway.

Alfert, F., Eckhoff, R.K., Fuhre, K., 1989. Zündwirksamkeit von Glimmnestern und heissen Gegenständen in industriellen Anlagen. VDI-Berichte 701, 303–319 (VDI-Verlag GmbH, Düsseldorf).

Alfert, F., 1985. Sicherheitstechnische Kenngrössen von polynären Brennstoff/ Luft-Gemischen und selbstzerfallenden Stoffen (Dr.-Ing. thesis). University of Dortmund, Germany.

Amyotte, P.R., Goraya, A.U., Hendershot, D.C., Khan, F.I., 2007. Incorporation of inherent safety principles in process safety management. Process Safety Progress 26, 333–346.

Amyotte, P.R., Pegg, M.J., Khan, F.I., 2009. Application of inherent safety principles to dust explosion prevention and mitigation. Process Safety and Environmental Protection 87, 35–39.

Anonym, April 4, 1975. ICI paper considers questions raised by Flixborough. European Chemical News 27.

Anonym, September 1976. Rapport over de explosie bij DSM te Beek (L) 7 November 1975. Gaswolkexplosie in Naftakraker II, Deel 1: Tekst en bijlagen, Deel 2: Foto's en tekeningen. Uitgave van het Directoraat-Generaal van de Arbeid van het Ministerie van Sociale Zaken (in Dutch).

Arntzen, B.J., Ringdal, L., Steiner, M., Eckhoff, R.K., 2014. Is the safe performance of flame gaps in flameproof electrical apparatus deteriorated by rusting and mechanical damage? Part II: group IIB and IIC gases. Process Safety Progress 33 (1), 49–55.

ASTM, 2015a. Standard test method for concentration limits of flammability of chemicals (vapours and gases): method E-681-09. In: Annual Book of ASTM Standards, vol. 14.02. American Society for Testing and Materials, Philadelphia, USA, pp. 175–186.

ASTM, 2015b. Standard test method for autoignition temperature of liquid chemicals: method E659. In: Annual Book of ASTM Standards, vol. 14.02. American Society for Testing and Materials, Philadelphia, USA, pp. 146–152.

ATEX Directive, 1999. Directive 1999/92/EC of the European Parliament and of the Council of 16 December 1999 on Minimum Requirements for Improving the Safety and Health Protection of Workers Potentially at Risk from Explosive Atmospheres.

ATEX Directive, 2014. Directive 2014/34/EU of the European Parliament and of the Council, on the Harmonization of the Laws of the Member States Relating to Equipment and Protective Systems Intended for Use in Potentially Explosive Atmospheres.

Baev, V.K., Bazhaikin, A.N., Buzukov, A.A., Timoshenko, B.P., 1984. Induction time measurements for ignition of liquid fuel jets in air at high temperatures and pressures. Progress in Astronautics and Aeronautics 95, 554–562.

Baker, W.E., Tang, M.J., 1991. Gas, Dust and Hybrid Explosions. Elsevier Science Publishers, Amsterdam, ISBN 0-444-88150-6.

Bakke, J.R., Wingerden, K., van, Hoorelbeke, P., Brewerton, B., 2010. A study on the effect of trees on gas explosions. Journal of Loss Prevention in the Process Industries 23, 878–884.

Ballal, D.R., Lefebvre, A.H., 1978. Ignition and flame quenching of quiescent fuel mists. Proceedings of Royal Society of London A 364, 277–294.

Ballal, D.R., Lefebvre, A.H., 1979. Ignition and flame quenching of flowing heterogeneous fuel/air mixtures. Combustion and Flame 35, 155–168.

Ballal, D.R., December 1976. Ignition of Fuel Sprays, S.M.E. Memo 2. Cranfield Institute of Technology, School of Mechanical Engineering.

Bartknecht, W., 1971. Brenngas-und Staubexplosionen. Forschungsbericht F45. Bundesinstitut für Arbeitsschutz, Koblenz, F. R. Germany.

Bartknecht, W., 1978. Explosionen — Ablauf und Schutzmassnahmen. Springer-Verlag, ISBN 0-387-08675-7.

Bartknecht, W., 1987. Staubexplosionen — Ablauf und Schutzmassnahmen. Springer-Verlag, ISBN 0-387-16243-7.

Bartknecht, W., 1993. Explosionsschutz — Grundlagen und Anwendung. Springer-Verlag, ISBN 3-540-55464-5.

BBC, 1983. Explosion Protection Manual, 2nd, revised edition produced by H. Olenik, H. Rentzsch and W. Wettstein, published by Brown, Boveri & CIE AG, Mannheim, Germany, at Verlag W. Girardet, Essen, Germany. ISBN: 3-7786-0876-4.

Beck, H., Jeske, A., 1989. Prüfung fahrbarer Industriestaubsauger zum Einsatz in durch Staub explosionsgefährdeten Bereichen. VDIBerichte 701, 881–897. VDI-Verlag, Düsseldorf, Germany.

Berg, E., 1989. Private Communication to R.K. Eckhoff from E. Berg, Dyno Industries, Gullaug, Norway.

Bergstrøm, T.H., 1995. A review of molten metal/water reactions. In: Proceedings of the 52nd Electric Furnace Conference, Nov. 13–16 1994 in Nashville. USA, pp. 303–308. Published by the Iron and Steel Society, USA in 1995, ISBN: 0-93289-97-5.

Birk, A.M., Cunningham, M.H., 1994. The boiling liquid expanding vapor explosion. Journal of Loss Prevention in the Process Industries 7, 474–480.

Birk, A.M., Cunningham, M.H., 1996. Liquid temperature stratification and its effect on BLEVEs and their hazards. Journal of Hazardous Materials 48, 219–237.

Birk, A.M., van der Steen, J.D.J., 2002. On the transition from no-BLEVE to BLEVE for a 1.8 m^3 propane tank. American Society of Mechanical Engineers Pressure Vessels and Piping Division PVP 446, 143–152.

Birk, A.M., 1996. Hazards from propane BLEVEs: an update and proposals for emergency responders. Journal of Loss Prevention in the Process Industries 9, 173–181.

Bjerketvedt, D., Bakke, J.R., van Wingerden, K., 1997. Gas explosion handbook. Journal of Hazardous Materials 52, 1–150 (special issue).

Bjørkhaug, M., Danielsen, V., Håverstad, T.A., Jacobsen, Ø., Schmidt Pedersen, K., 1985. Aarsaksforhold Ved Utvikling Av Brann Og Eksplosjon Ombord I Boreriggen 'West Vanguard' Natt Til Den 7. Oktober 1985. SINTEF-report ST25 F85013, SINTEF. Norges Branntekniske Laboratorium, Trondheim (in Norwegian).

Blything, K.W., Reeves, A.B., 1988. An Initial Prediction of the BLEVE Frequency of a 100 Tonnes Butane Storage Vessel. UK Atomic Energy Authority, Safety and Reliability Directorate. Report SRD/R488, 88 pp.
BMIIB, 2008. The Buncefield Incident 11 December 2005. The Final Report of the Major Incident Investigation Board, vol. 1, ISBN 978 07176 6270 8. Crown Copyright (2008).
Borisov, A., Gelfand, B., 1989. Private Communication to R.K. Eckhoff from A. Borisov and B. Gelfand, Russian Academy of Science, Moscow.
Bowen, P.J., Cameron, L.R.J., 1999. Hydrocarbon aerosol explosion hazards. Transactions of IChemE 77, 22–30.
Bowen, P.J., Shirvill, L.C., 1994. Pressurised atomization of high flash point liquids – Implications for hazardous area classification. In: Paper Presented at Symposium on European Advances in Process Safety, UMIST, Manchester 19–21 April, 1994.
Braaten, T.S., 1985. Investigation of Silo Plant at Kvalaberget, Stavanger, Norway, after Explosion on 22nd November, 1985. Norwegian Factory Inspectorate. Internal Report, 27. November (in Norwegian).
Brandsjö, K., 1988. I Brandmannaminne. Winbergs Förlag AB, Hudiksvall, Sweden. SBF, Stockholm, Sweden (in Swedish).
Bull, D.C., McLeod, M.A., Mizner, G.A., Aug. 1979. Detonation of unconfined fuel aerosols. In: Paper Presented at 7th ICGDERS, Göttingen, Germany.
Burgoyne, J.H., Cohen, L., 1954. The effect of drop size on flame propagation in liquid aerosols. Proceedings of Royal of Society A225, 375–392.
Burgoyne, J.H., 1963. The flammability of mists and sprays. In: Proceedings of 2nd Symposium on Chemical Process Hazards. Institution of Chemical Engineers, UK, pp. 1–5.
Cardillo, P., Anthony, E.J., 1978. The Flammability Limits of Hybrid Gas und Dust Systems. La Rivista dei Combustibili XXXII, 390–395.
Carson, D., 1996. A methodological approach to the spontaneous combustion of agricultural dusts. In: Proc. 7th Internat. Coll. Dust Explosions, (June 23–26), Bergen, Norway. Published by GexCon AS (previously CMR), Bergen, Norway.
Cashdollar, K.L., Hertzberg, M., 1985. 20 litre explosibility test chamber for dusts and gases. Review of Scientific Instruments 56, 596–602.
CCPS, 1985. Hazard evaluation Procedures. Center for Chemical Process Safety of the American Institute of Chemical Engineers, New York, ISBN 0-8169-0399-9.
CCPS, 1992. Guidelines Forhazard Evaluation Procedures, second ed. Center for Chemical Process Safety of the American Institute of Chemical Engineers, New York.
CCPS, 1999. Guidelines for Consequence Analysis of Chemical Releases. Center for Chemical Process Safety, American Institute of Chemical Engineers, New York.
CEN, August 1997. Explosive Atmospheres – Explosion Prevention and Protection – Part 1: Basic Concepts and Methodology. European Standard EN 1127–1. European Committee for Standardization, B-1050 Brussels.
CEN, June 2002a. Dust Explosion Venting Protective Systems, European Draft Standard PrEN 14491. European Committee for Standardization, Rue de Stassart, 36, B-1050 Brussels.
CEN, March 2002b. Explosion Suppression Systems, European Draft Standard PrEN 14373. European Committee for Standardization, Rue de Stassart, 36, B-1050 Brussels.
CEN, July 2003. Explosion Venting Devices, European Draft Standard CEN/TC 305 N 0388. European Committee for Standardization, Rue de Stassart, 36, B-1050 Brussels.
CEN, February 2004a. Guidance on Inerting for the Prevention of Explosions. European Technical Report CEN/TC 305 N 0428. European Committee for Standardization, Rue de Stassart, 36, B-1050 Brussels.
CEN, January 2004b. Explosion Isolation Systems, European Draft Standard CEN/TC 305 N 0427. European Committee for Standardization, Rue de Stassart, 36, B-1050 Brussels.
CENELEC, March 1997. Installations of Electrical Apparatus on Aread Endangered by Explosive Substances. Draft Europ. standard prEN 50273. CENELEC Central Office, Brussels.

CENELEC, September 1998. Electrical Apparatus for Use in the Presence of Combustible Dust. Part 1—2: Selection, Installation and Maintenance. European standard EN 50281-1-2. CENELEC Central Office, Brussels.

CENELEC, September 2002. Electrical Apparatus for Use in the Presence of Combustible Dust. Part 3: Classification of Areas Where Combustible Dusts Are or May Be Present. European Standard EN 50281—3. CENELEC Central Secretariat, Brussels.

Chan, K.-K., Jou, C.-S., 1988. An experimental and theoretical investigation of the transition phenomenon in fuel spray deflagration. Part 1: the experiment. Fuel 67, 1223—1227.

Cooper, P.W., Kurowski, S.R., 1993. An Introduction to the Technology of Explosives. Unpublished manuscript. Copyright Registered TXU 580 651.

Corradini, M.L., Kim, B.J., Oh, M.D., 1988. Vapour explosions in light water reactors: a review of theory and modeling. Progress in Nuclear Energy 22 (1), 1—117.

Cox, R.A., 1986. Appraisal of the utility of risk analysis in the process industry. In: Proceedings of 5th International Symposium on Loss Prevention and Safety Promotion in the Process Industry, in Cannes, France, September 1986. Paper No. 14.

Cox, R.A., 1987. An overview of hazard analysis. In: Woodward, J.L. (Ed.), Proceedings of the International Symposium on Prevention of Major Chemical Acidents, February 1987. Center for Chemical Process Safety, Washington DC.

Cozzani, V., Gubinelli, G., Salzano, E., 2006. Escalation thresholds in the assessment of domino accidental events. Journal of Hazardous Materials A129, 1—21.

Cozzani, V., Tugnoli, A., Salzano, E., 2009. The development of an inherent safety approach to the prevention of domino effects. Accident Analysis and Prevention 41, 1216—1227.

Crawley, F.K., 1982. The Effect of Ignition of a Major Fuel Spillage, the Assessment of Major Hazards (Report from the Institution of Chemical Engineers, Rugby, UK).

CSB, 1998. Explosives Manufacturing Incident, Sierra Chemical Company, Mustang, Nevada, 1998. Report No. 98-001-I-NV. U.S. Chemical Safety and Hazard Investigation Board (CSB), Washington DC, USA.

CSB, 2003. Dust Explosion, West Pharmaceutical Services, Inc.. Report No. 2003-07-I-NC Chemical Safety and Hazard Investigation Board (CSB), Washington DC, USA.

CSB, September 2009. Sugar Dust Explosion and Fire. Report No. 2008-05-I-GA. Chemical Safety and Hazard Investigation Board (CSB), Washington DC, USA.

CSB, May 2014. Catastrophic Rupture of Heat Exchangers in the Tesoro Anacortes Refinery. Report No. 2010-08-1-WA. U.S. Chemical Safety and Hazard Investigation Board (CSB), Washington DC, USA.

Cutler, D.P., 1974. The ignition of gases by rapidly heated surfaces. Combustion and Flame 22, 105—109.

Cutler, D.P., 1978. Further studies of the ignition of gases by transiently heated surfaces. Combustion and Flame 33, 85—91.

Cybulski, W., 1975. Coal Dust Explosions and Their Suppression. English Translation from Polish. Published for the Bureau of Mines, US Department of the Interior and the National Science Foundation, Washington DC, by the Foreign Scientific Publications Department of the National Center for Scientific, Technical and Economic Information, Warsaw, Poland.

Czujko, J. (Ed.), 2001. Design of Offshore Facilities to Resist Gas Explosion Hazard. Engineering Handbook, first ed. CorrOcean, Sandvika (Oslo), Norway, ISBN 82-996080-0-7.

Dahn, C. J. 1986. Contribution of low-level flammable vapour concentrations to dust explosion output. Proceedings of 2nd International Colloquium on Dust Explosions, Technical University of Warsaw, November 1986.

Davenport, J.A., March 1977. A study of vapor cloud incidents. In: Paper Presented at the 83rd National Meeting of American Institute of Chemical Engineers, Houston, Texas. Paper 24a.

Davenport, J.A., 1984. A Study of Vapor Cloud Incidents — an Update, pp. 245—504. The Institution of Chemical Engineers, Symposium Series No. 80, Rugby, UK.

de Bruyn, P.C.A.M., Weerheijm, J., Karelse, J.W., van Wees, R.M.M., 2002. The fireworks disaster in Enschede. Part 1: overview and reconstruction. In: Proc. 6th Internat. Conf. Probabilistic Safety Assessment and Management, 23–28 June 2002, San Juan Porto Rico, USA.

Dorsett, H.G., Jacobson, M., Nagy, J., 1960. Laboratory equipment and test procedures for evaluating explosibility of dusts. Report of Investigation No. 5424. US Bureau of Mines, Washington DC, USA.

Dybdahl, K., 1989. Vanndamp og gass sprengte digelen (Water vapour and gas burst the melting pot). Teknisk Ukeblad/Teknikk 136 (4), 24.

Eckhoff, R.K., Pedersen, G.H., 1988. Can Mechanical Sparks Ignite Gases? A Phase 1 Literature Study with Identification of Needs for Further Research. CM1 Report No. 02702–1. Chr. Michelsen Institute (now GexCon AS), Bergen, Norway.

Eckhoff, R.K., Thomassen, 1994. Possible sources of ignition of potential explosive gas atmospheres on offshore process installations. Journal of Loss Prevention in the Process Industries 7, 281–294.

Eckhoff, R.K., Fuhre, K., Guirao, C.M., et al., 1984. Venting of turbulent gas explosions in a 50 m^3 chamber. Fire Safety Journal 7, 191–197.

Eckhoff, R.K., Fuhre, K., Pedersen, G.H., December 1985. Vented Maize Starch Explosions in a 236 m^3 Experimental Silo. Report No. 843307–2. Chr. Michelsen Institute (now GexCon AS), Bergen, Norway.

Eckhoff, R.K., Arntzen, B.J., Opsvik, H.E.Z., Grov, A., Solheim, F., 2013. Is the safe performance of flame gaps in flameproof electrical apparatus deteriorated by rusting and mechanical damage? Part 1: group IIA gases. Process Safety Progress 32 (1), 49–56.

Eckhoff, R.K., 1977/1978. Pressure development during explosions in clouds of dusts from grain, feedstuffs and other natural organic materials. Fire Research 1, 71–85.

Eckhoff, R.K., 1983. Suppression of gas explosions on offshore gas and oil installations. In: Executive Summary of Proceedings of an Informal Seminar at Chr. Michelsen Institute Bergen, Norway (Now GexCon AS, Bergen, Norway) 5th and 6th October 1982. CMI report No. 833402–3.

Eckhoff, R.K., June 1991. Generation, Ignition, Combustion and Explosion of Sprays and Mists of Flammable Liquids in Air. A Literature Survey. Report No. CM1-91-A25014. Chr. Michelsen Institute (now GexCon AS), Bergen, Norway.

Eckhoff, R.K., 1994. Dust explosion hazards in the ferro-alloys industry. In: Proc. 52. Electric Furnace Conf., Nashville, TN, USA, November 13–16 (1994). Iron & Steel Soc. Inc., Warrendale, PA, USA, pp. 129–146.

Eckhoff, R.K., 1996a. Dust, mist and gas explosion accidents: causes and effects. In: Survey Paper Presented at the '2nd World Seminar on the Explosion Pheomenon and on the Application of Explosion Protection Techniques in Practice', 4–8 March, Gent, Belgium. Organized by Europex/Stuvex, Kontich, Belgium.

Eckhoff, R.K., 1996b. Electrical Equipment for Hazardous Areas. TAPIR, Trondheim, Norway, ISBN 82-519-1211-3 (in Norwegian).

Eckhoff, R.K., 2002. Minimum ignition energy (MIE) – a basic ignition sensitivity parameter in design of intrinsically safe electrical apparatus for explosive dust clouds. Journal of Loss Prevention in the Process Industries 15, 305–310.

Eckhoff, R.K., 2003. Dust explosions in the process industries, third ed. Gulf Professional Publishing/Elsevier, ISBN 0-7506-7602-7. 1st edition published in 1991, 2nd edition in 1997.

Eckhoff, R.K., 2004. Partial inerting – an additional degree of freedom in dust explosion protection. Journal of Loss Prevention in the Process Industries 17, 187–193.

Eckhoff, R.K., 2006. Differences and similarities of gas and dust explosions: a critical evaluation of the European "ATEX" directives in relation to dusts. Journal of Loss Prevention in the Process Industries 19, 553–560.

Eckhoff, R.K., 2009. Inadequate treatment of dust explosions and fires in ATEX. A critical view on resulting standards for electrical apparatus. Bulk Solids & Powder – Science and Technology 4 (3), 1–11.

Eckhoff, R.K., Arntzen, B.J., Opsvik, H.E.Z., Grov, A., Solheim, F., 2013. Is the safe performance of flame gaps in flameproof electrical apparatus deteriorated by rusting and mechanical damage? Part 1: Group IIA gases. Process Safety Progress 32, 49–56.

Eckhoff, R.K., 2014. Boiling liquid expanding vapour explosions (BLEVEs): a brief review. Journal of Loss Prevention in the Process Industries 32, 30–43.

Eckhoff, R.K., 2016a. Water vapour explosions – A brief review. Journal of Loss Prevention in the Process Industries 40, 188–198.

Eckhoff, R.K., 2016b. Ignition of dust clouds by capacitive electric sparks of short discharge times. In: Paper presented at the 11th International Symposium on Hazards, Prevention and Mitigation of Industrial Explosions (11th ISHPMIE), 24–29 July 2016 in Dalian, P.R. China.

Eichhorn, J., 1973. Mists Can Cause Explosions. Chapter in "Fire Protection Manual for Hydrocarbon Processing Plants", second ed. Gulf Publishing Company, Houston, Texas, USA, pp. 190–193 (Identical article in 3rd edition (1985), pp. 254–257).

Engler, C., June 1885. Beitrage zur Kenntniss der Staubexplosionen. Chemische Industrie 171–173.

Engler, C., 1907. Einfacher Versuch zur Demonstration der gemischten Kohlenstaub- und Gasexplosionen. Chemiker-Zeitung (28), 358–359.

Epstein, M., Fauske, H.K., Theofanous, T.G., 2000. On the mechanism of aluminium ignition in steam explosions. Nuclear Engineering and Design 201, 71–82.

Faeth, G.M., 1987. Mixing, transport and combustion in sprays. Progress in Energy Combustion and Science 13, 293–345.

Fire and Police Authorities of Bremen, 1979. Brand- und Explosionsschaden Bremer Rolandsmiffile am 6. Februar 1979. Eine Dokumentnation. Issued by the Fire and Police Authorities of Bremen (in German).

Fire Investigation Report, 1995. Pages 101–107 Were Kindly Made Available to R. K. Eckhoff by Mr. Jong-kil Lee, Manager of Korea Gas Safety Cooperation (Kyunggi-Do, South Korea).

Fischer, K., 1957. Ablauf von Kohlenstaubexplosionen. Zeitschrift für Elektrochemie 61, 685–692.

Fletcher, D.F., Anderson, R.P., 1990. A review of pressure-induced propagation models of the vapour explosion process. Progress in Nuclear Energy 23 (2), 137–179.

Förster, H., 1990. Private Communication to R.K.Eckhoff from H. Förster, PTB, Braunschweig, Germany.

Förster, H., 2000. Eigenschaften brennbarer Nebel und Schäume. Chapter in Steen (2000).

Franke, H., 1978. Bestimmung der Mindestzündenergie von Kohlenstaub/Methan/Luft-Gemischen (Hybride Gemische). VDI-Berichte 304. VDI-Verlag GmbH, Düsseldorf, pp. 69–72.

Frank-Kamenetskii, 1955. Diffusion and Heat Exchange in Chemical Kinetics. Princeton University Press, Princeton, USA.

Freytag, H.H., 1965. Handbuch der Raumexplosionen. Verlag Chemie, Weinheim, Germany.

Gardner, B.R., Winter, R.J., Moore, M.J., 1986. Explosion development and deflagration-to-detonation transition in coal dust/air suspensions. In: Proc. of 21st Symp. (Internat.) on Combustion. The Combustion Institute, pp. 335–343.

Gibson, N., Lloyd, F.C., 1965. British Journal of Applied Physics 16, 1619.

Gibson, N., Lloyd, F.C., Perry, G.R., 1968. Fire hazards in chemical plant from ignition sparks involving the thermite reaction. In: IChemE Symposium Series No. 25. 1ChemE, London, UK, p. 26.

Giesbrecht, H., Hess, K., Leukel, W., Maurer, B., 1981. Analysis of explosion hazards on spontaneous release of inflammable gases in the atmosphere: part 1: propagation and deflagration of vapour clouds on the basis of bursting tests on model vessels. German Chemical Engineering 4, 305–314.

Glarner, T., 1984. Mindestzündenergiё– Einfluss der Temperatur. VDI-Berichte 494, 109–118 (VDI-Verlag GmbH, Düsseldorf).

Glor, M., 1988. Electrostatic Hazards in Powder Handling. Research Studies Press Ltd, John Wiley & Sons Inc.

Goral, P., Klemens, R., Wolanski, P., 1988. Mechanism of gas flame acceleration in the presence of neutral particles. Progress in Astronautics and Aeronautics 113, 325–335.

Gould, J., Glossop, M., Ioannides, A., 2000. Review of Hazard Identification Techniques. Report HSL/2005/58 from Health and Safrty Laboratory. Broad Lane, Sheffield, S37HQ, UK.

Gow, H.B.F., 1991. Community Documentation. Centre on Industrial Risk, Bulletin No. 4. Institute for Systems Engineering and Informatics. Commission of the European Communities, SER Division, Joint Research Centre, Ispra.

Graves, C.B., et al., 1986. Ignition of a fuel spray by a hot surface. AIAA Journal 24, 1994–2001.

Green, H.L., Lane, W.R., 1964. Particulate Clouds: Dusts, Smokes and Mists. Chapters 2 and 5, second ed. E. & F.N. Spon Ltd, London.

Greiner, H., 2001. IP Degrees of Protection. Report SD 101 E issued by Danfoss Bauer GmbH, Esslingen, Germany.

Gubinelli, G., Cozzani, V., 2009. Assessment of missile hazards: identification of reference fragmentation patterns. Journal of Hazardous Materials 163, 1008–1018.

Gugan, K., 1979. Unconfined Vapour Cloud Explosions. The Institution of Chemical Engineers (UK), in association with George Godwin Ltd, Rugby/London.

Halstrick, V., 1995. Protego Fundamentals. Protego report No. 04, 3 March 1995. Published by Braunschweiger Flammenfilter GmbH, Germany.

Handbook of Chemistry and Physics, 1959/60. Chemical Rubber Publishing Co., Cleveland, Ohio, USA, p. 2512.

Hansen, O.R., Wilkins, B., 2004. An experimental study on oil mist explosions. In: Proc. Internat. ESMG Symposium on Process Safety and Industrial Explosion Protection, March 2004. ESMG, Hamm, Germany, ISBN 3-980-7567-3-4.

Harrold, P.J., April 1992. Probabilistic Assessment of Radio Frequency Ignition Hazards. Report No. SRD/HSE R579. Safety and Reliability Directorate, Culcheth, Cheshire, UK.

Heinrich, H.-J., 1989. Ablauf von Gas-und Staubexplosionen – Gemeinsamkeiten und Unterschiede. VDI-Berichte 709, 93–112 (VDI-Verlag GmbH, Düsseldorf, Germany).

Hendershot, D.C., 2006. An overview of inherently safer design. Process Safety Progress 25, 98–107.

Hjertager, B.H., 1981. Simulation of transient compressible turbulent reactive flows. In: Paper Presented at the International Meeting EUROMECH 139: Uncontrolled Blasts and Explosions in Industry, Aberystwyth, Wales, UK, 12–15 April, 1981.

Holden, P.L., Reeves, A.B., 1985. Fragment hazards from failures of pressurized liquefied gas vessels. Chemical Engineering Symposium Series 93, 205–217.

Hopkinson, B., 1915. British Ordnance Board Minutes No. 13565, p. 220.

IEC, 1975a. Electrical Apparatus for Explosive Gas Atmospheres. Part 4: Method of Test for Ignition Temperature. IEC Publication 79-4, second ed. Central office of International Electrotechnical Commission, Geneva, Switzerland.

IEC, 1975b. First Supplement IEC 79-1A to IEC 79-1(1971): Electrical Apparatus for Explosive Gas Atmospheres. Part 1: Construction and Testing of Flameproof Enclosures of Electrical Apparatus. Appendix D: Method of Test for Ascertainment of Maximum Experimental Safe Gap. Central office of International Electrotechnical Commission, Geneva, Switzerland.

IEC, February 1999. Electrical Apparatus for Explosive Gas Atmospheres. Part 11: Intrinsic Safety "i". IEC International Standard 60079-11, fourth ed. Central Office of International Electrotechnical Commission, Geneva, Switzerland.

IEC, 2001. Degrees of Protection provided by Enclosures (IP Code), 2nd edition of IEC International Standard 60529. Central office of International Electrotechnical Commission, Geneva, Switzerland.

International Standardization Organization, 1985. Explosion Protection Systems. Part 1: Determination of Explosion Indices of Combustible Dusts in Air. International Standard ISO/DIS 6184/1. International Standardization Organization, Geneva, Switzerland.

Johnsen, J.N., Holte, T., 1973. Hva er aarsaken til Tonstad-ulykken? Report on an oil spray/air explosion in a hydroelectric power station in Norway (in Norwegian). Elektro-Elektronisk Tidsskr 86 (22), 23–27.

Johnson, D.M., 2010. The potential for vapour cloud explosions – lessons from the Buncefield accident. Journal of Loss Prevention in the Process Industries 23, 921–927.

Kaesche-Krischer, B., Zehr, J., 1958. Untersuchungen an Staub/Luft-Flammen. Zeitschrift für Physikalische Chemie Neue Folge 14, 5/6.

Kaesche-Krischer, B., 1959. Untersuchungen an vorgemischten, laminarer Staub/Luft-Flammen. Staub 19, 200–203.

Kalkert, N., Schecker, H.-G., 1979. Theoretische Überlegungen zum Einfluss der Teilchengrösse auf die Mindestzündenergie von Staüben. Chemie-Ingenieur-Technik 51, 1248–1249.

Kauffman, C.W., Ural, E., Nicols, J.A., et al., 1982. Detonation waves in confined dust clouds, Proceedings of the 3rd Internat. School of Explosibility of Dusts, Turawa, Poland.

Kauffman, C.W., Hubbard, R.F., May 1984. An Investigation of Fourteen Grain Elevator Explosions Occurring between January 1979 and April 1981. Occupational Safety and Health Administration (OSHA), Washington DC.

Kauffman, C.W., Wolanski, P., Arisoy, A., et al., 1984. Dust, hybrid and dusty detonations. Progress in Astronautics and Aeronautics 94, 221–240.

Kauffman, C.W., 1982. Agricultural dust explosions in grain handling facilities. In: Lee, J.H.S., Guirao, C.M. (Eds.), Paper in Fuel-air Explosions. University of Waterloo Press, Canada, pp. 305–347. ISBN: 88898-035-3.

Kauffman, C.W., 1989. Recent dust explosion experiences in the US grain industry. ASTM Special Techn. Publ. 958. In: Cashdollar, K.L., Hertzberg, M. (Eds.), Industrial Dust Explosions. ASTM, Philadelphia, USA, pp. 243–264.

Khan, F.I., Abbasi, S.A., 1998. Techniques and methodologies for risk analysis in chemical process industries. Journal of Loss Prevention in the Process Industries 11, 261–277.

Khan, F.I., Amyotte, P.R., 2003. How to make inherent safety practice a reality. Canadian Journal of Chemical Engineering 81, 2–16.

Kim, B., 1985. Heat Transfer and Fluid Flow Aspects of Small-scale Single Droplet Fuel-coolant Interactions (PhD thesis). University of Wisconsin, Madison, USA.

Kirytopoulosa, K., Konstandinidou, M., Nivolianitoub, Z., Kazaras, K., 2014. Embedding the human factor in road tunnel risk analysis. Process Safety and Environmental Protection 9, 329–337.

Kjerpeseth, E., 1990. Private Commun. To R.K. Eckhoff, from E. Kjerpeseth Elkem-bremanger. Svelgen, Norway.

Kletz, T.A., November 2003. Inherently safer design – its scope and future. Transactions of IChemE 81 (Part B) (Institution of Chemical Engineers, UK).

Kohl, D., Kelleter, J., 1996. Multi-sensor system for partial pressure ratios of gas species. Sensors Actuators B 33, 134–136.

Kong, D., Eckhoff, R.K., Alfert, F., 1995. Auto-ignition of CH_4/air, C_3H_8/air, CH_4/C_3H_8/air and CH_4/CO_2/air using a 1 litre ignition bomb. Journal of Hazardous Materials 40, 69–84.

Kornai, T., Satsuki, U., Minoru, U., 1994. Safety Science 17, 91.
Krause, U., Hensel, W., 1996. Zündgefahren lagernder Staubschüttungen — Neue Hilfsmittel für ihre Bewertung. VDI-Berichte 1272, 183—201 (VDI-Verlag GmbH, Düsseldorf, Germany).
Krause, U., Schmidt, M., 2001. Untersuchungen zur Zündung und Ausbreitung von Schwelbränden in Stäuben und Schüttgütern. VDI-Berichte 1601, 397—410 (VDI-Verlag GmbH, Düsseldorf, Germany).
Krause, U., Schmidt, M., Lohrer, C., 2006. A numerical model to simulate fires in bulk materials and dust deposits. Journal of Loss Prevention in the Process Industries 19, 218—226.
Krause, U., 2009. Fires in Silos — Hazards, Prevention and Fire Fighting. Wiley-VCH Verlag & Co. KGaA, Weinheim, Germany, ISBN 978-3-527-31467-6.
Krause, U., Schmidt, M., Ferrero, F., 2009. Investigation of the development of conflagration of solid materials via analysis of coupled heat, mass and momentum transport. Chemical Engineering & Technology 32, 292—305.
Kroken, E., 1997. Fire in a Fish Meal Storage Silo at Vedde Fish Meal Factory, Norway September/October 1992. Letter report to R.K. Eckhoff from Erik Kroken, Vedde, Langevåg, Norway, dated dated 27 February, 1997 (in Norwegian).
Kuchta, J.M., 1985. Investigation of Fire and Explosion Accidents in the Chemical, Mining and Fuel Related Industries. A Manual. US Bureau of Mines. Bulletin 680.
Kühnen, G., 1978. Möglichkeiten zur Vermeidung von Zündquellen. VDI-Berichte 304, 97—102. VDI-Verlag, Düsseldorf, Germany.
L'Abbé, R.L., Crupi, V.G., April 1991a. Blast effects. Part I: overpressure injury. EuropEx Newsletter, pp. 4—6. Edition 15.
L'Abbé, R.L., Crupi, V.G., September 1991b. Blast effects. Part II: fragmentation injury. EuropEx Newsletter, pp. 11—12. Edition 16.
L'Abbé, R.L., Crupi, V.G., December 1991c. Blast effects. Part III: acceleration/deceleration type injuries. EuropEx Newsletter, pp. 8—12. Edition 17.
L'Abbé, R.L., Crupi, V.G., April 1992. Blast effects. Part IV: heat and heat stress. EuropEx Newsletter, pp. 8—10. Edition 18.
Lannoy, A., 1984. Analyse des explosions air-hydrocarbure en milieu libre: Etudes déterministe et probabiliste du scenario d'accident, Prévision des effets de suppression. Bulletin Direct. Etudes et Recherches EDF A4, 1984.
Larsen, Ø., Eckhoff, R.K., 2000. Critical dimensions of holes and slits for transmission of gas explosions. Some preliminary results for propane/air and cylindrical holes. Journal of Loss Prevention in the Process Industries 13, 341—347.
Larsen, Ø., Hagen, J.H., van Wingerden, K., Eckhoff, R.K., 2001. Ignition of dust clouds by brush discharges in oxygen enriched atmospheres. Gefahrst. — Reinhalt. Luft 61, 85—90.
Larsen, Ø., 1998. Schlieren Photo Taken as Part of Work for the Degree of cand.Scient. (M.Sc.) of the University of Bergen.
Laurendeau, N.M., 1982. Thermal ignition of methane/air mixtures by hot surfaces: a critical examination. Combustion and Flame 46, 29—49.
Lee, J.H.S., Frost, D.L., 1987. Steam explosions — major problems and current status. In: Paper Presented at the 11th ICDERS in Poland in August 1987.
Lee, J.H.S., 1984. Physics of Explosions. Lecture notes. MeGill University, Montreal.
Lees, F.P., 1996. Loss Prevention in the Process Industries. Hazard, Identification, Assessment and Control, second ed. Butterworth-Heinemann, Oxford, UK, ISBN 0 7506 1547 8.
Lefebvre, A.H., 1989. Properties of sprays. Particle and Particle System Characterization 6, 176—186.
Leuschke, G., 1965. Beitrage zur Erforschung des Mechanismus der Flammenausbreitung in Staubwolken. Staub 25, 180—186.

Li, G., Ji, T., 2015. A severe accidental water vapour explosion in China in 2012 during casting of a large stainless-steel object in a sand mould. In: Paper Prepared for Publication in an International Journal.

Li, G., Yang, H., Yuan, C., Ge, S., Eckhoff, R.K., 2015. A catastrophic aluminium-alloy dust explosion in China in 2014. Journal of Loss Prevention in the Process Industries paper in press.

Long, G., 1957. Explosions of molten aluminium in water — cause and prevention. Metal Progress 71, 107—112.

Lövsland, A., 2015. Private Communication to Rolf K. Eckhoff in July 2015.

Lütolf, J., 1971. Untersuchungen von Stäuben auf Brand- und Explosionsgefahr. Staub-Reinhalt. Luft 31, 93—97.

Lüttgens, G., Glor, M., 1989. Understanding and Controlling Static Electricity. Expert Verlag, Ehningen bei Böblingen, Germany.

Lüttgens, G., Wilson, N., 1997. Electrostatic Hazards. Butterworth-Heinemann, Oxford, UK.

Makhviladze, G.M., Roberts, J.P., Yakush, S.E., 1999. Combustion of two-phase hydrocarbon fuel clouds released into the atmosphere. Combustion and Flame 118, 583—605.

Marshall, V.C., 1987. Major Chemical Hazards, Elis Horwood Series in Chemical Engineering. Ellis Horwood Ltd/Halsted Press: a division of John Wiley & Sons, Chichester/New York, ISBN 0-470-20813-9.

Maskow, H., 1950. VDI-Fachberichte 14, 144—147 (Verein deutscher Ingenieure, Düsseldorf, Germany).

McQuaid, J., March 14, 1990. Know Your Enemy. The Science of Accident Prevention. The Vernon Clancey Memorial Lecture 1990. City University, London.

Meignen, R., Raverdy, B., Picchi, S., Lamome, J., 2014. The challenge of modeling fuel-coolant interaction: Part II: steam explosion. Nuclear Engineering and Design 280, 528—541.

Meyer, R., 1987. Explosives, third ed. VCH Verlagsgesellschaft, Weinheim, Germany. ISBN: 3-527-26599-6/0-89573-600-4.

Minetti, R., Carlier, M., Ribaucour, M., Therssen, E., Sochet, L.R., 1995. A rapid combustion machine investigation of oxidation and auto-ignition of n-heptane: measurement and modelling. Combustion and Flame 102, 298—309.

Minkhorst, J.H., 1957. Crank Case Explosions in Diesel Engines. Original report in Dutch published by Studiecentrum TNO voor Scheepsbouw en Navigatie, The Netherlands. Norwegian translation by J. Brun-Gulbrandsen, SFI, Trondheim (1962).

Moen, I.O., Lee, J.H.S., Hjertager, B.H., Fuhre, K., Eckhoff, R.K., 1982. Pressure development due to turbulentflame propagation in large-scale methanelair explosions. Combustion and Flame 47, 31—52.

Moore, P.E., 1981. Propagation and Suppression of Gas and Dust Explosions (PhD thesis). University of Surrey, UK.

Moore, P.E., Bartknecht, W., 1987. Extending the limits of explosion suppression systems, Staub-Reinhalt. Luft 47, 209—213.

Moore, P.E., Cooke, P.L., 1988. Suppression of metal dust explosions. Report N. 88/49. Britisk Materials Handling Board, Ascot, Berkshire, UK.

Mullen, J.W., Fenn, J.B., Irby, M.R., 1949. The ignition of high velocity streams of combustible gases by heated cylindrical rods. In: Proceedings of the Third International Symposium on Combustion and Flame and Explosion Phenomena. Williarns & Wilkins Co., Baltimore, USA, pp. 317—329.

Nagy, J., Portman, W.M., 1961. Explosibility of Coal Dust in an Atmosphere Containing a Low Percentage of Methane. Rep. Inv. 5815. US Bureau of Mines, Washington.

National Fire Protection Association, 1997. Recommended Practice for the Classification of Combustible Dusts and of Hazardous (Classified) Locations for Electrical Installations in Chemical Process Areas. Publication No. 499. National Fire Protection Association, Quincy, MA, USA.

Nesis, E.I., 1966. Boiling of liquids. Soviet Physics USPEKHI 8 (6), 883–907.
Nilsson, E., 1981. Gasmolnolyckan Göteborg 1981-05-08. Report from SAI (Swedish Explosives Inspectorate).
Nordahl, R., 1991. Private Communication to R.K. Eckhoff, from R. Nordahl, DnV, Oslo, Norway.
Nordtest, 1982. Dust and Powder Layers: Electric Spark Sensitivity. Nordtest Method NT Fire 016. Nordtest, Espoo, Finland.
Oberhagemann, D., 1989. Zündtemperaturen von Ein- und Mehr-komponentsystemen. Experimentelle Bestimmung und Vorherberechnung. VDI Fortschrittberichte, Reihe 3: Verfahrenstechnik No. 185. Verein Deutsche Ingenieure Verlag, Düsseldorf, Germany.
Olsen, W., Arntzen, B.J., Eckhoff, R.K., 2015. Electrostatic dust explosion hazard – towards a <1 mJ synchronized-spark generator for determination of MIEs of ignition sensitive transient dust clouds. Journal of Electrostatics 74, 66–72.
Olsen, O., 1989. Private Communication to R.K. Eckhoff, from O. Olsen, Stavanger Port Silo, Norway.
Palmer, K.N., Tonkin, P.S., 1971. Coal dust explosions in a large-scale vertical tube apparatus. Combustion and Flame 17, 159–170.
Palmer, K.N., 1973. Dust Explosions and Fires. Chapman and Hall, London. ISBN: 412 09430 4.
Parker, R.J., Pope, J.A., Davidson, J.F., Simpson, W.J., 1975. The Flixborough Disaster. Report of the Court of Inquiry to the Secretary of State for Employment. Her Majesty's Stationery Office, London, UK.
Parker, S.J., August 1985. Electric Spark Ignition of Gases and Dusts (PhD thesis). City University, London, UK. Research work carried out at Chr. Michelsen Institute, (presently: GexCon AS) Bergen, Norway.
Paterson, S., 1939. The ignition of inflammable gases by hot moving particles. The London, Edinburgh and Dublin Philosophical Magazine and Journal of Science, Ser. 7 28 (186), 1–23.
Paterson, S., 1940. The London, Edinburgh and Dublin Philosophical Magazine and Journal of Science 30, 437.
Patrtrick-Yeboah, J.R., Reid, R.C., 1981. Superheat-limit temperatures of polar liquids. Industrial and Engineering Chemistry Fundamentals 20, 315–317.
Pedersen, G.H., Eckhoff, R.K., December 1986. Initiation of Gas Explosions by Heat Generated during Single Impact between Solid Bodies. CMI Report No. 863302–1. Chr. Michelsen Institute (now GexCon AS), Bergen. Norway.
Pedersen, G.H., Eckhoff, R.K., 1987. Initiation of grain dust explosions by heat generated during single impact between solid bodies. Fire Safety Journal 12, 153–164.
Pedersen, G.H., Wilkins, B.A., 1988. Explosibility of Coal Dusts and Coal Dust/Limestone Mixtures at Elevated Initial Pressures. Report No. 88/02101. Chr. Michelsen Institute (now GexCon AS), Bergen, Norway.
Pellmont, G., 1979. Explosions- und Zündverhalten von hybriden Gemischen aus brennbaren Stäuben und Brenngasen. Doctorate Thesis No. 6498. ETH, Zürich.
Phillips, H., 1987. A comparison of "standard" methods for the determination of maximum expetrimental safe gaps (MESG). In: Proceedings of the International Symposium on the Explosion Hazard Classification of Vapors, Gases and Dusts. National Academy Press, pp. 83–108. Organized by the Committee on Studies of Hazardous Substances, National Materials Advisory Board, Publication NMAB-447.
Pilch, M., Erdman, C.A., Reynolds, A.B., 1981. Acceleration Induced Fragmentation of Liquid Droplets. Report NUREG/CR-2247 prepared for the Nucl. Reg. Commission, USA.
Pinkwasser, T., 1985. On the extinction of smouldering fires in pneumatic conveyors. In: Proc. Internat. Symp. Control of Risks in Handling and Storage of Granular Foods, (April). APRIA, Paris.

Pinkwasser, T., 1986. On the ignition capacity of free-falling smouldering fires. In: Euromech Colloquium 208. Explosions in Industry, (April).

Planas-Cuchi, E., Salla, J.M., Casal, J., 2004. Calculating overpressure from BLEVE explosions. Journal of Loss Prevention in the Process Industries 17, 431–436.

Powell, F., Quince, B.W., 1972. The kinetics of impact in drop-weight experiments. Journal of Physics D: Applied Physics 5, 1540–1553.

Powell, F., 1984. Ignition of gases and vapours by hot surfaces and particles – a review. In: Paper Presented at the 9th International Symposium on Prevention of Occupational Accidents and Diseases in the Chemical Industry, Section on Safety against Explosions, Lucerne, Switzerland, June 1984.

Proust, C., 1996. Ignition of dust-air mixtures by a laser beam. In: Proc. 7th Internat. Coll. Dust Explosions, (June 23–26), Bergen, Norway. Published by GexCon AS (previously CMR), Bergen, Norway.

Proust, C., 2002. Laser ignition of dust clouds. In: Paper Presented at 4. Int. Symp. On Hazards, Prevention and Mitigation of Industrial Explosions, Bourges, France, October, 21–25, 2002. J. Phys. IV France –Proceedings, Vol. 12, Pr7 79-88. Editor: I. Sochet.

Public Commission of Investigation, 1986. Ukontrollert Utblåsning Pa Boreplatffiormen West Vanguard 6. Oktober 1985. Norges Offentlige Utredninger, NOU 1986:16. Universitetsforlaget AlS, Oslo (1986). (in Norwegian.).

Rao, K.V.L., Lefebvre, A.H., 1976. Evaporation characteristics of kerosene sprays injected into a flowing air stream. Combustion and Flame 26, 303–309.

Rasmussen Report, 1974. Reactor Safety Study «WASH-1400» (Commonly Known as the Rasmussen Report). published by the US Nuclear Regulatory Commission.

Richards, G.A., Lefebvre, A.H., 1989. Turbulent flame speeds of hydrocarbon fuel droplets in air. Combustion and Flame 78, 299–307.

Rickenbach, H., 1983. Explosion barriers in conveyor installations for dust-air mixtures. Swiss Chemistry 5 (9a).

Ritter, K., 1984. Die Zündwirksamkeit mechanisch erzeugter Funken gegenüber Gas/Luft- und Staub/Luft-Gemischen. Doctorate Dissertation. Universität Fridericiana, Karlsruhe, Germany.

Roberts, A.F., Pritchard, D.K., 1982. Blast effects from vapour cloud explosions. Journal of Occupation Accidents 3, 231–247.

Sadee, C., Samuels, D.E., O'Brien, T.P., 1976/77. The characteristics of the explosion of cyclohexane at the Nypro (UK) Flixborough plant on 1. June 1974. Journal of Occupation Accidents 1, 203–235.

Salla, J.M., Demichela, M., Casal, J., 2006. BLEVE: a new aproach to the superheat limit temperature. Journal of Loss Prevention in the Process Industries 19, 690–700.

Schjelderup, S., 1990. Private Communication to R.K. Eckhoff from S. Schjelderup. Sira-Kvina Kraftselskap, Norway.

Schmitt, D.W., 1973. Compressor Test Facility Explosion, Chapter in 'Fire Protection Manual for Hydrocarbon Processing Plants', second ed. Gulf Publishing Company, Houston, Texas, USA, pp. 59–63.

Scholl, E.W., 1989. Vorbeugender Explosionsschutz durch Vermeiden von wirksamen Zündquellen. VDI-Berichte 701, 477–489 (VDI-Verlag GmbH, Düsseldorf).

Schuber, G., 1988. Zünddurchschlagverhalten von Staub/Luft Gemischen und Hybriden-Gemischen. In: Publication series' Humanisierung des Arbeitslebens, vol. 72. VDI-Verlag, Düsseldorf.

Schuber, G., 1989. Ignition breakthrough behavious of dust/air and hybrid mixtures through narrow gaps. In: Proc. 6th International Symp. On Loss Prevention and Safety Promotion in the Process Industries, Oslo, Norway.

Schulz-Forberg, B., Droste, B., Charlett, H., 1984. Failure mechanisms of propane-tanks under thermal stresses including fire engulfment. In: Proceedings of the International Symposium on Transport and Storage of LPG and LNG, Antwerp.

Sejlitz, I., 1987. Explosivämneskunskap. Tryckeri AB Knappen, Karlstad, Sweden.
Selway, M., 1988. The Predicted BLEVE Frequency of a Selected 2000 m^3 Butane Sphere on a Refinery Site. UK Atomic Energy Authority, Safety and Reliability Directorate. Report SRD/R492, 57 pp.
Shariff, A.M., Wahab, N.A., Rush, R., 2014. Application of inherent safety concept in preliminary design stage eliminate or minimize the potential accident due to BLEVE. In: Proceedings of 17th Annual International Symposium, October 28–30, 2014. Mary O'Connor Process Safety Center, College Station, Texas, pp. 433–456.
Silver, R.S., 1937. The ignition of gaseous mixtures by hot particles. The London, Edinburgh and Dublin Philosophical Magazine and Journal of Science 23 (156 Suppl.), 633. Ser. 7.
Skjold, T., 2003. Selected Aspects of Turbulence and Combustion in 20-litre Explosion Vessels, Cand. Scient. thesis. Department of Physics and Technology, University of Bergen, Norway.
Smeets, G., 1985. Spray detonation studies using laser doppler techniques. In: Bershader, D., Hanson, R. (Eds.), Proc. 15th Internat. Symp. On Shock Waves and Shock Tubes, pp. 547–554. Berkeley, CA, USA.
Smithsonian Physical Tables, 1959. Smithsonian Institution, Washington DC, USA, pp. 421–423.
Smoot, L.D., Horton, M.D., 1977. Propagation of laminar coal-air flames. Progress in Energy and Combustion Science 3, 235–258.
Sochet, I., Gardebas, D., Calderara, S., Marchal, Y., Longuet, B., 2011. Blast wave parameters for spherical explosives detonation in free air. Open Journal of Safety Science and Technology 1, 31–42.
Sochet, I., 2010. Blast effects of external explosions. In: Plenary Lecture at the 8th ISHP-MIE, September 5–10, 2010, Yokohama, Japan (Preliminary manuscript received from the author).
Spaeth, A., May 15, 1995. Accidents awaiting. Time 37.
Special Working Group, 1981. Propanexplosionen I Arendal (G6teborg) 8. Maj 1981. Gothenburgh, December (1981). (in Swedish).
Srinivasan, R., Natarajan, S., 2012. Development of inherent safety: A review of the progress 2001-2011 and opportunities ahead. Process Safety and Environmental Protection 90, 389–403.
SS 421 08 24, 1988. Swedish Standard SS 421 08 24 Classification of Hazardous Areas Endangered by Explosives (in Swedish). Produced by SEK (The Swedish Electrotechnical Commission). Published by SIS (The Swedish Standadization Commission).
Statoil, 2004. Performance Standards for Safety Systems and Barriers. TR1055 Version 1, valid from 17.08.2004, produced by the Statoil Group.
Steen, H., 2000. In: Steen, H. (Ed.), Handbuch des Explosionsschutzes. Wiley-VCH Verlag GmbH, D-69469 Weinheim, Germany.
Stokes, D.A., 1986. Fording Coal Limited Silo Explosion, pp. 56–60. CIM Bulletin 79 No. 891.
Storli, K., 1976. Private Communication to R.K. Eckhoff, from K. Storli, Norwegian Factory Inspectorate, Oslo.
Tam, V.H.Y., 2002. Barrier methods for gas explosion control. In: Lecture Given at 'Course on Explosion Prediction and Mitgation', at University of Leeds, Energy and Resources Research Institute, School of Process, Environmental and Materials Engineering, 12–15 November, 2002.
Tang, M.J., Nicholls, J.A., Lin, Z.C., Sichel, M., Kauffman, C.W., 1986. Detonation velocity in heterogeneous liquid decane/gas systems. Progress in Astronautics and Aeronautics 106, 461–473.

Templin, G., 1990. Private Communication to R. K. Eckhoff from G. Templin, Nord Mills, Malmö, Sweden.
The Institute of Petroleum, August 2002. Area Classification Code for Installations Handling Flammable Fluids, second ed. The Institute of Petroleum, London.
Thowle, C., October 2000. Optical sources of ignition in hazardous areas. Power Engineering Journal 238–242.
Török, Z., Ajtai, N., Turcu, A.-T., Ozunu, A., 2011. Comparative consequence analysis of the BLEVE phenomena in the context of land use planning. Case study: the Freyzin accident. Process Safety and Environmental Protection 89, 1–7.
Tugnoli, A., Gubinelli, G., Landucci, G., Cozzani, V., 2014. Assessment of fragment projection hazard: probability distributions for the initial direction of fragments. Journal of Hazardous Materials 279, 418–427.
Tulis, A.J., 1984. Initiation and propagation of detonation in unconfined clouds of aluminium powder in air. In: Proc. of 9th Int. Semin. Pyrotechnics.
Ural, E.A., 2004. Memo to ASTM E27 Main Committee Members.
van den Berg, A.C., van der Voort, M.M., Weerheijm, J., Versloot, N.H.A., 2006. BLEVE blast by expansion-controlled evaporation. Process Safety in Progress 25, 44–51.
van den Berg, A.C., 1985. The multi-energy method. A framework for vapour cloud explosion blast prediction. Journal of Hazardous Materials 12, 1–10.
van den Bosch, C.J.H., Weterings, R.A.P.M., 1997. Methods for the Calculation of Physical Effects. Committee for the Prevention of Disasters, CPR 14E (TNO 'Yellow Book'), The Hague, The Netherlands.
van der Wel, P.G.J., Lemkowitz, S.M., Leschonski, S., et al., 1994. Ignition of dust clouds using pulsed laser beams. In: Xufan, D., Wolanski, P. (Eds.), Proc. 6th Internat. Coll. Dust Explosions, pp. 125–140 (August 29–September 2), Shenyang, P. R. China.
van Laar, G.F.M., Zeeuwen, J.P., 1985. On the minimum ignition energy of dust/air mixtures. Archivum Combustionis 5, 145–159.
van Wingerden, K., Salvesen, H.-C., Perbal, R., 1995. Simulation of an accidental vapour cloud explosion. Process Safety Progress 14 (3), 173–181.
Verein deutscher Ingenieure, 1988. Untersuchungsmethoden zur Ermittlung von sicherheitstechnischen Kenngrössen von Stäuben. VDI Richtlinie 2263, (Draft). VDI-Verlag GmbH, Düsseldorf.
Verhaegen, H., 1989. Safety, a management task. In: Proceedings of 6th Internat. Symp. Loss Prevention and Safety Promotion in the Process Industries. (June), Norwegian Society of Chartered Engineers, Oslo, Norway.
von Müffling, L., 1946. XXXX. In: Jost, W. (Ed.), Explosion and Combustion Processes in Gases, first ed. McGraw-Hill Book Co. Inc., New York/London, p. 32.
Wakeshima, H., Takata, K., 1958. On the limit of superheat. Journal of the Physical Society of Japan 13, 1398–1403.
Walls, W.L., 1978. What is a BLEVE ? Fire Journal 31, 46–47.
Walls, W.L., 1979. The BLEVE -Part 1. Fire Command 17, 35–37.
Warner, F., 1975. The Flixborough, p. 77. Chemical Engineering Progress, 71, September (1975).
Weerheijm, J., de Bruyn, P.C.A.M., van Wees, R.M.M., Karelse, J.W., 2002. The fireworks disaster in Enschede. Part 21: safety analysis. In: Proc. 6th Internat. Conf. Probabilistic Safety Assessment and Management, 23–28 June, 2002, San Juan Porto Rico, USA.
Welzel, M.M., Schenk, S., Hau, M., Cammenga, H.K., Bothe, H., 2000. Ignition of combustible/air mixtures by small radiatively heated surfaces. Journal of Hazardous Materials A72, 1–9.
Wibbelhoff, H., 1984. Fahrbare Gross-Staubsauger in explosionsgeschütster Aussführung. Steine und Erden 168–169.

Wikipedia, 2015. Internet Publication Found March 2015 by Seeking Hazop on Google.
Williams, A., 1990. Combustion of Liquid Fuel Sprays, Butterworths (1990).
Witte, L.C., Cox, J.E., Bouvier, J.E., February 1970. The vapour explosion. Journal of Metals 39—44.
Wolanski, P., 1990. Deflagration and Detonation Combustion of Dust Mixtures. Paper published by the American Institute of Aeronautics and Astronautics, pp. 3—31.
Wolfhard, H.G., Bruszak, A.E., 1960. The passage of explosions through narrow cylindrical channels. Combustion and Flame 4, 149—159.
Xie, M.M., 2007. Thermodynamic and Gas Dynamic Aspects of a BLEVE. Draft report No.: 04—200708, Dept. Multi-Scale Physics. Faculty of Applied Sciences, Delft University of Technology, Delft, The Netherlands.
Xu, B., et al., 1988. The Model of Explosion Accident Determined by the Seismic Record (Copy of unpublished English manuscript concerning the Harbin Linen Textile Plant explosion, given by Xu Bowen to R.K. Eckhoff).
Xu, B., January 1988. The explosion accident in the Harbin Linen Textile plant. EuropEx Newsletter 2—3. Edition 6.
Young, C., 2004. LPG Tank Anti-BLEVE and Automotive Tank Safety System. US Patent 4,680,937.
Yu, Y.F., 1985. On the electrostatic charging of some finely divided materials in modern agricultural pneumatic transport systems. Journal of Electrostatics 16, 209—217.
Zabetakis, M.G., 1965. Flammability Characteristics of Combustible Gases and Vapors. US Bureau of Mines. Bulletin 627.
Zebel, G., 1966. Coagulation of aerosols, Chapter II. In: Davies, C.N. (Ed.), Aerosol Science. Academic Press, London, pp. 31—58. LCCCN: 66-16696.
Zehr, J., 1957. Anleitung zu den Berechnungen über die Zündgrenzwerte und die maximalen Explosionsdrücke. VDI-Berichte 19, 63—68.
Zehr, J., 1965. Eigenschaften brennbarer Stäube und Nebel in Luft. In: Freytag, H.H. (Ed.), Handbuch der Raumexplosionen. Verlag Chenfie GmbH, Weinheim, pp. 164—186.
Zevenbergen, J.F., 2002. Laser Ignition of Single Particles and Particle Clouds (Doctorate thesis). Technical University of Delft, ISBN 90-6464-409-8.
Zhang, J., Laboureur, D., Yi, L., Mannan, M.S., 2014. A Supercritical pressure BLEVE in Nihon Dempa Kogyo Crystal Inc.. In: Proceedings of 17th Annual International Symposium, October 28—30, 2014. Mary O'Connor Process Safety Center, College Station, Texas, pp. 385—395.
Zhu, H., 1988. Investigation of the Dust Explosion in Harbin Linen Factory (Unpublished English manuscript given to R.K. Eckhoff by Zhu Hailin).
Zockoll, C., 1996. Früherkennung von Bränden durch CO-Detektion. VDI-Berichte 1272, 411—427 (VDI-Verlag GmbH, Düsseldorf, Germany).
Zwahlen, G., 1989. Deflagrationsfähigkeit — ein gefährliches Zersetzungsverhalten abgelagerter Stäube. VDI-Berichte 701, 167—186 (VDI-Verlag GmbH, Düsseldorf).

INDEX

'*Note*: Page numbers followed by "f" indicate figures and "t" indicate tables.'

E

F

H

I

U

V

W

Z

CPI Antony Rowe
Chippenham, UK
2016-12-08 12:15